T0178360

This book, together with its companion volume *The science of crystallization: microscopic interfacial phenomena*, make up a complete course that will teach an advanced student how to understand and analyze scientifically any of the phenomena that are observed during natural or technological crystallization from any medium and via any technique. It is an advanced text that goes into considerable detail concerning the many elements of knowledge needed to understand both quantitatively and qualitatively a crystallization event.

This particular volume, having briefly reviewed the important findings of the companion volume, then deals specifically with convection, heat transport and solute transport to describe both steady state and transient solute distributions in bulk crystals, small crystallites of various shapes and thin films. The author then integrates all these factors to describe interface stability for interfaces of different shapes plus the dominant morphological characteristics found in crystals during either single phase or polyphase crystallization. The concepts are extended to embrace biological crystallization and the connecting links between the morphological features of polymer crystallization and simple organic molecule crystallization. The generation of physical and chemical defects is treated for both bulk crystals and thin films, to show both the origins of these faults and some procedures for eliminating them. A variety of mathematical examples are utilised to help the student gain a quantitative understanding of this topic area. Both the present book and its companion volume are much more broadly based and science oriented than other available books in this field, and are therefore more able to address any area of application, ranging from the production of dislocation-free single crystals in bulk or film form, at one extreme, to structurally sound large metal ingots, at the other.

This book and its companion can be used independently of each other, and together they provide the basis for advanced courses on crystallization in departments of materials science, metallurgy, electrical engineering, geology, chemistry, chemical engineering and physics. In addition the books will be invaluable to scientists and engineers in the solid state electronics, optoelectronics, metallurgical and chemical industries involved in any form of crystallization and thin film formation.

The science of crystallization:
macroscopic phenomena and defect generation

The science of crystallization: macroscopic phenomena and defect generation

William A. Tiller
Department of Materials Science and Engineering
Stanford University

CAMBRIDGE UNIVERSITY PRESS

Cambridge
New York Port Chester
Melbourne Sydney

CAMBRIDGE UNIVERSITY PRESS
Cambridge, New York, Melbourne, Madrid, Cape Town, Singapore, São Paulo

Cambridge University Press
The Edinburgh Building, Cambridge CB2 2RU, UK

Published in the United States of America by Cambridge University Press, New York

www.cambridge.org
Information on this title: www.cambridge.org/9780521381390

First published 1991

A catalogue record for this publication is available from the British Library

Library of Congress Cataloguing in Publication data

Tiller, William A.
  The science of crystallization: macroscopic phenomena and defect
generation / William A. Tiller.
      p.   cm.
    ISBN 0-521-38139-8. – ISBN 0-521-38828-7 (pbk.)
    1. Crystallization.   I. Title.
    QD921.T519   1992
    548'.5–dc20   91-26320   CIP

ISBN-13 978-0-521-38139-0 hardback
ISBN-10 0-521-38139-8 hardback

ISBN-13 978-0-521-38828-3 paperback
ISBN-10 0-521-38828-7 paperback

Transferred to digital printing 2006

To
My Loving Wife
Jean

Contents

Preface

This book first reviews the important findings described in the companion volume *The science of crystallization: microscopic interfacial phenomena*. It then deals specifically with convection, heat transport and solute transport to describe both steady state and transient solute distributions in bulk crystals, small crystallites of various shapes and thin films. It integrates all these factors to describe interface instability for interfaces of different shapes and predicts the dominant morphological characteristics found in crystals during either single phase or polyphase crystallization. These concepts have been extended to embrace biological crystallization and the connecting links between the morphological features of polymer crystallization and simple organic molecule crystallization.

These concepts are put to work to treat the generation of physical and chemical defects in both bulk crystals and thin films showing both the origin of these faults as well as some procedures for eliminating them.

The present book and its companion are the culmination of over 35 years of thought and personal study concerning modern-day science and technology of the crystallization process. A special perspective and a special approach have been developed for understanding the intricacies of *any* phase transformation in a fundamental and scientific way so that technological utilization can be readily engineered.

This pair of books differs from others available in the "crystal growth" area in that they are sufficiently broad-ranging to provide the scientific basis for understanding and treating any application area – single crystal growth, chemical crystallizers, film formation, ingots and castings, welding, frozen food processing, cryogenic organ storage, geological sleuthing, etc. The present book, and its companion, are advanced texts from which a number of university courses at the senior, masters or PhD level can be drawn by Departments of Materials Science, Metallurgy, Electrical Engineering, Geology, Chemistry, Chemical Engineering and Physics. In addition, technical personnel in the solid state electronics, optoelectronics, metallurgical and chemical industries involved in any form of crystallization or thin film formation should find them insightful and useful, if not indispensable.

I owe a great debt to many individuals for the completion of this work. First, to my PhD students over the past two and a half decades for helping me to refine and understand the concepts set forth here and in the companion text. To Irv Greenfield who provided me with the opportunity to make a good start on this version of the book (1980) after two earlier false starts (1964 and 1970). To my Stanford classroom students in this topic area who, over the years, helped me to correct my many inconsistencies with respect to symbols and logic. To my secretary, Miriam Peckler, who provided much support for the earlier versions. To Jerri Rudnick who has converted my handwritten manuscript to TEX. To my colleague and friend Guy Marshall Pound (deceased) for reading and helpfully criticizing the text. To Bruce Chalmers who first introduced me to this subject. To John Dodd who first made me want to learn. To my wife Jean for her loving patience in putting up with both the many years of distraction to this task and the perpetually littered surfaces of our home with working notes and papers.

Stanford, California   William A. Tiller
1990

Symbols

1 Arrangement

The most often used subscripts and superscripts are given immediately below and this is followed by the list of symbols. English symbols are listed first and Greek second with the symbols being arranged in alphabetical order. For each letter, printed capitals are listed first, script capitals second and lower case third.

2 Subscripts

a - adsorption site	β - phase	C - solute
$\hat{c}$ - crystal	c - critical value	E - excess
$\tilde{E}$ - eutectic	e - electronic	F - fusion
f - face	i - interface	i^* - intrinsic
∞ - far field value	K - kinetic	k - kink
L - liquid	ℓ - ledge	M - melting
N - nutrient	o - reference condition	P - pressure
Φ - electrostatic	S - solid	$\tilde{S}$ - solution
s - surface	s^* - sublimation	T - thermal, temperature
V - vapor	v^* - vaporization	

3 Superscripts

a - adsorbate	e - electron	j - species
M - major solute	m - minor solute	$\hat{v}$ - vacancy
$*$ - equilibrium value		

4   English symbols

A - area

A_{sh} - strain hardening factor

A_0 - area occupied by a growth unit in a crystal

$\hat{A}_0$ - interface stability parameter

A^* - activated state

$\tilde{A}$ - Hamaker constant for colloid interaction

$\mathcal{A}_0$ - area of surface with unit probability for a new nucleation event

$\mathcal{A}$ - aspect ratio for spheroids

a, a', a'' - lattice distances in different directions

$\hat{a}$ - thermodynamic activity

a_0 - molecular radius

$a_{\hat{c}}$ - crystal radius

a_d - droplet area

a^* - convection parameter ($a^* = 2\tilde{U}_\infty/\pi R$)

a_* - dimensionless suction parameter

$\tilde{a}$ - sphere diameter

$\bar{a}$ - nucleation parameter ($\bar{a} = \pi h \gamma_E^2/3\Delta S_F \mathcal{R} T$)

B - magnetic flux

$\hat{B}$ - dynamic bond number ($\hat{B} = G_r/Re_M$)

$\mathcal{B}$ - non-planar field decay coefficient

$$\left(\mathcal{B} = \frac{1}{2}\left\{ 1 + \left[1 + \left(\frac{2D}{V}\omega\right)^2 \right]^{1/2} \right\} \right)$$

$\mathcal{B}_*$ - value of $\mathcal{B}$ at the marginal stability condition

b - growth rate parameter ($\dot{R} = bt^{-1/2}$); thermal convection parameter

$\tilde{b}$ - Henry's law constant

b_* - Burger's vector

$\hat{b}$ - diffusion parameter ($\hat{b} = a^2/4\alpha t_0$)

C - concentration

C_t - concentration of transition state

$\bar{C}$ - Laplace transform concentration

$\tilde{C}$ - electrolytic capacitance

$\hat{C}$ - Taylor constant

c - velocity of light (2.998×10^8 m s^{-1})

c_p - specific heat at constant pressure

c_v	- specific volume ratio ($c_v = v_0^j/v_0^0$)
$\tilde{c}$	- number of chemical elements in system
D	- diffusion coefficient
$\tilde{D}_T$	- transitional diameter for a dendrite array
$\hat{D}$	- dislocation Damköher number
$\mathcal{D}$	- an effective diffusivity
d_f	- film thickness
d_c	- critical crystal diameter
E	- energy
$\bar{E}$	- partial molar internal energy
$E_{\bar{T}}$	- total energy
E_j^f	- energy of the jth type of fault
E_g	- band gap energy
E_F	- chemical binding energy at the Fermi level
E_v	- valence band energy
E_c	- conduction band energy
E_B	- boundary energy
E^*	- Young's modulus
$\hat{E}$	- electric field
E_i	- exponential integral function
Erf	- error function
Erfc	- complementary error function
ΔE_r	- energy change due to roughening
ΔE^f	- formation energy
ΔE_b	- excess energy per bond
$E_{\hat{L}}$	- electron lattice energy
e_T	- dilatational transformation strain
F	- Helmholtz free energy
$\hat{F}$	- force between half spaces
F_*	- Cochran function
ΔF_r	- Helmholtz free energy change due to roughening
ΔF_{rec}	- Helmholtz free energy change due to surface reconstruction
$\underline{f}$	- Helmholtz free energy density
$\vec{f}$	- volume force
$\tilde{f}, \tilde{f}'$	- unspecified general mathematical functions
$\hat{f}$	- fraction of active sites

$\bar{f}_{mn}$	- surface stress tensor
f_*	- correlation factor
$\tilde{f}_*$	- degrees of freedom
G	- Gibbs free energy
G^*	- Gibbs free energy of the activated state
G_*	- Cochran function
$\tilde{G}$	- shear modulus
G_+	- Green's function
ΔG	- Gibbs free energy driving force
ΔG_{sv}	- driving force consumed by the thermodynamic state variable change (∞ to i)
ΔG_N	- free energy change per unit volume transformed
ΔG_A	- activation energy
ΔG_p^f	- pseudo kink formation free energy
ΔG_m	- free energy of mixing
$\Delta \tilde{G}_0^{\,j}$	- free energy change of j due to an interface field
$\bar{\Delta} G_+, \bar{\Delta} G_-$	- free energy change for transitions to right or to left
ΔG_S	- free energy of separation
ΔG_0	- bulk free energy change
ΔG^{uh}	- free energy change for unhindered rotation
δG_{C^*}	- free energy change for surface ledge creation
δG_{A^*}	- free energy change for surface ledge annihilation
δG_{I^*}	- free energy change for surface ledge interaction
$\mathcal{G}$	- temperature gradient
$\mathcal{G}_n$	- temperature gradient in the normal direction
g	- Gibbs free energy density
$\tilde{g}_1 \tilde{g}^1$	- unspecified general mathematical functions
$\hat{g}$	- geometrical factor
g_*	- fraction solidified
H	- enthalpy, distance
$\bar{H}$	- partial molar enthalpy
H_*	- Cochran function
H^*	- Biot number
$\tilde{H}$	- height of cavity
$\hat{H}$	- magnetic field strength
ΔH	- heat content change
$\Delta H'$	- heat change per molecule
ΔH_m	- heat of mixing

$\Delta H''$	- thermodynamic state with two degrees of freedom factored out
ΔH_g	- molar heat of fusion
h	- ledge height
h_c	- transport coefficient
h^f	- facet height
$h', \bar{h}'$	- vector distances
$\tilde{h}$	- heat transfer coefficient
h^*	- Planck's constant
I	- nucleation frequency
$\tilde{I}$	- moment of inertia of a cluster
$\hat{I}$	- rate of elementary process
I^*	- Bessel function of the first kind (modified)
$\mathcal{I}$	- kinetic amplification factor in spinodal decomposition
i	- number of molecules in a cluster
i^*	- number of molecules in a critical size cluster
J	- flux of heat or matter
$\mathcal{J}$	- Bessel function
j	- species
K	- thermal conductivity
$\hat{K}$	- equilibrium constant
$\hat{K}_R$	- reaction constant
K_*	- transmission coefficient
K_f	- drag coefficient
$\bar{K}_0$	- gross equilibrium solute partition coefficient for a polycomponent system
$\tilde{K}$	- surface curvature parameter ($\tilde{K} = (\gamma/\mathcal{G}\Delta S_F)^{1/2}$)
K^*	- Bessel function of the second kind (modified)
$\mathcal{K}$	- surface curvature
$k_0, \tilde{k}_0$	- phase diagram solute partition coefficient $(C_s/C_L, X_S/X_L)$
k_i	- interface solute partition coefficient
$\hat{k}_i$	- net interface solute partition coefficient
k	- effective solute partition coefficient
$k_{\hat{E}}$	- transformed solute distribution coefficient
$\bar{k}$	- rate constant

$\bar{k}_p$	- parabolic rate constant
$\bar{k}_\ell$	- linear rate constant
$\bar{k}_+, \bar{k}_-$	- rate of transition to right or to left
k_*	- suction parameter
L	- length
$\bar{L}$	- dimensionless length $(\bar{L} = Vt/a_{\hat{c}})$
L_*	- crucible size
$\tilde{L}$	- $\tilde{H}/\ell$ - dimensionless height
$\hat{L}_i$	- a thermodynamic potential component
$\mathcal{L}$	- a lumped material parameter for dendrite growth
$\mathcal{L}_\pm$	- mathematical functions for initial transient solute redistribution
ℓ	- distance
ℓ_{tw}	- length of twin plane
M	- resistive torque
$\hat{M}$	- atomic mobility
$\hat{M}_{\hat{E}}$	- effective ionic mobility
$\hat{M}_{ij}$	- a thermodynamic potential component
M_a	- Marangoni number
M_*	- hydromagnetic interaction parameter
$\tilde{M}$	- biaxial elastic modulus
m	- mass
m_r	- reduced mass
m_L	- liquidus slope
$\bar{m}$	- average liquidus slope component
$\hat{m}_L$	- effective liquidus slope
$\tilde{m}$	- degree of a kink; effective stress exponent in dislocation mobility
m_*	- log spiral convection parameter
m^*	- number of excess vacancies involved in dislocation climb
$m^\wedge$	- slope in stirring/field strength plot
N^j	- number of j species
N_a	- number of adsorption sites per square centimeter
$\tilde{N}$	- number of nearest neighbour sites in a crystal
N_k	- number of heterogeneous nuclei per cubic centimeter

N_*	- transport number
$\hat{N}_{ij}$	- thermodynamic potential function component
Nu	- Nusselt number
$\mathcal{N}_A$	- Avogadro's number
n^j	- number of moles of j
n_j^f	- number of j-type faults
$\hat{n}$	- distance in the interface normal direction
$\hat{n}_x, \hat{n}_y$	- projections of the unit normal vector on x and y
$\tilde{n}_1$	- number of nearest neighbour sites in the interface
n_{bf}	- number of broken bonds in the face
$n_{\hat{r}}$	- number of rearrangements
n_n	- average number of foreign nuclei
n_D	- number of dislocations per square centimeter
$n_{\hat{v}}^*$	- equilibrium concentration of vacancies
N_I^*	- equilibrium concentration of interstitials
$n_{I\alpha}$	- number of interstitial sites per unit cell
P	- pressure
$\tilde{P}$	- dimensionless pressure $(\tilde{P} = P/\rho U_\infty^2)$
$\hat{P}$	- Péclet number $(\hat{P}_T = VR/2D_T, \hat{P}_C = VR/2D_C)$
Pr	- Prandtl number $(Pr = \nu/D_T)$
P_*	- Cochran function
P_p	- purification per pass
p, p_+, p_-, p_0	- probabilities
p_f	- stacking fault formation probability
p_{sp}	- spiral dislocation formation probability
$\tilde{p}$	- number of coexisting phases
p_B	- dislocation breeding efficiency
$\bar{Q}$	- total energy content
$\tilde{Q}$	- centrifugal fan volume
Q_*	- non-conservative system effect on solute content
$\hat{Q}$	- parameter, magnetic Taylor number
q	- charge
$\bar{q}$	- energy distribution
$\tilde{q}_\omega$	- heat flux
q_*	- rate factor for non-conservative system
$\hat{q}$	- canonical partition function
R	- radius

R_c	- crystal radius
R_C	- critical radius for sphere instability
R_*	- crucible radius
R^*	- critical nucleus radius
$\tilde{R}_1, \tilde{R}_2$	- principal radii of curvature
$\hat{R}_0, \hat{R}$	- metallic radii
$\bar{R}_n^2$	- mean square displacement
Re	- Reynold's number $(Re = U\ell/\nu)$
Re_D	- rotational Reynold's number
Re_M	- Marangoni Reynold's number
Ra	- Rayleigh number
$\mathcal{R}$	- gas constant per gram mole, 8.134 J K^{-1} mole^{-1}, 1.987 cal K^{-1} mole^{-1}
r	- radial coordinate
$\hat{r}$	- surface roughness
r_0	- equilibrium separation of atoms
r_D^c	- dislocation core radius
r^*	- critical radius of two-dimensional embryo
S	- entropy
$\bar{S}$	- partial molar entropy
S_b	- entropy per bond
$\tilde{S}_{ij}$	- compliance
S_*	- sticking coefficient
$\hat{S}$	- Laplace transform variable
Sc	- Schmidt number $(Sc = \nu/D_C)$
S_p^q	- charged species parameter
ΔS	- entropy change
$\Delta S''$	- entropy change for states with two degrees of freedom factored out
ΔS_r	- entropy change on roughening
$\mathcal{S}$	- stability function
$\mathcal{S}_*$	- value of $\mathcal{S}$ at the marginal stability condition
s	- surface entropy density, surface coordinate
$\hat{s}$	- interface shape
$\hat{s}_*$	- optimum surface shape
s_{c^*}, s_{A^*}	- coordinates for surface creation, annihilation regions
$\tilde{s}$	- crucible rotation parameter

T	- temperature
$\overset{\bullet}{T}$	- cooling rate
T_L	- liquidus line, liquidus surface
T_s	- heat source temperature
$T_{s'}$	- heat sink temperature
$\tilde{T}_s$	- substrate temperature
T_g	- glass transition temperature
T_t	- transition temperature
T_w	- wall temperature
$\hat{T}$	- Taylor number
ΔT	- thermal driving force (temperature potential difference)
$\widehat{\Delta T}$	- superheat
$\Delta T^*, \Delta T^{**}, \Delta T^{***}$	- special transition driving forces for attachment kinetics
ΔT_k^{fr}	- driving force for attachment at the root of a facet
ΔT_{NP}	- driving force due to non-planar isotherms
ΔT_c	- critical supercooling
$\widetilde{\Delta T}_s$	- non-equilibrium temperature range of solidification
δT	- temperature oscillation amplitude
t	- time
t_k	- relaxation time to enter a kink
$U, u, \vec{u}$	- fluid velocity
$\hat{U}, \hat{u}$	- intermolecular potential
$\hat{U}'_{dd}$	- potential for interacting layers
U_*	- potential driving a fluid
U_M	- Marangoni velocity
U_g	- buoyancy driving velocity
$\bar{U}_R$	- return flow velocity
$\bar{U}_\infty$	- uniform velocity at ∞
U_N	- natural convection velocity
U_D	- axial forced flow velocity
$\hat{u}_j$	- multiplicity factor
u_f	- fluid velocity in a filamentary array
$\vec{u}_R$	- relative velocity
$\tilde{u}_j$	- dimensionless velocity in the jth coordinate direction

V	- growth velocity
V_*	- optimum velocity
V_y	- velocity of surface in the y-direction
V_ℓ	- ledge velocity
$\tilde{V}$	- Volta potential, solid velocity
$\bar{V}_c$	- critical velocity for loss of transport communication
$\bar{V}_c^*$	- critical velocity for loss of interface equilibrium
$V_{\hat{E}}$	- transformed frame velocity
V_{max}^{CSC}	- maximum velocity without formation of constitutional supercooling
V_0	- climb velocity
$\underline{V}$	- macropotential, voltage
$\mathcal{V}$	- nucleation velocity limit
v	- volume
$\tilde{v}$	- partial molar quantity
v_m	- molar volume of a material
dv/v_0	- fraction of volume transformed
Δv_t	- volume transformed
$\hat{v}$	- vacancies
$\hat{W}$	- a general potential
W_a	- work of atomic adsorption rearrangement
W_e	- work of electron rearrangement
$W_{\alpha\beta}$	- work of separation of two phases
W^f	- facet width
$\bar{w}$	- a parameter
$\tilde{w}$	- possible number of complexions
X	- mole fraction
X_t	- fraction transformed
X^*	- optimum density of roughened states
$\bar{X}_k$	- average kink spacing in a'' units
X_d	- fraction of droplets
$\bar{X}_{\hat{s}}$	- mean surface diffusion distance in a' units
X_{ox}	- oxide thickness
$\hat{X}$	- position of the interface
x_j	- experimental coordinate of the j-type
$\tilde{x}$	- coordinate on face relative to a crystal corner
Y	- Young's modulus ($Y = E^*/(1 - \sigma^*)$)

$\bar{Y}_e$ — average ledge spacing

$\hat{Y}_{em}$ — harmonic test function

$\tilde{Y}_j$ — dimensionless coordinate (y_j/ℓ)

y — coordinate direction

Z_3 — three-body interaction parameter

$\tilde{Z}$ — Zeldovich factor

$Z_{\hat{s}}$ — z-coordinate variation on the surface shape

z — distance

z_* — number of equivalent jumps

$\tilde{z}_1$ — number of nearest neighbors

$\hat{z}^j$ — valence of the j species

5   Greek symbols

α - Jackson α-parameter, phase, grouped material parameter for dislocation loop formation

$\alpha_1, \alpha_2, \alpha_3$ - parameters

α^*, α^{A_j} - thermodynamic α-factors

α_k - vaporization fraction

α' - monomer in cluster formation process

$\tilde{\alpha}$ - field width at the interface

$\bar{\alpha}$ - growth rate parameter ($V = \bar{\alpha} t^{1/2}$)

$\bar{\alpha}^*$ - thermal expansion coefficient, interface attachment parameter at the onset of instability

α_D - dislocation parameter

α_β - interface breakdown parameter

β - phase, mathematical decay coefficient, parameter for conversion rate of species

$\beta_0, \beta_1, \beta_2$ - parameters used in the interface attachment process

$\bar{\beta}$ - surface energy per unit area of low index projected face

β' - embryo in cluster formation process

β_* - fraction dissociated to tetramer state

β_* - reaction rate parameter ($\beta_* = -\bar{k}_+ C_A / \kappa T$)

β^* - volume expansion coefficient

$\bar{\beta}$ - magnitude of field strength at the interface

β_u - convective flow parameter

β_r - chemical reaction parameter

Γ^j - surface excess density of j

$\tilde{\Gamma}$ - surface capillarity parameter ($\tilde{\Gamma} = \gamma / \Delta S_f$)

$\hat{\Gamma}$ - heat transfer parameter

γ - surface Gibbs free energy, phase

$\hat{\gamma}$ - thermodynamic activity coefficient

γ_ℓ - ledge free energy

γ_f^o - face free energy relative to a standard condition

γ_e - electronic contribution to the surface free energy

γ_{tw} - twin boundary energy

γ_0'' - surface torque term ($\partial^2 \gamma / \partial n_x^2$)

γ_c - quasi-chemical contribution

γ_t - transitional diffuseness contribution

γ_d - dislocation/strain contribution

γ_a - adsorption contribution

γ'	- interfacial free energy per molecule
γ_g	- molar surface tension
$\Delta\gamma^{\hat{R}}$	- free energy change in forming a ridge
$\tilde{\Delta}$	- surface creation parameter $(\lambda''/\omega^2\delta)^{1/2}$
δ	- boundary layer thickness, off-stoichiometry fraction, phase, strain
$\tilde{\delta}$	- amplitude of surface undulation
$\dot{\tilde{\delta}}$	- time rate of change of amplitude
δ_*	- half-width of activation barrier
δ_B	- dislocation breeding factor
ε	- an effective bond energy
ε_n	- pair bonds to the nth neighbor
ε_a	- adsorption energy per bond
ε_0	- binding energy
$\hat{\varepsilon}_j$	- parameter of the jth type
$\tilde{\varepsilon}_{mn}$	- strain tensor
ε_*	- dielectric permeability
$\tilde{\varepsilon}$	- lattice parameter strain
$\tilde{\varepsilon}_e, \tilde{\varepsilon}_p$	- elastic and plastic strain
$\bar{\varepsilon}$	- emissivity
$\Delta\varepsilon$	- energy fluctuation
ζ	- Fermi energy
$\tilde{\zeta}$	- a dimensionless coordinate
$\eta, \tilde{\eta}$	- expanded electrochemical potential, dimensionless coordinate $(y/\delta_m, z/(4\nu t)^{1/2}, V_z/D, D_s\tau_s/\lambda^2$, etc.)
$\bar{\eta}$	- viscosity
η^*	- a parameter
$\hat{\eta}$	- a quantity
$\tilde{\eta}_*$	- flow field variable
θ	- angle
$\hat{\theta}$	- surface coverage
$\hat{\theta}^a$	- surface coverage of adatoms
$\hat{\theta}^{\hat{v}}$	- surface coverage of vacancies
$\tilde{\theta}$	- dimensionless temperature $((T - T_A)/(T_M - T_a))$
θ_*	- diffusion parameter $((Vx/D)^{1/2})$

$\Delta\theta$	- non-dimensional supercooling or Stefan number $(c\Delta T_\infty/\Delta H_F)$
κ	- Boltzmann's constant
κ_*	- a drag coefficient
$\hat{\kappa}_0, \hat{\kappa}_1, \hat{\kappa}_2$	- potential function components
κ_D	- inverse Debye length (λ_D^{-1})
λ	- grid spacing
$\hat{\lambda}$	- $\gamma_{max}/\gamma_{min}$
λ_D	- Debye length
λ_D^e	- Debye length for electrons
λ_*	- de Broglie wavelength
$\tilde{\lambda}$	- width of interface field
λ^*	- optimum perturbation wavelength
$\Delta\lambda/\lambda$	- lattice parameter change
μ	- chemical potential
μ_0	- standard state chemical potential
μ_0^*, μ_*	- magnetic permeabilities
$\Delta\mu_b$	- excess chemical potential per bond
$\bar{\mu}$	- viscosity
ν	- kinematic viscosity $(\nu = \bar{\eta}/\rho)$
$\hat{\nu}$	- field frequency
$\check{\nu}$	- vibrational frequency
$\bar{\nu}$	- a surface roughness parameter $(2\varepsilon_1/kT^*)$
$\tilde{\nu}_r$	- normal frequency of reaction
ξ	- distance coordinate in a moving frame of reference
$\hat{\xi}$	- a surface roughening parameter
ρ	- density, radius of curvature of a filament
ρ_c	- radius of curvature of the tip of a filament
ρ_m	- mass density
σ	- stress
$\hat{\sigma}$	- supersaturation
σ^*	- Poisson's ratio
σ_*	- electrical conductivity
$\bar{\sigma}_*$	- electrical charge density

$\tilde{\sigma}$	- Stefan–Boltzmann constant
$\bar{\sigma}$	- supersaturation number
Φ	- electrostatic potential, Galvani potential
ϕ	- angle, velocity potential, electrostatic potential
$\hat{\phi}$	- dimensionless velocity potential
$\tilde{\phi}$	- dissipation function
$\bar{\phi}$	- dimensionless concentration (C/C_∞)
ϕ_*	- Fourier spectrum of perturbations
τ	- time constant
τ_k	- average time for two adjacent kinks to grow together
τ_{R_1}	- time for a surface pillbox disc to grow to radius R_1
τ_1^a	- jumping time of adsorbate
$\hat{\tau}_\ell$	- ledge overgrowth time
$\hat{\tau}_f$	- face overgrowth time
$\bar{\tau}_0$	- modulus parameter
$\tau_{\hat{s}}$	- adatom lifetime on surface
τ_σ	- shear stress
τ_{diff}	- effective solute redistribution time
τ_{rise}	- effective rise time to achieve a steady state solute distribution
$\tilde{\Psi}$	- electron work function
ψ	- angle
$\hat{\psi}$	- nucleation time lag factor
Ω	- molecular volume
$\hat{\Omega}$	- angular velocity
ω	- frequency, angular velocity
ω_u	- frequency of configurational rearrangements
ω_D	- frequency of diffusive transport
$\omega_{\bar{\eta}}$	- frequency of viscous flow events
ω_t	- frequency of transition
ω_{cr}	- crucible rotation rate
χ	- electron affinity
$\tilde{\chi}_{\hat{s}}$	- surface potential
χ_E^*, χ_M^*	- electric and magnetic susceptibilities

1

Introduction

1.1 Coupling equations

Crystallization is a many-variable, many-parameter interaction event in Nature and in technology and scientific understanding of this area feeds the very large application field illustrated in Fig. 1.1. For the territory lying within the scientific understanding box, it is necessary to treat the crystallization event as a dynamic system with many interacting parts or subsystems. These are illustrated in Fig. 1.2 for a typical case. The unique morphology and special phenomena associated with the crystallization event arise out of the conjunction of these subsystems interacting with each other in pairs, triplets, etc.

It is in the far-field domain of a growing crystal where the macroscopic or global thermodynamic state variables, $(C, T, P, \phi) \equiv (C_\infty, T_\infty, P_\infty, \phi_\infty)$ are fixed or changed dynamically according to some particular program $(\dot{C}_\infty, \dot{T}_\infty, \dot{P}_\infty, \dot{\phi}_\infty)$. However, it is mainly at the interface of the growing crystal that all the morphological, chemical segregation and embryonic physical defect phenomena arise in the crystallization event. Thus, scientific understanding of the overall process requires a description in terms of the thermodynamic state variables and their gradients at the growing interface; i.e., (C_i, T_i, P_i, ϕ_i) and $(\nabla_i C, \nabla_i T, \nabla_i P, \nabla_i \phi)$. The connection between the interface conditions and the far-field conditions occurs via a set of partial differential equations and their boundary value constraints. In the most general case, these are:

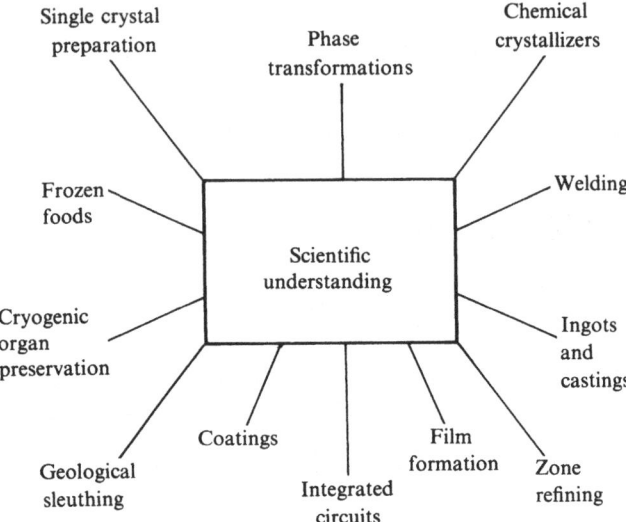

Fig. 1.1. A wide variety of application areas nourished by scientific understanding of crystallization.

(1) the transport of heat equation,
(2) the hydrodynamic flow equations,
(3) the transport of matter equation,
(4) the electrical potential equation,
(5) the stress and strain equations and
(6) the perturbation response of the interface equations.

The overall thermodynamic potential, ΔG_∞, driving a crystallization event is consumed, in part, in setting up these various fields, ΔG_{sv}, (since they all store energy) and, in part, at the interface itself, ΔG_i, where the latter energy is consumed in driving both the interface reaction leading to a change of state, ΔG_K, and the energy storage reactions in (i) new surface creation, ΔG_γ, (ii) strain field generation, ΔG_σ, (iii) physical defect formation, ΔG_d, (iv) metastable phase formation, ΔG_p, and non-equilibrium point defect formation, ΔG_{PD}. The free energy conservation equation for this process is

$$\Delta G_\infty = \Delta G_{sv} + \Delta G_i \qquad (1.1a)$$

$$= \Delta G_{sv} + \Delta G_K + \Delta G_E \qquad (1.1b)$$

$$= \Delta G_{sv} + \Delta G_K + \Delta G_\gamma + \Delta G_\sigma + \Delta G_d + \Delta G_p + \Delta G_{PD} \qquad (1.1c)$$

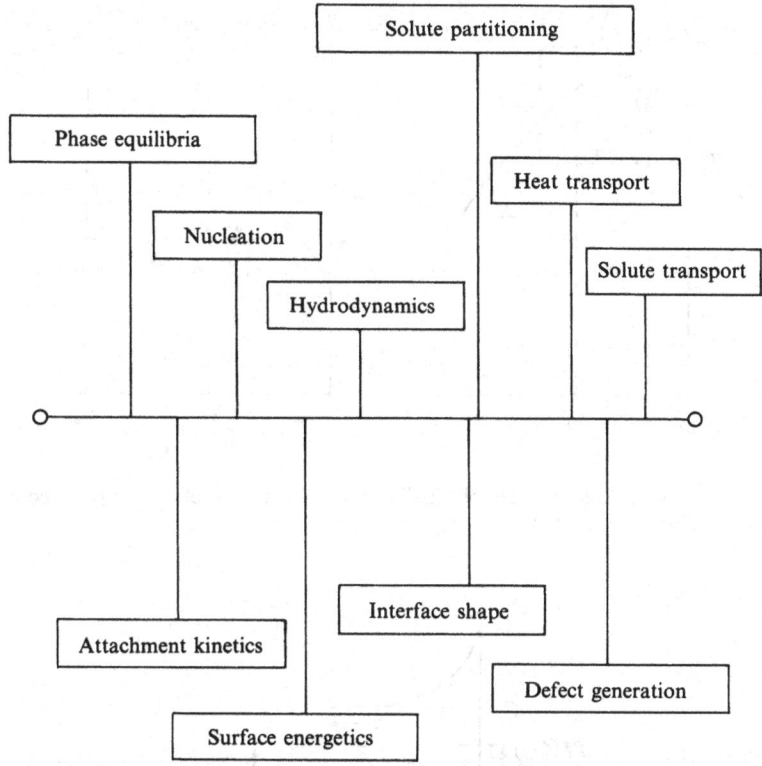

Fig. 1.2. System of interacting scientific considerations involved in a typical crystallization event.

In many crystal growth situations, it is not necessary to use the ΔG-coupling equation and, for simplicity, surrogate driving force equations may be used; i.e.,

$$\Delta T_\infty = \Delta T_C + \Delta T_T + \Delta T_E + \Delta T_K \qquad (1.2a)$$

or

$$\Delta C_\infty = \Delta C_C + \Delta C_T + \Delta C_E + \Delta C_K \qquad (1.2b)$$

or

$$\Delta P_\infty = \Delta P_C + \Delta P_T + \Delta P_E + \Delta P_K \qquad (1.2c)$$

where the subscripts C and T refer to matter (concentration) and heat (temperature), respectively. The partitioning associated with Eqs. (1.2a) and (1.2b) is illustrated using the phase diagrams of Fig. 1.3 for the two different modes of phase change:

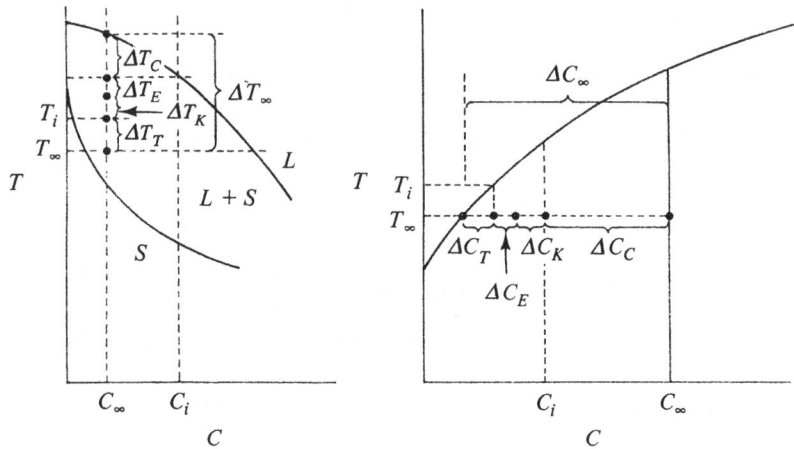

Fig. 1.3. Typical phase diagram illustrations of the coupling equations for ΔT and ΔC.

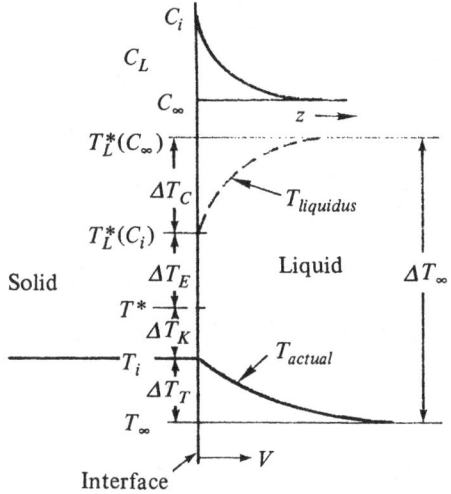

Fig. 1.4. Solute and temperature distributions plus key temperatures for unconstrained crystallization.

(1) *unconstrained crystallization*, where the surrogate far-field driving force in the nutrient phase, ΔT_∞, is greater than the interface surrogate driving force, ΔT_i, (see Fig. 1.4) and

(2) *constrained crystallization*, where the surrogate far-field driving force in the nutrient phase, ΔT_∞, is less than the interface surrogate driving force, ΔT_i.

Although in the past, the tendency has often been to consider only one of these driving forces to be dominant and to neglect the rest, it can be readily shown that all of the primary quantities enter every crystallization event and each may dominate the growth process at different times. This is perhaps best illustrated by considering the unconstrained growth of a crystal in its nutrient from the onset of nucleation to the fully grown state. For this case, the velocity of crystallization, V, is as given in Fig. 1.5(a). At time $t = 0$, all of the driving force goes into forming the surface needed to comprise the nucleation event so that $V = 0$ and $\Delta G_\infty = \Delta G_E$ at $t = 0$. At long times, all of the driving force is consumed in the transport of matter and heat and in the storage of excess energy (for solid state transformation) so that $\Delta G_\infty = \Delta G_C + \Delta G_T + \Delta G_E$. At early to intermediate times, all processes simultaneously consume free energy. The variation of subprocess free energy with crystal size, R, is shown in Fig. 1.5(b) for the case of a crystal growing from a pure melt ($\Delta G_\sigma = \Delta G_d = \Delta G_p = \Delta G_{PD} = 0$). A number of specific subprocess dominance examples have been treated in Chapter 1 of the companion book.[1]

1.2   Interface attachment kinetics

For any system, the ΔG_j contribution of ΔG_E and the ΔG_K contribution are intimately coupled. For almost all cases of our experience, we are dealing with a terrace/ledge/kink (TLK) model of the interface on an atomistic level so that the free energy of the interface per unit area, γ, can be given in terms of the face (terrace) energy, γ_f, and the ledge energy, γ_ℓ, plus the density of each. Structural roughening of the ledges to generate kinks increases the enthalpy of the ledges but it increases the configurational entropy of the ledges even more so that γ_ℓ decreases as ledge roughening increases up to some optimum level which depends upon the enthalpy per bond between the crystal and the nutrient phase. These values of γ_f and γ_ℓ are generally also strongly decreased by atomic bond reconstruction on the terrace and ledge so that a marked interaction between these atomic displacement changes on the terrace, as well as kink formation changes on both the ledge and terrace, occurs.[1] Using the oversimplified nearest-neighbor-only dangling bond model, Fig. 1.6 gives a plot of the computed equilibrium

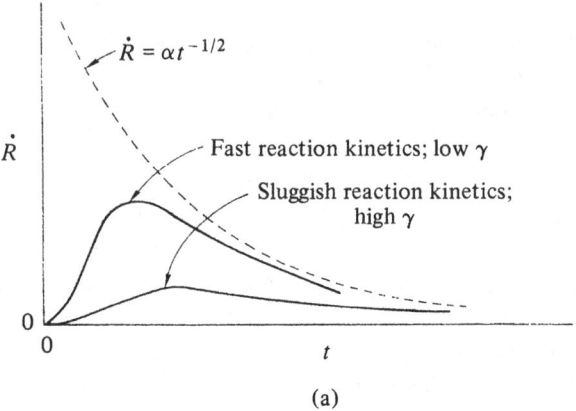

(a)

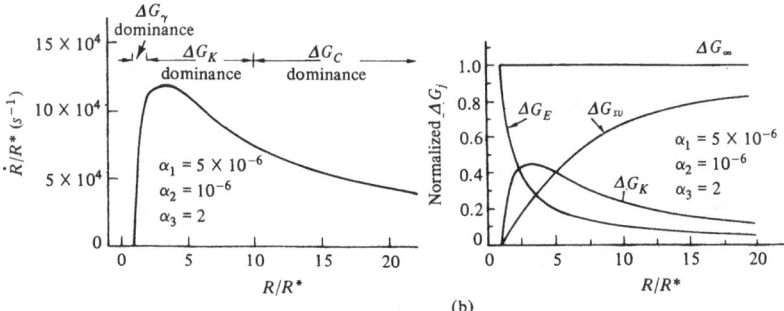

(b)

Fig. 1.5. (a) Schematic representation of particle growth velocity, $\dot{R}$, versus time, t, illustrating the effect of interface reaction kinetics plus the asymptotic approach to the transport dominance regime, and (b) relative spherical crystal velocity, $\dot{R}/R^*$, and the normalized driving force for the various subprocesses as a function of the relative crystal size, R/R^* ($R^* =$ nucleus radius whereas α_1, α_2 and α_3 are a particular set of parameters defined in the mathematical description of the overall growth process[1]).

fraction of roughened states, X^*, on the (100) face of a simple cubic crystal as a function of the important bonding parameter $\varepsilon_1/\kappa T$, where ε_1 is the nearest-neighbor bond energy across the interface. Because of the many approximations involved in such a calculation, the error bars are expected to be quite large and only qualitative significance should be placed on such a plot.[1]

The kinetics of crystallization depends strongly on the density of kinks as well as on the interface driving force. At very small driving

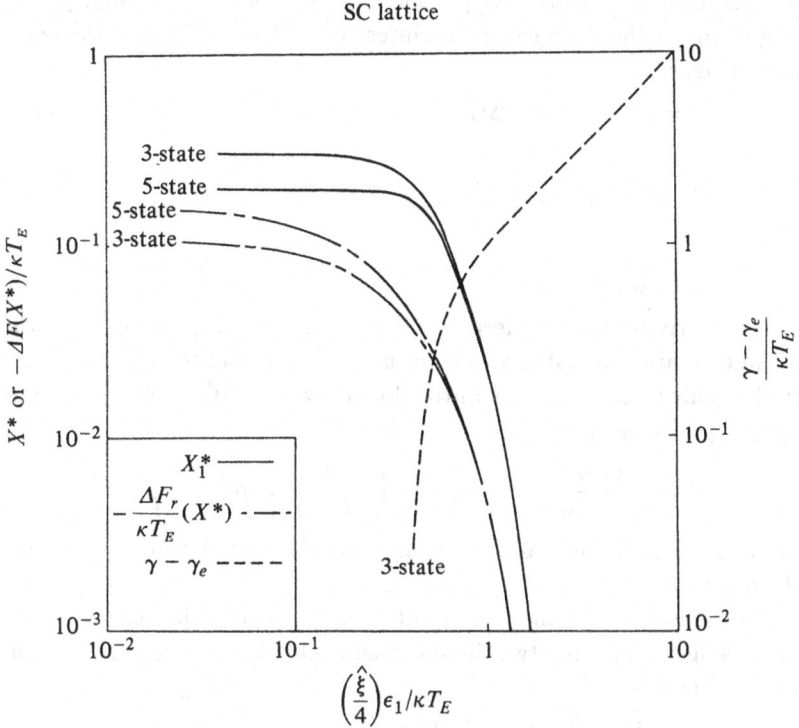

Fig. 1.6. Plot of rougnening factors for a simple cubic lattice as a function of the key parameter $\vec{v}_1/\kappa T_E$ ($\hat{\xi}$ is a universal factor): (i) equilibrium population of +1 states for a 3- and 5-state model, (ii) decrease in free energy, $\Delta F(X^*)$, due to roughening for these two models and (iii) the interfacial energy, γ, for the 3-state model.

forces, $(\Delta G_K < (\mathrm{d}\gamma_\ell/\mathrm{d}x)_{max})$ phase change can only occur at kink sites on the ledges so the interface moves forward by the movement of kinks on ledges and the lateral passage of ledges over the surface. At intermediate driving forces $((\mathrm{d}\gamma_\ell/\mathrm{d}x)_{max} < \Delta G_K < (\mathrm{d}\gamma_f/\mathrm{d}z)_{max})$, the ledges become more diffuse (higher kink density over more levels) because kinks can now form spontaneously on the ledge. At very large driving forces $(\Delta G_K > (\mathrm{d}\gamma_f/\mathrm{d}z)_{max})$, kinks can also form spontaneously on the terrace so that phase change can occur anywhere on the interface. Qualitatively, the smaller is $\varepsilon_1/\kappa T$, the greater is the kink density, the smaller are γ_ℓ and γ_f, the smaller is the anisotropy in γ_ℓ or γ_f and the greater is the crystallization rate for a given value of ΔG_K.

For those cases where a continuous supply of ledges occurs easily, the velocity of crystallization is proportional to both the density, ρ_ℓ, and height, h_ℓ, of these ledges on the interface and the rate of movement of these ledges; i.e.,

$$V = h_\ell \rho_\ell V_\ell = \beta_K \Delta G_K \qquad (1.3a)$$

where

$$\beta_K = 2h_\ell a \beta_0 / \lambda_\ell \lambda_k \qquad (1.3b)$$

and

$$\lambda_k = a'/X'^* \qquad (1.3c)$$

at small driving forces. Here, a and a' represent atomic spacings while λ_ℓ and λ_k are the distances between ledges and between kinks, respectively. This formula applies in the domain where $\Delta G_K / \mathcal{R}T \ll 1$. When $\Delta G_K / \mathcal{R}T \widetilde{>} 1$, we must use

$$V = \left(\frac{2h_\ell aa'B}{\lambda_\ell \lambda_k} \right) \exp \left(\frac{-\Delta G_A}{\mathcal{R}T} \right) \left[1 - \exp \left(\frac{-\Delta G_K}{\mathcal{R}T} \right) \right] \qquad (1.3d)$$

From Eq. (1.3d), one can readily see how the kink density (λ_k^{-1}) enters the expression for V.

For the case where the velocity of crystallization is limited by the supply of ledges, via either two-dimensional nucleation or screw-dislocations, we find that

(a) two-dimensional nucleation:

$$V = V_0 \exp[-(\pi a''\gamma_\ell^2 / \mathcal{R}T \Delta G_i)] \qquad (1.4a)$$

where

$$V_0 = a'' A I_0 \qquad (1.4b)$$

(b) screw dislocations:

$$V = \beta_K' |\Delta G_K| \Delta G_K \qquad (1.4c)$$

where

$$\beta_K' = h_\ell a \beta_0 / 2\pi \gamma_\ell \lambda_k \qquad (1.4d)$$

this formula applies only for the $\Delta G_K / \mathcal{R}T \ll 1$ domain. For the general case where both two-dimensional nucleation of layers and the lateral spreading of layers are equally important to the interface attachment kinetics, we find that[1]

$$V = \alpha_1 \exp[-\alpha_2/\kappa T][1 - \exp(-\overline{\Delta G}_K / \mathcal{R}T)]^{2/3} \qquad (1.5a)$$

where

$$\alpha_1 = \pi^{1/3} a'' I_0^{1/3} (2aa'B/\lambda_k)^{2/3} \exp(-2\Delta G_A/3\mathcal{R}T) \qquad (1.5b)$$

$$\alpha_2 = \pi a'' \gamma_\ell^2 / 3\overline{\Delta G_i} \qquad (1.5c)$$

The topography of such a surface is determined by the relative rates of the two processes defined by α_1 and α_2. When the nucleation rate is low (α_2 small) and the lateral spreading rate is high (α_1 large), the surface remains flat. When the nucleation rate is small to intermediate and the lateral spreading rate is very slow, the surface develops large undulations even though ΔG_K is constant. If impurities are also present and solute partitioning occurs at such undulations, ΔG_K varies from the center to the ledge of the undulation so that the undulations can be significantly increased in amplitude.

For the unconstrained growth of a crystal, we recognize that, from a simple dangling bond viewpoint, different ledges will require different driving forces at which uniform ledge attachment occurs ($\Delta G_K^{(hk\ell)} > (d\gamma_\ell^{(hk\ell)}/dx)_{max}$) so that the attachment kinetics in a particular crystallographic direction may take sudden "spurts" depending upon the overall magnitude of the local ΔG_K. From a more realistic viewpoint, where surface and ledge reconstruction are factored into the process, the kink density is not expected to follow such a simple relationship and some ledges with high γ_ℓ may roughen more readily than ledges with lower γ_ℓ because the former beneficially allow terrace reconstruction to proceed while the latter do not.[1] Finally, one will also find that, if the crystal is extrinsically conducting at the growth temperature, the population of charged kinks is greatly increased just as are the population of charged point defects in the volume of a crystal and the population of charged jogs on dislocation lines in the crystal. Such an extrinsic conduction factor could be partially or completely offset by impurity adsorption to poison kink sites. At this point in time, few specific experimental data are available to quantify either the surface and ledge reconstruction factor or the extrinsic conduction factor with respect to kink formation in specific crystal systems or specific orientations in these systems. This is territory for new research and is most likely to be extremely important in the area of thin film formation (where $\varepsilon_1/\kappa T$ is large, generally).

Returning to the overly simplistic viewpoint before ending this section, we note the baseline fact that kink formation is easiest when γ_ℓ is small so that rapid attachment kinetics should correspond with low surface energetics. This often makes it extremely difficult to decide upon the key factors determining a crystal's morphology.

1.3   Surface energetics

For a static surface, it is well known that γ is a function of the crystallographic angles (θ, ϕ) of the surface normal relative to the closest packed crystallographic plane in the crystal. It is also well known that a particular surface (θ_1, ϕ_1) is thermodynamically predicted to decompose into two or more lower energy surfaces at (θ_2, ϕ_2) and (θ_3, ϕ_3), etc., when the γ-plot energetics are such that

$$\left\{ \gamma + \left[\left(\frac{\partial^2 \gamma}{\partial \theta^2} \right) + \left(\frac{\partial^2 \gamma}{\partial \phi^2} \right) \right] \right\}_{\theta_1, \phi_1} < 0 \tag{1.6}$$

What is not so well appreciated is that, because of surface reconstruction effects, $\gamma(\theta, \phi)$ is also a function of ledge height (or terrace length) even for a static surface so we should begin to think in terms of $\gamma(\theta, \phi, h_\ell)$. Further, since terrace reconstruction is a thermally activated process and ledge movement requires terrace unreconstruction at the bottom of the ledge as well as terrace reconstruction at the top of the ledge, it is important to recognize that

$$\gamma \equiv \gamma(\theta, \phi, h_\ell, V_\ell) \tag{1.7a}$$

so that faceting and crystal form development also depend upon h_ℓ and V_ℓ. This also applies to layer formation via two-dimensional nucleation or screw dislocations. Finally, since the kink population on ledges in the intermediate driving force range will also depend upon ΔG_K and since γ_ℓ is a function of the ledge kink population, we extend Eq. (1.7a) to become

$$\gamma \equiv \gamma(\theta, \phi, h_\ell, V_\ell, \Delta G_K) \tag{1.7b}$$

Certainly facet formation and crystal form development will follow Eq. (1.6) utilizing Eq. (1.7b) as the proper basis for describing γ. However, this approach uses equilibrium thermodynamic reasoning for shape change but a dynamic description of γ for implementation. Since we are concerned here with dynamic faceting and surface morphology adjustments during crystal growth, it would be preferable to be more inclusive in our thermodynamic reasoning and include the departure from equilibrium at the growing crystal faces as part of an "effective" surface energy, γ_{eff}, i.e.,

$$\gamma_{eff} = \gamma(\theta, \phi, h_\ell, V_\ell, \Delta G_K) + \Delta G_K(\theta, \phi, h_\ell, V_\ell, \Delta G_K) \tag{1.8a}$$

Now, the criterion for dynamic facet formation and morphology adjustments during crystal growth becomes, from Eq. (1.6),

$$\gamma_{eff} + \left[\frac{\partial^2 \gamma_{eff}}{\partial \theta^2} + \frac{\partial^2 \gamma_{eff}}{\partial \phi^2} \right] < 0 \tag{1.8b}$$

Finally, in such a description, we must remember to include extrinsic conduction effects on the creation of kink sites and adsorbate poisoning effects on the annihilation of kink sites since both factors alter the energetics of both ledges and terraces.

1.4 Surface creation

One must distinguish the excess chemical potential of an atom or molecule on a static surface that is curved from the excess driving force that must be available in the nutrient phase at a dynamically expanding surface to allow it to change its local surface to volume ratio. In the former case, we have

$$\Delta\mu_1 = \Omega \left[\left(\gamma + \frac{\partial^2 \gamma}{\partial n_x^2} \right) \mathcal{K}_x + \left(\gamma + \frac{\partial^2 \gamma}{\partial n_y^2} \right) \mathcal{K}_y \right] \tag{1.9a}$$

where $\mathcal{K}_j$ is the curvature in the j-direction and n is the surface normal. In the latter case, $\Delta\mu$ must include Eq. (1.9a) as well as a contribution that accounts for the change in surface to volume ratio, i.e.,

$$\Delta\mu_2 = \Delta\mu_1 + \Omega \frac{\Delta(\dot{\gamma}A)}{\Delta \dot{v}} \tag{1.9b}$$

where $\Delta\dot{A}$ and $\Delta\dot{v}$ are the rate of change of local surface area and local volume, respectively. Most of the current literature in crystal growth neglects the second term of Eq. (1.9b) but, to show that it is non-zero, one has only to consider Fig. 1.7 which represents the development of a periodic interface undulation. Across the interface, the curvature reverses sign periodically so that the average curvature is zero giving $\overline{\Delta\mu_1} = 0$; however, the area of the interface is clearly growing as the amplitude grows.

The evaluation of $\Delta\dot{A}/\Delta\dot{v}$ depends on the model chosen to describe the interface.[1] For the convenient but overly simplistic model of a continuum interface of constant γ moving at velocity V with small harmonic departure from smoothness of amplitude δ, we have

$$\frac{\Delta\dot{A}}{\Delta\dot{v}} = \frac{\omega^2 \delta\dot{\delta} \cos^2 \omega x}{V_y^0} + \cdots \tag{1.10a}$$

For large amplitudes, this relationship is not correct and should be replaced by the approximate relationship

$$\frac{\Delta\dot{A}}{\Delta\dot{v}} \approx \frac{2\dot{\delta}}{\lambda V} \tag{1.10b}$$

where λ is the average wavelength of the undulations. For a more re-

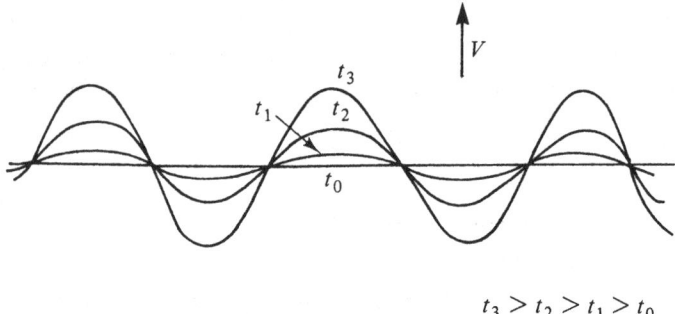

$$t_3 > t_2 > t_1 > t_0$$

Fig. 1.7. Illustration of time-dependent growth in amplitude for a harmonic shape distortion on a macroscopic front with average growth velocity, V.

alistic model of the interface; i.e., a TLK model, *all* the surface creation occurs within the plane of the ledge movement. These ledges are initially created by either two-dimensional nucleation or at dislocation sources and increase their ledge area as they expand so that $\Delta\dot{A}$ involves the ledge expansion. Additional energetic contributions involve (i) the change of γ_ℓ with λ_ℓ and h_ℓ as well as position relative to a crystal ridge or corner and (ii) the annihilation of layer edges between adjacent sources or in the formation of a ridge or corner. A variety of formulae for expressing $\Delta(\gamma\dot{A})/\Delta\dot{v}$ can be readily developed for specific TLK cases. In particular, for a flat interface with layers flowing at an angle ϕ to the interface plane and beginning to develop periodic undulations of amplitude δ, we find that

$$\frac{\Delta(\gamma\dot{A})}{\Delta\dot{v}} = \left(\frac{\partial\gamma_\ell}{\partial\theta}\right)_\phi \frac{\omega\dot{\delta}\cos\omega x}{V_\phi} \tag{1.10c}$$

This formula assumes that h_ℓ remains constant during the undulation development.

This TLK requirement of ledge creation, ledge interaction and ledge annihilation leads to the development of flats, ridges and corners on real crystals as well as to the wide variety of crystallization anisotropies seen in nature and technology. In particular, it is the symmetrical placement of two-dimensional nucleation layer sources in the meniscus region during Czochralski (CZ) crystal growth that accounts for the cross-sectional shape of the pulled crystals as well as for the location of surface ridge-lines and flats. Likewise, it is the symmetrical placement of two-dimensional nucleation layer sources on the cap of a growing dendrite that accounts for both its preferred growth direction and the onset of side-branching.[1]

1.5   Surface adsorption layer

Although the interfacial adlayer has not been considered to be important in melt growth of crystals, it has certainly been recognized as having a significant impact on solution-grown crystals because of the propensity for kink site poisoning by specific adsorbates. However, at least three other important crystal growth considerations are traceable to the interfacial adlayer: (1) chemical reactions and polymerization, (2) interface mobility enhancement and (3) interface field development. These considerations are especially relevant to film growth from the vapor phase via chemical vapor deposition (CVD) or molecular beam epitaxy (MBE) techniques.

Although CVD reactions can occur homogeneously in the vapor phase and "rain-down" particles onto a growing film surface, many important film-forming reactions, such as Si from $SiH_{4-n}C\ell_n$ in H_2, occur heterogeneously at the surface to form a variety of saturated chemical intermediates, that have a short lifetime in the adlayer before diffusing away into the gas phase, and a variety of radicals (such as $SiH_2, SiC\ell_2$, etc.), which are the film-forming species, have a long lifetime in the adlayer because of one or more covalent bonds formed with the substrate surface. These are the species that diffuse to the ledges where they undergo dissociation and phase change reactions ($SiH_2(ad) \rightarrow Si(s) + H_2(ad)$, for example). Under certain conditions, they can also form polymerization reactions, $(SiH_2)_n$ or $(CH_2)_n$, etc., which tend to impede the crystallization reaction. For chemical reactions occurring in the adlayer, to determine the equilibrium species population, it is necessary to include the free energies of adsorption along with the usual free energies of the gaseous species. Since such adsorption generally alters the surface reconstruction state of the substrate, this factor must be included in the free energies of adsorption which may be expected to be a strong function of surface concentration.

The growth rate of the ledges is a strong function of the mobility of the film-forming species in the adlayer and this, in turn, is a strong function of the adlayer constitution and concentration. For example, the effective surface diffusion coefficient of Cu on a Cu substrate may be increased by factors of $\sim 10^5$ with the addition of $C\ell_2$ vapor at partial pressures $\sim 0.1 - 0.3$ of the saturation pressure. Here, we see the effect of a foreign element with a reasonable adsorption energy to the surface that aids the movement of Cu on the surface leading to what is probably a mobile adsorption isotherm in this Cl_2 partial pressure range. It is probable that a Cu–Cl complex forms on the surface which has less binding to the surface than a bare Cu atom and moves more easily in the adlayer medium, perhaps much like chemical vapor transport phenomena via the gas phase. In the $SiH_{4-n}Cl_n$ case, it is the Si–H or Si–Cl, etc., complex that moves readily through the adlayer to the ledges where the Si is deposited. However, in this case, the movement can be limited if too much chemisorbed H or Cl is present on the surface. Thus, at too high a temperature, the surface mobility is limited by insufficient adsorbed H or Cl to provide a good surface concentration of Si–H or Si–Cl complexes and a continuous adlayer medium. At the other extreme of too low a temperature, too much chemisorbed H or Cl is present so that any Si-bearing complexes that exist on the surface are limited in their movement by the availability of surface vacancies in the adlayer. In practice, adjustment of the total H_2 or Cl_2 pressure is often used to restore mobility to the adlayer at low temperatures (low pressure CVD (LPCVD)). Likewise, reduction of the binding energy of the key adsorbate involved in the chemical reaction restores mobility to the adlayer at low temperutres (metal organic CVD (MOCVD)).

The presence of an adlayer alters the reconstruction and the electrical state of the interface and this, in turn, produces a built-in electric field that may extend some distance into the solid because the interface potential becomes screened by a redistribution of free electrons, charged point defects and impurity ions. Of course, an interface electric field is present even when no adlayer, due to foreign species, has developed. Even in this case, surface dipole, charge state, intermolecular potential and stress tensor effects exist so that relatively long range fields of both stress and electrical nature exist in the vicinity of the interface. An important short range (1–5 lattice planes) field at the interface which influences initial heterogeneous nucleation at substrates is the intermolecular potential field. It is this field that, in large part, determines whether the Stranski–Krastanov, the Volmer–Weber or the van der Merwe mechanism of film formation occurs.[1]

1.6   Solute partition coefficients

It is the solute partition coefficient between the crystal and nutrient phases that allows the redistribution and manipulation of doping levels in the crystal. For the growth of crystals from convectively stirred melts, the two most familiar partition coefficients, k and k_0, are related by the equation

$$k = \frac{k_0}{k_0 + (1 - k_0)\exp(-V\delta_C/D)} \tag{1.11a}$$

where k_0 is the phase diagram partition coefficient defined as the ratio of solubility on the solidus and the liquidus curves of the phase diagram at the interface temperature, T_i, k is the effective distribution coefficient defined as the ratio of the solute content in the crystal just behind the interface to that in the bulk liquid and δ_C is the thickness of the solute boundary layer at the interface. We will see in Chapter 3 that the parameter $V\delta_C/D$ must be altered to describe properly an essentially one-dimensional solute distribution in a CZ crystal pulled from an essentially two-dimensional fluid flow field. Used in Eq. (1.11a) form, $D = D_{\text{eff}}$ where $D_{\text{eff}} \sim (3\text{--}5)D$.

Eq. (1.11a) is inaccurate in a second and perhaps even more important way for real systems where an interface field is present in either the crystal, the nutrient phase or both. For those cases where the effective thickness of the interface field, ΔZ_j (where $j = S$ or L), is much less then δ_C, Eq. (1.11a) should be replaced by

$$k = \frac{\hat{k}_i}{\hat{k}_i + (1 - \hat{k}_i)\exp(-V\delta_C/D)} \tag{1.11b}$$

where $\hat{k}_i$, which is generally a function of crystal orientation, is defined as the ratio of the concentration in the crystal a distance ΔZ_S from the interface to the concentration in the liquid phase a distance ΔZ_L from the interface (see Fig. 1.8).[1] When V is small, $\hat{k}_i \rightarrow k_0$ because solute transport across the distance $\Delta Z_S + \Delta Z_L$ is rapid compared to the rate of movement of the interface and essentially thermodynamic equilibrium prevails over this dimension. For interface velocities that are large relative to effective transport velocities but small relative to the maximum transition velocity for solute attachment and detachment at the interface, V^*, $\hat{k}_i \rightarrow k_i$, the equilibrium interface partition coefficient defined as the concentration ratio immediately at the interface and illustrated in Fig. 1.8 for the four general combinations of interface field. The qualitative variations of $\hat{k}_i$ with V for these four types of interface field are shown in Fig. 1.9.[1] The transition of $\hat{k}_i$ to $k_0 + \Delta k_i$ occurs when transport communication is lost across the field zone in the solid.[1] The further transition of $\hat{k}_i$ to k_i occurs when transport communication is also lost across the field zone in the liquid. Finally, the

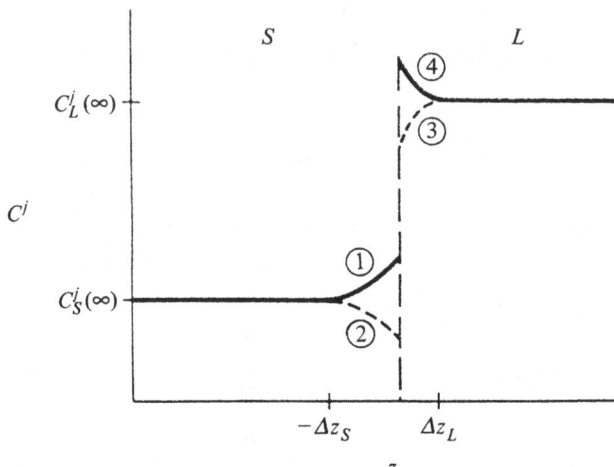

Fig. 1.8. Solute concentration profiles in the solid, C_S, and liquid, C_L, for a static interface which exhibits all four possible types of distributed binding potentials.

transition of $\hat{k}_i$ to unity occurs when $V > V^*$ because $k_i \to 1$ and the solute becomes trapped.

When one applies these considerations to film formation, interface fields exist both normal to the ledges in the plane of the adlayer and normal to the bulk film/adlayer interface. This is one of the reasons why it is so difficult to dope crystals during film growth. It is in such cases that Fig. 1.8 (substituting the adlayer for the nutrient phase) and Fig. 1.9 are expected to apply most strongly, although there are also many geological and solution growth as well as melt growth crystal examples.[1]

This chapter has been essentially a review of the most important findings discussed in the companion book.[1] There, one will also find the background science on both the thermodynamics of bulk media as well as surfaces and the kinetics of chemical reactions as well as nucleation. Thus, we are in position to integrate these microscopic interfacial phenomena with the macroscopic aspects of crystal growth via the coupling equations (see Eqs. (1.1) and (1.2)). Chapters 2, 3 and 4 allow us to appreciate quantitatively ΔG_{sv} and the surrogates $\Delta T_T, \Delta T_C, \Delta C_T$ and ΔC_C while Chapters 5 and 6 allow us to investigate the morphology of crystals via use of the coupling equations. Finally, Chapter 7 shows us how these various solute, temperature and stress profiles interact to generate defects in our growing and grown bulk crystals. Chapter 8 extends the exposition of Chapter 7 to thin films.

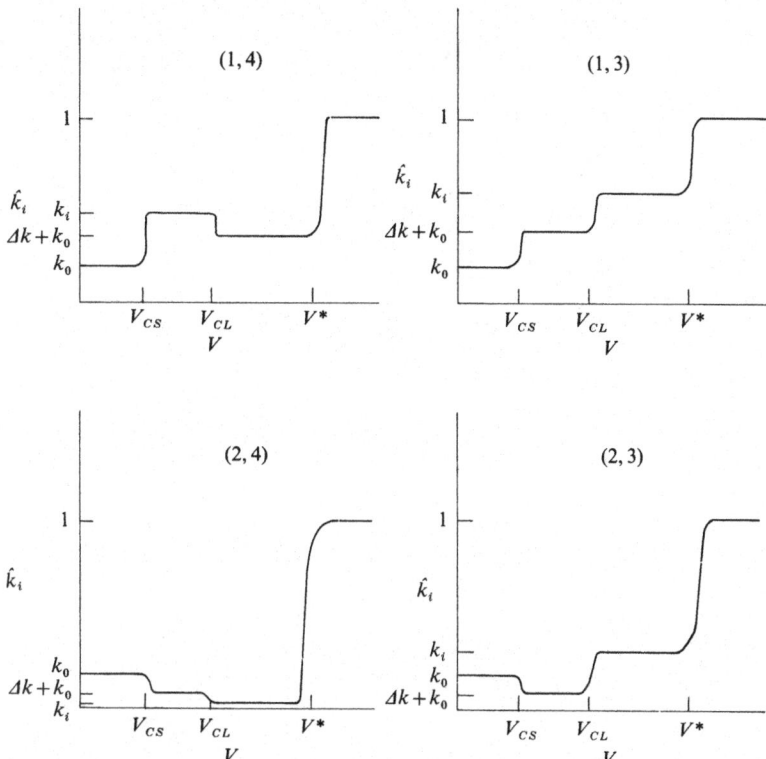

Fig. 1.9. Schematic variation of the net solute partition coefficient, $\hat{k}_i$ as a function of interface velocity, V, for the four different kinds of interface field given in Fig. 1.8.

2

Convection and heat transfer

In the following sections on convection, the intent is to apply all of this understanding and analysis to the growth of crystals from a fluid or gaseous nutrient. Most of this convection information comes from studies on fluids in contact with rigid and non-reacting boundaries. However, a crystal/nutrient interface is a reacting interface which can store heat and matter and thus can act as a type of "storage ballast" for the energy contributions imbedded in the flow patterns of the adjacent fluid. Thus, the numerical details of the fluid flow in a box with *reacting walls* may be somewhat different from those in a box with *non-reacting* walls. However, for pedagogical simplicity, most of the chapter will deal with flows adjacent to non-reacting walls so that the author can simply "lift" work from many other studies and then weave it into a fabric that has relevance to crystal growth. At the end of the convection section, we shall return to the "reacting wall" consideration and present some data illustrating the importance of this factor.

2.1   Convection

The state of a moving incompressible fluid is fully described if, for each point in space and each instant of time, the following four quantities can be defined: the three components of the fluid velocity $\vec{u}$ and the pressure p. Conservation of matter is expressed for an incompressible fluid by

$$\operatorname{div} \vec{u} = 0 \tag{2.1}$$

and the equations of motion for an element of fluid are given by the Navier–Stokes equation

$$\rho\left(\frac{\partial \vec{u}}{\partial t} + \vec{u} \cdot \nabla \vec{u}\right) = \rho \vec{f} - \nabla p + \nabla(\bar{\eta}\nabla \vec{u}) \tag{2.2}$$

Here, the left side comprises the product of the mass of the unit and its acceleration while the right side represents the sum of the external forces acting on the unit.

Following Levich,[1] the term $\nabla(\bar{\eta}\nabla \vec{u})$, where $\bar{\eta}$ is the viscosity of the fluid, accounts for the effect of viscous forces. The internal friction due to the viscous nature of the fluid is manifested only when one region of fluid moves relative to another. Faster moving layers entrain slower moving ones so that momentum is transferred from the faster to the slower layers. This unique value of volume force arises in those fluids where the transfer obeys Newton's laws of motion. Such fluids are called Newtonian fluids which include water, aqueous solutions of inorganic and many organic substances, a number of organic liquids, alcohols, hydrocarbons, liquid metals, glycerine, glasses, gases and certain resins. In most examples, we shall use the parameter $\nu = \bar{\eta}/\rho$ which is the kinematic viscosity of the fluid (ρ = density).

The negative of the pressure gradient is that volume force which acts on the fluid element when the pressure changes from point to point. It is not the pressure itself but only its gradient which is required in the equation of motion (since we want the net force on an element of volume) while the local pressure may be reckoned from an arbitrary datum.

The vector $\vec{f}$ represents the volume force exerted on the element of fluid. Gravity is one example of a volume force while the electromagnetic Lorentz force due to induced currents is another. In general, any gradient of extended chemical potential can be considered as such a body force. The gravity effect is of great importance in natural convection which is driven by a buoyancy density difference and impeded by the viscous inertia of the fluid. The viscosity dissipates kinetic energy while the buoyancy force releases internal energy. One example of $\vec{f}$ due to spatial temperature and concentration differences is

$$\vec{f} = -g\rho[\beta_T^*(T - \bar{T}) + \beta_C^*(C - \bar{C})] \tag{2.3a}$$

$$\beta_T^* = -(\partial \ell \mathrm{n}\rho/\partial T)_C \tag{2.3b}$$

$$\beta_C^* = -(\partial \ell \mathrm{n}\rho/\partial C)_T \tag{2.3c}$$

where $\bar{T}$ and $\bar{C}$ refer to the average values and g is the gravitational constant. We shall see later that this gives rise to Bénard convection. While this is a volume drive, we can also think in terms of a surface or interface drive due to the variation of γ with temperature and concentration along the surface, i.e.,

$$\vec{f}_s \propto \left[\left(\frac{\partial \gamma}{\partial T} \right) \nabla_{\hat{s}} T + \left(\frac{\partial \gamma}{\partial C} \right) \nabla_{\hat{s}} C \right] \tag{2.4}$$

We shall see later that this gives rise to Marangoni convection.

Numerous experimental studies of the flow of Newtonian liquids past the surface of a solid body wetted by them have established that the layer of fluid immediately adjacent to the surface remains motionless. Velocity measurements have shown that the thickness of this stationary layer is quite small, $\sim$ several molecular layers. The absence of slip past the surface is highly important to fluid flow, in general, and the boundary condition that is generally assumed at solid/liquid interfaces is

$$\vec{u}_R = 0 \tag{2.5}$$

where $\vec{u}_R$ is the relative velocity since the solid may be moving. Thus, the fluid exerts on each unit area of the solid a force which is numerically equal to the rate of momentum transfer across the surface. For example, defining p_{xz} as the x-component of momentum transfer across a unit surface which is perpendicular to the z-axis,

$$p_{xz} = (\rho u_x) u_z + \bar{\eta} \left(\frac{\partial u_x}{\partial z} + \frac{\partial u_z}{\partial x} \right) \tag{2.6}$$

The first term on the right side of Eq. (2.6) is the x-component momentum transfer accompanying physical transfer (convection) of fluid volume across a surface perpendicular to the y-axis. The second term represents the momentum transfer caused by fluid viscosity. The viscous properties of the fluid assure the transfer of a portion of the momentum from regions of greater velocity to regions of lesser velocity.

In this book we wish to make use of the similarity of different flows of a viscous fluid. For this, the equations for flow must be expressed in dimensionless form so that all dimensional variables that appear in the hydrodynamic equations must be expressed in terms of factors characteristic of these variables. Defining ℓ as our characteristic dimension (size of crystal around which flow occurs or tube in which flow occurs) and U_∞ as our characteristic velocity (far-field stream velocity, for example), all linear dimensions and velocities take the form $Z_i = z_i/\ell$ and $U_i = u_i/U_\infty$, respectively. Expressed in dimensionless form, the *steady flow* of an incompressible fluid without body force yields equations of the form[1]

$$U_k \frac{\partial U_i}{\partial Z_k} = -\frac{\partial P}{\partial Z_k} + \frac{1}{Re} \frac{\partial^2 U_i}{\partial Z_k^2} \qquad (2.7a)$$

$$\frac{\partial U_k}{\partial Z_k} = 0 \qquad (2.7b)$$

where $P = p/\rho U_\infty^2$ is the dimensionless pressure and Re is the Reynolds number

$$Re = \frac{U_\infty \ell}{\nu} \qquad (2.7c)$$

In this example of the steady flow of an incompressible fluid there is only one controlling number, the Reynolds number. Thus, the dimensionless shear force, τ, acting on one square centimeter of surface past which the fluid streams is equal to

$$\tau = f(Re) \qquad (2.8)$$

and is completely determined by the magnitude of Re. In more complex cases such as for unsteady flow or for flow in the presence of an external field of volume forces, etc., other controlling parameters enter along with the Reynolds number. In these cases the flows are similar if the geometrical conditions are similar, the initial and boundary conditions are identical and all the controlling dimensionless numbers have the same respective numerical values. As an additional example to illustrate the point, if we consider the free fall of a sphere of density ρ and diameter a in a fluid of density difference $\Delta\rho$ and viscosity $\bar{\eta}$, the terminal velocity, U, of the sphere is completely determined by two dimensionless numbers, $Re = Ua/\nu$ and $Gr = g(\Delta\rho/\rho)a^3/\nu^2$ where Gr is called the Grashof number. This terminal velocity is governed by the dimensionless relation

$$Re = f(Gr) \qquad (2.9)$$

At high Reynolds numbers, $Re \gg 1$, viscosity effects play a secondary role and, if they are eliminated, the Navier–Stokes equation transforms to the Euler equation (first order differential equation) for an ideal fluid with dimensional velocity potential ϕ given by $\vec{u} = \text{grad } \phi$ having a solution

$$\frac{U}{\rho} + \frac{\partial\phi}{\partial t} + \frac{u^2}{2} + \frac{p}{\rho} = \text{constant} \qquad (2.10)$$

where $\vec{f} = -\text{grad } U$ and yields Bernoulli's theorem if the flow is steady $(\partial\phi/\partial t = 0)$. This theorem is analogous, in some degree, to the energy principle in ordinary mechanics showing that, as we go from regions of higher flow velocity to regions of lower velocity, the pressure of the fluid changes in the opposite direction.

To show that the ideal fluid approximation is inadequate for some problems even at very high Reynolds numbers, we need only consider the example of a sphere moving steadily through an ideal fluid. Reasoning based on the Bernoulli equation shows that the force acting on the front hemisphere is exactly matched by the force acting on the rear hemisphere so that the sphere shouldn't move. Clearly, then, the viscosity exerts a very significant influence in the region immediately adjacent to the surface of a solid body. Combining these two features, one comes to the conclusion that, near the surface of a solid body, there must be a thin zone in which the tangential velocity component undergoes a very abrupt change from a high value at the outer border of the zone to zero at the solid surface. The retardation of the fluid in this boundary layer is caused by viscous forces alone and, although the boundary layer occupies an extremely small volume, it exerts a significant influence on the motion of the fluid. The phenomena that take place in the boundary layer are the sources of hydrodynamic resistance to the motion of solids through fluids.

In a thin boundary layer, all quantities change rapidly in the direction perpendicular to the wall while their tangential rate of change is comparatively small. Moreover, provided the dimensions of the body are large compared to the thickness of the boundary layer, the flow in the boundary layer may be regarded as laminar. Simplistically, the entire zone of motion may be roughly subdivided into two regions: a region of inviscid motion, where Euler's equation applies, and a boundary layer region of thickness δ_m, where viscosity plays an important part. The transition from viscous flow in the boundary layer to inviscid flow in the main stream is smooth and gradual. The quantity δ_m represents the thickness of the region across which the principal change in velocity from zero to U_∞ takes place. By comparing magnitudes of the various terms in the Navier–Stokes equation for a body of dimensions ℓ where the change in velocity along the z-axis takes place over distances of the order of δ_m, one finds that[1]

$$u_z \sim \frac{\delta_m}{\ell} u_x \ll u_x \tag{2.11a}$$

$$\delta_m \sim \ell/Re^{1/2} \tag{2.11b}$$

and

$$\frac{\partial p}{\partial z} \sim \frac{\delta_m}{\ell} \frac{\partial p}{\partial x} \sim \frac{\delta_m U_\infty^2}{\rho \ell^2} = \frac{\nu^2}{\rho} Re^{3/2} \tag{2.11c}$$

Thus, for $Re^{1/2} \gg 1, \delta_m/\ell \ll 1$ and the pressure does not change much in the normal direction to the surface but remains essentially equal to the pressure outside the boundary layer. Therefore, the pressure variation in the x-direction within the boundary layer is determined by the change in pressure outside and this may be determined from Bernoulli's theorem.

2.1.1   Boundary layer relationships

The boundary layer equations allow an exact solution for the case of flow past a semiinfinite plate when its leading edge encounters a fluid moving at velocity U_∞ (see Fig. 2.1(a)). With x referring to the coordinate along the plate and z being perpendicular to the plate, we define a new dimensionless variable $\eta^* = \frac{1}{2}(U_\infty/\nu x)^{\frac{1}{2}} z$ and use the stream function ψ defined by

$$u_x = \frac{\partial \psi}{\partial z}; \quad u_z = -\frac{\partial \psi}{\partial x} \tag{2.12a}$$

Letting ψ have the form

$$\psi = (\nu U_\infty x f)^{1/2} (\eta^*) \tag{2.12b}$$

the mathematical solution yields

$$f = \frac{\alpha \eta^{*2}}{2!} - \frac{\alpha^2 \eta^{*5}}{5!} + \frac{11\alpha^3 \eta^{*8}}{8!} + \cdots \tag{2.12c}$$

where $\alpha = 1.33$. The drag force, F_x, acting on one side of the plate of thickness b is given by

$$F_x = \int_0^b \int_0^\ell \bar{\eta} \left(\frac{\partial u_x}{\partial z} \right)_{z=0} dx\, dz = \frac{\alpha \rho U_\infty^2}{2} b\ell \left(\frac{\nu}{U_\infty \ell} \right)^{1/2} \tag{2.12d}$$

This converts readily to an often used "drag coefficient" defined as

$$K_f = \frac{F_x}{\rho \left(U_\infty^2/2 \right) b\ell} = \frac{\alpha}{Re^{1/2}} \tag{2.12e}$$

In turn, if δ_m is defined as the distance from the surface of the plate to the point where u_x attains a value equal to 90% of the mainstream velocity, U_∞, then one finds numerically that

$$\delta_m = 4.6 \left(\frac{ux}{U_\infty} \right)^{1/2} \tag{2.12f}$$

Qualitatively, the formulae given here for the plate are applicable to an arbitrary crystal surface having small macroscopic curvature. In Fig. 2.1(b), the boundary layer thickness and shear stress ($\tau = K_f$) are given as functions of x along the plate. Finally, we find that

$$\frac{u_z}{U_\infty} = \frac{3}{2} \left[\frac{z}{\delta_m(z)} \right] - \frac{1}{2} \left[\frac{z}{\delta_m(z)} \right]^3 \tag{2.12g}$$

When the plate-like surface is a growing or dissolving crystal, then we

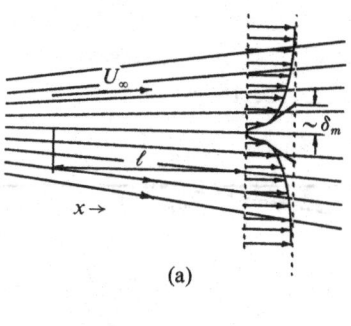

(a)

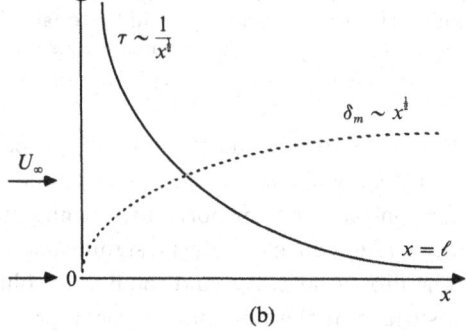

(b)

Fig. 2.1. (a) Schematic illustration of fluid flow past a plate of length ℓ, and (b) boundary layer thickness, δ_m, and shear stress, τ, variations with distance x along the plate.[1]

are not only concerned with the velocity distributions in the fluid but also with the temperature and solute distributions as well. For these state variables, the boundary layer approximation can also be used so that, at a crystal surface, there are three boundary layers formed in the adjacent fluid over which the transport character changes sharply. These are the momentum, δ_m, thermal, δ_T, and solute, δ_C, boundary layers and they are schematically illustrated in Fig. 2.2. A set of scaling relationships exists between these boundary layer thicknesses; i.e.,

$$\delta_C/\delta_m \sim Sc^{-1/3} = (D_C/\nu)^{1/3};$$
$$\delta_T/\delta_m \sim Pr^{-1/3} = (D_T/\nu)^{1/3} \qquad (2.13)$$

where Sc and Pr refer to the Schmidt number and Prandtl number, respectively, and where D_C and D_T are the solute and thermal diffusivity, respectively. Equation (2.11b) or (2.12f) can be used to provide δ_m. In Table 2.1, physical properties are given for four fluids and we note that,

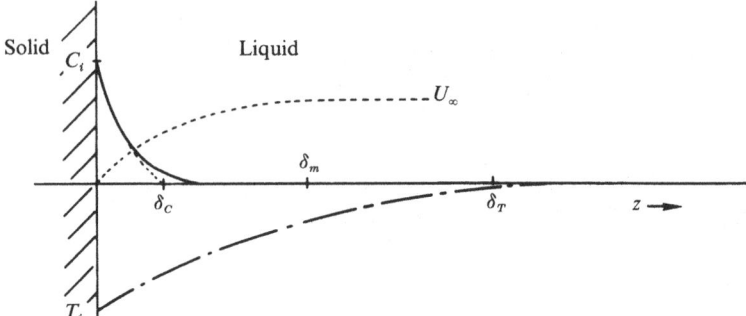

Fig. 2.2. Schematic plots vs. distance, z, for concentration, C, temperature, T, and fluid velocity, U, at a solid/liquid interface illustrating the relationship between the respective boundary layer thicknesses, δ_C, δ_T and δ_m.

comparing H_2O to Hg, $Pr \sim 1$–$\sim 10^{-2}$ while $Sc \sim 1$–10^3, respectively. Thus, for liquid metals $\delta_T >> \delta_m$ while $\delta_C << \delta_m$ so that mild convection has a very strong effect on matter transport but a negligible effect on heat transport. For water, mild convection has a significant influence even on heat transport. The flux of heat, J_T, and the flux of solute, J_C, to or from such a crystal surface in the boundary layer approximation are given by

$$|J_T| = K_L \frac{|T_i - T_\infty|}{\delta_T} \approx 0.22 K_L |T_i - T_\infty| Pr^{1/3} (U_\infty/\nu x)^{1/2}$$

(2.14a)

$$|J_C| = D_{CL} \frac{|C_i - C_\infty|}{\delta_C} \approx 0.22 D_{CL} |C_i - C_\infty| Sc^{1/3} (U_\infty/\nu x)^{1/2}$$

(2.14b)

2.1.2   Natural convection

In the CZ crystal growth technique we encounter both forced and natural convection modes of flow. The rotating crystal produces a fan effect which draws the fluid upwards along the central axis and radially *outwards* in a spiral-like fashion along the upper region of the melt. This driven flow meets the natural convection flow due to the hot crucible wall that causes liquid to rise upward along the crucible wall and *inwards* along the upper region of the melt. The latter flow is generally stronger than the former in most practical situations. These two flows meet generally to produce coupled rotating cells of fluid and, under certain conditions, one type of flow completely dominates over

Table 2.1. *Physical properties of four fluids.*

Fluid			Air	Light oil	H_2O	Hg
Temperature	T	(°C)	27	60	20	100
Density	ρ	(kg m^{-3})	1.28	860	1000	13,400
Dynamic viscosity	$\bar{\eta}$	$\times 10^3$(kg m^{-1} s^{-1})	0.0185	72.2	1.0	1.25
Kinematic viscosity	ν	$\times 10^6$(m^2 s^{-1})	15.7	84	1.006	0.093
Thermal conductivity	K	(W m^{-1} K^{-1})	0.0262	0.14	0.60	10.5
Specific heat	c_p	$\times 10^{-3}$(J kg^{-1}K^{-1})	1.006	2.05	4.18	0.137
Thermal diffusivity	$D_T = K/\rho c_p \times 10^6$(m^2 s^{-1})		22.1	0.080	0.144	5.72
Prandtl number	$Pr = \nu/D_T$ (dimensionless)		0.711	1058	7.01	0.0163
Thermal expansivity	β_T	$\times 10^3$ (K^{-1})	3.33	0.70	0.18	0.182
Buoyancy parameter	$g\beta_T/\nu^2$	$\times 10^9$ (K^{-1} m^{-3})	0.133	0.00097	1.75	206
Buoyancy parameter	$g\beta_T/\nu D_T$	$\times 10^9$ (K^{-1} m^{-3})	0.095	1.03	12.2	3.36
Molecular diffusivity	D_C	$\times 10^6$ (m^2 s^{-1})	2	1	10	50
Schmidt number	$Sc = \nu/D_C$ (dimensionless)		1	10^2	10	500

the other. In this section, we shall concentrate on situations where only natural convection flows occur and shall choose examples that will be helpful to our understanding of practical crystal growth geometries.

Case 1 (Isothermal vertical plate): As a first approximation to the type of flow that one might find near the wall of a large crucible used in CZ crystal growth, let us consider the case of a thin vertical plate of uniform temperature T_w placed in a fluid of infinite extent and of temperature $T_\infty < T_w$. The initial fluid is at rest but, as fluid layers near the plate are heated, they expand to become lighter than layers far from the plate and rise. Here, we consider the steady state buoyancy-driven convective flow (see Fig. 2.3) which is confined to a boundary layer. The upflow near the plate must be compensated by the downflow far from it; however, because of the large extent of fluid, this downflow rate can safely be put at almost zero.

Confining attention to the range of $Pr \sim 1$, we may assume $\delta_T \approx \delta_m$ (see Fig. 2.4 for exact solutions). In this case, defining $\tilde{\eta} = z/\delta_m(x)$, the flow and temperature distributions are given by

$$u(x, z) = u_1 \tilde{\eta}(1 - \tilde{\eta})^2 \qquad (2.15a)$$

$$T(x, z) - T_\infty = (T_w - T_\infty)(1 - \tilde{\eta})^2 \qquad (2.15b)$$

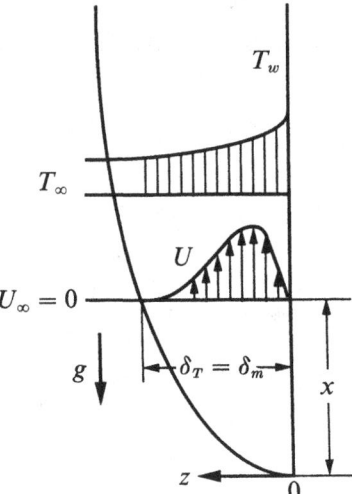

Fig. 2.3. Natural convection boundary layers over an isothermal vertical plate ($Pr \sim 1$).

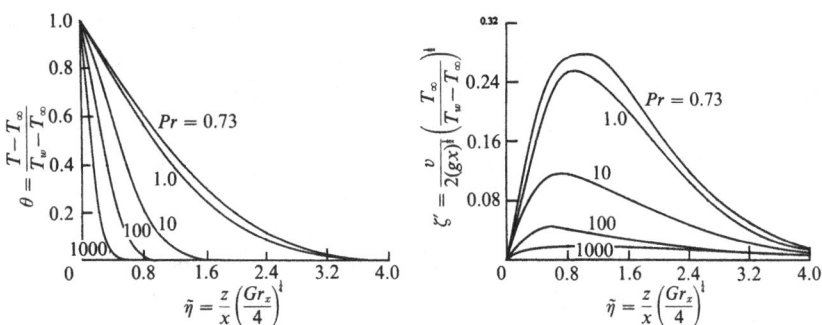

Fig. 2.4. Dimensionless temperature, θ, and velocity, ζ', boundary layers for natural convection near a hot vertical plate.

where

$$u_1(x) = \left(\frac{80}{3}\right)^{1/2} \left(\frac{Gr_x}{\frac{20}{21} + Pr}\right)^{1/2} \frac{\nu}{x} \tag{2.15c}$$

and

$$\frac{\delta_m(x)}{x} = \left[\frac{240\left(1 + \frac{20}{21}Pr\right)}{PrGr_x}\right]^{1/4} \tag{2.15d}$$

In Eqs. (2.15), $Gr_x = g\beta_T(T_w - T_\infty)x^3/\nu^2$ is the local Grashof number.

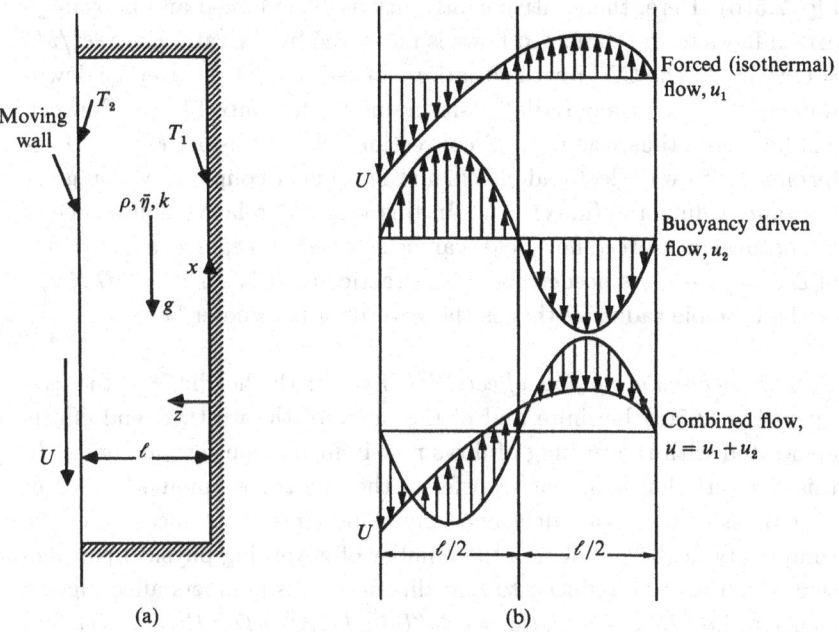

Fig. 2.5. (a) Geometrical and thermal constraints for combined forced and buoyancy-driven flows, and (b) relative velocity profiles for (a).[2]

We note that $\delta_m(x) \sim x^{1/4}$ so that the boundary layer approximation is appropriate for large values of Rayleigh number, $Ra_x = Pr Gr_x$, and laminar flow is expected for $Ra_x < 10^8$–10^9.

Case 2 (Forced and buoyancy-driven flows combined): Consider a fluid confined in a long vertical slot with the vertical sidewalls being spaced a distance ℓ apart and kept at temperatures T_1 and T_2 as illustrated in Fig. 2.5(a). One wall moves downwards at constant velocity U and can be thought to approximate the condition of the forced convection cell in the CZ setup. Here, the fluid velocity u is given in terms of the linear sum of two flows, u_1, a forced isothermal flow and u_2, the buoyancy-driven flow. The solution in this case is

$$u_1(z) = U \left[2\left(\frac{z}{\ell}\right) - 3\left(\frac{z}{\ell}\right)^3 \right] \tag{2.16a}$$

and

$$u_2(z) = \frac{-g\beta_T \Delta T \ell^2}{\nu} \left[\frac{z}{\ell} - 3\left(\frac{z}{\ell}\right)^2 + 2\left(\frac{z}{\ell}\right)^3 \right] \tag{2.16b}$$

The superposition of Eqs. (2.16) leads to the combined flow shown in

Fig. 2.5(b). Here, the relative contribution of the forced and buoyancy-driven flows to the combined flows is measured by the ratio $g\beta_T\delta_T\ell^2/\nu U$ $= Gr/Re$. In the CZ growth situation, an estimate of the average downward velocity $\bar{U}$ at any radial position due to the central forced flow cell can be made, thus, the natural convection cell dimension, ℓ, can be determined (if we neglect end effects and non-linear coupling) which gives the same value of $u_2(\max) \approx \bar{U}$. In this way, the relative dominance of the natural and forced flow fields can be assessed. Of course, a knowledge of $\Delta T = T_1 - T_2$ is needed for this. Practically, it is $\Delta T/\ell \approx (\partial T/\partial r)_R$ at the crucible radius R that is the governing parameter here.

Case 3 (Recirculating flow effects): [2] If we let the height $\tilde{H}$ of the cavity in Fig. 2.5(a) be finite and of the order of the width ℓ, end effects associated with the turning of flows now influence the central region. In this case, friction and heat transfer in the end zones amount to significant parts of the total effect and cannot be ignored. A measure of the complexity of the problem is the number of governing physical parameters which may be reduced to four dimensionless numbers affecting the solution; i.e., $Re = U\ell/\nu, Ra = g\beta_T(T_2 - T_1)\tilde{H}^3/\nu D_T; Pr = \nu/D_T$ and $\tilde{L} = \tilde{H}/\ell$. To demonstrate this complexity, consider Figs. 2.6(a) and (b) $(\theta = T - T_w)$ which show numerical solutions to two simpler problems: the cavity flow without buoyancy and the buoyancy-driven flow in a cavity. The gross behavior of the driven cavity flow is the recirculating fluid motion portrayed by the closed streamlines in the bulk of the cavity. However, at the corners, far from the moving wall, small recirculation zones (corner vortices) exist at the Reynolds numbers shown. The associated temperature distributions of this flow, portrayed by the isotherms in Fig. 2.6(a), show how the thermal stratification, from being confined to the hot moving wall at low Reynolds numbers, is pushed towards the stationary cold wall as Re is increased leaving a nearly isothermal core of rotating fluid.

The example of buoyancy driven flow in a square cavity (Fig. 2.6(b)) illustrates the effect of increasing the Rayleigh number on the streamline patterns and the isotherms. The flow regimes represented here range from that of conduction ($Ra \sim 100$) via that of transition ($Ra \approx 2500$) to the boundary layer regime ($Ra \approx 25 \times 10^3$–$100 \times 10^3$).

Such non-parallel flows as illustrated by Fig. 2.6 arise from the complexity of the full Navier–Stokes equations which are non-linear partial differential equations of the elliptic-type whose solutions depend upon the magnitude of Re. For increasing Reynolds number, the non-linear

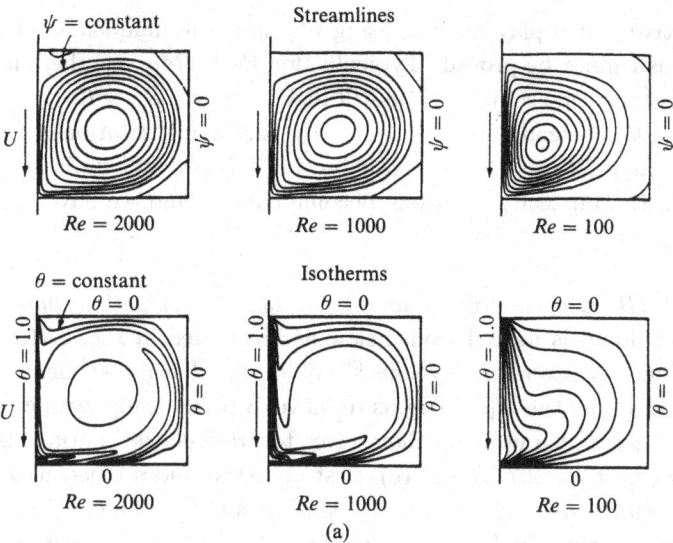

(a)

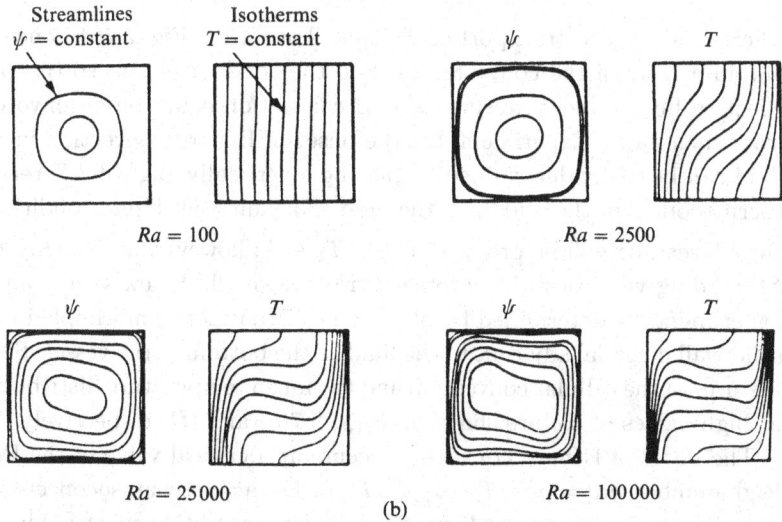

(b)

Fig. 2.6. (a) Effect of Reynolds number, *Re*, on driven square cavity flow; computed streamlines and isotherms ($Pr = 1$) for $\theta = 1$ on the moving wall and $\theta = 0$ on the other walls, and (b) effect of Rayleigh number, *Ra*, on buoyancy-driven flow in a square cavity, having isothermal sidewalls and insulated top and bottom walls ($Pr = 0.73$) and giving streamlines, ψ, and isotherms, *T*.

convective terms play an increasing role and the uniqueness of steady solutions cannot be proved. Typically, for $Re > Re_1$, say, there may exist more than one solution. For still higher values, $Re > Re_2$, say, there may exist no steady solution but only oscillating solutions. For still higher values, $Re > Re_c$, which is beyond the stability limit of laminar motion, the unsteady solutions become chaotic and we have turbulent flow.

Case 4 (Bénard convection and the Rayleigh instability condition): When a liquid is heated from below as illustrated in Fig. 2.7(a) ($T_2 > T_1$), a density inversion develops.[3] When the density gradient exceeds a critical value, the liquid begins to move in periodically arranged vertical columns which close on themselves to produce the continuous loops of flow (see Figs. 2.7(b) and (c)) first observed and studied by Bénard. This abrupt onset of convection leads to an abrupt departure of the Nusselt number, Nu, from unity at the critical Rayleigh number, Ra^c, as indicated in Fig. 2.7(d). The Nusselt number, Nu, is defined as the ratio of the total heat flux, q_w, to the conducted heat flux; i.e.,

$$Nu = \frac{q_w}{-K_L \left(\partial T/\partial z\right)_w} = \frac{\tilde{h}(T_2 - T_1)}{-K_L \left(\partial T/\partial z\right)_{T_2}} \approx \frac{\tilde{h}d}{K_L} \qquad (2.17)$$

where $\tilde{h}$ is the heat transport coefficient. Thus, from Fig. 2.7(d), one can obtain a ratio of the convected heat to the conducted heat so that the effective fluid velocity during Bénard convection could be deconvolved from such data. Experimentally, the onset of Bénard convection in the fluid can be easily determined by placing a vertically aligned differential thermocouple in the middle of the fluid. For quiescent liquid conditions, the δT reading will be proportional to $T_2 - T_1$; however, at $Ra > Ra^c$, the δT reading will abruptly become extremely small. Likewise q_w can be experimentally determined by placing a differential thermocouple in the solid wall material adjacent to the fluid at the bottom of the vessel. Some features of the cellular convection and the mean temperature distribution at high values of Ra are shown in Figs. 2.7(e) and (f), respectively.

The onset of Bénard convection occurs at a critical value of the Rayleigh number, $Ra = g\beta_T (T_2 - T_1)d^3/D_T \nu$. For a quiescent isoconcentrate fluid in a temperature gradient, the following condition must hold:

$$\mathcal{G}_L = \frac{\Delta T}{d} < \mathcal{G}_L^c = \frac{Ra^c D_T \nu}{g\beta_T d^4} \qquad (2.18a)$$

where $Ra^c = 1700 \pm 140$. It is well known that a density inversion due to a solute gradient in a solvent can also produce this mode of convection

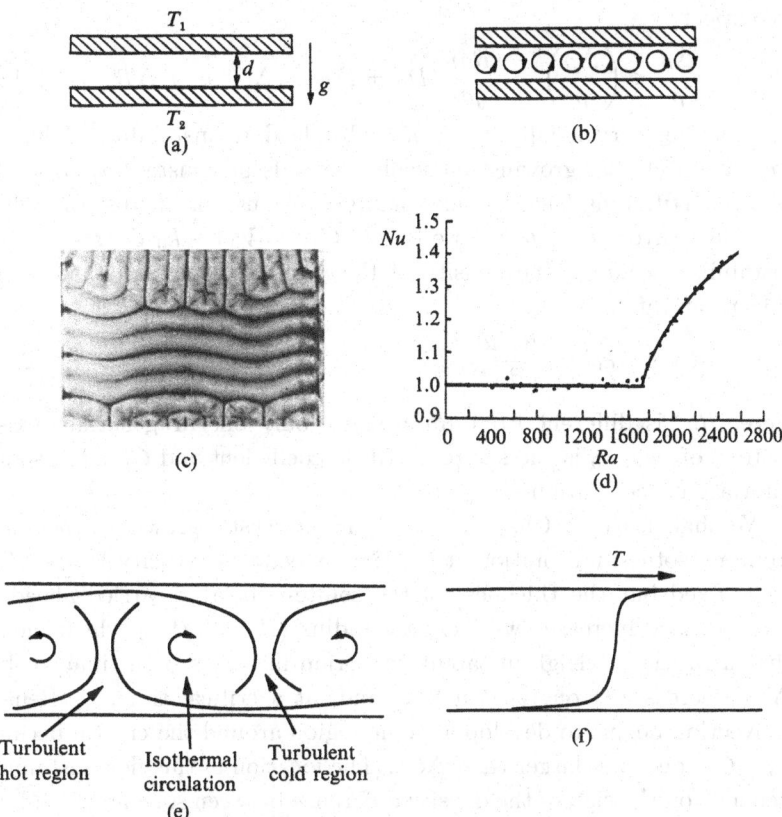

Fig. 2.7. (a) Definition sketch for Bénard convection, (b) convective roll cells at Ra slightly above Ra^c, (c) Bénard convection at $Ra = 3 \times 10^3$ and $Pr = 367$, (d) variation of Nusselt number with Ra showing the increase of heat transfer for $Ra > Ra^c$, (e) principal features of cellular convection in the boundary layer regime, and (f) the general form of mean temperature profile for Bénard convection at large Ra.

(see Eq. (2.3a)). For a quiescent isothermal fluid in an inverse density gradient produced by a concentration difference, the following condition must hold

$$\left(\frac{\partial C}{\partial z}\right) = \frac{\Delta C}{d} < \left(\frac{\partial C}{\partial z}\right)^c = \frac{Ra^c D_C \nu}{g \beta_C d^4} \qquad (2.18b)$$

Combining these two conditions, for a non-isothermal and non-isoconcentrate fluid, one would expect that it is a critical density gradient that sets the fluid in motion so that, for a quiescent liquid condition to hold, we expect

$$\frac{\Delta\rho}{\rho} < \left(\frac{\Delta\rho}{\rho}\right)^c = \frac{Ra^c\nu}{gd^3}(D_C \pm D_T) = \Delta C^c \beta_C \pm \Delta T^c \beta_T \quad (2.18c)$$

In practice, very small values of $\Delta C/d$ lead to instability. Thus, if we have a crystal growing vertically upwards at constant velocity, V, solute partitioning (see the next chapter) tells us that, across the solute boundary layer, $d \approx D_C/V$, we have $\Delta C/d \approx V(1 - k_0)C_i/D_C$ so that instability occurs if the density of the solute is less than that of the solvent and if

$$C_i(1 - k_0) > \frac{Ra^{c\prime}\nu V^3}{g\beta_C D_C^2} \quad (2.19)$$

Here, $Ra^{c\prime}$ is different from Ra^c because only one drag surface exists instead of two, k_0 is the solute partition coefficient and C_i is the solute interface concentration.

We shall learn in Chapter 4 that, for a crystal growing spherically from an isothermal solution at the time-dependent velocity $V = \varepsilon t^{-1/2}$, C_i is fixed but the thickness of the solute-rich layer carried ahead of the interface increases with crystal radius ($R = 2\varepsilon t^{1/2}$) which means that d in the Rayleigh instability criterion increases with time so that ΔC^c decreases strongly with time and, at a critical time, t_c, Bénard convection begins to develop at some region around the crystal because $C_i - C_\infty$ becomes larger than ΔC^c. The location of this flow instability depends on the sign of the density difference between solvent and solute. The abrupt onset of such flow generally leads to conditions favoring the onset of dendritic growth for such crystals. Space-grown crystals do not suffer from this type of convective instability so that relatively large polyhedral crystals of "difficult to grow materials" like isocitrate lyase have been grown in low gravity satellite orbits.

Using these concepts, a convection-free growth cell illustrated in Fig. 2.8 was designed for the growth of crystals. Bulk liquid is doped from the vapor and stirred completely while the porous plug restricts the convection to the large outer region of liquid. The seed temperature is programmed and d is calculated to be less than that which would produce Rayleigh instability.

Case 5 (Marangoni convection): Reactive processes, adsorption–desorption phenomena, evaporation, deformation of interfaces and local temperature gradients are all mechanisms that can induce surface tension inhomogeneities leading to convection in adjacent fluid phases. On some

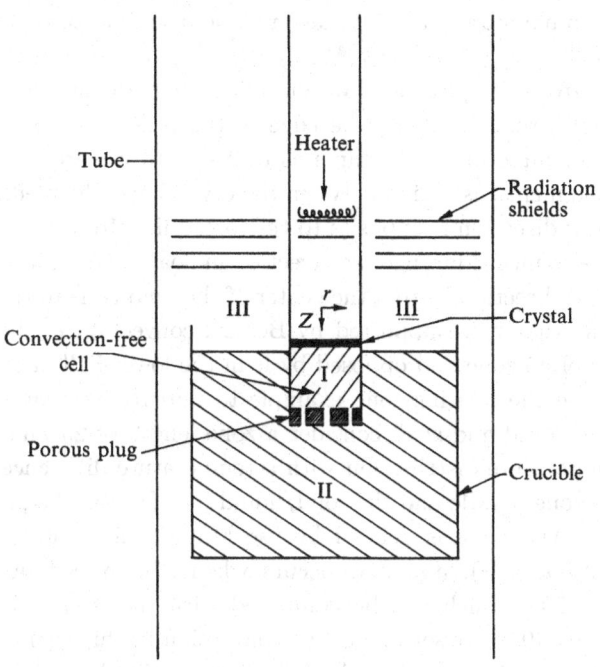

Fig. 2.8. Schematic illustration of convection-free cell for crystal growth (Region I) imbedded in a well-mixed alloy melt volume (Region II) in contact with the gas phase (Region III).

occasions, it is sufficient to have merely the coupling of phenomena at an interface regardless of whether or not each of the processes is destabilizing by itself. Sometimes the origin of the instability can be a large disparity in time or space scales of the coupled processes such as one finds in oscillating chemical (or biochemical) reactions. For our purposes here, we will restrict our considerations to surface variations of temperature and composition as indicated in Eq. (2.4).

The basic consideration which introduces Marangoni convection is the shear stress balance at the interface between two fluid phases 1 and 2; i.e.,

$$\bar{\eta}_1 \frac{\partial u_{x1}}{\partial z} - \bar{\eta}_2 \frac{\partial u_{x2}}{\partial z} = \frac{\partial \gamma}{\partial x} = \frac{\partial \gamma}{\partial T} \frac{\partial T}{\partial x} + \frac{\partial \gamma}{\partial C} \frac{\partial C}{\partial x} \qquad (2.20)$$

For the majority of pure liquids $\partial \gamma / \partial T < 0$ and, since surface atoms move from positions of high chemical potential to lower chemical potential, we should expect the surface diffusion to occur from regions of high γ (low T) to regions of lower γ (high T). Thus, the compensating surface flow to maintain a flat interface will be from hot (low γ) to cold (high γ).

In crystal growth applications we meet Marangoni convection in CZ crystal growth primarily at (i) the edge of the melt with the crucible where the flow direction is the same as that due to Bénard convection and (ii) in the meniscus region between the crystal and the melt surface where the flow direction is opposite to the forced flow from the rotating crystal. We also meet Marangoni convection in float-zone crystal growth where the flow direction is from the center of the fluid zone to the cooler zone edges so that it is supported by Bénard convection in the upper half of the molten zone and opposed by it in the lower half of the zone.

We shall use the floating-zone example to illustrate the major considerations involved and shall consider a zone length of $2h$ and a symmetrical temperature distribution with a temperature difference of ΔT between the zone middle and the solid/liquid interface on the periphery of the zone. The force is defined by the Marangoni number, $Ma = -(\partial\gamma/\partial T)(\Delta T h/\bar{\eta}D_T)$, and flow occurs when Ma exceeds a critical value $Ma^c = 79.6$ which can be compared with the critical Rayleigh number $Ra^c = 669$ for onset of Bénard convection in this geometry. To compare these two flows we use the following relationship coming from basic definitions:

$$Ra = \frac{g\beta_T \Delta T h^3}{D_T \nu} = \frac{g\beta_T \rho h^2}{|\partial\gamma/\partial T|} Ma \tag{2.21}$$

Since $\partial\gamma/\partial T \sim -0.1$ to -0.5, $g \sim 10^3$ and $\beta_T \sim 10^{-5}$, Eq. (2.21) shows that $Ra \ll Ma$ for most systems unless $h \gg 1$ cm. Stated another way, we can see that Marangoni convection drives will dominate in small systems while Bénard convection drives will dominate in large systems. For our float-zone example with $\Delta T = 1$ °C, $h = 1$ cm and $\partial\gamma/\partial T = -0.4$, $Ma \sim 10^3 >> Ma^c$ whereas $Ra \equiv 10^2 << Ra^c$. Thus, Marangoni convection rather than Bénard convection drives the fluid motion in this case.

On earth, a coupled flow of thermal Marangoni convection and buoyancy-driven convection is present in floating zones and is governed by a new similarity number, the dynamic Bond number, $\hat{B}$, defined as

$$\hat{B} = \frac{Gr}{Re_M} = \frac{\rho g D_T (2r)^2}{|\partial\gamma/\partial T|} \tag{2.22a}$$

where

$$Re_M = \rho \left|\frac{\partial\gamma}{\partial T}\right| \frac{\Delta T 2r}{\bar{\eta}^2} \tag{2.22b}$$

is the Marangoni Reynolds number and r is the radius of the floating zone. Each of these drives gives rise to a certain flow velocity; i.e., the Marangoni velocity, U_M, is given by

$$U_M = |\partial\gamma/\partial T|\Delta T/\bar{\eta} \qquad (2.23a)$$

while the buoyancy-driven velocity, U_g, is given by

$$U_g = gD_T\Delta T(2r)^2/\nu \qquad (2.23b)$$

We find that the dynamic Bond number is related to the velocity ratio U_g/U_M as follows:

$$U_g/U_M = \hat{B}^{1/2}Re_M^{-1/6}(2r)^{-5/6} \qquad (2.23c)$$

Equation (2.23c) makes it easy to decide when flow domination is by thermal Marangoni convection ($U_g/U_M \widetilde{<} 0.1$) and when it is by buoyancy-driven convection ($U_g/U_M > 10$).

When one exceeds a critical Marangoni number $Ma^{c^*} \sim 10^4$, typically, the steady thermal Marangoni convection in the zone shows a transition to an oscillatory state of convection in the low frequency range ($f \sim 1$–10 Hz). The disturbance is a traveling rather than a standing wave with the circumference of the zone being an integral number of wavelengths. These oscillatory flow disturbances generate temperature oscillations in the fluid which, in turn, influence the crystallization velocity such that velocity oscillations develop which lead to periodic impurity striations in the crystal. If we consider CZ crystal growth for a moment and focus attention on the meniscus region, we can ask if $Ma > Ma^{c^*}$. We shall use the definition of Ma in the form

$$Ma = |\partial\gamma/\partial T|\,(\partial T/\partial y)\,h^2/\bar{\eta}D_T \qquad (2.24)$$

Since $\bar{\eta} \sim 10^{-2}$, $D_T \sim 10^{-1}$ and $|\partial\gamma/\partial T| \sim 0.4$, if $h \sim 0.1$–0.2 cm and $\partial T/\partial y \sim 10$ °C cm^{-1} then $Ma^c < Ma < Ma^{c^*}$. However, at higher rotation rates, the interface rises higher in the meniscus so that h increases to ~ 0.5–1 cm and $\partial T/\partial y$ increases to ~ 100 °C cm^{-1} and now $Ma \widetilde{>} Ma^{c^*}$ in the meniscus region.

Case 6 (Temperature oscillations due to natural convection): As already mentioned, the rolling fluid due to natural convection loops leads to thermal fluctuations in the liquid. The magnitude of the temperature fluctuations may be small in metals, $\delta T \widetilde{<} 1$ °C, but can be quite large in insulators, $\delta T \widetilde{<} 50$ °C, and thus has a significant influence on crystal growth. Perhaps one of the best examples which illustrates this effect is shown in Fig. 2.9 where we see, for a liquid metal, the effect of interface orientation on the buoyancy force and fluctuations. At the left,

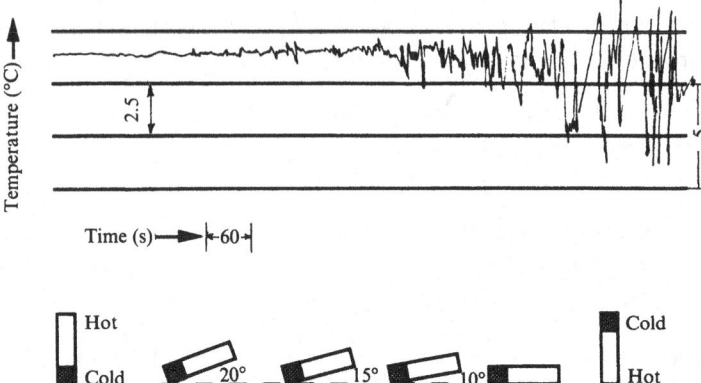

Fig. 2.9. Effect of interface orientation in buoyancy-force and temperature fluctuations.

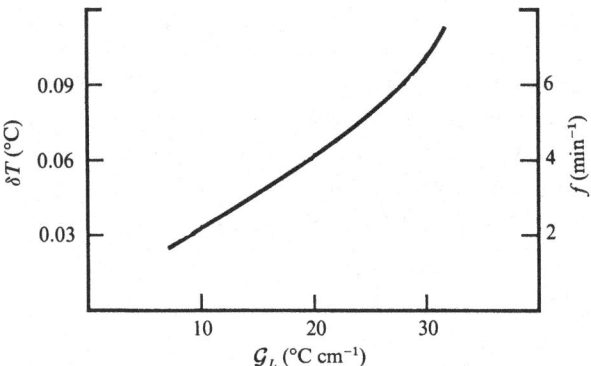

Fig. 2.10. Magnitude, δT, and frequency, f, of melt temperature oscillations as a function of temperature gradient, $\mathcal{G}_L$, in a horizontal crucible for an Mn–Bi melt.

no convection occurs because there is no buoyancy-driving force; as the force increases, the flow increases and reaches a maximum on the right for heating from below. We note that the amplitude of the temperature oscillations also increases as the flow increases.

As a second example, during the growth of a BiSb crystal in a horizontal crucible at $V \sim 10^{-5}$ cm s^{-1} and a temperature gradient, $\mathcal{G}_L$, bands of concentration excess aligned parallel to the interface and with a periodic spacing ~ 5–10 μm were found to form in the crystal. From this spacing and interface velocity, the banding frequency, $f \sim 1$ min^{-1}. A measure of f and δT as a function of $\mathcal{G}_L$ is given in Fig. 2.10.

To understand the effect of different material parameters on this fluid oscillation phenomenon, we can think of a rotating tube of fluid connecting the hot region (T_h) with the cold region (T_c). Half of it tends to pick up excess heat while it is in the hotter region and to lose heat while in the cooler region. The ability to transfer heat from the tube (reduce δT) thus increases as K_L increases, as ν increases and as f decreases.

One can stop these fluid oscillations and suppress Bénard convection in electrically conducting systems by the application of a strong DC magnetic field. This magnetohydrodynamic effect is significant for magnetic fields $\sim 10^3$–2×10^3 G. One can also stop fluid oscillations by rotating the fluid and crucible even for non-conducting systems. In this case ~ 10 rpm is slightly more effective than a 1 kG magnet on an electrically conducting system. Several years ago, Japanese investigators used a DC magnetic field ($H \sim 1$ kG) with CZ growth of Si crystals and found that the O concentration of the crystals was significantly reduced, presumably due to the greatly reduced fluid flow velocity and thus the reduced drag on the SiO$_2$ crucible containing the melt leading to less crucible dissolution into the melt.

In a reference frame rotating at angular velocity $\hat{\Omega}$, the Coriolis force acts as a body-force on the Navier–Stokes equations. One finds that the critical Rayleigh number for the onset of Bénard convection increases with the Taylor number, $\hat{T} \equiv 4\hat{\Omega}^2 d^4/\nu^2$ where d is the radius of the fluid cylinder. In the limit of large $\hat{T}$, the relationship is simply given by

$$Ra^c = C\hat{T}^{2/3} \tag{2.25}$$

where C is a constant that depends slightly but definitely on the boundary conditions ($C = 8.7$ for the two free boundary case). In the magnetic field case, the parameter $\hat{Q}/\pi^2$ plays the same role as $\hat{T}/\pi^4$ in the foregoing where $\hat{Q} \equiv \mu^{*2}\hat{H}^2\sigma_*d^2/\rho\nu$. Here, μ^* is the magnetic permeability, σ_* is the electrical conductivity and $\hat{H}$ is the magnetic field strength. In Figs. 2.11(a) and (b) the theoretical relationships between Ra^c and either $\hat{T}$ or $\hat{Q}$ plus experimental data for Hg are given.[4]

Case 7 (Vertical Bridgeman configurations): Buoyancy-driven convection in closed vertical cylinders heated from below and crystal growth in corresponding configurations (vertical Bridgeman technique with top seeding) have been studied both theoretically and experimentally and the results strengthen our understanding of natural convection phenomena.[5] In this case, three dimensionless numbers characterize the hydro-

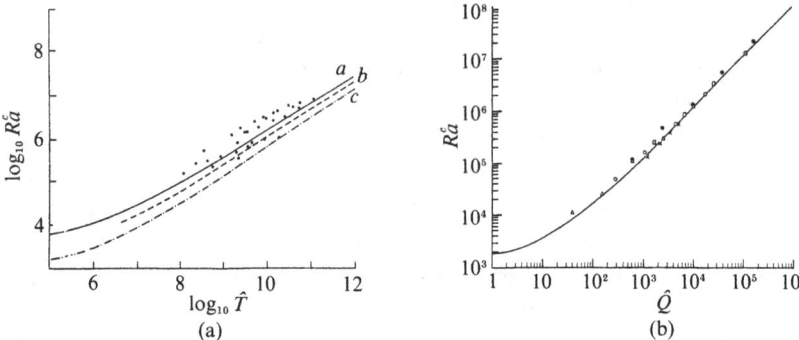

Fig. 2.11. Critical Rayleigh number, Ra^c for Hg as a function of Taylor number, $\hat{T}$, (a) Curves aa, bb and cc are theoretical for (i) both bounding surfaces rigid, (ii) one rigid and one free and (iii) both free at $Pr = 1.025$. The solid circles are the experimental data points for Hg at $Pr = 1.025$. (b) Full line curve is theory while solid circles, squares, open circles and triangles are experimental points with a large magnet for liquid layers at depths of $d = 6, 5, 4$ and 3 cm respectively while the four crosses are for a small magnet ($H = 1500$ G) at the same depths, respectively.[4]

dynamic state of the fluid inside the cavity, Ra, Pr and $\tilde{L}$, the aspect ratio of the crucible of height $\tilde{H}$ and diameter ℓ. Flow experiments were carried out in test cells with H_2O and Ga before crystals of GaSb were grown.

The experimental flow results for H_2O and Ga are summarized in the Ra versus $\tilde{L}$ plots of Fig. 2.12. For H_2O, convective flow sets in at Ra^{c1} which increases with $\tilde{L}$ and a similar dependence is found for the transition from steady convective flow to the unsteady turbulent state at Ra^{c3} ($Ra^{c3} \approx (50\text{--}300)Ra^{c1}$). Different flow patterns, drawn on Fig. 2.12(a), were observed depending on Ra and $\tilde{L}$. For Ga, steady flow sets in at values of $Ra^{c1}(L)$ which are very close to those obtained for H_2O but unsteady convection sets in at much lower Ra values (see Fig. 2.12(b)). A periodic flow range for $Ra^{c2} < Ra < Ra^{c3}$ is observed which was not found in H_2O and the transition to unsteady turbulent convection occurs at a small value of Ra^{c3} ($Ra^{c3} \approx (1.5\text{--}13)Ra^{c1}$). In addition, a striking result is the extreme extension of the laminar convective range for $\tilde{L} \approx 1$ ($Ra^{c2} \sim 8Ra^{c1}$) compared to the $\tilde{L} \neq 1$ configurations. The symmetries of the different convective flows for Ga were found to be very similar to those observed in H_2O; i.e., axisymmetric (ax) for $\tilde{L} = 0.5$

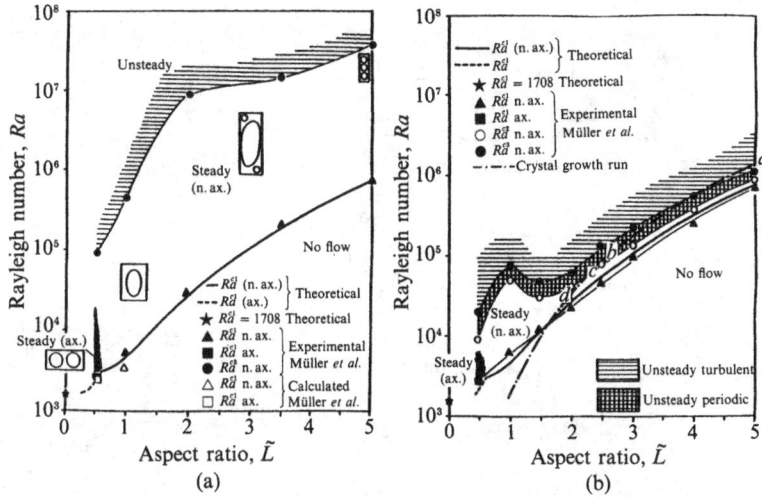

Fig. 2.12. Rayleigh number, Ra, versus crucible aspect ratio, $\tilde{L}$, for (a) H_2O and (b) Ga systems. (Courtesy of G. Müller.)

and non-axisymmetric (n.ax) for $\tilde{L} > 1$. Theoretical calculations of the different flow modes and symmetries were in remarkably good correlation with the experimental data of Fig. 2.12.

Te-doped GaSb crystals were grown with top seeding in a 10 mm diameter and 60 mm long vertical Bridgeman arrangement. The axial temperature gradient was approximately 8 K cm^{-1} and temperature was measured with four protected thermocouples positioned at various heights in the thin quartz tube. Crystal growth from the seed was achieved by a downward movement of the furnace at 0.1 mm min^{-1}. Typical temperature results were observed with decreasing melt length: non-periodic temperature fluctuations for tall melt columns ($\tilde{H} \approx 60-30$ mm), a transition to periodic temperature oscillations (period ≈ 30 s) for melt heights below 30 mm ($\tilde{L} \lesssim 3$) and no temperature fluctuations at all for $\tilde{H} < 24$ mm ($\tilde{L} < 2.4$). These results are shown in Fig. 2.13 with the correlation to the observed striation patterns. The four typical results for one growth run are depicted in Figs. 2.13(a)–(d) and we note that a clear correlation exists between the temperature fluctuations in the melt and the growth striations in the crystal. A quantitative correlation of these samples with the convective flow patterns is shown in Fig. 2.12(b) via the points a-d on the dashed-dotted curve. We note that the pairs of values of Ra and $\tilde{L}$ corresponding to Figs. 2.13(a)–(d) can be

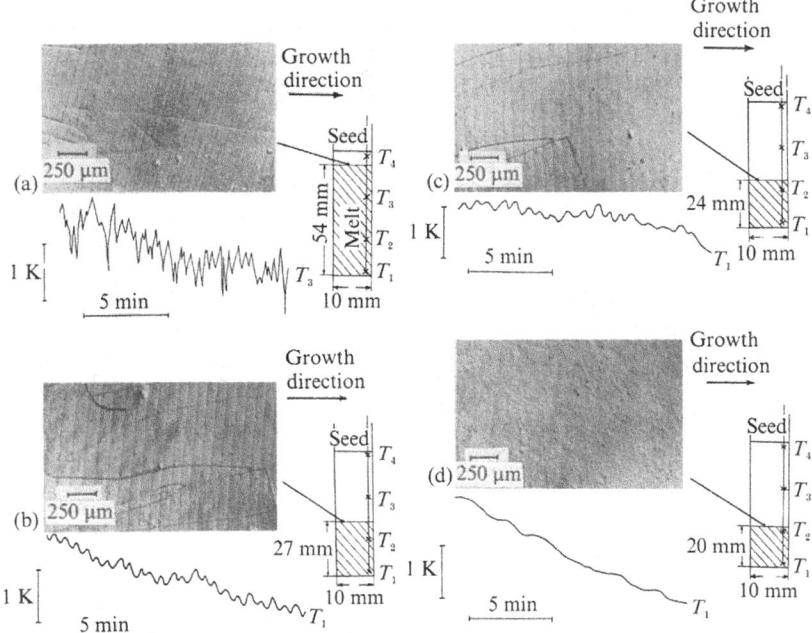

Fig. 2.13. Temperature fluctuations and correlated grown-in striations for the InSb system during downward growth in a verticle crucible
at different residual melt heights, (a) 5.4 cm, (b) 2.7 cm, (c) 2.4 cm
and (d) 2.0 cm. (Courtesy of G. Müller.)

clearly correlated to the ranges of turbulent convection (a), oscillatory
convection (b), to the transition from oscillatory to steady convection
(c) and to steady convection (d), respectively. We can expect all flow
configurations to be partitioned into different regimes as depicted in Fig.
2.14.

2.1.3 Forced convection

Among the advantages of forced convection in the crystallization process are thought to be: (1) homogenization of the liquid, (2)
development of thermal and solutal symmetry in the crystal, (3) control of interface shape, (4) influence of the crystallization velocity via
control of the heat balance at the interface, (5) manipulation of solute
partitioning at the crystallization front, and (6) the level of force can
be adjusted at the crystal grower's desire. One of the obvious disad-
vantages is the dissolution of the crucible containing the nutrient phase.
Perhaps the most famous forced convection example in crystal growth

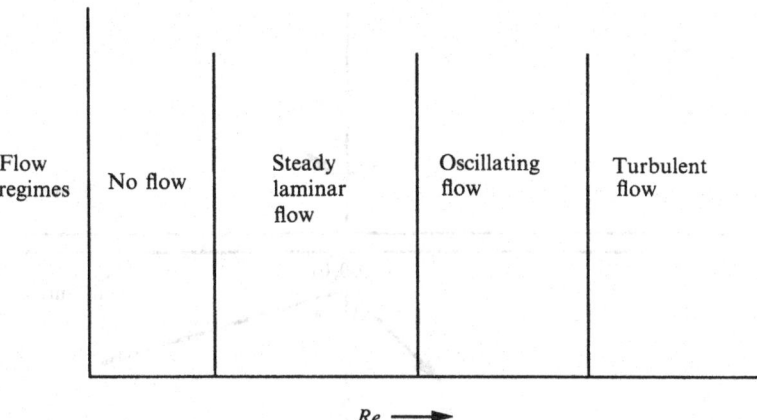

| Flow regimes | No flow | Steady laminar flow | Oscillating flow | Turbulent flow |

$Re \longrightarrow$

Fig. 2.14. The four typical fluid flow regimes as a function of Reynolds number, Re.

is that present in the CZ technique. We know a great deal about this flow situation because, to first order, it is commonly thought to approximate the Cochran flow due to an infinite disc rotating on a semiinfinite fluid. We shall begin our treatment of laminar forced convection by a description of the Cochran analysis and its consequences for CZ crystal growth.

Case 1 (Laminar motion produced by a rotating disc): [6] We consider an infinite plane disc at $z = 0$ rotating in a viscous medium around the axis $r = 0$ with constant angular velocity ω and study the motion of the fluid in the half-space $z > 0$ (see Fig. 2.15(a)). The boundary conditions of the problem have the form:

$$u_r = 0, \quad u_\phi = r\omega, \quad u_z = 0 \quad \text{for} \quad z = 0 \qquad (2.26a)$$

$$u_r = 0, \quad u_\phi = 0 \qquad\qquad \text{for} \quad z = \infty \qquad (2.26b)$$

The velocity component $u_z \neq 0$ occurs at $z = \infty$ because the disc acts as a centrifugal ventilator and causes suction resulting in a negative value of u_z at infinity, as well as a radial motion away from the center near the disc.

Since the momentum boundary layer thickness $\delta_m \sim (\nu/\omega)^{1/2}$ in this case, it is appropriate to replace z by a dimensionless distance $\zeta = z/\delta_m = z(\omega/\nu)^{1/2}$ and to make the following changes in variables

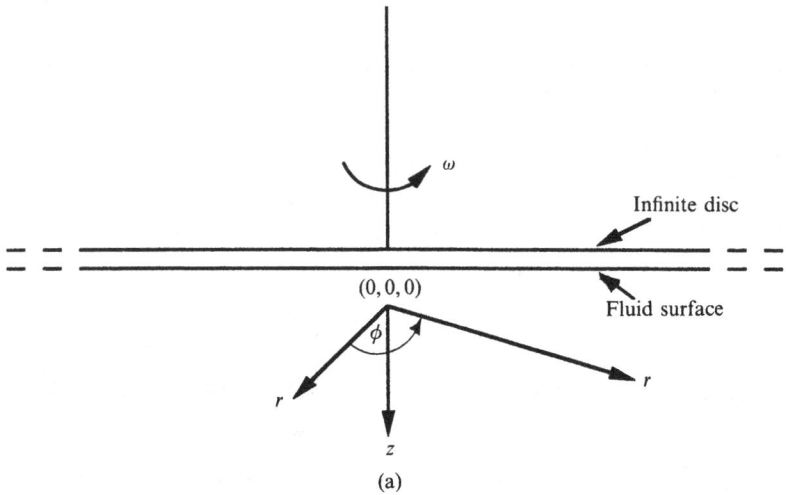

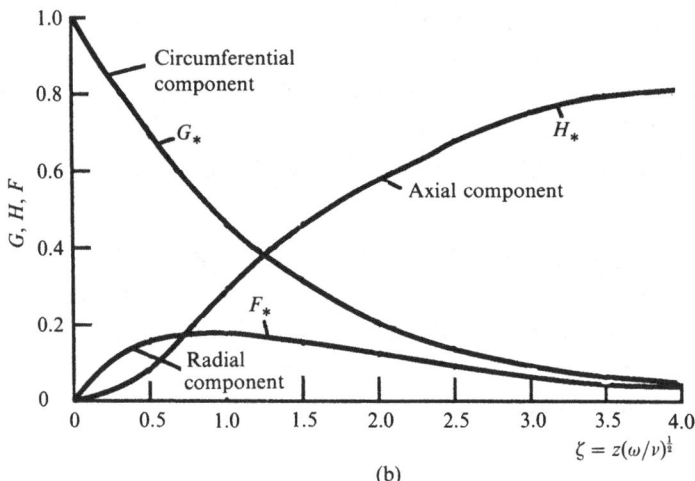

Fig. 2.15. (a) Schematic representation of the infinite radius rotating disc used in Cochran's analysis and (b) his theoretical velocity distribution functions G_*, H_* and F_* near a disc rotating on a liquid initially at rest.[6]

$$u_r = r\omega F_*(\zeta), \quad u_\phi = r\omega G_*(\zeta),$$
$$u_z = (\nu\omega)^{1/2} H_*(\zeta), \quad p = -\rho\nu\omega P_*(\zeta) \qquad (2.26c)$$

Cochran showed that, if $H_* \to -c$ for $\zeta \to \infty$, F_*, G_*, and H_* could be expanded in powers of $\exp(-c\zeta)$, to satisfy the $\zeta \to \infty$ condition, as well as in powers of ζ to satisfy the boundary conditions at $\zeta \to 0$.

In this way, he was able to determine the functional forms for F_*, G_* and H_* (see Fig. 2.15(b)) which are given in detail in Table 2.2. Near $\zeta = 0$ we find that $F_* \approx 0.51\zeta$, $G_* \approx 1 - 0.62\zeta$ and $H_* = -0.51\zeta^2$ so that $u_z \approx 0.51\omega^{3/2}\nu^{-1/2}z^2$, $u_r \approx 0.51r\omega^{3/2}\nu^{-1/2}z$ and $u_\phi = r\omega(1 - 0.62\omega^{1/2}\nu^{-1/2}z)$. The graphs in Fig. 2.15(b) show that the distance from the disc at which the circumferential component of the flow velocity drops to half its value at the disc is $\delta_{0.5} \sim (\nu/\omega)^{1/2}$ which is ~ 0.1 cm for $\nu \sim 10^{-2}$ and $\omega = 60$ rpm. Both F_* and G_* are independent of r and $c = 0.886$ in the Cochran analysis. For these same values of ν and ω, we see from Table 2.2 that H_* is very close to $-c$ at $z \approx 0.5$ cm; thus, even for fluid depths $\stackrel{\sim}{<} 2$ cm, this analysis should give a reasonable approximation to the actual forced flow conditions. We might expect that the effect of a finite rather than an infinite disc radius would involve comparable transition regions; i.e., the outer few millimeters of the disc might begin to show a radial dependence in F_*, G_* and H_*. However, the effect of an opposing natural convection flow at the outer edge of a finite radius disc will certainly influence the radial dependence of the flow.

The volume of fluid pumped outwards by the rotating disc per second as a result of the centrifugal fan effect, $\tilde{Q}$, is given by

$$\tilde{Q} = 2\pi R \int_0^\infty u_r \mathrm{d}z = 0.886\pi R^2 (\nu\omega)^{1/2} \qquad (2.26d)$$

which is just the axial flow from $\zeta \to \infty$ over this disc area. The average flow velocity of this fluid is $0.886\nu\omega^{1/2}$ cm s^{-1} so that, in a crucible of radius R_* with a crystal of radius R_c, the average return velocity of the flow, $\bar{U}_R$ that would be competing with the natural convection flow is

$$\bar{U}_R \approx 0.886(\nu\omega)^{1/2} \left(\frac{R_c^2}{R_*^2 - R_c^2} \right) \qquad (2.27)$$

In addition, the shearing stress on the disc due to the fluid is given by $\tau_w \sim \rho r\omega^2 \delta_m \sim \rho r\omega(\nu\omega)^{1/2} = \rho r\nu^{1/2}\omega^{3/2}G'(0)$ and the resistive torque, M, of a disc of radius R_c, which is the product of the shearing stress, area and moment arm, will be

$$M \sim \tau_w R_c^3 \sim \rho R_c^4 \omega(\nu\omega)^{1/2} = 0.308\pi\rho R_c^4(\nu\omega^3)^{1/2} \qquad (2.28)$$

Finally, the transition to turbulence for this type of flow occurs for a Reynolds number, $Re = R_c^2\omega/\nu \sim 3 \times 10^5$, so that, even for $R_c \approx 12.5$ cm, ω would need to be extremely high for CZ Si crystal growth to generate turbulent flow due to forced convection.

To consider the effect of crystal growth on these flows, people have considered the effect of a uniform suction condition over the surface of

Table 2.2. *Values of the functions determining the distribution of velocities and pressure near a disc rotating in a fluid at rest.*[6]

$\zeta = z(\omega/\nu)^{1/2}$	F_*	G_*	$-H_*$	$-P_*$	F'_*	$-G'_*$
0	0	1.000	0	0	0.510	0.616
0.1	0.046	0.939	0.005	0.092	0.416	0.611
0.2	0.084	0.878	0.018	0.167	0.334	0.599
0.3	0.114	0.819	0.038	0.228	0.262	0.580
0.4	0.136	0.762	0.063	0.275	0.200	0.558
0.5	0.154	0.708	0.092	0.312	0.147	0.532
0.6	0.166	0.656	0.124	0.340	0.102	0.505
0.7	0.174	0.607	0.158	0.361	0.063	0.476
0.8	0.179	0.561	0.193	0.377	0.032	0.448
0.9	0.181	0.517	0.230	0.388	0.006	0.419
1.0	0.180	0.468	0.266	0.395	−0.016	0.391
1.1	0.177	0.439	0.301	0.400	−0.033	0.364
1.2	0.173	0.404	0.336	0.403	−0.046	0.338
1.3	0.168	0.371	0.371	0.405	−0.057	0.313
1.4	0.162	0.341	0.404	0.406	−0.064	0.290
1.5	0.156	0.313	0.435	0.406	−0.070	0.268
1.6	0.148	0.288	0.466	0.405	−0.073	0.247
1.7	0.141	0.264	0.495	0.404	−0.075	0.228
1.8	0.133	0.242	0.522	0.403	−0.076	0.210
1.9	0.126	0.222	0.548	0.402	−0.075	0.193
2.0	0.118	0.203	0.572	0.401	−0.074	0.177
2.1	0.111	0.186	0.596	0.399	−0.072	0.163
2.2	0.104	0.171	0.617	0.398	−0.070	0.150
2.3	0.097	0.156	0.637	0.397	−0.067	0.137
2.4	0.091	0.143	0.656	0.396	−0.065	0.126
2.5	0.084	0.131	0.674	0.395	−0.061	0.116
2.6	0.078	0.120	0.690	0.395	−0.058	0.106
2.8	0.068	0.101	0.721	0.395	−0.052	0.089
3.0	0.058	0.083	0.746	0.395	−0.046	0.075
3.2	0.050	0.071	0.768	0.395	−0.040	0.063
3.4	0.042	0.059	0.786	0.394	−0.035	0.053
3.6	0.036	0.050	0.802	0.394	−0.030	0.044
3.8	0.031	0.042	0.815	0.393	−0.025	0.037
4.0	0.026	0.035	0.826	0.393	−0.022	0.031
4.2	0.022	0.029	0.836	0.393	−0.019	0.026
4.4	0.018	0.024	0.844	0.393	−0.016	0.022
∞	0	0	0.886	0.393	0	0

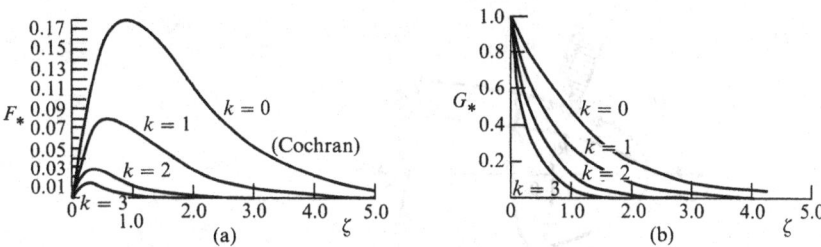

Fig. 2.16. (a) Radial velocity distribution function, F_*, near a rotating disc with uniform suction parameter, k, as a function of reduced distance ζ, and (b) circumferential velocity distribution function, G_*.[6]

the rotating disc. They solved the hydrodynamic equations with $F_* = 0$, $G_* = 1$, $H_* = -k$ on $\zeta = 0$ and $F_* = G_* = 0$ on $\zeta = \infty$. For small values of k, around $\zeta = 0$ they found

$$\left.\begin{array}{l} F_* = a_1\zeta + \cdots \\[4pt] G_* = 1 + b_1\zeta + \cdots \\[4pt] H_* = -k - a_1\zeta^2 + \cdots \end{array}\right\} \qquad (2.29)$$

For $k = 1$, they found $a_1 = 0.389$, $b_1 = -1.175$ and $c = 1.295$ instead of 0.886 as in the Cochran analysis. More complete data for F_* and G_* are given in Fig. 2.16. From this figure, we see that the effect of suction (freezing for metals) at the disc or crystal surface for materials that contract on freezing is not large. In fact, it is quite small for most metals. We can see this by letting $u_z = (\nu\omega)^{1/2}H_* = (\Delta v/v)V \approx 10^{-3}$ cm s$^{-1} \approx 10^{-1}k$ for $\omega \approx 60$ rpm. Thus, $k \stackrel{\sim}{<} 10^{-2}$ for CZ crystal growth of metals and, from Fig. 2.16, we note a small effect. For CZ Si, the volume expands upon freezing as the sign of k is negative but the effect is still likely to be extremely small and can be neglected.

In the normal crystal growth situation, two effects other than a finite crystal radius influence the flow: (1) a curved rather than a flat interface and (2) competing Cochran and natural convection flows. For the case of a curved interface on a rotating crystal, a schematic illustration of the flow is shown in Fig. 2.17 where curvilinear coordinates are used. Here, the x-axis is along the meridional section of the solid surface at any point, the y-axis is along a section of the surface by a plane perpendicular to the axis of revolution, while the z-axis is perpendicular to the tangent plane. Since we are generally interested in only the small curvature case (small departure from flatness), the approximate solution for the rotation of a sphere in a medium at rest is of real interest.

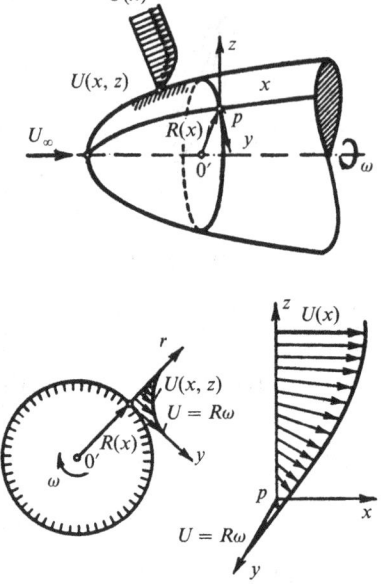

Fig. 2.17. Schematic illustration of the flow around a rotating axially symmetric solid in an axial stream.[6]

Taking into account that, for a sphere of radius R_m, we have

$$\psi = \frac{x}{R_m}; \quad R(x) = R_m \sin \psi; \quad \frac{\mathrm{d}R}{\mathrm{d}x} = \cos \psi \qquad (2.30a)$$

For small values of ψ the results are quite close to those for a rotating disc; i.e.,

$$\delta_m = 2.57 \left(\frac{\nu}{\omega} \right)^{1/2}; \quad \left(\frac{\partial \hat{u}}{\partial z} \right)_0 = 0.61 \frac{\omega^{3/2} R_m \psi}{\nu^{1/2}};$$

$$\left(\frac{\partial \hat{v}}{\partial z} \right)_0 = -0.58 \frac{\omega^{3/2} R_m \psi}{\nu^{1/2}} \qquad (2.30b)$$

Thus, the important result, since $R_m \psi = x$, is that $\hat{u} \approx u_r$ in the Cochran analysis but with x substituted for r. In addition, $\hat{v}$ substitutes for u_ϕ in the Cochran analysis. Because the curvature is so small and we are in the lamellar flow regime, the change of surface contour has a negligible effect on the flow characteristics of the fluid near the central region of the crystal. Following the same line of reasoning, it should be possible to show that the same result holds for an interface concave downwards with a radius of curvature $-R_m$ and Cochran type flow.

In Cochran flow, $u_r \propto r$ and this is what leads to the constant value of δ_m along a rotating disc while, for flow along a flat plate in Fig. 2.1 at $u =$ constant, $\delta_m \propto x^{1/2}$. In most real CZ crystal growth situations, the net fluid flow field is the superposition of Cochran flow and natural convection flow which oppose each other along the radial coordinate. If the natural convection flow velocity is strong then u_r will no longer increase as r but as r^n where $0 \overset{\sim}{<} n \overset{\sim}{<} 1$ which means that δ_m will increase as r^m where $1/2 \overset{\sim}{>} m \overset{\sim}{>} 0$. The important consequence of this increase of δ_m with r is that both δ_T and δ_C will also increase with r. This leads to small radial heat flux effects and, more importantly, for $k_0 < 1$, the crystal solute content will increase with increase of r.

Perhaps one of the most neglected considerations in CZ Si growth is the transient start-up and stopping times.[6] To evaluate the transient cases, we use the variable $\tilde{\eta} = z/2(\nu t)^{1/2}$ and the start-up flow solution is found in terms of the power series

$$\left. \begin{aligned} F_* &= \omega t\, f_1(\tilde{\eta}) + (\omega t)^3 f_3(\tilde{\eta}) + \cdots \\ G_* &= 1 + f_0(\tilde{\eta}) + (\omega t)^2 f_2(\tilde{\eta}) + \cdots \end{aligned} \right\} \qquad (2.31a)$$

and

$$u_z = -4\omega^2 t(\nu t)^{1/2} \int_0^{\eta} f_1(\tilde{\eta}')d\tilde{\eta}' = -4\omega^2 t^{3/2}\nu^{1/2} f_1^*(\tilde{\eta})$$

Values for all these functions are given in Fig. 2.18(a) where we can deduce that the stationary state is reached only for $\tilde{\eta} \approx 1.5$. Thus, for $\nu \sim 5 \times 10^{-3}$ and $z \sim$ crucible size, $L^* \sim 25$ cm, $t_t \sim \nu^{-1}(L^*/3)^2 \sim 4$ hr. This is a very long time, suggesting that present batch CZ Si crystallization, with a continuously falling liquid level, is a totally transient process which never fully attains a stationary state. This is the reason why continuous CZ crystal growth of Si is commercially destined to replace the present batch process. This is probably also the reason why a shallow, large diameter crucible is commercially favored over a deep, smaller diameter crucible. For the shallow crucible, one can reduce L^* by a factor of ~ 4 and thus t_t by a factor of 16. In addition, the liquid drop rate will be reduced so that a stationary state can be more readily attained.

For the alternate transition of the sudden stopping of rotation, the solution is of the form

$$\left. \begin{aligned} F_* &= \omega t\, f_1(\tilde{\eta}) + (\omega t)^3\, f_3(\tilde{\eta}) + \cdots \\ G_* &= f_0(\tilde{\eta}) + (\omega t)^2\, f_2(\tilde{\eta}) + \cdots \\ u_z &= -4\omega^2 t(\nu t)^{1/2}\, f_1^*(\tilde{\eta}) \end{aligned} \right\} \qquad (2.31b)$$

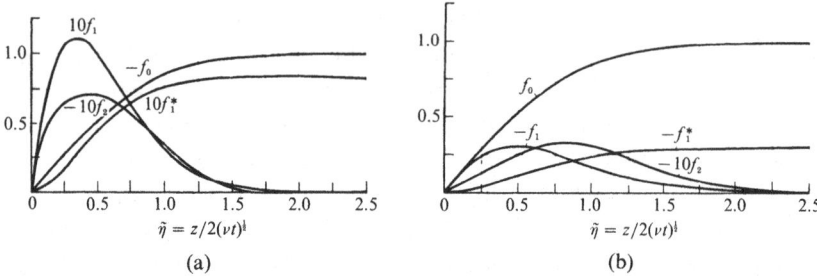

Fig. 2.18. Values of the functions, f_0, f_1, f_1^* and f_2 versus reduced distance $\tilde{\eta}$ for (a) sudden acceleration of a rotating disc on a fluid and (b) sudden stopping.[6]

with the functions being plotted in Fig. 2.18(b). Again we come to the same general conclusion about the stopping time transients; i.e., they are almost as large as the starting transients.

During the transient periods, the relative lines of fluid flow are logarithmic spirals determined from the differential equation for start-up

$$\frac{dr}{rd\phi} = \frac{F_*}{G_* - 1} = m_* \qquad (2.32a)$$

Since the right side of Eq. (2.32a) is independent of r we obtain the logarithmic spiral flow pattern

$$r = \exp(m_*\phi) \qquad (2.32b)$$

For stopping, $m_* = F_*/G_*$. For long times after start-up, $m_* \rightarrow -0.73\omega t/[1 + 0.23(\omega t)^2]$ whereas theoretical and experimental values are given in Figs. 2.19(a) and (b). For sudden stopping, values of m_* are given in Fig. 2.19(c) and, for long times after stopping, $m_* = -1.24 wt/[1 - 0.046(\omega t)^2]$. The qualitative point to be noted here is that the flow is spiral-like, as illustrated in Fig. 2.19(d), with the pitch changing as a function of time because F_* and G_* are functions of time during the transient start-up and stopping periods. An important conclusion to be drawn is that the spiral segregation lines observed in CZ Si crystals have their origin in this long term transient flow field of the fluid. Because of the continuously changing liquid level, the transient flow mode is maintained, m_* continues to change with time and the splines of the crystal segregation pattern are continuously shifting.

Quantitative studies have been made of the laminar flow fields between a rotating and a stationary infinite disc at finite separation. The streamline pattern is as expected from the earlier qualitative deductions and is as shown in Fig. 2.20(a) for a meridional plane. One finds that, for

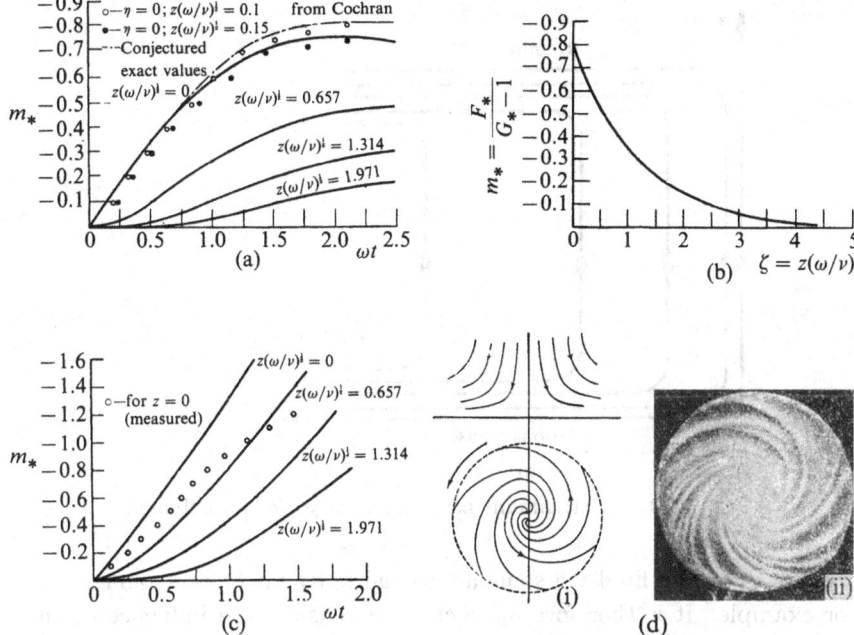

Fig. 2.19. (a) Slope coefficient, m_*, of the relative streamlines in sudden disc acceleration, (b) value of m_* for the stationary motion of a disc in a medium at rest, (c) value of m_* for sudden stopping, and (d) distribution of streamlines at the surface of a rotating disc (i) schematic and (ii) etched lines on the rotating disc.[1,6]

$\widetilde{Re} > 10^3$, flow occurs in which boundary layers are formed on both walls and a fluid core rotates as a solid body. When the problem of laminar flow of fluid in a cylindrical vessel with a plane stationary base and with a rotating plane lid was analyzed, it was found necessary to separate out a circulatory potential flow $u_\phi = B/r$ which must be joined continuously to the flow at the boundary layers on the walls. Examples of calculated streamlines in the meridional plane are given in Figs. 2.20(b) (i) and (ii) for the aspect ratios of $\tilde{L} = 2$ and $\tilde{L} = 0.25$ respectively.[6] These patterns are in accord with our qualitative expectations.

Case 2 (Influence of a flow perpendicular to the surface of a rotating disc):[6] In the previous section, Eq. (2.27) provided a measure of the average forced flow velocity, $\bar{U}_R$, near the wall region of a typical CZ crucible where it must compete with the natural convection flow. In order to enhance $\bar{U}_R$ without increasing the crystal rotation rate, one might

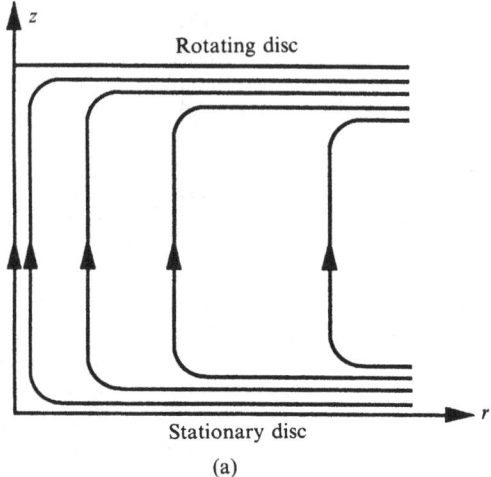

(a)

Fig. 2.20. (a) Streamline pattern for large Re and unbounded discs.

wish to drive the fluid via some additional body force, electromagnetic for example. It is therefore instructive to consider the influence of an axially symmetric fluid current impinging on the disc at right angles to the surface.

Assuming a circular disc of radius R and a uniform flow at infinity with velocity W_∞ at the outer limits of the boundary layer, the axially symmetric potential flow in the space $z>0$ is described by the equations

$$u_r = ar, \quad u_\phi = 0, \quad u_z = -2az \qquad (2.33a)$$

where

$$a = \frac{2}{\pi}\frac{W_\infty}{R} \qquad (2.33b)$$

The ratio of the impinging flow velocity, W_∞, to the rotational velocity, $R\omega$, is determined by the quantity a/ω as

$$\frac{W_\infty}{R\omega} = \frac{\pi}{2}\frac{a}{\omega} \qquad (2.33c)$$

For steady flow we require that Eq. (2.25a) hold but that $u_r = ar$ replace $u_r = 0$ in Eq. (2.25b). The following transformation is made to reduce the equations to dimensionless form

$$\zeta = z\left(\frac{a+\omega}{\nu}\right)^{1/2}, \quad u_z = [\nu(a+\omega)]^{1/2}H_*(\zeta),$$

$$u_\phi = r(a+\omega)G_*(\zeta) \qquad (2.34)$$

and the solution is given in Fig. 2.21. In the limiting case of $a/\omega = 0$,

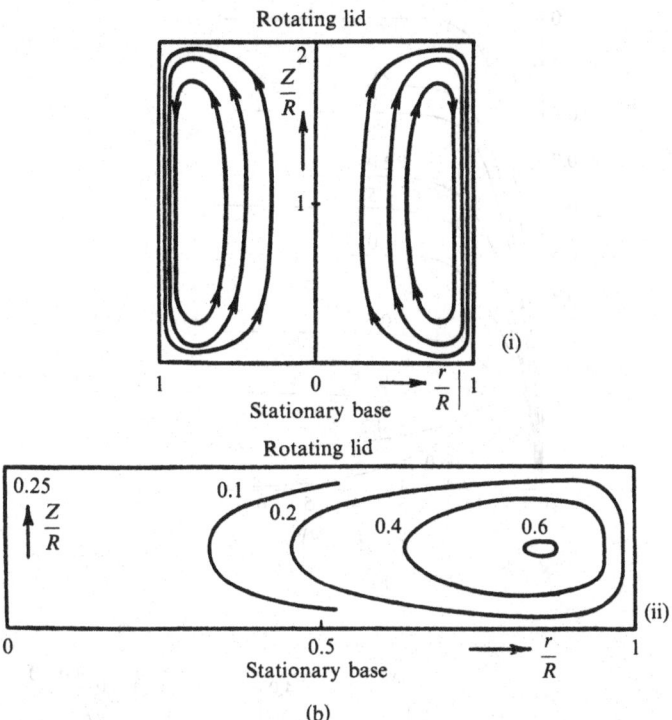

Fig. 2.20. (b) Streamlines in a meridional plane for the rotation of a flat lid on a circular cylinder containing fluid: (i) for $\tilde{L} = 2R$ and (ii) for $L = 0.25R$.[6]

the solution agrees with the Cochran analysis while, in the other limiting case, $a/\omega = \infty$, one obtains good agreement with the solution for axially symmetric flow in the vicinity of a stagnation point. We note that u_r increases while u_ϕ decreases as a/ω increases and, as expected, u_z also increases. Since we have a continuous range of solutions, an axial flow drive seems to add synergistically to the rotational drive and offers the possibility of greatly altering the solute "swirl" effect since u_ϕ can be greatly reduced as a/ω increases. From Fig. 2.21(a), more than an order of magnitude increase in $\bar{U}_R$ is expected in going from $a/\omega = 0$ to $a/\omega \tilde{>} 1$. No crystal growth experiments have yet been carried out to test this forced flow opportunity.

Case 3 (Ekman layer spiral shear flow): One of the most important examples of forced convection arises in the accelerated crucible rotation

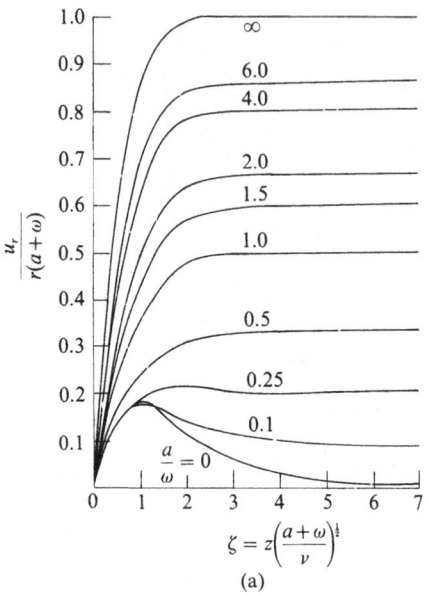

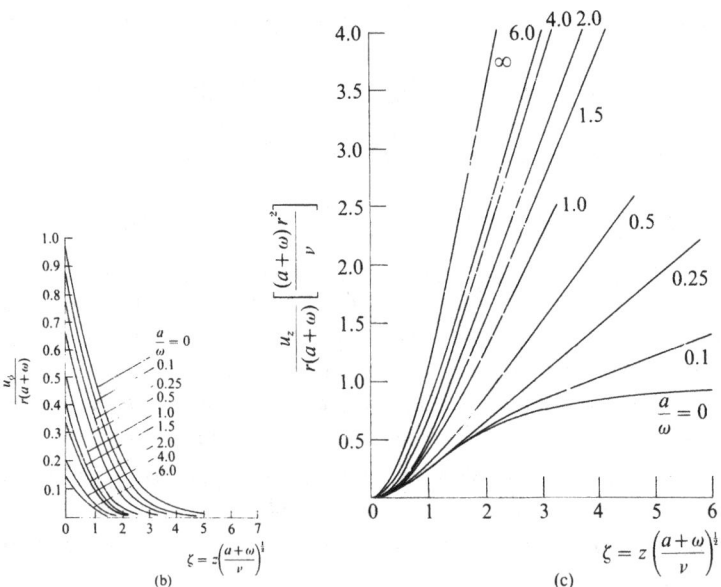

Fig. 2.21. (a) Radial velocity distribution, μ_r, near a rotating disc in a flow impinging axially, (b) circumferential velocity distribution, u_ϕ, and (c) axial velocity distribution, u_z.[6]

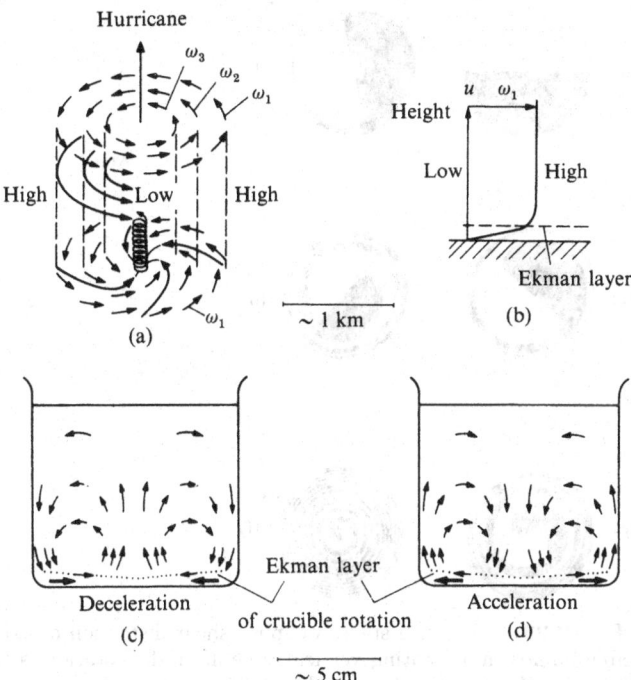

Fig. 2.22. Ekman layer flow in a hurricane (a), with its distribution of the peripheral angular velocities (b), and in decelerated (c), and accelerated (d) containers.[7]

technique (ACRT) for crystal growth. In ACRT, the crucible rotation is periodically accelerated and decelerated. It is the Ekman layer flow that is most important in ACRT mixing just as it is in hurricanes. There, a rapid suction of air occurs in the momentum boundary layer (Ekman layers) because the differential pressure, ΔP, between the outside of the hurricane and the center is not balanced by centrifugal forces. This is illustrated in Fig. 2.22 for both a hurricane and a crucible.[7] Surprisingly, it takes only a time $\tau_1 \approx \omega^{-1}$ (a fraction of a revolution) to set up an Ekman layer of thickness $\delta_E = (\nu/\omega)^{1/2}$ with a radial flow velocity of $U_r = \omega r$ and it requires a time $\tau_2 = R/(\omega\nu)^{1/2}$ to complete spin-up or spin-down in either a crucible of radius R or a hurricane. The Ekman layer flow is inwards in a flat-bottomed crucible during deceleration and it is outwards during acceleration (see Fig. 2.22). Everyone is familiar with the settling of stirred tea leaves which is controlled by Ekman flow. For a crucible of $R = 10$ cm, $\omega = 2\pi$ s^{-1} (1 rps) and $\nu = 10^{-2}$ cm^2 s^{-1},

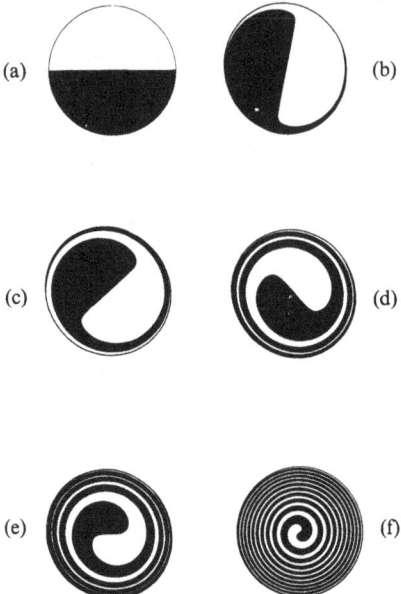

Fig. 2.23. Successive stages of spiral shear distortion of two immiscible liquids in a rotating vertical tube after the rotation is suddenly stopped. (Courtesy of H. Scheel.)

we find that $\delta_E = 4 \times 10^{-2}$ cm after $\tau_1 = 0.2$ s with U_r increasing to ~ 60 cm s^{-1} and the spin-up period lasting $\tau_2 \approx 50$ s.

ACRT also generates a second flow mode; i.e., a spiral shearing distortion which one might expect from the earlier discussion of the Cochran analysis and which is somewhat less efficient as a mixing agent than Ekman flow. This flow mode is illustrated in Fig. 2.23 where we consider two immiscible fluids (differing only in color) vertically aligned in an infinite cylinder which is uniformly rotated and has the initial cross-sectional geometry shown in Fig. 2.23(a). At a sudden cessation of the rotation, the angular velocity of the fluid near the wall decreases rapidly due to wall drag whereas the center decelerates more slowly due to inertia. The shearing occurs between annular liquid parts of various velocities leading to the steady distortion into the spiral pattern shown in Fig. 2.23(f).

Fig. 2.24 shows the decrease in relative rotation rate, ω/ω_0 of the fluid as a function of relative radius, r/R, in this spiral shearing distortion for a number of dimensionless times, $\tau = \nu t/R^2$. The liquid rotation rate in the center of the tube is reduced to $\sim \omega_0/2$ after a time $t \sim$

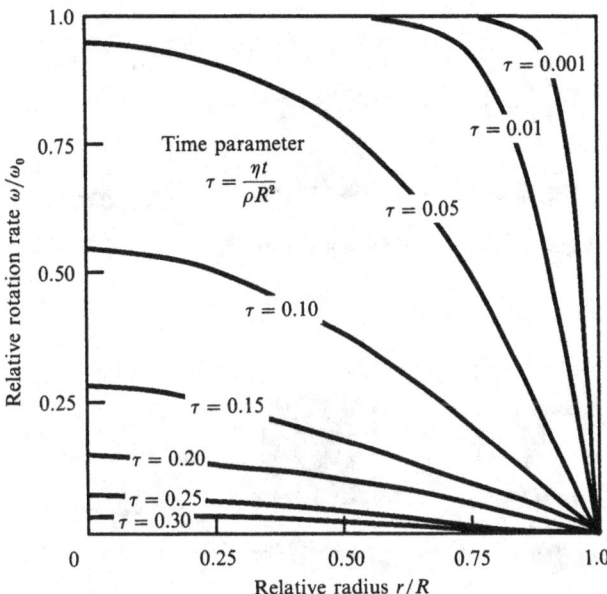

Fig. 2.24. Change of the relative rotation rate versus radius during spiral shear distortion for various times, τ.

$0.1R^2/\nu \sim 10^3$ s for $R \sim 10$ cm and $\nu \sim 10^{-2}$ cm s^{-1} so the acceleration period should be of this order to achieve optimum mixing (for long, tall containers where spiral shear flow may be dominating). In this case, the spiral arm separation is $\sim 10^{-3}$ cm so that homogenization of immiscible liquids by thermal convection and diffusion will occur in ~ 0.1 s.

A demonstration/simulation experiment was created by Scheel[7] to show the excellent efficiency of ACRT mixing (see Fig. 2.25). Three glass containers containing dilute sulfuric acid are heated from below to establish natural convection. Then, equal amounts of KMnO$_4$ (coloring agent) are placed on the surface of each to float, dissolve and mix. In Fig. 2.25, for the ACRT sample, as the KMnO$_4$ dissolved, it was homogeneously distributed after 1–2 ACRT cycles and, after 15 minutes, all the KMnO$_4$ had dissolved to yield a homogeneous solution. In contrast, the uniformly rotated container showed unmixed regions with no coloration at all while the stationary container exhibited even less mixing. Clearly, the ACRT is an excellent technique for initial mixing of an alloy melt.

An even more effective fluid mixing technique has recently been utilized by Feigelson *et al.*[8] They clamped a vertical cylindrical container

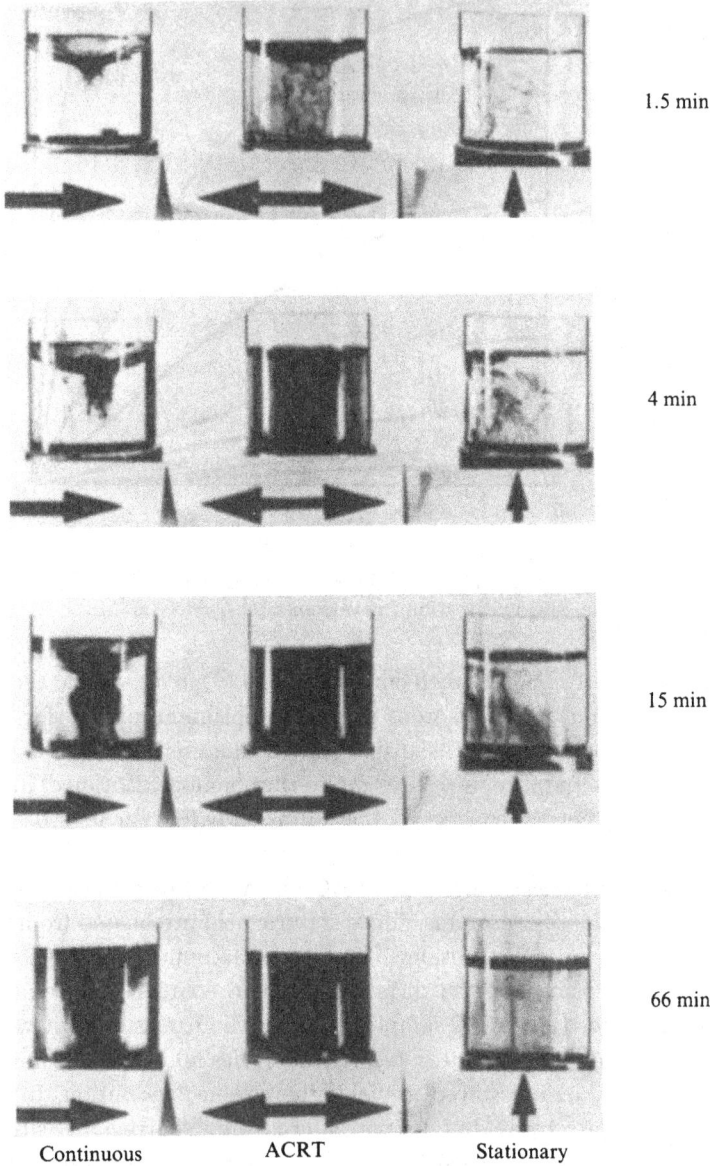

1.5 min

4 min

15 min

66 min

Continuous   ACRT   Stationary
rotation

Fig. 2.25. Successive mixing stages for continuous rotation (at left), ACRT (middle) and stationary (right) containers heated from below. (Courtesy of H. Scheel.)

containing water to a low frequency vibration table and, by injecting drops of dye, were able to show that complete mixing in the fluid required an order of magnitude less time than was found by using the ACRT technique. This technique has been applied successfully to the control of solute partitioning during alloy crystal growth from the melt.

2.1.4   Extension to film formation

It should be clear that a flow velocity will exist in a vapor from which a crystal is growing whenever the vapor/crystal transition involves a change in the number of vapor phase molecules, i.e., either a change in volume or pressure. This flow will set in motion convection patterns in the gas phase independent of Bénard or Marangoni flows. For the simplest case of solid $\rightarrow$ vapor $\rightarrow$ solid film formation, there is a flow velocity away from the source material towards the growing crystal. This is called Stefan flow and has been described in useful detail by Faktor and Garrett.[9] For the simplest case of a long silica capsule containing a solid Ag source at one end and an Ag seed crystal at the other plus sufficient inert gas to yield a total pressure P, let us consider the gaseous transport.

The Ag vapor is carried down the capsule by both flow of the whole volume of gas at a velocity U and also by diffusion. The inert gas is also carried down the capsule at flow velocity U and diffuses back at such a rate as to give no net flow. It can be shown that $U = (\mathcal{R}T/P)J_{Ag} = (P/P - P_{Ag})dP_{Ag}/dz$ where z refers to distance along the length of the capsule. Thus, since $V \propto J_{Ag}$ for the film, one can see that the importance of the Stefan velocity factor to a particular film growth situation will depend upon the ratio P_{Ag}/P. For $P_{Ag}/P \ll 1$, the effect is small but, as $P_{Ag}/P \rightarrow 1$, the effect becomes very large. The important point, for our purposes here, is that gas phase convection can also be an extremely important consideration during film growth.

2.1.5   Combined forced and natural convection flows

In the CZ crystal growth technique, the buoyancy-driven flow (combined with some Marangoni flow) yields an upwardly rising natural convection velocity near the crucible wall of $U_N = Cg\beta_T\bar{\mathcal{G}}\ell^3/\nu$ where C is a geometry-dependent numerical factor, typically $\sim 10^2$, $\bar{\mathcal{G}}$ is the temperature gradient normal to the wall and ℓ is a typical crucible dimension ($\ell \approx$ crucible radius). Thus, for $\bar{\mathcal{G}} \approx 1\,°C\ cm^{-1}$ and $\ell \approx 10$ cm, $U_N \approx 10$ cm s^{-1} which is a very strong flow velocity near the wall. This flow velocity decays with distance from the wall in a manner analogous to Fig. 2.4(b).

The competing flow from the Cochran drive has a velocity $\bar{U}_R$ given by Eq. (2.27) and this may be much smaller than U_N, especially at radii $R_* \gg R_c$. If we define $U_N(r)$ and $\bar{U}_R(r)$ as

$$U_N(r) \approx U_N(\text{max}) \exp[-b(R_M - r)] \qquad (2.35a)$$

and

$$U_R(r) \sim 0.886(\nu\omega)^{1/2} \left(\frac{R_c^2}{r^2 - R_c^2} \right) \qquad (2.35b)$$

where R_M is the radius at which $U_N = U_N(\text{max})$ and b is a constant, then the value of r, r_B, at which the natural convection velocity and the forced flow velocity are equal can be determined; i.e.,

$$\left(\frac{R_c^2}{r_B^2 - R_c^2} \right) \exp[b(R_M - r_B)] \approx \frac{8}{7} \frac{U_N(\text{max})}{(\nu\omega)^{1/2}} \qquad (2.35c)$$

This will determine the relative sizes of the zones of natural convection and forced flow in the CZ technique. Figs. 2.26(a)–(c) illustrate, from Carruther's work,[10] the decrease in natural convection dominance as the crystal rotation rate, ω, is increased.

When one adds the additional factor of crucible rotation, as in Figs. 2.26(d) and (e), then the spiral Cochran-type flow at the bottom of the crucible is in the same upward/downward pattern as the natural convection flow but modifies the natural flow into a spiral pattern as long as the spiral shear does not have too large a pitch. In such a case (high crucible rotation rates), the flow pattern splits into a central forced flow-dominated pattern and an outer natural convection pattern perturbed into an almost turbulent state by the spiral forced flow. The degree of complexity of these Taylor–Proudman-type flow cells for different cases of crystal/crucible isorotation and counterrotation is illustrated by Fig. 2.27.[10]

In the floating zone technique, rotation of the rods is often applied to induce forced convection in the floating zone so as to (i) manipulate the steady Marangoni convection to a more advantageous pattern and (ii) suppress the oscillatory Marangoni convection so as to avoid crystal striations. Iso, single and counter-rotations of the two coaxial rods have been studied. Single rotation of the lower rod induces a radial outward moving secondary flow in the vicinity of the rotating lower rod and a radial inward moving flow in the vicinity of the upper stationary rod. The whole secondary flow is a toroidal vortex whose rotation is just counterdirected to the interacting Marangoni convection due to the

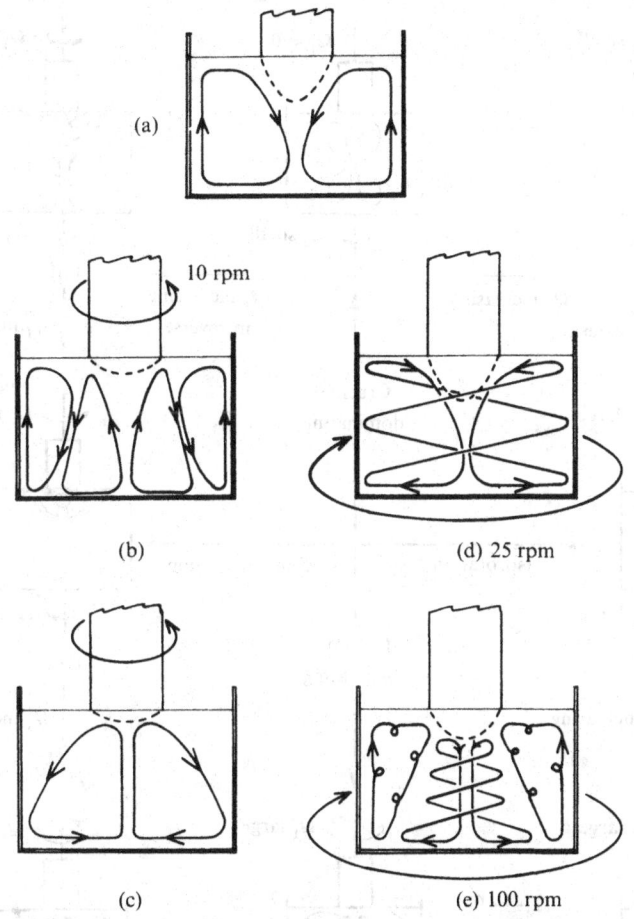

Fig. 2.26. Schematic CZ flow patterns from a transparent fluid model (a) thermal convection only, (b),(c) crystal rotation and (d),(e) crucible rotation. (Courtesy of J. Carruthers.)

zone heating. These flow situations are indicated schematically in Fig. 2.28(a).[11] In the large figure to the left, on the left side is given the secondary flow induced by single rotation alone while, on the right side, the Marangoni convective flow alone is given. The combined secondary flows are shown at the right for different values of ω_2 and different ratios of the rotational and Marangoni Reynolds numbers given by

$$\frac{Re_D}{Re_M} = \frac{(\omega r^2/\nu)}{\rho|\partial\gamma/\partial T|\Delta T (2r/\bar{\eta}^2)}$$

(2.36)

Owing to the competition between the two counterdirected flows, it

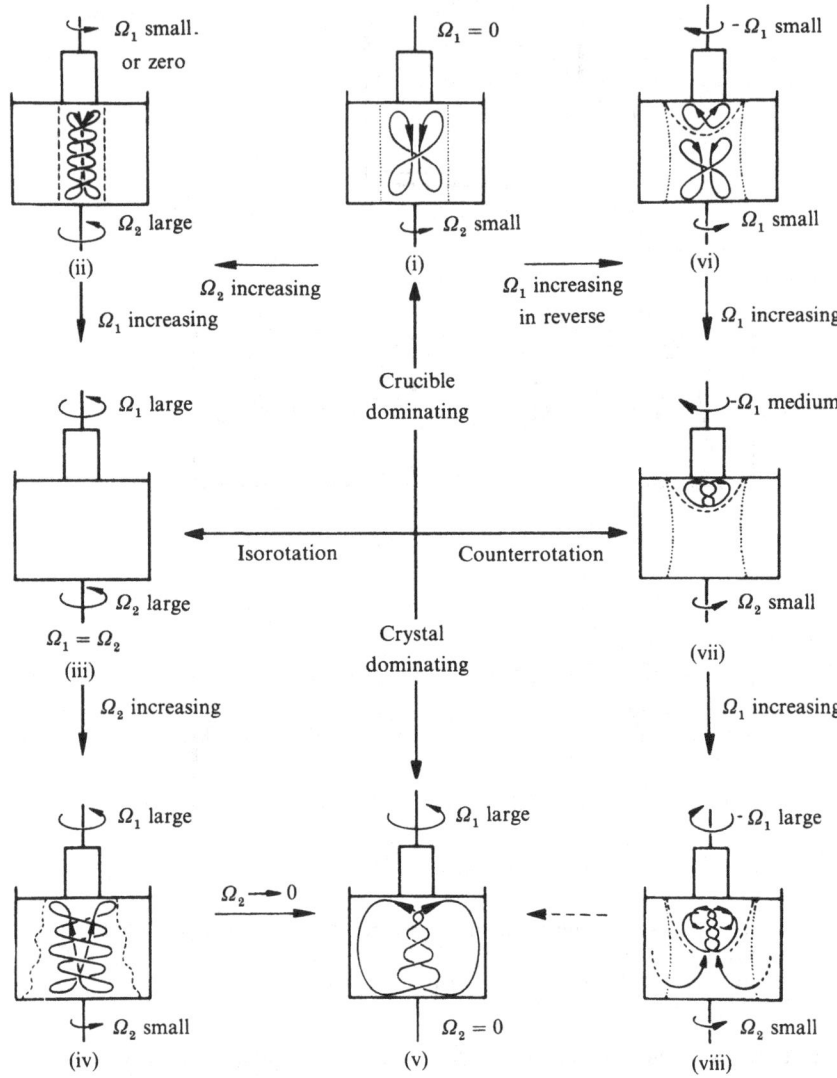

Fig. 2.27. Taylor–Proudman convection cell formation in a melt as a function of crystal or crucible rotation (iso or counter). (Courtesy of J. Carruthers.)

is possible to reduce the toroidal Marangoni convection by increasing the rotation rate of the lower rod. At $Re_D/Re_M = 0.75$, the ratio of the axial forced flow velocity, V_D, due to rotation to the velocity U_M of the Marangoni convection, is $V_D/U_M = 0.21$. At higher rotation rates, V_D/U_M increases and the rotating flow becomes more dominant. If the upper rod is rotated while the lower one remains at rest, the Marangoni convection and the rotation-induced secondary flow will be isodirectional, resulting in an intensification of the coupled flow.

Most important, the oscillatory Marangoni convection can be suppressed by applying rotation to only the lower rod. Fig. 2.28(b) clearly demonstrates the capability. The upper diagram shows the temperature oscillation measured at $Re_D = 87.0$. Its first harmonic frequency is 2.6 Hz according to Fourier analysis. After superimposing the single rotation of the lower rod at 1000 rpm ($Re_D/Re_M = 1.5$), the temperature oscillation is completely suppressed. The amplitude is reduced to a negligible value compared to a low rpm case and Fourier analysis shows only the frequency of the rotation speed (1000 rpm $\equiv$ 16.66 Hz). This fact confirms successful suppression of the flow oscillation.

Before closing this section, it is beneficial to consider the results by Kuroda, Kozuka and Takano[12] *re* the effect of temperature oscillations at the growth interface on crystal perfection. Dislocation-free crystals were grown via the CZ technique under various growth conditions, including changed heater position, pull rate plus crystal and crucible rotation rates. The density of microdefects was found to decrease as the amplitude of the temperature oscillation, δT, was reduced. This microdefect density decreased with higher heater position, increased pulling rate, decreased crystal rotation rate and increased crucible rotation rate. Let us now see how the velocity of natural convection, U_N, correlates with these observations since one expects δT to increase as U_N increases.

We saw earlier that U_N decreases as the average temperature gradient at the crucible wall, $\bar{G}$, decreases and as the crucible rotation rate increases ($\hat{T}$ effect). In addition, $\bar{G}$ is determined by the average upward heat loss from the melt by radiation, conduction and convection and by any heat loss from the bottom of the crucible; i.e., $K_L\bar{G}A_\omega \approx$ heat loss–radiation input where A_ω is the crucible wall/melt contact area. As the heater is raised, the radiation input increases and $\bar{G}$ decreases. As the crystal rotation rate decreases, the convected heat transfer into the crystal is reduced so that a smaller $\bar{G}$ is needed to maintain the upward heat loss via the crystal "wick" effect. And, as the crystal pull rate increases, the latent heat evolution saturates the crystal's capacity to act as a "wick" for heat conduction from the melt so $\bar{G}$ decreases. Thus, qualitatively at least, all of these observations are consistent with the postulate that reducing the natural convection velocity, U_N, reduces the temperature oscillation amplitude, δT, and thus increases the micro-

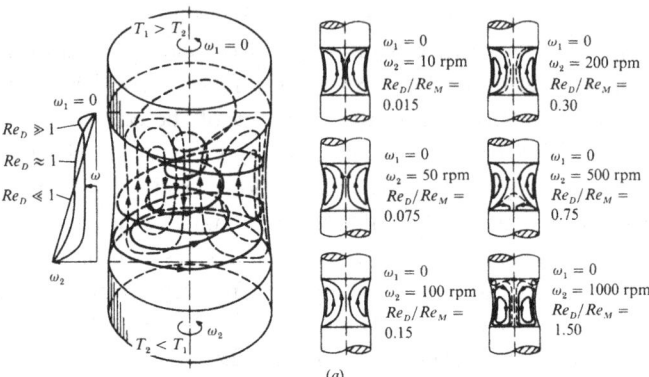

Fig. 2.28. (a) Interaction of Marangoni convection with forced convection due to single rotation. (Courtesy of D. Schwabe.)

perfection of the crystal. Since we saw earlier that $U_N \propto \bar{\mathcal{G}}\ell^3$, as the crucible size increases, U_N will strongly increase as will δT so that the microperfection of the crystal will strongly decrease with increase of ℓ unless ω_c or $\hat{H}$ also increase to constrain the natural convection.

Since the early 1980s, commercial CZ crystal pulling of Si has involved the control of a number of key process parameters. These are pull rate, V, crystal rotation rate, ω, temperature gradient in the liquid, $\mathcal{G}_L$, crucible rotation rate, ω_c, and axial magnetic field flux, B_0. A general similarity flow solution in terms of this collection of parameters has been obtained[13] and it is a useful conclusion to this mixed-flow section to describe some of these results.

Following the same approach as Cochran, for steady state flow but with a suction parameter, $k_* = V/(\nu\omega)^{1/2}$, one obtains Eq. (2.26c) except that p is given by

$$p = -\rho\nu\omega P_*(\zeta) + \frac{1}{2}\rho(\tilde{s}^2 + bM_* - b^2)\omega^2 r^2 \qquad (2.37a)$$

where

$$b = \left(\frac{\beta_T g \mathcal{G}}{2\omega}\right)^{1/2}, \quad M_* = \frac{\sigma_* B_0^2}{\rho\omega}, \quad \tilde{s} = \frac{\omega_c}{\omega} \qquad (2.37b)$$

The important growth conditions are incorporated into four parameters; i.e., the suction parameter, k_*, the hydromagnetic interaction parameter, M_*, the crucible rotation parameter, $\tilde{s}$, and the thermal convection parameter, b. Unfortunately, the similarity solutions for the non-isothermal case ($b \neq 0$) exist only when the melt has no solid rotation at infinity. Although numerical solutions can be easily obtained for all possible com-

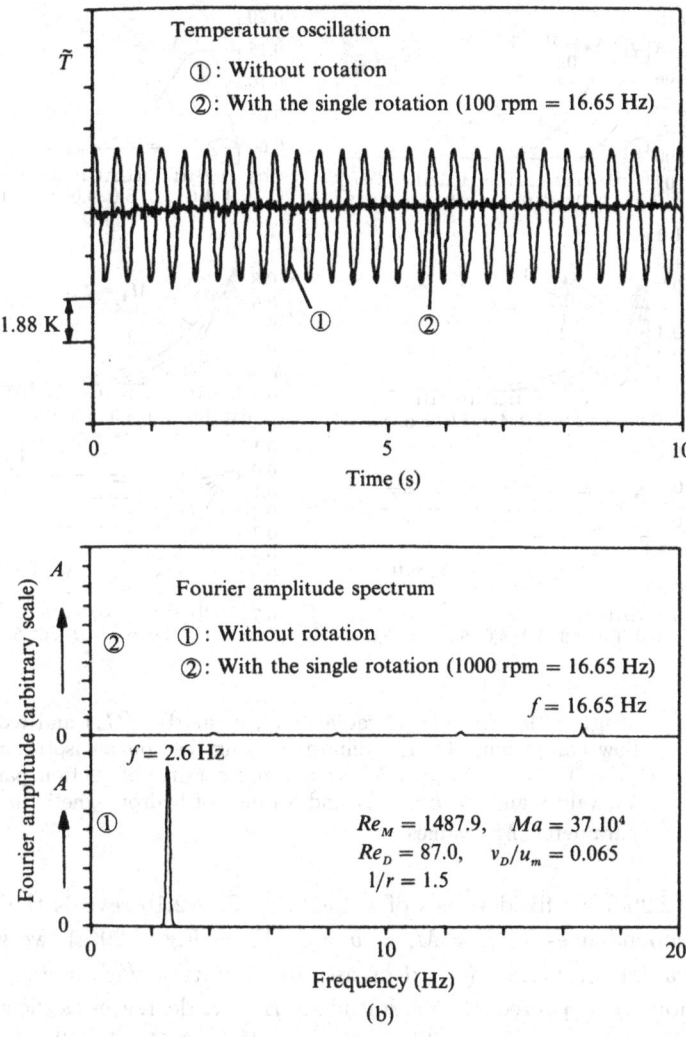

Fig. 2.28. (b) Suppression of temperature oscillations in the fluid by applying single rotation. (Courtesy of D. Schwabe.)

binations of k_*, $\tilde{s}$, M_* and b, only a few results will be given to focus on the influence that individual parameters have on the resultant melt flow configuration.

We first consider the isothermal fluid case so that $b = 0$ and consider the variation in suction parameter, k_*, which combines both growth velocity and rotation rate, on the key flow functions F_*, G_* and H_* of

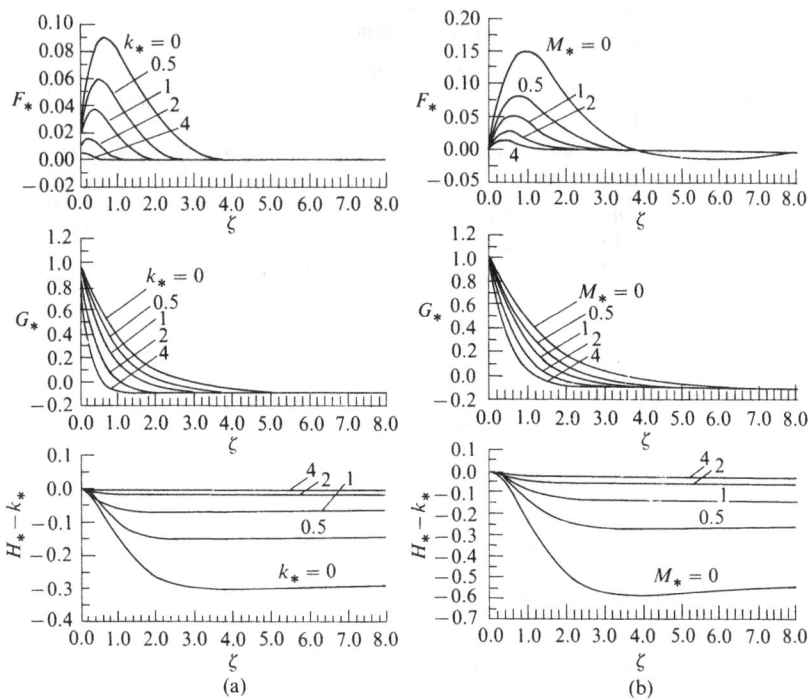

Fig. 2.29. Calculated radial (F_*), azimuthal (G_*) and axial (H_*) flow components for an infinite rotating disc on an isothermal fluid $(b = 0, \tilde{s} = -0.1)$. (a) $M_* = 0.5$ and a range of suction parameter, k_*, values and (b) $k_* = 0.1$ and a range of hydromagnetic interaction parameter, M_*, values.[13]

Eq. (2.26c) for fixed values of $\tilde{s}$ and M_*. Fig. 2.16 reveals the F_* and G_* dependences for $\tilde{s} = M_* = b = 0$. From Fig. 2.29(a), we see that the radial component (F_*) decreases in magnitude with increase of the suction. As expected, the magnitude of $H_* - k_*$ decreases as the strength of the suction increases. The values for F_* and G_* also illustrate that the width of the momentum boundary layer decreases with increasing k_*.

If an axial magnetic field is applied, the effect of the magnetohydro-dynamic interaction on the flow configuration can be seen from Fig. 2.29(b). In a similar fashion to the effect of V, increasing M_* reduces the extent of the forced convection caused by the rotating crystal. At fixed M_*, as the crucible rotation parameter, $\tilde{s}$, is changed, important fluid flow configurations are generated.

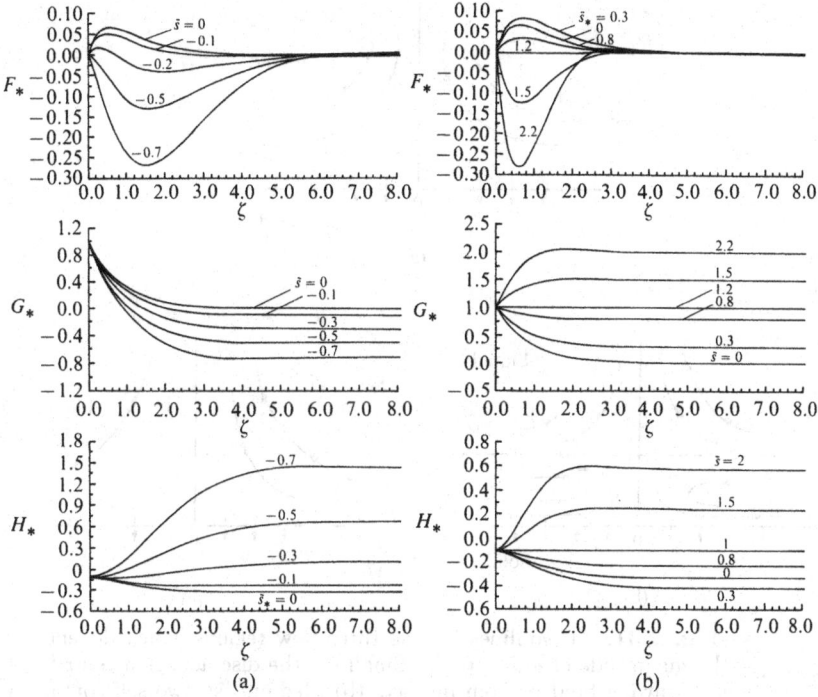

Fig. 2.30. Calculated radial (F_*), azimuthal (G_*) and axial (H_*) flow components for an infinite rotating disc on an isothermal fluid ($b = 0$, $k_* = 0.1$, $M_* - 1.0$), (a) a range of negative values for the crucible rotation parameter, $\tilde{s}$, and (b) a range of positive values for the crucible rotation parameter, $\tilde{s}$.[13]

When the fluid at infinity is rotating in the same sense as the disc, physically meaningful solutions exist for all values of $\tilde{s}$. On the other hand, if the fluid at infinity is rotating in the opposite sense to that of the disc, numerical integration gives solutions that diverge for some large values of $|\tilde{s}|$ and, for such cases, the extrapolation method is used to describe the flow velocity qualitatively.[13] Fig. 2.30(a) shows the solutions for $k_* = 0.1$, $M_* = 1.0$ and a range of $\tilde{s} \leq 0$ while Fig. 2.30(b) treats the $\tilde{s} \geq 0$ domain. We can note that, from G_* in Fig. 2.30(a), the flow configuration eventually divides into three different regimes as illustrated in the streamline diagram of Fig. 2.31. For the cases where $\tilde{s} > 0$, Fig. 2.30(b) shows the results. When $0 < \tilde{s} < 1$, the magnitudes of the outward radial flow and the upward axial flow just increase with increase of $\tilde{s}$ until a maximum flow velocity is reached. After that,

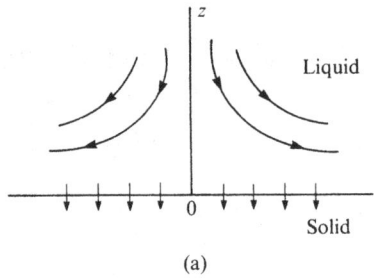

(a)

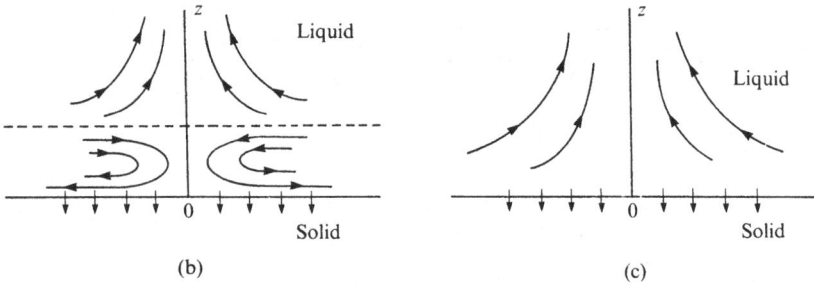

(b)   (c)

Fig. 2.31. Streamlines for the three flow regions which depend on the magnitude of $\tilde{s}$, $\tilde{s} < 0$. (a) Small $|\tilde{s}|$, the disc acts as a centrifugal fan drawing fluid in from infinity. (b) Medium $|\tilde{s}|$, two self-contained cells are formed where the disc still acts like a centrifugal fan locally. (c) Large $|\tilde{s}|$, the fluid flow is laterally inward near the disc and flows away toward infinity axially.[13]

the fluid motion gets progressively weaker with increase of $\tilde{s}$ until fluid motion ceases at $\tilde{s} = 1$. When $\tilde{s} > 1$, the flow configuration becomes the one shown in Fig. 2.31(c). There is an inward radial flow near the disc surface which is balanced by an axial flow towards infinity.

The very important non-isothermal flow regime occurs when $\mathcal{G}_L \neq 0$. Fig. 2.32 presents the results for these $b \neq 0$ cases. Fig. 2.32(a) gives plots for $\tilde{s} = 0$, $k_* = 0.1$ and $M_* = 0.5$ for a range of b. We see that the radial outward flow (F_*) near the disc decreases in magnitude and reverses direction beyond a critical distance that decreases as b increases. At the limit of very large b, the radial component of the flow velocity will be completely inward as in Fig. 2.31(c). Similarly, the axially upward flow (H_*) near the disc surface (which is induced by disc rotation) decreases in magnitude with increase of b and the flow direction changes from upwards to downwards at a specific distance from the disc which moves closer to the disc as b increases (see Fig. 2.31(b)). When

$b \to \infty$, the flow will be downwards everywhere except in a tiny layer at the disc surface required by the suction (see Fig. 2.31(c)).

The flow configuration also has a strong dependence on the suction parameter, k_*. This can be clearly seen in Fig. 2.32(b). As shown, due to the presence of thermal convection, the radial flow (F_*) far away from the disc will be in an inward direction $(F_* \to -b)$. If the suction is small, the direction of the axial component maintains near the disc but has a decreasing strength as k_* increases. At the limit of large k_*, the flow will become completely inwards. The influence of M_* on the melt flow configuration is seen in Fig. 2.32(c). With increase of M_*, not only does the amplitude of the radially outward flow (F_*) near the disc decrease but also the range of this zone becomes smaller. Similarly, the induced upward axial flow (H_*) decreases in magnitude with increasing M_* and the range of this upward flow shrinks.

During CZ crystal growth, the fluid flow configuration is likely to be like that illustrated in Fig. 2.31(b) produced by some effective convection parameter, b, associated with the particular crystal growing furnace and growth program. We cannot expect to obtain the true value of b from purely theoretical calculations and, thus, recourse must be made to experimental procedures. One such procedure will be discussed in Chapter 4 when we are dealing with transient solute distributions. One sees that slavish adherence to the simple boundary layer approximations can lead to great errors and that full three-dimensional solutions are often needed to produce *qualitatively* correct representations of heat and mass flow in such applications. It is also interesting to note that, even in the description of solute profiles during vertical Bridgeman crystal growth, one can have near-interface roll cells of convection that transfer solute laterally to alter the radial segregation pattern while, at the same time, no significant long range axial convection pattern develops to produce end-to-end segregation in the crystal. Thus, the three-dimensional character of convection can impart very anisotropic transport characteristics to the nutrient phase and must be taken into consideration in any quantitative attempt at overall modelling of crystal growth systems.

2.1.6   Interaction of flows with the crystal/melt interface

Glicksman, Coriell and McFadden[9] have described several cases wherein the reactive nature of the solid/melt interface has led to the development of hydrodynamic instabilities in the fluid at Grashof or Rayleigh numbers well below those observed for vessels with non-reacting (rigid) walls. Their work clearly shows that a strong coupling can occur

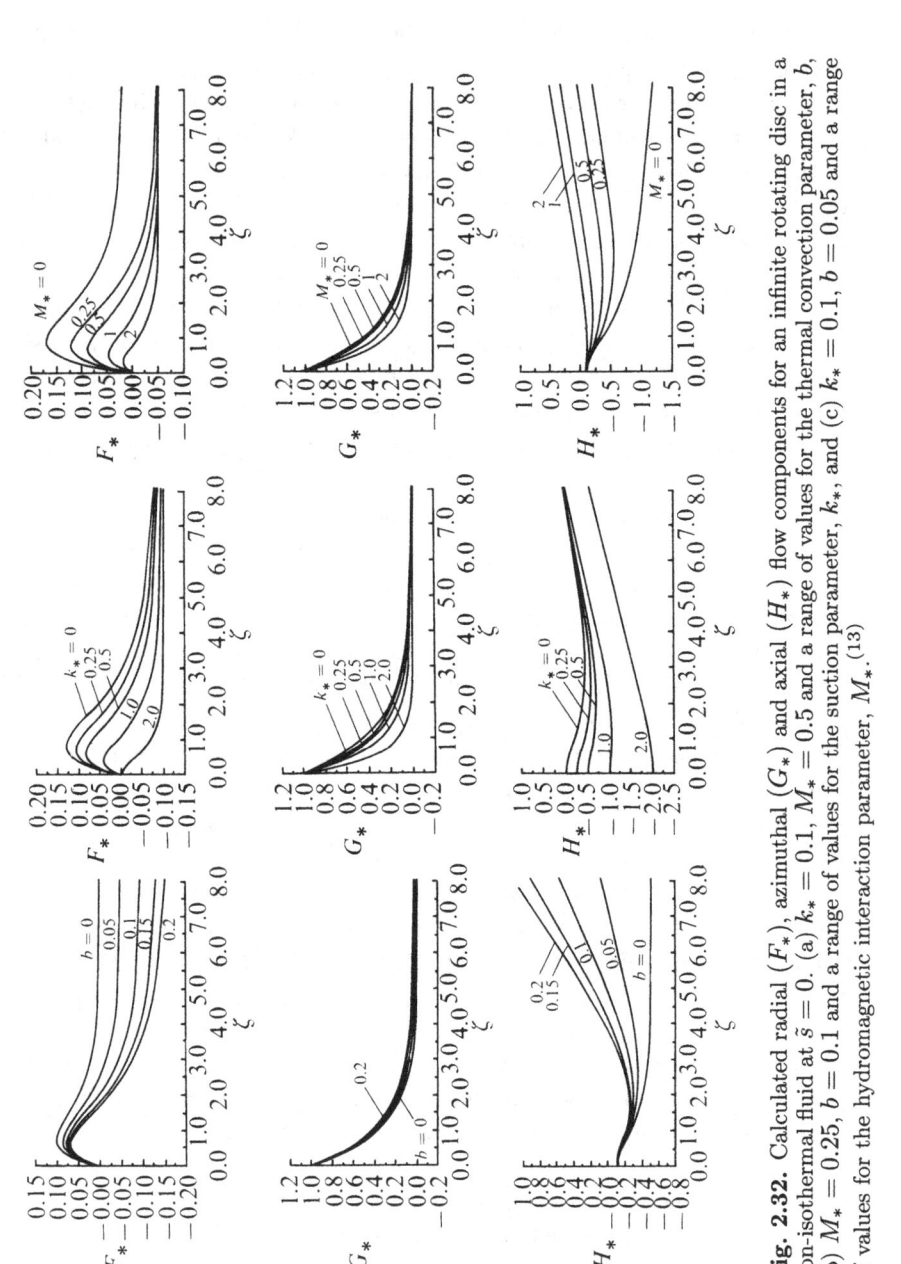

Fig. 2.32. Calculated radial (F_*), azimuthal (G_*) and axial (H_*) flow components for an infinite rotating disc in a non-isothermal fluid at $\tilde{s} = 0$. (a) $k_* = 0.1$, $M_* = 0.5$ and a range of values for the thermal convection parameter, b, (b) $M_* = 0.25$, $b = 0.1$ and a range of values for the suction parameter, k_*, and (c) $k_* = 0.1$, $b = 0.05$ and a range of values for the hydromagnetic interaction parameter, M_*.[13]

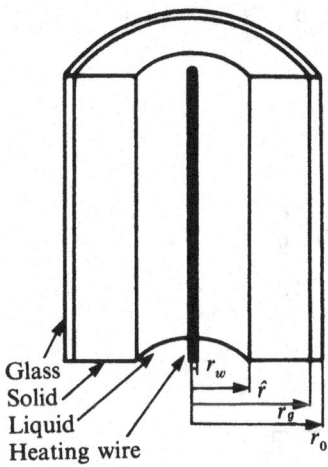

Fig. 2.33. Section of annular experimental arrangement for studying convection flow patterns.

between hydrodynamic flows and a phase change and that neglect of this important coupling can sometimes lead to erroneous conclusions.

One particularly interesting example involved the radial movement of a solid/liquid interface held in a tall vertical annulus. A representative section of this experimental arrangement is illustrated in Fig. 2.33. The heating wire generates a constant heat flux and the outer glass envelope is maintained at a constant temperature below the melting point of the material. For a small temperature difference between the wire and the solid/liquid interface, a steady state interface position is maintained in the presence of a molten zone wherein steady natural convection occurs. The convection is generally in the form of a long unicell torus with warm fluid rising near the axial heating wire and cooler fluid descending adjacent to the solid/fluid interface.

The experimental material was succinonitrile (SCN) whose molten phase has a Prandtl number of $Pr = 23$. Below a critical Grashof number, $Gr^* \approx 150$, steady flow existed. However, above Gr^*, the steady state flow was replaced by a periodic state wherein the interface developed into a helical form that slowly rotated about the wire. The period of this rotation ranged from several minutes to hours depending upon the radial distance, ℓ, between the wire and the interface. The axial wavelength scaled with ℓ while the rotational period scaled with ℓ^5 and the helical waves travelled up the interface opposite to the direction

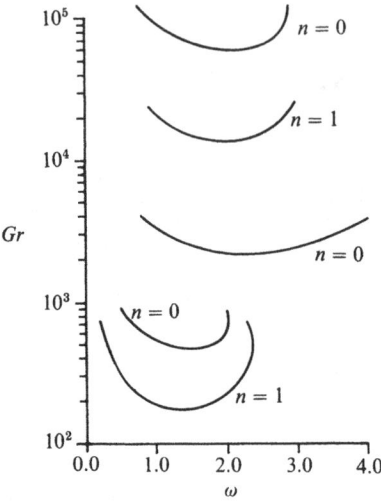

Fig. 2.34. Grashof number for the onset of hydrodynamic instabilities as a function of the axial wave number, ω, of a sinusoidal perturbation ($Pr = 22.8$) for different azimuthal wave number, n. The lowest test curves occur only in the presence of a crystal/melt interface.[14]

of the local flow. The wave speed was found to be about 1% of the maximum unperturbed flow velocity and scaled as ℓ^{-4}.[14]

The hydrodynamic instability of the fluid in this configuration for both non-reacting and reacting walls has been studied to yield the marginal stability curves for the coupled system. In Fig. 2.34, the Grashof number at the onset of instability is given as a function of the axial wave number, ω, of a sinusoidal shape perturbation on the solid/liquid interface. The upper three curves, from top to bottom, are the axisymmetric shear mode ($n = 0$), the asymmetric shear mode ($n = 1$) and the symmetric buoyant mode ($n = 0$). These levels correspond within 1% to the modes found for a rigid wall. The lower two modes represent instabilities that result from a coupling of the fluid flow with a deformable (reacting) interface. The lowest mode is asymmetric ($n = 1$), with a minimum Grashof number of $Gr^* = 176$, and this mode corresponds to the experimentally observed instability. The second mode is axisymmetric with a minimum Gr^* of 464. The wavespeeds for these two modes are predicted to be only $\sim 1\%$ of the wavespeeds for the three higher modes and, at onset, the disturbance travels in the upward direction. Clearly, this type of coupling phenomenon has great relevance for crystal growth

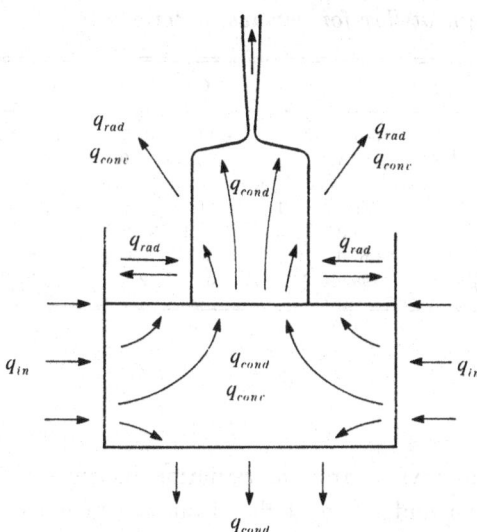

Fig. 2.35. Schematic representation of heat flow during CZ crystal growth.

in the absence of forced convection! This is particularly relevant for geological crystallization.

2.2   Heat transport

A schematic illustration of the various heat fluxes occurring during conventional CZ crystal growth is given in Fig. 2.35. The heat supplied to the crucible is transferred to the heat sinks around the furnace in various ways. The heat is partly conducted, convected and radiated directly from the crucible to the melt and is partly radiated from the crucible to the crystal. The heat flowing in the melt is transferred to the interface and the melt surface by both conduction and convection. Heat is dissipated from the melt/free surface to the atmosphere by both gaseous convection and by radiation. If the crystal material is transparent, then q_{rad} is also transferred in the crystal. In addition, for a growing crystal, latent heat is generated at the solid/liquid interface. Because this is a batch process, the factors which must vary during the growth of a crystal are crystal size, height of the melt and meniscus shape and they are determined by the time-dependent heat transfer. Thus, a time-dependent heat input and pull rate is necessary to try and balance these factors so as to pull a constant diameter crystal.

Table 2.3. *Comparative heat flux for various materials.*[15]

	Cu	W	Ge	Si	SBN	Cr$_2$O$_3$
Melting point T_M(K)	1356	3653	1212	1685	1470	2538
Thermal conductivity K_S(W cm^{-1} K^{-1})	3.3	0.895	0.174	0.220	0.008	0.04
Emissivity $\bar{\varepsilon}_S$	0.056	0.36	0.59	0.46	0.88	0.8
Crystal radius^a a(cm)	1	1	1	1	1	1
Heat flux^b q_L(W cm^{-2})	1.07	363	7.17	21.0	23.3	188
q_S(W cm^{-2})	62.0	945	34.7	80.0	14.8	124
Total heat flux $Q_S = \pi a^2 q_S$(W)	195	3060	109	248	46.5	388

^{a}Assumed.
$^b\varepsilon_L = \varepsilon_S$ is assumed

For high temperature crystal growth, radiation becomes the prime heat transfer mechanism and the heat flux from the melt to the surroundings (neglecting back radiation) is

$$q_L = \bar{\varepsilon}_L \tilde{\sigma} T_L^4 \qquad (2.38a)$$

where $\bar{\varepsilon}_L$ is the emissivity of the liquid, $\tilde{\sigma}$ is the Stefan–Boltzmann constant ($\tilde{\sigma} = 1.37 \times 10^{-12}$) and T_L is the temperature of the liquid. If the heat conducted into the crystal is only dissipated by radiation into the surroundings, then a one-dimensional crystal approximation can be made and the heat flux in the crystal at the interface, q_S, is approximately given by

$$q_S = \left(\frac{2\bar{\varepsilon}_S \tilde{\sigma} T_M^5}{3aK_S} \right)^{1/2} \qquad (2.38b)$$

where K_S is the thermal conductivity in the solid at the temperature T_M and a is the radius of the crystal. In Table 2.3,[15] values of T_M, $\bar{\varepsilon}_S$, K_S, q_L and q_S are given for a range of materials. From such values, an estimate of the needed heat input for melt and crystal stabilization for various crucible and crystal sizes can be readily obtained.

2.2.1   Basic equations

For an isotropic opaque solid, the heat transfer equation is given by

$$\rho c_p \frac{\partial T}{\partial t} = \nabla(K \nabla T) \qquad (2.39a)$$

where ρ is the density and c_p is the specific heat at constant pressure. If the solid is moving with a velocity $\vec{V} = (V_x, V_y, V_z)$, Eq. (2.39a) becomes

$$\rho c_p \left(\frac{\partial T}{\partial t} + \vec{V} \cdot \nabla T \right) = \nabla(K \nabla T) \qquad (2.39b)$$

For a fluid at constant pressure or for fluids with a density independent of temperature, the heat transfer equation is given by

$$\rho c_p \left(\frac{\partial T}{\partial t} + \vec{V} \cdot \nabla T \right) = \nabla(K \nabla T) + \bar{\eta} \phi \qquad (2.39c)$$

where $\bar{\eta}$ is the viscosity and ϕ is the dissipation function given by

$$\phi = 2 \left[\left(\frac{\partial V_x}{\partial x} \right)^2 + \left(\frac{\partial V_y}{\partial y} \right)^2 + \left(\frac{\partial V_z}{\partial z} \right)^2 \right] + \left(\frac{\partial V_y}{\partial x} + \frac{\partial V_x}{\partial y} \right)^2$$

$$+ \left(\frac{\partial V_z}{\partial y} + \frac{\partial V_y}{\partial z} \right)^2 + \left(\frac{\partial V_x}{\partial z} + \frac{\partial V_z}{\partial x} \right)^2 - \frac{2}{3}(\nabla \cdot \vec{V})^2 \qquad (2.39d)$$

At the boundary surfaces the temperature may be specified by various conditions: (1) Prescribed surface temperature; if the temperature of the crucible is controlled to be a constant, it may be specified by $T = T_c$. (2) Prescribed heat flux across the surface; this is equivalent to specifying the temperature gradient. (3) Heat flux transferred by fluid convection; sometimes this flux at a solid/gas or liquid/gas boundary is related to the temperature difference between the boundary and the bulk fluid ($q = \tilde{h}(T - T_b)$) where $\tilde{h}$ is the heat transfer coefficient and T_b is the temperature of the bulk fluid. If heat is transferred by free convection in a gaseous fluid, the heat transfer coefficient from the surface is usually given by $\tilde{h} L_c / K = C_0 Ra^{1/n}$ where L_c is a characteristic length, C_0 is a constant and Ra is the Rayleigh number. If the convection is laminar, $n = 4$ and, if turbulent, $n = 3$. (4) Heat flux transferred by blackbody radiation; a body at the absolute temperature T surrounded by a blackbody at temperature T_a will radiate at the rate $q = \bar{\varepsilon}\tilde{\sigma}(T^4 - T_a^4) \approx 4\bar{\varepsilon}\tilde{\sigma}T_a^3(T - T_a)$ when the temperature difference $T - T_a$ is not too large. (5) Continuities of T and q at the boundary between two media; at the solid/liquid interface $T_S = T_L$ and $q_{n,S} = q_{n,L} + \Delta H_F V_n$ where n is the surface normal direction, ΔH_F is the latent heat per unit volume and V is the growth rate.

2.2.2 One-dimensional heat transport approximation

A one-dimensional model is very useful for a rough understanding of heat transfer during solidification. The model assumes a flat interface, neglects any radial or angular variations in the temperature and thus assumes flat isotherms. It is useful in that it allows an easy calculation of approximate values for V and $\mathcal{G}_S(0)$, the temperature gra-

Table 2.4. Heat transfer properties for various materials.[15]

	Cu	W	Ge	Si	SBN	Cr$_2$O$_3$
Melting point T_M(K)	1356	3653	1210	1685	1470	2538
Thermal conductivity						
$\quad K_S$(W cm^{-1} K^{-1})	3.3	0.895	0.174	0.220	0.008	0.04
Emissivity $\bar{\varepsilon}_S$	0.056	0.36	0.59	0.46	0.89	0.8
Crystal radius^a a(cm)	1	1	1	1	1	1
Biot number $H^* = \bar{\varepsilon}_S \bar{\sigma} T_M^3 a / K_S$	0.000240	0.111	0.0341	0.0567	1.98	1.85
Interfacial temperature gradient						
$\quad \tilde{\mathcal{G}}(0) = (0.8H^*)^{1/2}$	0.0139	0.298	0.165	0.213	1.26	1.22
$\quad \mathcal{G}_S = \frac{T_M}{a}\tilde{\mathcal{G}}(0)$(K cm^{-1})	18.8	1090	200	359	1850	3090
Characteristic crystal length						
$\quad \bar{L} = (0.8H)^{-1/2}$	72.7	4.53	6.06	4.70	0.794	0.821
$\quad z^c = a\bar{L}$(cm)	72.2	4.53	6.06	4.70	0.794	0.821

^{a} Assumed

dient in the solid at the interface. The radial heat loss at the surface of a short slice of the crystal is averaged over the volume of that slice of crystal to become a uniform heat loss contribution. Thus, heat balance over a differential volume element in a cylindrical crystal gives

$$\rho_S c_{pS}\left(\frac{\partial T}{\partial t} + V\frac{\partial T}{\partial z}\right) = \frac{\partial}{\partial z}\left(K_S\frac{\partial T}{\partial z}\right) - \frac{2}{a}q_{dis} \qquad (2.40)$$

where q_{dis} is the heat dissipation from the surface of the crystal of radius a and z is the axial distance into the crystal from the interface; i.e., the coordinate system is tied to the interface. In Table 2.4, typical values for some thermal properties are given for several materials.[15]

Case 1 (Linear heat dissipation): $q_{dis} = \tilde{h}_a(T - T_a)$: For this case we define the dimensionless variables

$$\theta = \frac{T - T_a}{T_M - T_a}, \quad Z = \frac{z}{a}, \quad \tau = \frac{Vt}{a} \qquad (2.41a)$$

where T_a is the temperature of the surroundings and T_M is the melting point. The dimensionless crystal length, for $V = $ constant, is $\bar{L} = Vt/a = \tau$. With these new variables, Eq. (2.40) becomes

$$\frac{\partial^2 \theta}{\partial Z^2} - 2\hat{P}_T\frac{\partial \theta}{\partial Z} - 2H^*\theta = 2\hat{P}_T\frac{\partial \theta}{\partial \tau} \qquad (2.41b)$$

where $H^* = \tilde{h}_a a / K_S = $ the Biot number, and $\hat{P}_T = V\rho_S c_{pS}a/2K_S = Va/2D_{TS} = $ the Péclet number. In simple terms, the thermal Péclet number is roughly the ratio of heat carried by movement of the crystal to that conducted in the crystal, while the Biot number is roughly the heat dissipated from the crystal surface to the axial conduction in the crystal. The boundary conditions are

$$\theta(0) = 1; \quad \left(\frac{\partial \theta}{\partial Z}\right)_{\bar{L}} = -H^*\theta \tag{2.41c}$$

During initial seed necking and crystal shoulder development, $\tilde{h}_e = \tilde{h}_a$ is used ($H_e^* = \tilde{h}_e a/K_S$).

Case 2 (Small Péclet number, steady state): For slow growth, the Péclet number is small so that terms containing $\hat{P}_T$ in Eq. (2.41b) can be neglected and we have

$$\frac{\partial^2 \theta}{\partial Z^2} = 2H^*\theta \tag{2.42}$$

which can be readily solved to yield θ and the reduced temperature gradient at the interface, $\tilde{\mathcal{G}}(0) = -(d\theta/dZ_0)$; i.e.,

$$\theta = \frac{(H_e^*/H^*)\sinh[(2H^*)^{1/2}(\bar{L} - Z)] + (2/H^*)^{1/2}\cosh[(2H^*)^{1/2}(\bar{L} - Z)]}{(H_e^*/H^*)\sinh[(2H^*)^{1/2}\bar{L}] + (2/H^*)^{1/2}\cosh[(2H^*)^{1/2}\bar{L}]} \tag{2.43a}$$

$$\tilde{\mathcal{G}}(0) = (2H^*)^{1/2}\frac{(H_e^*/H^*)\cosh[(2H^*)^{1/2}\bar{L}] + (2/H^*)^{1/2}\sinh[(2H^*)^{1/2}\bar{L}]}{(H_e^*/H^*)\sinh[(2H^*)^{1/2}\bar{L}] + (2/H^*)^{1/2}\cosh[(2H^*)^{1/2}\bar{L}]} \tag{2.43b}$$

A plot of $\tilde{\mathcal{G}}(0)/(2H^*)^{1/2}$ is given in Fig. 2.36 as a function of crystal length $(2H^*)^{1/2}\bar{L}$. As might be expected, $\tilde{\mathcal{G}}(0)$ varies as $\bar{L}$ increases except for the special case of $H^* = H_e^* = 2$.

For a short crystal, $\tilde{\mathcal{G}}(0)$ depends on the thermal conditions at the cold end and is approximately given by

$$\left.\begin{array}{ll} \tilde{\mathcal{G}}(0) = 2H^*L & \text{for} \quad H_e^* = 0 \\ = H^*[1 + (2 - H^*)\bar{L}] & \text{for} \quad H_e^* = H^* \\ = \bar{L}^{-1} & \text{for} \quad H_e^* = \infty \end{array}\right\} \tag{2.44}$$

For a long crystal, θ and $\tilde{\mathcal{G}}(0)$ are given by

$$\theta = \exp[-(2H^*)^{1/2}Z]; \quad \tilde{\mathcal{G}}(0) = (2H^*)^{1/2} \tag{2.45}$$

The practical utility of these results, for fixed H^*, H_e^* and $\bar{L}$, comes first from noting that $\tilde{\mathcal{G}}(0)$ is completely specified, which means that the ability of the crystal to extract latent heat plus conducted heat from the liquid is also fixed; i.e., we have the following constraint,

$$K_S\tilde{\mathcal{G}}(0) = \frac{aK_S}{(T_M - T_a)}\left(\frac{dT}{dz}\right)_0 = \left(\bar{q}_L + \Delta H_F \frac{d\bar{v}_c}{dt}\right)\left(\frac{a}{T_M - T_a}\right) \tag{2.46a}$$

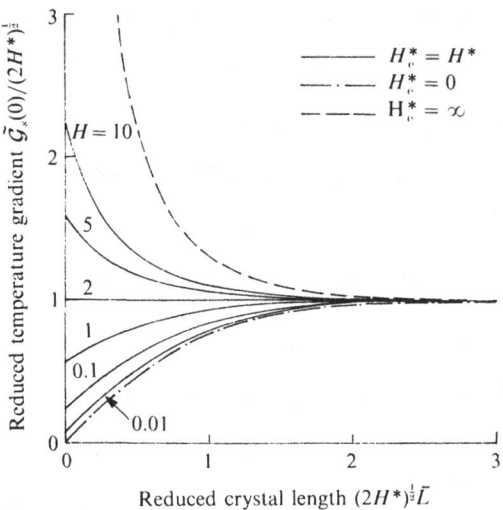

Fig. 2.36. Reduced temperature gradient at the interface as a function of reduced crystal length.

with

$$\frac{d\bar{v}_c}{dt} = \pi a^2 V \left(\frac{dV}{V} + \frac{2da}{a} \right) \tag{2.46b}$$

Eqs (2.46) tell us that a fixed interface heat extraction rate of the crystal, $K_S\tilde{\mathcal{G}}(0)$, and a fixed average heat flux from the liquid, $\bar{q}_L$, predict the change in volume of the crystal with time, $d\bar{v}_c/dt$. Thus, for a varying pull rate, the crystal radius will either be expanding, $da/dt > 0$, or shrinking, $da/dt < 0$. As the crystal rotation rate, ω, is increased, δ_m and thus δ_T decrease so that $\bar{q}_L$ increases and $d\bar{v}_c/dt$ must decrease in accordance with Eqs. (2.46). During the crystal seeding operation, it is Eqs. (2.43)–(2.46) that determine the rate of diameter increase for a given reduced crystal length $\bar{L}$. For long crystals, $\tilde{\mathcal{G}}(0)$ becomes a constant if H^* is constant, so that constant diameter crystals can be grown provided $\bar{q}_L$ is constant. We shall see later that, in a real situation, the effective H^* is not constant (H^* may be negative near $z \approx 0$ and positive for larger z) and heater power adjustments are needed so that $\bar{q}_L$ can be altered to balance the change in $\tilde{\mathcal{G}}(0)$ for long crystals. If one wishes, Eq. (2.43b) can be combined with Eq. (2.46a) to obtain a full constraint equation on the process with parameters $H_e^*, H^*, \bar{q}_L, d\bar{v}_c/dt$ and $\bar{L}$.

Considering Si in the long crystal limit, since $H_{Si}^* \sim 0.057$, we find that $\tilde{\mathcal{G}}(0) \sim 0.34$ which gives $(\mathrm{d}T/\mathrm{d}z)_0 \approx 20 ~°\mathrm{C}~\mathrm{cm}^{-1}$ for $a \approx 7.5$ cm and $T_M - T_a \approx 450~°\mathrm{C}$. For this case, the maximum growth velocity $(\bar{q}_L = 0)$ is just $V_{max} \approx (K_S/\Delta H_F)(\mathrm{d}T/\mathrm{d}z)_0 \approx 2 \times 10^{-3}~\mathrm{cm~s}^{-1}$. It is important to recognize here that infinitely rapid interface attachment kinetics have been assumed throughout this section. The special case of sluggish attachment kinetics will be returned to later.

Case 3 (Finite Péclet number): For this case, the heat carried by the movement of the crystal is now considered and, for a long crystal,

$$\theta = \exp\left[\hat{P}_T - (\hat{P}_T^2 + 2H^*)^{1/2}\right] Z \tag{2.47a}$$

$$\tilde{\mathcal{G}}(0) = \left[(\hat{P}_T^2 + 2H^*)^{1/2} - \hat{P}_T\right] \tag{2.47b}$$

These results reduce to Eq. (2.45) as $\hat{P}_T$ shrinks towards zero.

Case 4 (Steady state rotating disc flow effects): To the Cochran analysis described in Eqs. (2.26), using the flow functions $F_*(\zeta), G_*(\zeta)$ and $H_*(\zeta)$ where $\zeta = z(\omega_s/\nu)^{1/2}$, we can add the heat transport equation

$$\theta'' - PrH_*(\zeta)\theta' = 0 \tag{2.48a}$$

where

$$\theta = (T - T_M)/\Delta T \tag{2.48b}$$

and Pr is the Prandtl number, ΔT is the temperature difference between the rotating disc and the bottom of the crucible, ω is the crystal rotation rate and the primes refer to differentiation with respect to ζ. The BCs are

$$\theta = 0 \quad \text{at} \quad \zeta = 0; \quad \theta = 1 \quad \text{at} \quad \zeta \to \infty. \tag{2.48c}$$

Eq. (2.48a) is easily integrated so that the temperature and the interfacial temperature gradient are given respectively by

$$\theta = \frac{1}{\Delta T} \int_0^\zeta \exp\left[Pr \int_0^{\zeta_1} H_*(\zeta_2)\mathrm{d}\zeta_2\right] \mathrm{d}\zeta_1 \tag{2.49a}$$

$$\theta' = \frac{1}{\Delta T} \tag{2.49b}$$

where

$$\Delta_T - \delta_T \left(\frac{\omega}{\nu}\right)^{1/2} = \int_0^\infty \exp\left[Pr \int_0^\zeta H_*(\zeta')\mathrm{d}\zeta'\right] \mathrm{d}\zeta \tag{2.49c}$$

and δ_T is the thermal boundary layer thickness which is constant over

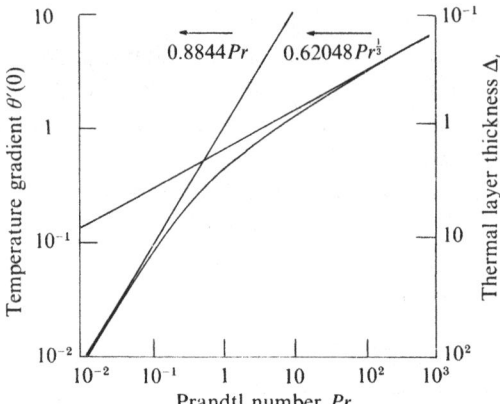

Fig. 2.37. Reduced temperature gradient and thermal layer thickness as a function of Prandtl number.[15]

an infinite rotating disc. The asymptotic expression for $\theta'(0)$ is given by

$$\theta'(0) = H_*(\infty)Pr \qquad \text{for} \quad Pr \ll 1 \qquad (2.50a)$$
$$= [-H_*''(0)/6]^{1/3}Pr^{1/2}\hat{T}^{-4/3} \quad \text{for} \quad P_r \gg 1 \qquad (2.50b)$$

where $\hat{T}$ is the Taylor number. The exact solution, for the case of a non-rotating crucible, was derived numerically by Cochran and is explicitly given by

$$\theta'(0) = 0.844\,Pr \quad \text{for} \quad Pr \ll 1 \qquad (2.51a)$$
$$= 0.620\,Pr^{1/3} \quad \text{for} \quad Pr \gg 1 \qquad (2.51b)$$

and shown graphically in Fig. 2.37.

To compare the effects of crystal versus crucible rotation on heat transfer to the crystal, we can use Kobayashi's calculated results[15] for the thermal film thickness, δ_f, defined by

$$\delta_f = \frac{T_b - T_M}{(\partial T/\partial z)_0} \qquad (2.52)$$

where T_b is the bulk melt temperature. In the presence of crystal rotation alone, δ_f is independent of radial position as ω increases with values as shown in Fig. 2.38(a). In the presence of crucible rotation alone, δ_f depends on the radial position as the crucible rotation rate, ω_c, increases as illustrated in Fig. 2.38(b). When the crystal and crucible rotate in the same sense, the thermal film thickness changes with ω in the same way as in the $\omega_c = 0$ case. When the crystal and crucible rotate in the opposite sense, ω_f at the center becomes a minimum at a value of $\omega \sim \omega_c/3$. For flow induced by combined crystal and crucible rotation, the rotation

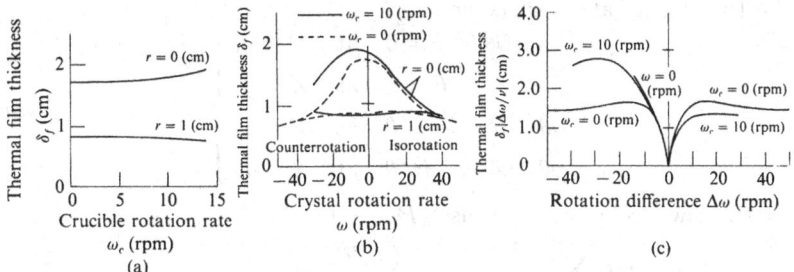

Fig. 2.38. Thermal film thickness as a function of (a) crystal rotation rate, (b) crucible rotation rate and (c) rotation difference, $\Delta\omega$, for $Pr = 0.31$.[15]

difference, $\Delta w = \omega - \omega_c$, should be used in place of the crystal rotation rate, ω. The quantity $\delta_f |\Delta w/\nu|^{1/2}$ is shown in Fig. 2.38(c) and it tends to become independent of Δw for large values of Δw. Using these data, a value for $\bar{q}_L = K_L(\partial T/\partial z)_0$ in Eq. (2.46a) can be readily evaluated so that the complete one-dimensional heat transfer approximation is described. However, it should be remembered that the actual fluid flow has a long initial transient for large crucible diameters and deep melts, so time-dependent factors will alter the simple one-dimensional result. Perhaps more important is to note that Fig. 2.31 reveals that the simple Cochran fluid flow analysis is *not* a good approximation to the actual flow during CZ crystal growth and this will significantly influence the heat transport result.

2.2.3   Three-dimensional heat transport approximation

Especially for high temperature crystal growth, a steep radial temperature distribution is easily introduced and the one-dimensional model is no longer valid because the Biot number is usually large. Although it is not totally correct to do so, the angular dependence of T is generally neglected so one obtains only a ϕ-average effect. However, let us remind ourselves again that it is the ϕ-variation that leads to the "swirl" defects to be discussed later in Chapter 4. We shall neglect the ϕ-variation here. Introducing the dimensionless radius, $R = r/a$, Eq. (2.41b) is replaced by

$$2\hat{P}_T \left(\frac{\partial \theta}{\partial \tau} + \frac{\partial \theta}{\partial Z} \right) = \frac{\partial^2 \theta}{\partial Z^2} + \frac{1}{R}\frac{\partial}{\partial R}\left(\frac{\partial \theta}{\partial R} \right) \tag{2.53a}$$

For very low Péclet numbers, this just becomes the Laplace equation.

The BCs for a flat interface are

$$\left.\begin{array}{rcl} \partial\theta/\partial R = 0 & \text{at} & R = 0; \\ \theta = 1 & \text{at} & Z = 0 \\ \partial\theta/\partial R = -H^*\theta & \text{at} & R = 1; \\ \partial\theta/\partial Z = -H_e^*\theta & \text{at} & Z = \bar{L} \end{array}\right\} \qquad (2.53b)$$

For the low Péclet number case ($\hat{P}_T \to 0$),

$$\theta = 2H^* \sum_{n=1}^{\infty} \frac{1}{H^{*2} + \gamma_n^2} \frac{J_0(\gamma_n R)}{J_0(\gamma_n)}$$

$$\times \left\{ \frac{\gamma_n \cosh[\gamma_n(\bar{L} - Z)] + H^* \sinh[\gamma_n(\bar{L} - Z)]}{\gamma_n \cosh(\gamma_n \bar{L}) + H^* \sinh(\gamma_n \bar{L})} \right\} \qquad (2.54a)$$

where γ_n is the nth root of

$$\gamma_n J_1(\gamma_n) - H^* J_0(\gamma_n) = 0 \qquad (2.54b)$$

For a long crystal with a flat interface, the curly bracket in Eq. (2.54a) reduces to $\exp(-\gamma_n z)$. Apart from the radial dependence, the temperature is approximately that given by the one-dimensional model.

If H^*/a is small (expected values are $\ll 1$ cm^{-1} for most systems), then

$$\gamma_0 \sim (2H^*)^{1/2}; \; \gamma_1 \sim 3.8; \; \gamma_n \sim [7.02 + 3.15(n - 2)] \quad \text{for} \quad n > 2. \qquad (2.54c)$$

Using these values, it can be seen that Eq. (2.54a) is dominated by the $n = 0$ term and that the contributions by the other terms, which are of alternating sign, are only a few per cent. Thus, for $\bar{L} - Z > \gamma_0^{-1}$, the temperature distribution at a position Z is independent of $\bar{L}$ and the following approximate relations apply:

$$\theta \sim \left(\frac{1 - H^* R^2}{1 - H^*/2} \right) \exp[-(2H^*)^{1/2} Z] \qquad (2.55a)$$

$$\frac{\partial\theta}{\partial Z} \sim -(2H^*)^{1/2} \left(\frac{1 - H^* R^2}{1 - H^*/2} \right) \exp[-(2H^*)^{1/2} Z] \qquad (2.55b)$$

$$\frac{\partial\theta}{\partial R} \sim -\left(\frac{2H^* R}{1 - H^*/2} \right) \exp[-(2H^*)^{1/2} Z] \qquad (2.55c)$$

and

$$\frac{\partial^2\theta}{\partial R^2} \sim -\left(\frac{2H^*}{1 - H^*/2} \right) \exp[-(2H^*)^{1/2} Z] \qquad (2.55d)$$

All these approximations improve as Z increases while Eq. (2.55c) requires that $Z > \gamma_1^{-1}$. From Eq. (2.55d), we see that the curvature of the isotherms is proportional to $2H^*/(1 - 0.5H^*)$ which will be a measure of the planarity of the interface. Brice[16] carried out experiments on Ge crystals. He showed that both $a^{1/2}\partial\theta/\partial Z$ and $a\partial^2\theta/\partial Z^2$ were constant, in agreement with Eq. (2.55b) and its derivatives $(a^{-1/2}\partial\theta/\partial Z$ and $a^{-1}\partial^2\theta/\partial Z^2$, respectively).

Although the temperature distribution in a Ge crystal does not depend on the crystal length so long as this length is greater than $6a^{1/2}$, the approximation does not hold for metals because of their order of magnitude larger thermal conductivity. In such cases, much longer crystal lengths are required.

2.2.4 Interface shape considerations

Until now we have assumed that the solid/liquid interface shape is planar. However, in practice, the interface shape may be macroscopically concave, planar, convex or some combination of these, ranging from deeply concave to extremely convex. One determinant of the deviation of the interface shape from planarity is the Biot number. The interface becomes more concave to the liquid with increasing Biot number, unless other conditions are also varied. A convex interface is obtained if the heat is transferred mainly radially from the melt to the interface (strong natural convection mode) or if the crystal near the interface absorbs heat from the surroundings such as exposed crucible walls.

Analysis of the heat loss for a rotating disc in contact with a viscous fluid has shown that the heat transfer depends significantly upon the existence of a radial temperature gradient in the disc.[6] For a disc with a temperature distribution along the surface of

$$T(0,r) = T_{12} + T_{20}r^2 \tag{2.56a}$$

we can define a transport number, N_*, of the form

$$N_* = -rT'(0,r)/T(0,r) \tag{2.56b}$$

where the prime refers to differentiation with respect to $\zeta = z(\omega/\nu)^{1/2}$. Thus, neglecting dissipation of energy in the boundary layer,

$$\frac{N_*}{Re^{1/2}\phi(Pr)} = 0.616 - 0.217\frac{T_{10}}{T(0,r)} - \frac{3.83}{Re}\left[1 - \frac{T_{10}}{T(0,r)}\right] \tag{2.57}$$

Here, $\phi(Pr) \approx Pr^{0.45}$ so that, in the regions of Re of interest, the two extremes of $T_{20} = 0$ versus $T_{10} = 0$ show a 50% higher value of N_* when $T_{10} = 0$. From this we see that the radial temperature gradient in a crystal is a significant factor in the degree of heat transfer from a viscous melt to the growing crystal.

When considering the fluid flow aspects for a non-planar rotating interface, we saw earlier that the flow field is the same as for a disc at the surface when one replaces the radial coordinate by the surface coordinate, x, along any radial direction (see Fig. 2.17). Because of this, one would expect that such a curved surface would be almost an isotherm (pure system) and therefore Eq. (2.54a) might hold approximately on any cross-section of the crystal axis.

For the case of very slow growth (neglect of latent heat evolution), the interface shape, $Z = Z_i$ can be obtained from the continuity equation and an approximate condition of constant heat flux through the interface from the melt[17]

$$-K_S\frac{\partial T_S}{\partial n} = -K_L\frac{\partial T_L}{\partial n} + V_n\Delta H_F \approx q_0 \tag{2.58a}$$

The BCs are

$$\theta = 1 \text{ at } R = 1, \ Z = 0 \text{ and } -\frac{\partial\theta}{\partial Z} = Q_0 \text{ at } Z = 0 \tag{2.58b}$$

where Q_0 is the dimensionless heat flux

$$Q_0 = q_0 a / K_S(T_M - T_a) \tag{2.58c}$$

For a long crystal, the solution is very much like Eq. (2.54), i.e.,

$$\frac{\theta}{Q_0} = \sum_{n=0}^{\infty} \frac{1}{\gamma_n(H^{*2} + \gamma_n^2)} \frac{J_0(\gamma_n R)}{J_0(\gamma_n)} \exp(-\gamma_n Z) \tag{2.59a}$$

and

$$Q_0 = \frac{1}{\sum_{n=0}^{\infty} [\gamma_n(H^{*2} + \gamma_n^2)]^{-1}} \tag{2.59b}$$

The interface shape, $Z = Z_i$, is the isotherm of $\theta = 1$ which appears near the coordinate ($R = 1$, $Z = 0$) and is given approximately by

$$Z_i = -\frac{1}{\gamma_1}\ln\left[\frac{Q_0\gamma_1(H^{*2} + \gamma_1^2)J_0(\gamma_1)}{J_0(\gamma, R)}\right] \tag{2.59c}$$

except for the region $R \approx 1$.

Interface shapes for various positive Biot numbers are shown in Fig. 2.39 where we see that H^* is a good parameter for evaluating the deviation of the interface from planarity. Negative values of H^* would provide interface shapes that were convex to the melt.[17] From Fig. 2.39 we see that, for small Biot numbers ($H^* < 0.1$), the interface shape is nearly planar and the one-dimensional approximation becomes valid. From the basic definition of the Biot number, we expect the interface to become concave to permit the easier loss of heat to the surroundings so that the isotherm shape becomes strongly concave as the Biot number becomes

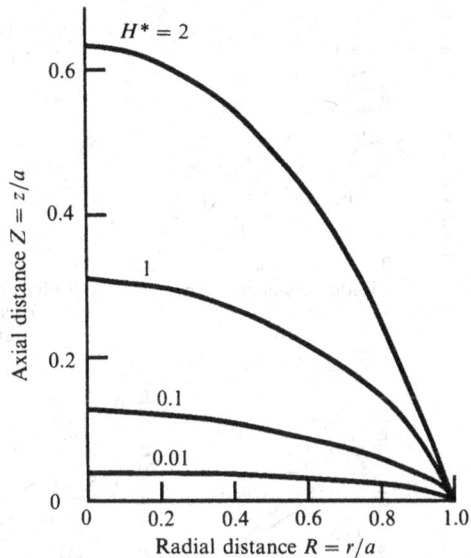

Fig. 2.39. Interface shapes for various Biot numbers.[17]

large, plus nearly planar when the Biot number is small and strongly convex when the Biot number is large and negative.

In the real world of CZ crystal growth, one must consider both radiation and convection from all important surfaces, crucible, melt and crystal, and the Biot number is now defined by

$$H^* = \left(\bar{\varepsilon}_S \tilde{\sigma} T_M^3 + \alpha_h T_M^{1/4} \right) a / K_S \qquad (2.60)$$

In Eq. (2.60), the $T_M^{1/4}$ term comes from gaseous convection with α_h being the heat transfer coefficient. A computer solution that takes into account heat transfer from the top surface of the crystal $(0 < r < a,\ Z = \bar{L})$, from the crystal side surface $(r = a,\ 0 \stackrel{-}{<} Z \stackrel{-}{<} \bar{L})$ and at the melt free surface $(a \stackrel{-}{<} r \stackrel{-}{<} R_*,\ Z = 0)$ allows one to gain qualitative insight concerning the effects of the ambient gas, crystal radius, crystal length, crucible radius, melt height and crucible temperature on the interface shape, $Z_i(r)$. These results, for Ge as a model system, have been collected in Fig. 2.40 and conform to one's qualitative expectations. To make the results more quantitative, it is necessary to include both the heat transfer view factors, f_{jk}, illustrated in Fig. 2.41 and the fluid flow in the melt. As one might expect, this greatly increases the complexity of the analysis.

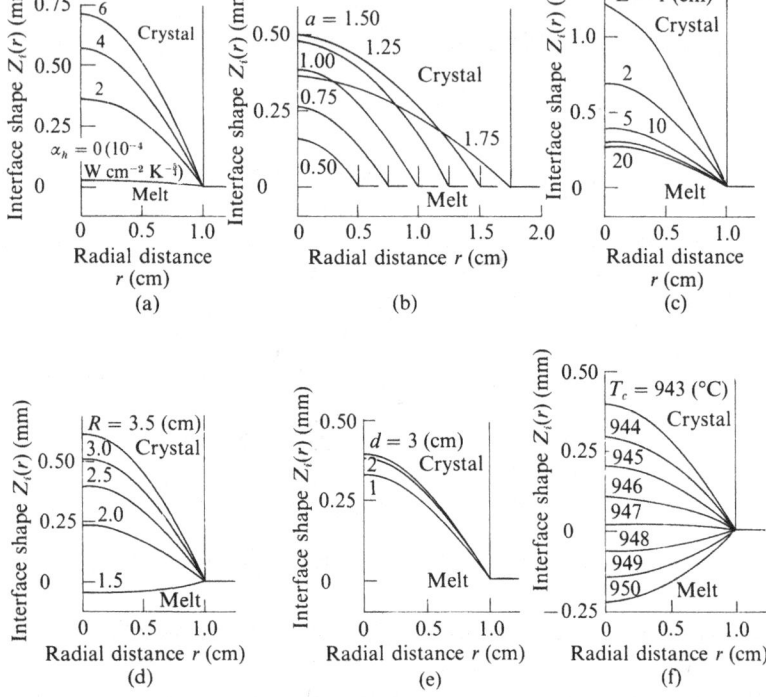

Fig. 2.40. Ge interface shapes, $Z_i(r)$, for (a) various heat transfer coefficients, (b) various crystal radii, (c) various crystal lengths, (d) various crucible radii, (e) various melt heights and (f) various crucible temperatures.

Experimental results show that the melt height markedly influences heat transfer through the exposed crucible wall and that the exposed crucible wall plays the role of an afterheater. Naturally, radiative heat exchange between the crystal and the exposed crucible wall becomes more effective if the radius of the crystal is near that of the crucible. The interface shape changes from concave to convex with increasing exposed crucible wall because H^* changes from positive to negative values. Often features thought of as crystal length and melt height effects are actually due to the effect of exposed crucible wall.

For high Pr melts, the heat flow from the melt to the interface sometimes changes during growth when the crystal diameter exceeds a critical value and sudden remelt of the crystal occurs. At the same time, the fluid flow in the melt usually varies from free convection to forced con-

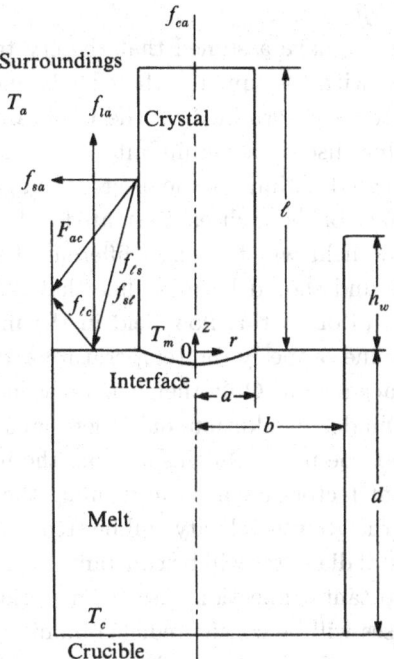

Fig. 2.41. Schematic illustration of radiative heat exchange during crystal growth.

vection due to crystal rotation and relatively abrupt inversion of the interface shape from extremely convex to planar occurs. A summary of the experimental findings on interface shape are:

(a) **To make the interface shape less convex (more concave):** (1) Use a large crystal radius, (2) increase the crystal rotation rate, (3) reduce the crucible rotation rate, (4) increase the growth rate, (5) thermally insulate the crucible bottom, and (6) use a flat-bottomed crucible.

(b) **To make the interface shape less concave (more convex):** (1) Cool the crucible base, (2) use a large crucible, (3) use an afterheater, and (4) use a shallow melt with a high crucible wall.

Although the foregoing heat transfer considerations have been directed to CZ crystal growth, they are readily applicable to both the vertical and horizontal Bridgeman methods and to other methods because the operating principles have all been well demonstrated. Very similar considerations are involved when one wishes to evaluate temperature profiles during film growth.

2.2.5   *Microscopic meniscus-lift picture*

In all of the foregoing, we have assumed that the crystal/melt interface is essentially coplanar with the upper surface of the melt and that the convective flow velocity is determined by the Cochran analysis for an infinite radius rotating disc on a semiinfinite fluid. Actually, the crystal/melt interface is located within the meniscus envelope a distance, z_i, above the upper surface of the melt as illustrated in Fig. 2.42. Thus, the local convective flow field will be quite different than that given by the Cochran analysis and should be more like that described by Fig. 2.31(b). A good description of this flow field in the meniscus region is urgently needed to define properly the temperature and solute distributions in the near interface region. Only then can a reasonable assessment of interface shape be made. In addition, only then can a careful prediction of interface transients be made. In this section, the intent is merely to outline the important factors involved in defining the microscopic interface state whereby the growth velocity will be steady state or transient and whereby the crystal diameter will be constant, shrinking or growing. To illustrate the important connection with the material of the companion volume,[18] two cases will be considered: (1) infinitely rapid interface attachment kinetics and (2) sluggish and strongly anisotropic interface attachment kinetics.

For case (1) the relevant diagram is that given in Fig. 2.43(a) where, because of the infinitely rapid attachment kinetics, the liquid contact point for steady state growth (R_c = constant) is at the juncture of the straight wall and the curved rim region[18] where the temperature is

$$T = T^* = T_M - \frac{\gamma_{SL}\mathcal{K}_{SL}}{\Delta S_F} \tag{2.61}$$

where $\bar{\mathcal{K}}_{SL}$ is the curvature of the solid/liquid interface at this junction point. For large values of R_c, the height of this junction point above the bulk liquid is given approximately by

$$h_m \approx \frac{2\gamma_{LV}}{\rho_L g} \tag{2.62a}$$

where γ_{LV} is the effective liquid/vapor interfacial energy, ρ_L is the liquid density and g is the gravitational constant. Here, γ_{LV} is treated as an effective energy because it depends somewhat on the magnitude of the local fluid flow velocity. For a quiescent liquid, γ_{LV} attains the true value of the liquid/vapor surface energy because the only lift force comes from the negative curvature of the surface. However, for Cochran-type flow in this meniscus region, an additional outwardly directed force develops which can be modeled via a larger value of γ_{LV}. For the reverse direction

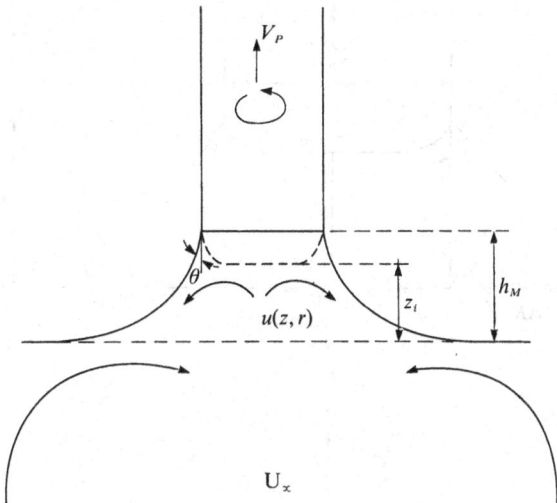

Fig. 2.42. Schematic illustration of general interface shape and fluid flow fields in the meniscus region.

of flow, a smaller value of γ_{LV} is needed. This type of flow contribution to the lift force on the meniscus is also expected to change the wetting angle, θ, to some degree. The other dimensions in Fig. 2.43(a) are

$$z_i \approx h_M - (\bar{\mathcal{K}}_{SL})^{-1} \qquad (2.62b)$$

and

$$\Delta R = h_M \tan\theta \qquad (2.62c)$$

where $\bar{\mathcal{K}}$ is the average curvature in the rim region ($\bar{\mathcal{K}}_{SL}^{-1} \sim 10\text{--}50\ \mu m$). Thus, all these geometrical features are known provided the effective value of γ_{SV} is known. This leads to $h_M \sim$ several millimeters while $h_M - z_i \sim 10^{-2} h_M$ and $\Delta R \sim$ several millimeters.

Energy conservation for this case yields

$$Q_{in} = \tilde{h}_{cr} A_{cr}(T_{cr} - T_a') + \tilde{h}_L A_L(T_L(0) - T_a)$$
$$+ \tilde{h}_m A_m(T_L(z) - T_a) + K_S A_c (dT_S/dz)_0 \qquad (2.63a)$$

where $\tilde{h}$ refers to heat transfer coefficient, A refers to area and the subscript cr refers to crucible. Thus, this heat balance leads, in steady state, to a value of $A_c = \pi R_c^2$, so the crystal size is determined. To change the crystal size we operate on Eq. (2.63a) at constant V_P to yield

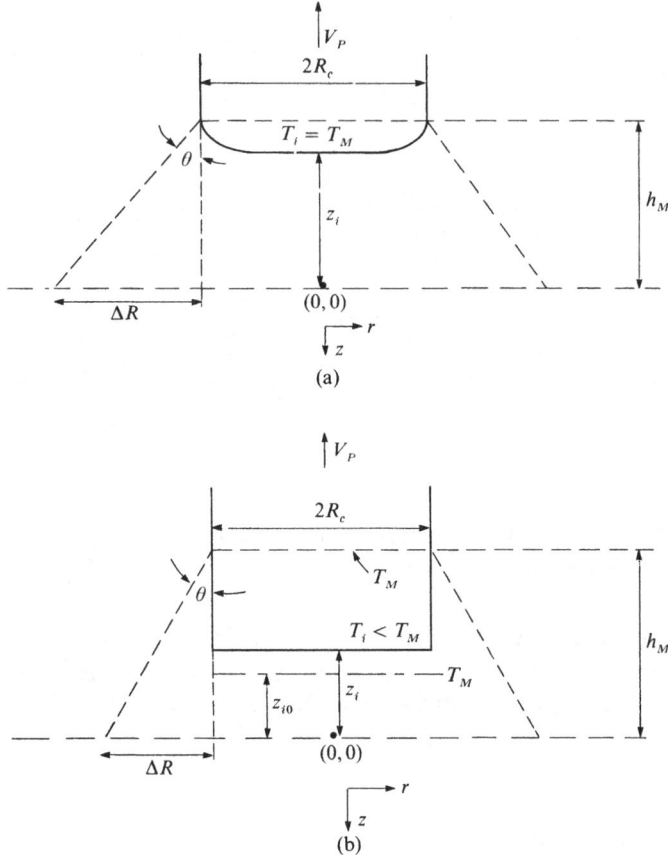

Fig. 2.43. Schematic illustration of local interface geometry and key temperatures in the meniscus region for (a) infinitely rapid interface attachment kinetics, and (b) sluggish and strongly anisotropic interface attachment kinetics.

$$\frac{\mathrm{d}Q_{in}}{\mathrm{d}t} \approx \left[v_M c_L - \left(\tilde{h}_{cr} A_{cr} + \tilde{h}_L A'_L + \frac{\tilde{h}_m A'_m}{2} \right) \right] \frac{\mathrm{d}T_\infty}{\mathrm{d}t}$$
$$+ K_S \left(\frac{\mathrm{d}T_S}{\mathrm{d}z} \right)_0 \frac{\mathrm{d}A_c}{\mathrm{d}t} \qquad (2.63b)$$

where v_M is the melt volume and T_∞ is the bulk melt temperature. Thus, the rate of crystal broadening is connected to the rate of melt temperature change (note that A'_L and A'_m are different than A_L and A_m, respectively, because $\mathrm{d}A_c/\mathrm{d}t \neq 0$). If we are allowed to neglect any

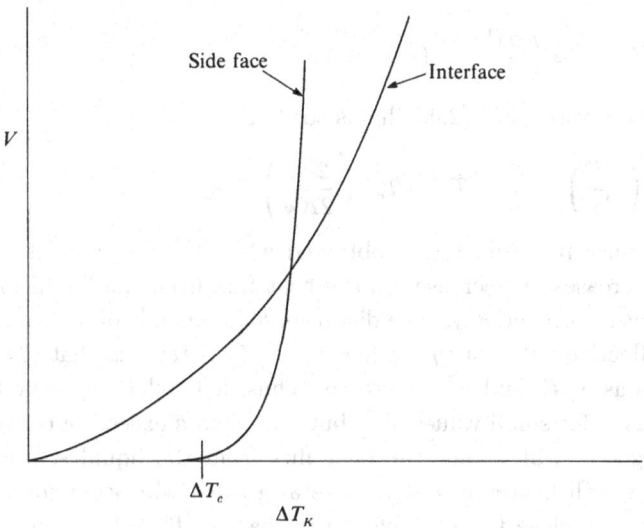

Fig. 2.44. Proposed attachment kinetic laws at the main crystal interface and at the side faces for the situation illustrated in Fig. 2.43(b).

changes in fluid flow velocity (doubtful) due to the changes in T_∞, we can expect the following balance to hold

$$\frac{K_L}{z_i}\frac{dT_\infty}{dt} \approx \Delta H_F V_P \frac{dA_c}{dt} \tag{2.63c}$$

Combining Eqs. (2.63c) and (2.63b) leads to a relationship between dT_∞/dt and dQ_{in}/dt. Of course, convective velocity changes during the transient will alter Eq. (2.63c).

For case (2) the relevant diagram is that given in Fig. 2.43(b). The solid/liquid interface is at temperature $T_i = T_M - \Delta T_{K_i}$ where the magnitude of ΔT_{K_i} depends on the limiting mechanism that is operating (see Eqs. (1.3)–(1.5)). For simplicity, let us assume that it is a layer source limitation associated with axial dislocations threading the interface. Let us also assume that the lateral face attachment mechanism is two-dimensional nucleation. Thus, the relationship between the particular face growth velocity, V, and face undercooling, ΔT_K, for both the interface and the side faces is indicated in Fig. 2.44. So long as $\Delta T_m = T_M - T_m < \Delta T_c$, no growth can occur on the sidewall faces and the crystal cannot broaden.

In practice, we expect

$$\Delta T_m \approx \left(\frac{dT}{dz}\right)_{T_m}(h_m - z_{i0}) \tag{2.64a}$$

while

$$\Delta T_{K_i} \approx \left(\frac{dT}{dz}\right)_{T_m} (z_i - z_{i0}) \qquad (2.64b)$$

For a long crystal, Eq. (2.45) holds so that

$$\left(\frac{dT}{dz}\right)_{T_m} = -(T_m - T_a)\left(\frac{2\tilde{h}_a}{2K_S}\right)^{1/2} \qquad (2.64c)$$

The interface position, z_{i0}, is obtained when $V_P = 0$ so that its magnitude increases or decreases as the heat flux from the liquid increases or decreases, respectively. The distance, h_m, depends on the magnitude of the effective value of γ_{LV} while Eq. (2.64c) tells us that $|dT/dz|_{T_m}$ increases as a, T_a and h_a^{-1} decrease. Thus, it is relatively easy to have $\Delta T_m > \Delta T_c$ for small values of a, but not when a exceeds a certain size. Reducing Q_{in} will reduce the heat flux from the liquid so that both z_{i0} and z_i will decrease, but no lateral growth will occur unless ΔT_m becomes very close to ΔT_c before the change. If ΔQ_{in} is sufficient to increase $(h_m - z_{i0})$ so that ΔT_m exceeds ΔT_c, then a burst of pillbox nucleation will occur on the side faces immediately adjacent to the meniscus and these new layers will propagate down the side faces to broaden the crystal. When they do, a increases slightly so that $(dT/dz)_{T_m}$ decreases slightly and the lateral growth process stops. If, by chance, this sidewall layer flow creates dislocations pointing laterally through a face, then the screw dislocation attachment mechanism will become operative and new layers can be spun out at a great rate and a great deal of broadening will occur rather abruptly. All the details of the foregoing will be altered by the shape of the isotherms in both the main interface and meniscus regions. Thus, once again a major need exists for a careful analysis of fluid flow, heat transport and matter transport in this general meniscus region for crystals with sluggish and anisotropic interface attachment kinetics.

Problems

1. Considering one-dimensional parallel flow between vertical plates as in Fig. 2.5(a), Eqs. (2.1), (2.2) and (2.38a) become

$$\partial u/\partial x = 0$$

$$\frac{\partial u}{\partial t} = -\frac{1}{\rho}\frac{d}{dx}(p - p_0) + g\beta_T(T - T_0) + \nu\frac{\partial^2 u}{\partial z^2}$$

$$\frac{\partial T}{\partial t} = D_T\frac{\partial^2 T}{\partial z^2}$$

For *steady* fluid motion with temperature difference, ΔT, between the walls and subject to the boundary conditions and mass conservation condition

$$u(0) = 0 \; ; \; u(\ell) = -U \; ; \; \int_0^\ell u(z)\mathrm{d}z = 0$$

show that, by considering the flow, $u(z)$, as the superposition of forced (isothermal) flow at $u_1(z)$ and buoyancy-driven flow at $u_2(z)$, Eqs. (2.16) are obtained (the unknowns p_0 and T_0 are determined from the conservation of mass condition).

2. For the crystal growth cell shown in Fig. 2.8, where the length of Region I is ℓ cm, develop a formula to define the maximum growth velocity, V, for the crystal when the system is isothermal and no convection develops in Region I.

3. Considering Fig. 2.7, show how the Nusselt number, Nu, can be used to define the thermal boundary layer thickness, δ_T. If the lower surface is a slowly growing crystal, what is the relationship between the solute boundary layer thickness, δ_C, and Nu?

4. Using the Cochran analysis for forced fluid flow during CZ-type growth of Si, what crystal rotation rate is needed to make $\delta_C \sim 10^{-2}$ cm ($\nu = 2 \times 10^{-2}$cm^2 s^{-1}, $D_C = 5 \times 10^{-5}$cm^2 s^{-1})?

5. For 15 cm diameter CZ growth of Si at 30 rpm rotation rate, what value of impinging flow velocity, W_∞, (see Eq. (2.33)) is needed to reduce the value of δ_C by about a factor of 5 compared to the $u_\infty = 0$ case?

6. During CZ crystal seeding at constant pull rate with an initial seed rotation rate of ω_1, how would you change ω (without a change in furnace power) to produce diameter broadening for the case where (a) no natural convection exists in the melt and (b) strong natural convection exists in the melt? What else can you do to enhance crystal broadening?

3

Steady state solute partitioning

For the considerations of this and the next chapter, two distinctly different geometrical arrangements of liquid and crystal will be considered: (1) the unidirectional crystallization of a channel of liquid and (2) the three-dimensional crystallization of differently shaped particles of solid that are entirely surrounded by liquid. Most of the chapter will deal with the former case which can be subdivided into two distinct modes of crystallization: (a) *normal freezing* wherein all of the charge is melted and then progressively solidified from one end as illustrated in Fig. 3.1(a) and (b) *zone melting*, wherein only a portion of the solid charge is melted and this fluid zone is moved through the remainder of the solid (see Fig. 3.1(b)). Each mode leads to its own characteristic solute redistribution patterns which can be altered or perturbed by changing the crystallization procedures in one of two ways: (i) by changing the concentration of the bulk liquid via changing the liquid volume, adding or removing a constituent from the vapor or by making the unmelted portion of the charge have a variable composition and (ii) by changing the character of the solute boundary layer at the interface through a change in the forces controlling mass transport; i.e., by changes in crystallization velocity (V), fluid velocity (u), or applied field ($\delta \widetilde{\Delta G_0}$).

3.1   Normal freezing

In this topic, the main considerations are the fraction of the total volume frozen and the interface rejected solute that accumulates

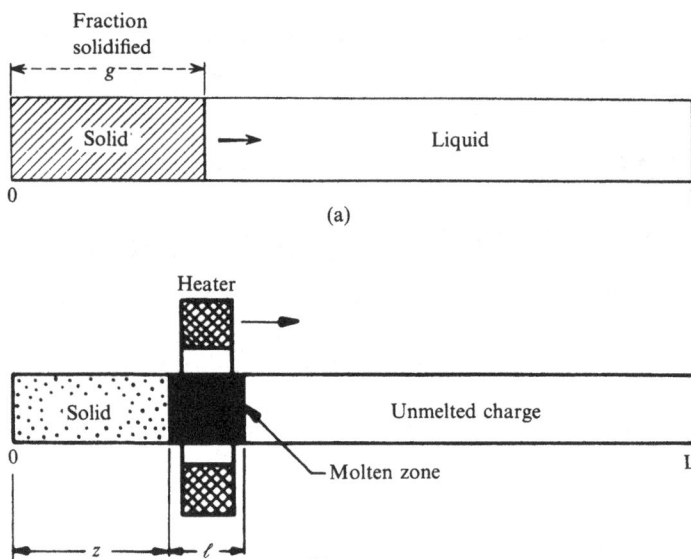

Fig. 3.1. Schematic illustration of the two common procedures of solidification from a melt: (a) normal freezing and (b) zone melting.

in the liquid as freezing progresses. This leads to an end-to-end segregation of solute along the length of the crystal which is called "normal segregation". Four distinct types of normal segregation profiles arise depending upon the crystallization conditions. These profiles are illustrated in Fig. 3.2 and are:

(a) *equilibrium freezing* wherein the freezing is so slow that diffusion in both the liquid and solid is complete so that all concentration gradients are eliminated and no chemical segregation develops (never occurs in practice);

(b) *complete mixing* wherein freezing is slow enough for liquid mixing to eliminate all concentration gradients in the liquid but fast enough that no diffusion occurs in the solid leading to the maximum possible end-to-end segregation (may occur in practice);

(c) *no mixing* wherein freezing is rapid enough that only liquid diffusion affects the solute distribution in the liquid leading to only slight segregation (may occur in practice); and

(d) *partial mixing* wherein the freezing rate and fluid mixing are such that the solute distribution in the liquid is affected by both

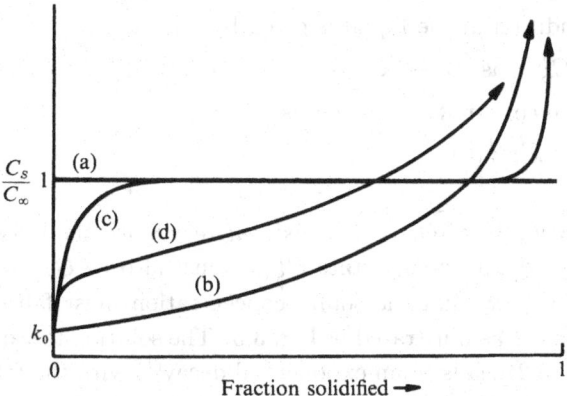

Fig. 3.2. Solute distribution in a crystal frozen from a long constant cross-section of initial concentration C_0, for (a) equilibrium freezing, (b) complete mixing, (c) no mixing and (d) partial mixing.

diffusion and convection leading to the general case of intermediate segregation (generally occurs in practice).

3.1.1 The steady state distributions (planar front, V = constant)

In one dimension, the general expression for solute flux, J^j, is

$$J^j = - \left(D^j \frac{\partial C^j}{\partial z} + \left\{ \frac{D^j}{\kappa T} \left[\frac{\partial}{\partial z} \delta \widetilde{\Delta G}_0^{\,j}(z) \right] - u(z) \right\} C^j \right) \quad (3.1a)$$

where $\delta \widetilde{\Delta G}_0$ is the interface field and u is the fluid velocity, so that the solute conservation condition in a fixed coordinate system without creation or annihilation of species is

$$\frac{\partial}{\partial z}(-J^j) = \frac{\partial C^j}{\partial t} \quad (3.1b)$$

However, this never yields a steady state solution so we transfer our coordinate system to the moving interface frame where we are able to reach a steady state solution. Solute conservation in a coordinate system moving at velocity V is given by

$$\frac{\partial}{\partial z}(-J^j_\beta + V C^j_\beta) = \frac{\partial C^j_\beta}{\partial t}$$

$$= D^j_\beta \frac{\partial^2 C^j_\beta}{\partial z^2} + \frac{\partial}{\partial z} \left\{ \left[V + \frac{D^j_\beta}{\kappa T} \left(\frac{\partial}{\partial z} \delta \widetilde{\Delta G}^{\,j}_{0\beta} \right) - u \right] C^j \right\};$$

$$\beta = S, L \quad (3.1c)$$

The interface conservation condition is given by

$$V(1 - k^j_i) C^j_L(0) = J^j_L(0) - J^j_S(0) \quad (3.2a)$$

the far-field condition in the liquid is given by

$$C_L^j \to C_\infty^j \quad \text{as} \quad z \to \infty \qquad (3.2b)$$

and the interface continuity condition is

$$C_S^j(0) = k_i^j C_L^j(0) \qquad (3.2c)$$

Case 1 ($D_{SC} = u_\infty = \delta \widetilde{\Delta G}_0 = 0$, *conservative*): Since the bulk liquid concentration is C_∞^j, at steady state $C_S^j(0)$ must also be C_∞^j and thus $C_L^j(0)$ must be C_∞^j / k_0^j and the solute concentration must fall off into the liquid for $k_0^j < 1$ as illustrated in Fig. 3.3. The solution to Eqs. (3.1) subject to Eqs. (3.2) leads to an exponential decay[1] with the following specific form

$$C_L(z) = C_\infty \frac{(1 - k_0)}{k_0} \exp\left(-\frac{Vz}{D_L}\right) + C_\infty \qquad (3.3)$$

Calculated profiles for $C_L(z)$ for various values of k_0 are given in Fig. 3.3 while Fig. 3.4 shows the results of a rapidly quenched $A\ell$ + Cu alloy crystal (the upward bending in the solid is probably a quenching artifact).[2] Eq. 3.3 also applies for $k_0 > 1$.

The characteristic width of the solute rich layer at the interface is $\sim D_L/V$ and, since $D_L \sim 2 \times 10^{-5}$ cm^2 s^{-1} this is $\sim$ 1–10^3 μm for V between 0.2 cm s^{-1} and 2 cm s^{-1}. Since the interface concentration in the liquid, $C_i = C_\infty / k_0$, this can give rise to extremely large local solute concentrations even when C_∞ is very small provided k_0 is also very small. Because of this, important interface reaction and precipitation effects may develop during crystallization that would not be anticipated from bulk fluid considerations alone. Let us consider a few examples: (1) *Ice cube bubbles* – The air that dissolves in water partitions at the interface and builds up an interface concentration sufficient to exceed the condition for bubble nucleation. Thus, bubbles nucleate on the interface and are fed by solute diffusion within the interface boundary layer. If the flux of gas molecules to the bubbles is large and the rate of freezing of the ice is low, then the bubbles grow with a continually expanding diameter. If the solute flux and the interface movement are well balanced, then long tubular pores develop in the ice. When the ice freezing rate is high relative to the gas flux, then the bubbles grow with a continually shrinking diameter. For ice cubes formed in an old-style refrigerator, an on–off cycle occurs which leads to an oscillation in V so that the bubbles look like a string of connected beads. (2) *Rimming action in steel* – In a ladle of molten steel, pouring is delayed until the bulk C + O $\overset{\rightarrow}{\leftarrow}$ CO formation reaction has stopped. However, after pouring and as

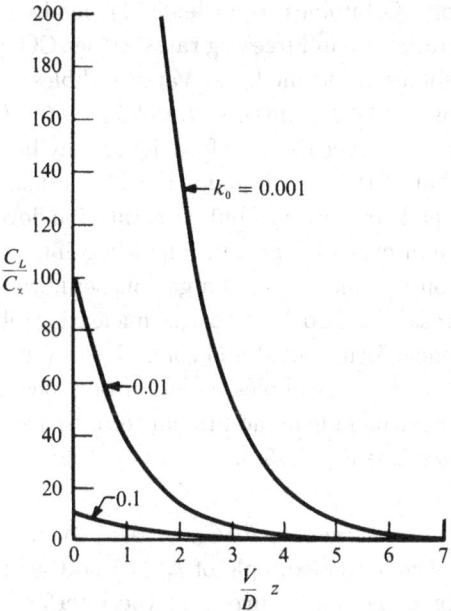

Fig. 3.3. Dimensionless solute distribution profile in the liquid, C_L/C_∞, as a function of the dimensionless distance parameter, Vz/D, for the "no mixing" case.

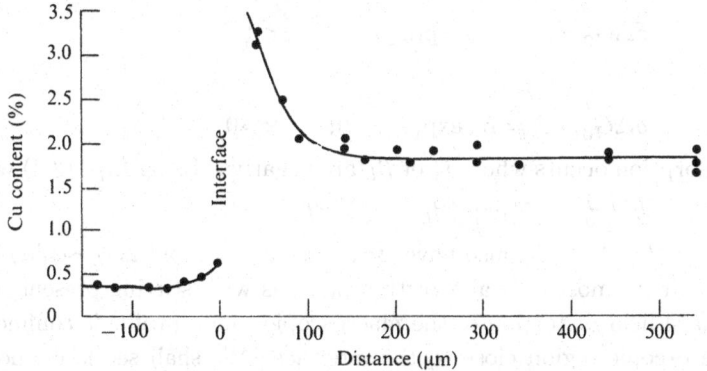

Fig. 3.4. Cu distribution in an Aℓ–Cu alloy quenched during solidification (resolution $\sim 1\ \mu m$ and $50\ \mu m$ to left and right of interface, respectively).[2]

freezing occurs in the mold, $C_i^O \to C_\infty^O/k_0^O$ and $C_i^C \to C_\infty^C/k_0^C$ so that renewed chemical reaction to form CO occurs in the interface boundary layer and it is sufficient for CO bubbles to nucleate. Depending on the relative supersaturation, transport and freezing rates, either CO channel pores or a string of CO bubbles are formed. (3) *Vapor bubbles* – During single crystal growth, if j is a volatile constituent and $C_i^j \to C_\infty^j/k_0^j$, the equilibrium partial pressure of j over the interface liquid may be greater than 1 atm even though that of the bulk liquid is much less than 1 atm. Thus, the interface liquid tends to boil with bubble formation developing and a "Swiss Cheese" type of crystal is grown. The foregoing examples dealt with bubble formation via nucleation of a gas phase; however, the interface enrichment process can also lead to the nucleation of liquid phases (oils in some geological formations) and solid phases (eutectic or peritectic phases). In these three examples, creation of a new species occurs so that Eq. (3.1c) must include an additional term to account for the annihilation of j-species (see Eq. 3.13a).

Case 2 ($D_{SC} = u_\infty = 0$, conservative, $\delta \widetilde{\Delta G}_0 \neq 0$): To understand this case, we shall first consider the example of $k_0 = 1$ so that there is no large solute build-up or depletion occurring at the interface. Thus, the effects associated with the interface field will be more clearly seen. For use in numerical evaluation later, a reasonable and simple choice of interface field is one that decays exponentially with distance from the interface so we shall assume the following forms

$$\delta \widetilde{\Delta G}_{0S}^{\ j}(z) = \bar{\beta}_S \exp(\alpha_S z); \qquad z<0 \tag{3.4a}$$

and

$$\delta \widetilde{\Delta G}_{0L}^{\ j}(z) = \bar{\beta}_L \exp(-\alpha_L z); \qquad z>0 \tag{3.4b}$$

Adsorption occurs when $\bar{\beta}_S$ or $\bar{\beta}_L$ are negative. Using Eqs. (3.4) we have

$$k_i^j/k_0^j = \hat{\gamma}^{*j} \exp[(\bar{\beta}_L - \bar{\beta}_S)/\kappa T] \tag{3.5}$$

with $\hat{\gamma}^{*j} \sim 1$.[3] We also have $\Delta z_S \sim 5\alpha_S \ll \delta_C$ and $\Delta z_L \sim 5|\alpha_L| \ll \delta_C$ so that, in most crystal growth situations with stirring present, the interface field distortion of the macroscopic solute profile is confined to a microscopic region close to the interface. We shall see later, however, that the effect of the interface field on the solute profile during crystallization can extend orders of magnitude beyond the Δz_L distance for some special range of growth parameters. One is thus led to ask, "What are the consequences of this microscopic solute profile on the net solute incorporation into the crystal?" However, before answering this question, it should be made clear that a variety of field dependencies on z

could be considered; i.e.,

$$\delta \widetilde{\Delta G}_0^j(z) = \bar{\beta}(1 - \alpha|z|) \quad ;\alpha|z| \widetilde{<} 1 \atop = 0 \qquad\qquad ;\alpha|z| > 1 \left.\right\} \tag{3.6}$$

is another reasonable possibility.

In what is to follow, we shall choose to approximate the interface fields as exponential functions as given in Eqs. (3.4). We shall choose $\bar{\beta}_S/\kappa T = \bar{\beta}_L/\kappa T = -5$ with $\alpha_L = 10^6$ cm^{-1} and $D_L = 10^{-5}$ cm^2 s^{-1} in order to determine the solute profile in the liquid as a function of V. The mathematical equations for the solute distributions will be given later so that we can focus first on the results. Fig. 3.5 illustrates the liquid solute profile results. Here, the $V = 0$ curve gives the equilibrium solute profile yielding the expected adsorption for this sign of $\bar{\beta}_L$. The $V > 0$ curves reveal a remarkable dip in the liquid profile in the region of 10^{-7} cm $< z < 10^{-2}$ cm and a significant drop in the interface concentration even at V as small as 10^{-3} cm s^{-1}. The dip occurs because the interface field is pushing the solute up-hill towards the interface to deplete the region at $z \sim \alpha_L^{-1}$ cm. Since solute cannot diffuse into this region fast enough to match the up-hill diffusion flux, a depletion zone develops. The in-filling of this depletion zone by diffusion spreads the disturbance out to $z \sim 10^2$–$10^3 \alpha_L^{-1}$ cm. The drop in interface concentration occurs because of the drop in liquid concentration at $z \sim \alpha_L^{-1}$ cm since a Boltzmann constant type of enrichment factor operates in this region as an effective bulk concentration; i.e., $C_i^j \approx C_L^j(\text{min}) \exp(\bar{\beta}_L^j/\kappa T)$. We note from Fig. 3.5 that this ratio $C_i/C_L(\text{min})$ begins to decrease for $V \widetilde{>} 10^{-1}$ cm s^{-1}; thus, we expect $V_{cL} \sim 10^{-1}$ cm s^{-1} (see Fig. 1.9). Finally we notice that, as V increases, the position of $C_L(\text{min})$ moves in towards the interface ($\Delta z_L \sim 10^{-6}$ cm) because there is not enough time for the diffusion to propagate much beyond the minimum as V increases (using $\Delta z^2 = Dt$ with $t = \Delta z/V$ leads to $\Delta z \sim 10^{-5}$ cm for $V = 1$ cm s^{-1}). Using Eq. (3.5) for k_i, one notes that $C_S(0) \approx C_L(0)$ for this case since $k_0 \approx 1$.

Fig. 3.6 shows the calculated C_S/C_∞ and C_L/C_∞ distributions for the same set of conditions as Fig. 3.5 except that now $k_0 = 0.1$ and the conditions for the solid have been added.[4] Comparing these two figures, one can readily separate out the $\exp(-Vz/D_L)$ portion from the short range field effect portion. We note that the two effects are roughly additive in the liquid. As can be seen, $k_i \sim k_0$ as expected from Eq. (3.5) and the distribution in the solid decays to C_∞ within $z = a$ few α_S^{-1} cm as predicted from our earlier qualitative discussion. The logarithmic

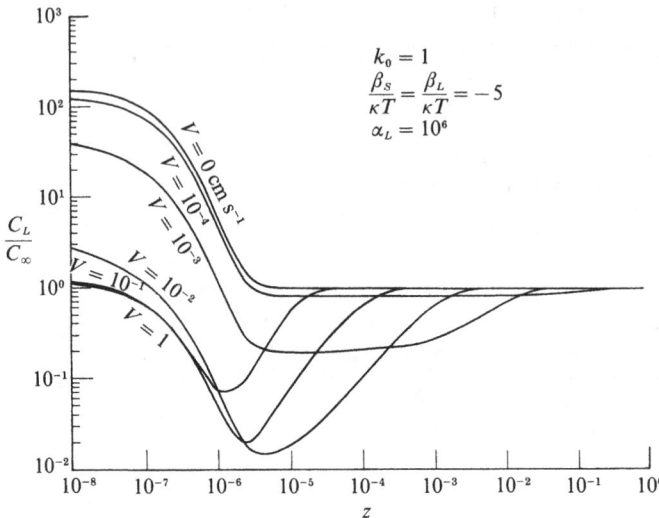

Fig. 3.5. Computed curves for C_L/C_∞ as a function of distance ahead of the interface, z, for a range of V with $k_0 = 1$ to illustrate that the dip below unity level comes from the interface field effect.[4]

scale has been used here to reveal the two very different scales of the phenomena operating in the liquid. Perhaps the most important point to note here is that the concentration depletion zone occurring at $z \sim +\Delta z_L$ was not considered in the development of Fig. 1.9 so that, in some cases, this will make it impossible to use the simple approach of replacing k_0 by $\hat{k}_i$ to translate results from the zero field case to the non-zero field case.

The general analytical solution to the solute transport equation in the steady state is

$$C = A_1 \exp\left(-\frac{V}{D}z\right) \exp\left\{-\left[\frac{B}{\alpha'} \exp(\alpha' z)\right]\right\}$$
$$+ \frac{A_2}{\alpha'} \left(-\frac{\alpha'}{B}\right)^{V/\alpha' D} \gamma(m, S) \exp\left(-\frac{V}{D}z\right) \exp\left\{-\left[\frac{B}{\alpha'} \exp(\alpha' z)\right]\right\}$$
$$(3.7a)$$

where

$$\gamma(m, S) = \int_0^S p^{m-1} \exp(-p)\mathrm{d}p; \quad m = \frac{V}{\alpha' D} > 0 \qquad (3.7b)$$

$$= \sum_{n=0}^{\infty} \frac{(-1)^n S^{m+n}}{(m+n)n!} \qquad (3.7c)$$

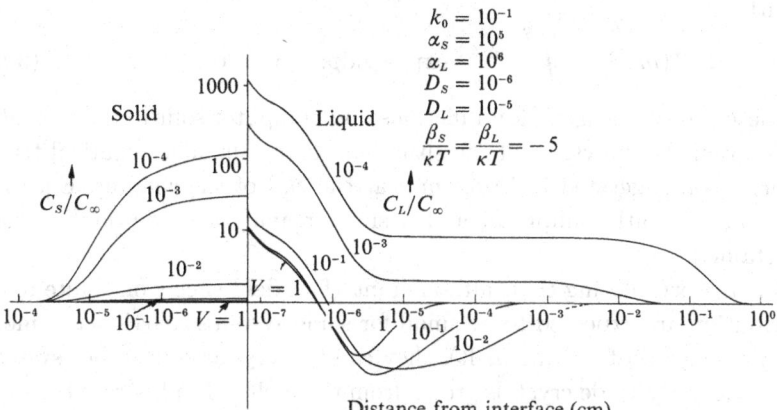

Fig. 3.6. Computed curves for C_S/C_∞ and C_L/C_∞ as a function of distance from the interface for several V, the fixed interface field parameters noted and $k_0 = 0.1$.[4]

and

$$\alpha' = \alpha_S \quad \text{for} \quad z < 0 \quad \text{while} \quad \alpha' = -\alpha_L \quad \text{for} \quad z > 0 \qquad (3.7d)$$

with $B = \alpha'\bar{\beta}/\kappa T$ and $\alpha_L > 0$. The actual solute distributions in the solid and liquid are thus given by

$$\frac{C_S}{C_\infty} = \frac{V}{\alpha_S D_S}\left(-\frac{\alpha_S}{B_S}\right)^{V/(\alpha_S D_S)}\gamma\left[\frac{V}{\alpha_S D_S}, \frac{B_S}{\alpha_S}\exp(\alpha_S z)\right]$$
$$\times \exp\left\{-\left[\frac{B_S}{\alpha_S}\exp(\alpha_S z)\right]\exp\left(-\frac{V}{D_S}z\right)\right\} \qquad (3.8a)$$

and

$$\frac{C_L}{C_\infty} = \exp\left(-\frac{V}{D_L}z\right)\exp\left\{\left[\frac{B_L}{\alpha_L}\exp(-\alpha_L z)\right]\right\}$$
$$\times\left(\frac{1}{k_i}\left\{\frac{V}{\alpha_S D_S}\left(-\frac{\alpha_S}{B_S}\right)^{V/(\alpha_S D_S)}\gamma\left(\frac{V}{\alpha_S D_S}, -\frac{B_S}{\alpha_S}\right)\right.\right.$$
$$\left.\times\exp\left[-\left(\frac{B_S}{\alpha_S}+\frac{B_L}{\alpha_L}\right)\right]\right\} + \frac{V}{\alpha_L D_L}\left(\frac{\alpha_L}{B_L}\right)^{-V/(\alpha_L D_L)}$$
$$\left.\times\left\{\gamma^*\left(-\frac{V}{\alpha_L D_L}, \frac{B_L}{\alpha_L}\right) - \gamma^*\left[-\frac{V}{\alpha_L D_L}, \frac{B_L}{\alpha_L}\exp(-\alpha_L z)\right]\right\}\right)$$

$$(3.8b)$$

with

$$\gamma^*(m, S) = \int_{S_1}^{S} \beta^{m-1} \exp(-p)\mathrm{d}p; \quad S_1 \neq 0 \tag{3.8c}$$

These are rather unwieldy functions and computer solutions are needed to reveal the "physics" of the situation. Using the mathematical transformation suggested in Problem 2 at the end of the chapter, Eq. (3.1) becomes greatly simplified and a simpler analytical solution is easily obtained.

It is worth noting that, although interface field effects on solute redistribution are expected to be small for some systems (particularly metal crystals grown from the melt), they can be very large for other systems (particularly oxide crystals grown from the melt). To illustrate the magnitude of effect that can occur, consider the case of a 1000 μm diameter TiO_2 crystal pulled from a pedestal melt containing 2.3 wt% SiO_2 at a rate of 4×10^{-5} cm s^{-1} and $\mathcal{G}_S \sim 10^4$ °C cm^{-1}. Melt convection is extremely strong in such a situation so that the interface is smooth rather than cellular. In Fig. 3.7(a), a typical longitudinal solute profile is shown while a longitudinal micrograph is presented in Fig. 3.7(b).

The longitudinal chemical analysis was made using the EDS (energy dispersion spectroscopy) technique on rectangular strips perpendicular to the specimen axis of 100 μm $\times$ 100 μm dimensions and spaced every 200 μm along the length of the crystal. In the photomicrograph we see a homogeneous single phase zone close to the interface of $\sim$ 500 μm width adjacent to a two-phase zone for the major portion of the crystal. We also note that the solid/liquid interface was smooth and slightly convex to the melt.

Returning to Fig. 3.7(a), we note that the interface concentration in the solid is small and $\sim$ 0.05 wt% while the interface concentration in the liquid is $\sim$ 10 wt% leading to $k_0 \approx 5 \times 10^{-3}$ for SiO_2 in TiO_2. With such a small value of k_0 we expect the liquid to be enriched in SiO_2 with time and the longitudinal concentration profile in the solid to be monotonically increasing and concave downwards from the seed end to the interface end.

Experimentally, we do see an enriched liquid but the solid concentration profile decreases near the interface. Such a solute profile can only arise from $k_0 < 1$ when a thermodynamic chemical potential field exists in the solid behind the freezing interface that pumps the SiO_2 up-hill into the bulk solid from the interface region. Here, the concentration enrichment in the bulk solid is a factor of $\sim$ 40. Similar behavior was

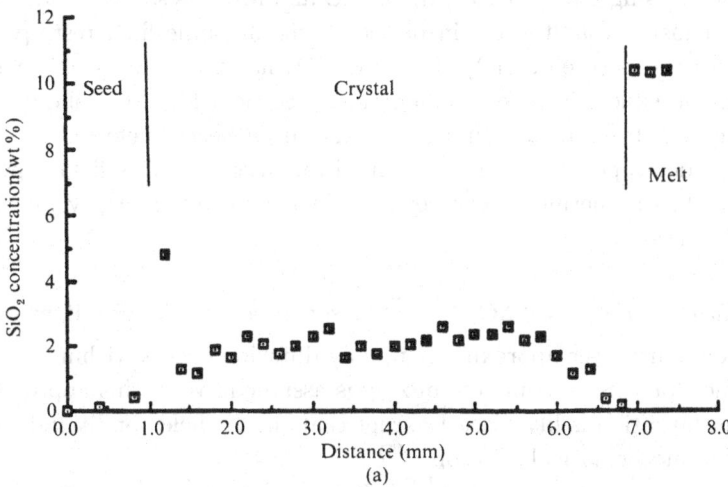

(a)

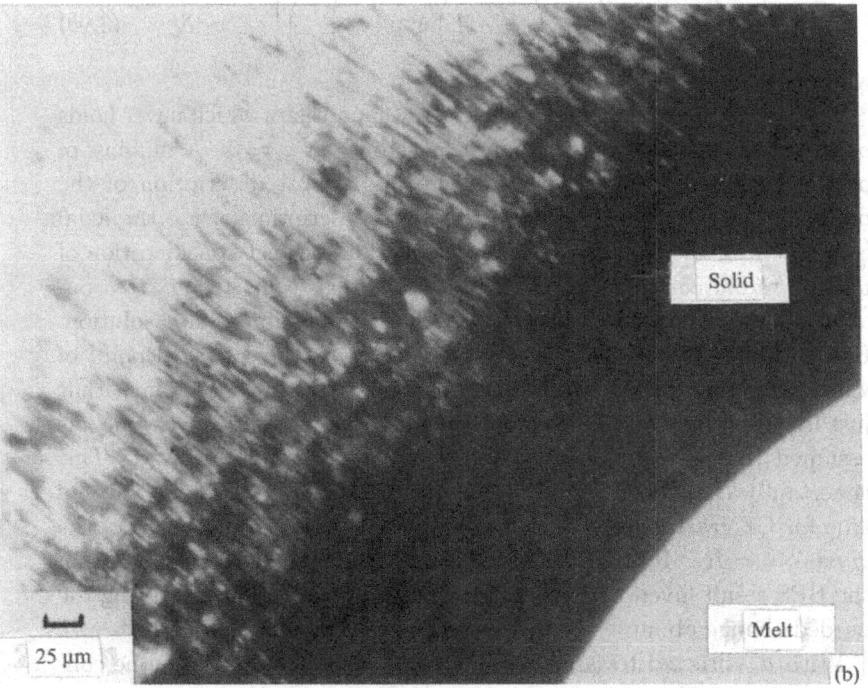

(b)

Fig. 3.7. TiO$_2$ crystal pulled from a TiO$_2$ + 2.3 wt% SiO$_2$ pedestal melt under $\mathcal{G}_S \sim 1000\,°C$ mm and $V \sim 4 \times 10^{-5}$ cm s^{-1}: (a) longitudinal solute profile in crystal and quenched melt and (b) longitudinal photomicrograph showing the smooth solid/liquid interface, the precipitation-free zone near the interface and the SiO$_2$ precipitation in the bulk of the crystal.

found using CaO, Aℓ_2O$_3$, MgO, FeO and MnO as solutes leading to the conclusion that the origin of the thermodynamic field rests primarily with the nature of TiO$_2$. Because TiO$_2$ has a large dielectric constant, the possible existence of a long range electric field in the solid was investigated. Preliminary studies showed that an electric field $\sim$ 1–10 V cm^{-1} was developed in the crystal behind the interface. This field is probably due to the thermoelectric power of TiO$_2$ and to the large value of $\mathcal{G}_S$ in the crystal.

Case 3 ($D_{SC} = \delta\widetilde{\Delta G}_0 = 0$, conservative, $u_\infty \neq 0$): If we use the boundary layer approximation, only diffusion occurs within $0 < z < \delta_C$ and, for $z > \delta_C$ complete mixing is assumed. With this approximation the situation is like Case 1 except that the far-field condition is altered. The mathematical solution is[5]

$$C_L(z) = \frac{C_\infty(1 - k_0)\exp\left(-Vz/D_L\right) + k_0 C_\infty}{k_0 + (1 - k_0)\exp\left(-V\delta_C/D_L\right)}; \quad z\bar{<}\delta_C \quad (3.9a)$$

$$= kC_\infty\left[1 + \left(\frac{1}{k_0} - 1\right)\exp\left(-\frac{Vz}{D_L}\right)\right]; \quad z\bar{<}\delta_C \quad (3.9b)$$

where k is given by Eq. (1.11a).

This result applies only for a perfectly stagnant film which never holds because $u = 0$ inside the boundary layer only at $z = 0$. A number of people have provided a more accurate and robust description of the boundary layer but they all involve only a description of u_z, the axial component of the convective fluid velocity, and neglect consideration of u_r, the radial component. On the surface, this seems to be a reasonable choice because the concern has been for a one-dimensional solution. However, if one considers the situation from a purely physical point of view, then there are real problems associated with an approach that just involves the u_z fluid flow component. To date, almost everyone has assumed that C is independent of r so that u_r is not involved. This is certainly true for an infinite diameter crystal but it is certainly *not* true for CZ crystal growth where C falls to C_∞ at the periphery of the crystal, $r = R_c$. It is this finiteness of R_c that requires a change from the BPS result given by Eq. (1.11a).[5] Let us now see how to bring the needed change about.

With u_z directed towards the interface, as in Cochran flow, the convective flux, $u_z C_\infty$, entering the boundary layer tends to increase the solute pile-up at the interface and reduce the boundary layer thickness. However, this flux alone is unable to account for the experimentally observed increase in the bulk liquid concentration, C_∞, with time. Only a flow combination involving the lateral convective flux, $u_r C_L$, can remove solute from the boundary layer and make it available for mixing with the bulk liquid. It is this lateral solute loss mechanism from the boundary layer that must be included in the solute transport equation in order to develop a one-dimensional equation that properly describes solute redistribution with convection.

Except for a small volume change on freezing, the volumetric flow, $\pi R_c^2 u_z(\delta_m)$, into the boundary layer must equal the lateral volumetric flow, $2\pi R_c \delta_m u_r(R_c)$, within the boundary layer. Thus, the net solute exchange is a loss, $2\pi R_c \delta_c u_r(R_c) \int_0^\delta C_L(z, R_c) dz$, laterally and a net gain, $\pi R_c^2 u_z(\delta_c) C_\infty$, axially. Combining these factors into an average loss rate, $-\dot{q}$, for a one-dimensional crystal much as was done in Eq. (2.40) for radial heat dissipation from a crystal surface, it can be shown that

$$-\dot{q} = \frac{-2u_r(R_c, z)}{R_c}[C_L(z) - C_\infty] \qquad (3.9c)$$

This is very similar to an annihilation/creation term, such as one would find for a chemical reaction in the melt, with a z-dependent reaction coefficient. For the variety of convective flows that one may wish to consider, it is perhaps simplest to approximate the $2u_r/R_c$ term as in Eq. (3.10a).

Let us suppose that the effect of lateral convection can be approximated by the z-dependent annihilation/creation term of Eq. (3.9c) to modify Eq. (3.1) in the following manner for the steady state case; i.e.,

$$D_L \frac{\partial^2 C_L}{\partial z^2} + V \frac{\partial C_L}{\partial z} - \beta_u'[1 - \exp(-p'z)](C_L - C_\infty) = 0 \quad (3.10a)$$

In this formulation, the effective annihilation/creation rate is assumed to be proportional to the velocity of the convective flux $\to 0$ at the interface and exponentially approaching a constant value, β_u', away from the interface. In terms of dimensionless coordinates, $C = C_L/C_\infty$ and $\tilde{\eta} = Vz/D$, Eq. (3.10a) becomes

$$\frac{\partial^2 C}{\partial \tilde{\eta}^2} + \frac{\partial C}{\partial \tilde{\eta}} - \beta_u[1 - \exp(-p'z)](C - 1) = 0 \qquad (3.10b)$$

where

$$\beta_u = \beta_u' D_L/V^2 \quad \text{and} \quad p = p' D_L/V \qquad (3.10c)$$

with the boundary conditions becoming

$$(1 - k_0)C = -\partial C/\partial \tilde{\eta} \qquad \text{at } \tilde{\eta} = 0 \qquad (3.10d)$$

$$C \to 1 \qquad \text{at } \tilde{\eta} \to \infty \qquad (3.10e)$$

With the incorporation of lateral convection, this model does not introduce a discontinuity in the solute concentration gradient as in the BPS model.[5] As a result, it can be utilized to study solute segregation behavior for a variety of static or dynamic growth situations.

One solution to Eq. (3.10b) is the following[6]

$$C(\eta) - 1 = \sum_{n=1}^{\infty} a_n \exp(-np\tilde{\eta}) \qquad (3.10f)$$

where

$$p = \frac{1 + (1 + 4\beta u)^{1/2}}{2} \qquad (3.10g)$$

$$a_n = \frac{-a_{n-1}}{n^2(p^2/\beta_u) - n(p/\beta_u) - 1} \qquad (3.10h)$$

$$= \frac{(-1)^{n+1}a_1}{\prod\limits_{m=2}^{n}\left[m^2\left(\dfrac{p^2}{\beta_u}\right) - m\left(\dfrac{p}{\beta_u}\right) - 1\right]}, \quad n = 2, 3, 4, \ldots \infty \quad (3.10i)$$

and

$$\frac{a_1}{(1-k_0)} = \left\{ (1-k_0) - p + \sum_{n=2}^{\infty}\left\langle \frac{(-1)^{n+1}[(1-k_0)-np]}{\prod\limits_{m=2}^{n}[m^2\left(\dfrac{p^2}{\beta_u}\right) - m\left(\dfrac{p}{\beta_u}\right) - 1]}\right\rangle \right\}^{-1}$$

$$(3.10j)$$

This result can be shown to revert to Eq. (3.3) when $\beta_u = 0$ since Eq. (3.10g) shows that $p = 1$ and $p' = V/D_L$ from Eq. (3.10c). Fig. 3.8 shows calculated solute profiles in the melt ahead of the growing interface for three values of k_0 and a number of β_u values. The $\beta_u = 0$ cases represent no mixing while $\beta_u \to \infty$ represents the complete mixing case. In between, one sees that the larger is β_u, the smaller is the amount of solute being carried ahead of the moving interface.

For general CZ crystal growth conditions, we can truncate Eq. (3.10f) after the a_2 term with an ultimate error of only 1%; i.e.,

$$C(\eta) = 1 + a_1 \exp(-p\tilde{\eta}) + a_2 \exp(-2p\tilde{\eta}) \qquad (3.10k)$$

with

$$a_1 = \frac{-(1-k_0)[4(p^2/\beta_u) - 2(p/\beta_u) - 1]}{(1 - k_0 - p)[4(p^2/\beta_u) - 2(p/\beta_u) - 1] - (1 - k_0 - 2p)}$$

$$(3.10l)$$

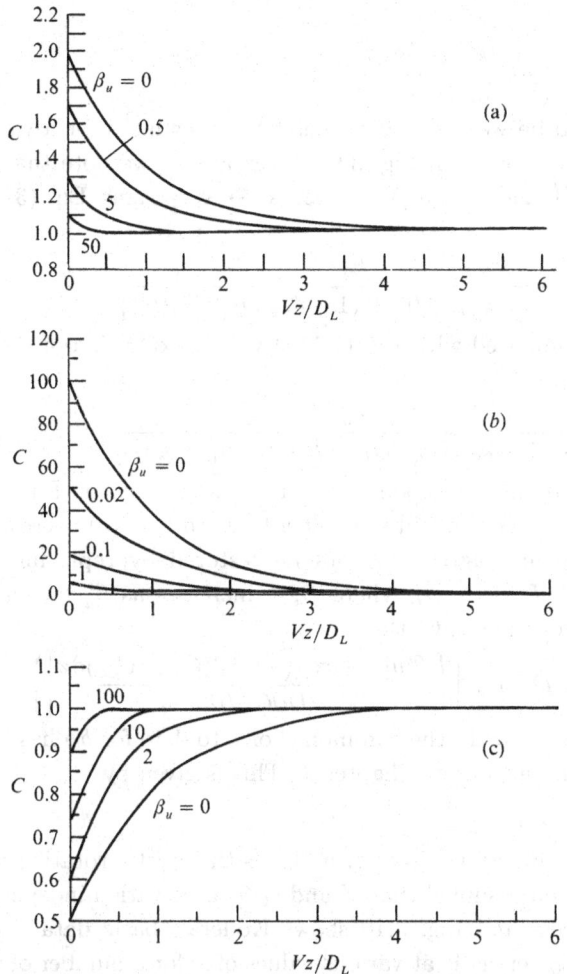

Fig. 3.8. Calculated steady state normalized solute profiles, $C = C_L/C_\infty$, in the melt as a function of the dimensionless distance parameter, $\tilde{\eta} = Vz/D$, at (a) $k_0 = 0.5$, (b) $k_0 = 0.1$ and (c) $k_0 = 2.0$ with several values of the convection parameter $\beta_u = \beta'_u D_L/V^2$.[6]

and

$$a_2 = \frac{(1 - k_0)}{(1 - k_0 - p)[4(p^2/\beta_u) - 2(p/\beta_u) - 1] - (1 - k_0 - 2p)}$$

(3.10m)

Since $k = k_0 C(0)$ in this notation, the effective segregation coefficient, k, can be readily calculated from this truncated solution and it is given by

$$k = \frac{k_0}{1 - (1 - k_0)2/(p + 1)} \qquad (3.11a)$$

The relationship between k and β_u has been calculated for a variety of k_0 and these are plotted in Fig. 3.9. As expected, we note that, when $\beta_u \to 0$, $k \to 1$ and, when $\beta_u \to \infty$, $k \to k_0$. Using Eq. (3.10c) in Eq. (3.11a),

$$k = \frac{k_0}{1 - (1 - k_0)\left\{4/[3 + (1 + 4\beta'_u D_L/V^2)^{1/2}]\right\}} \qquad (3.11b)$$

which can be compared with Eq. (1.11a) via an expansion of the exponential term; i.e.,

$$k = \frac{k_0}{1 - (1 - k_0)[V\delta_c/D_L - \frac{1}{2}(V\delta_c/D_L)^2 + \cdots]} \qquad (3.11c)$$

No simple one-to-one correspondence can be seen between Eqs. (3.11b) and (3.11c). However, we will see below that Eq. (3.11c) overestimates the value of δ_C and agreement can be restored by replacing D_L in Eq. (3.11c) with $D_{eff} > D_L$ where D_{eff} increases as β'_u increases. In fact, it can be readily shown that

$$D_{eff} = D\left\{1 + \left|\frac{\int \beta'u[1 - \exp(-p'z)](C_L - C_\infty)dz}{D\partial C_L/\partial z}\right|\right\} \qquad (3.11d)$$

For CZ crystal growth, the common choice to date for δ_C has been to use the Cochran analysis of Chapter 2. This is given by

$$\delta_C = 1.6 D_L^{1/3} \nu_L^{1/6} \omega^{-1/2} \qquad (3.12)$$

where ν_L is the kinematic viscosity and ω is the crystal rotation rate. It has generally been assumed that V and δ_C can be varied independently in the CZ process and Fig. 3.10 shows Kodera's basic data[7] on the variation of k/k_0 versus V at various values of ω for a number of solutes in Si. The best fit values of δ_C/D_L are also given from the use of Eq. (1.11a) and it is interesting to note that he does not obtain the same value of δ_C/D_L at the same V and ω for the different k_0. For example, with $\omega = 55$ rpm, we find $60 < \delta_C/D_L < 160$ and, for $\omega = 5$ rpm, $144 < \delta_C/D_L < 280$; i.e., about a factor of 2 variance. Kodera then proceeded to evaluate values of D_L for these different species and these values are presented in Table 3.1. The dilemma that one finds with these values of D_L is that they appear to be about a factor of 10 too large. This suggests that there is more convection present than is accounted for by the Cochran analysis and Eq. (1.11a). From the Cochran analysis, one expects $\delta_C \omega^{1/2}/D_L = $ constant but, from Fig. 3.10, this quantity ranges from 268 to 1185; i.e., more than a factor of 4 variance.

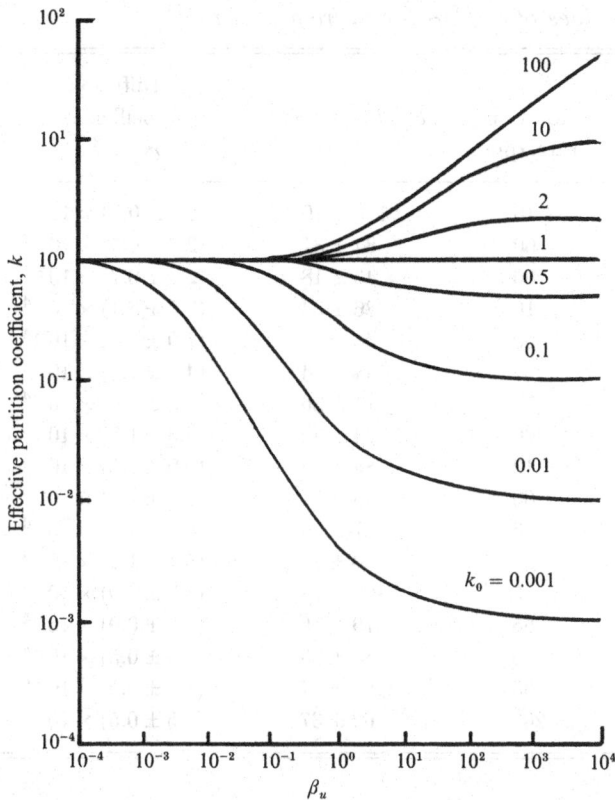

Fig. 3.9. Calculated values of the effective partition coefficient, k, as a function of the convection parameter, $\beta_u = \beta'_u D_L / V^2$ for a number of k_0.[6]

As an aspect of the β'_u effect of Eq. (3.10a), one should consider Fig. 3.11. From Chapter 2, one sees that the Cochran centrifugal fan effect produces a flow loop with a small velocity compared to the velocity of the natural convection loop. Thus, the natural convection loop tends to drag the Cochran loop causing it to spin faster and push it closer to the interface as in Fig. 2.31(b) and this would lead to a smaller value of δ_C than calculated by the Cochran analysis.

An alternate quantitative approach is to follow the Tifford and Chu flow analysis of Chapter 2 and compare it with the Cochran analysis. From the Cochran analysis, the centrifugal velocity that determines δ_C is given by $u_\phi = r\omega G(\zeta) \rightarrow u_\phi/r\omega = 0.5$ at which point the flow is almost at the stream value. Likewise, from the Tifford and Chu analysis,

Table 3.1. *Values of δ_C/D_C for Si from Kodera.*[7]

Impurity element	Rotation rate, rpm	δ_C/D_C, s cm^{-1}	Diffusion coefficient D_C, cm^2 s^{-1}
B	10	170 ± 19	$(2.4 \pm 0.7) \times 10^{-4}$
	60	84 ± 37	$(2.4 \pm 0.7) \times 10^{-4}$
	200	43 ± 18	$(2.4 \pm 0.7) \times 10^{-4}$
Aℓ	10	86 ± 34	$(7.0 \pm 3.1) \times 10^{-4}$
	60	40 ± 17	$(7.0 \pm 3.1) \times 10^{-4}$
Ga	5	144 ± 54	$(4.8 \pm 1.5) \times 10^{-4}$
	55	71 ± 26	$(4.8 \pm 1.5) \times 10^{-4}$
	200	24 ± 8	$(4.8 \pm 1.5) \times 10^{-4}$
In	10	84 ± 15	$(6.9 \pm 1.2) \times 10^{-4}$
	60	43 ± 5	$(6.9 \pm 1.2) \times 10^{-4}$
P	5	127 ± 36	$(5.1 \pm 1.7) \times 10^{-4}$
	55	60 ± 19	$(5.1 \pm 1.7) \times 10^{-4}$
As	5	190 ± 53	$(3.3 \pm 0.9) \times 10^{-4}$
	55	79 ± 16	$(3.3 \pm 0.9) \times 10^{-4}$
Sb	5	283 ± 55	$(1.5 \pm 0.5) \times 10^{-4}$
	55	157 ± 57	$(1.5 \pm 0.5) \times 10^{-4}$
	200	62 ± 37	$(1.5 \pm 0.5) \times 10^{-4}$

$u_\phi/r\omega \approx 0.05$ and $\delta_m \approx 4(\nu/\omega)^{1/2}$ at $\delta_m = \zeta^*_{a/w}(\nu/\omega)^{1/2}(1 + a/\omega)^{-1/2}$ with $a = 2W_\infty/\pi R$, where W_∞ is the upward plume velocity at the crucible ($z = \infty$) and R is the radius of the crystal. Here, $\zeta^*_{a/w}$ has been selected from Fig. 3.12 using $u_\phi/r\omega(1 + a/\omega) = 0.05/(1 + a/\omega)$. These values are given in Table 3.2.

The point with this approach is that, for any CZ growth process, the natural convection drive contributes a jet effect to the axial Cochran flow of magnitude W_∞. This leads to some value of a for the system that influences the determination of δ_m. For example, if $a = 4$ s^{-1} and $\omega = 1, 10, 20, 40, 60$ and 100 rpm, then $a/\omega = 40, 4, 2, 1, 0.6$ and 0.4, respectively, so that δ_m is reduced by the factor of $> 10, 6, 3.6, 2.3, 2.0$ and 1.7, respectively. Since the values of D_C in Table 3.1 seem to be in error by factors of 5–10, an appropriate choice is $a/\omega \sim 4$–6. This type of consideration becomes an important factor in Chapter 4 when one considers changes in δ_C caused by changes in V, ω or heater power

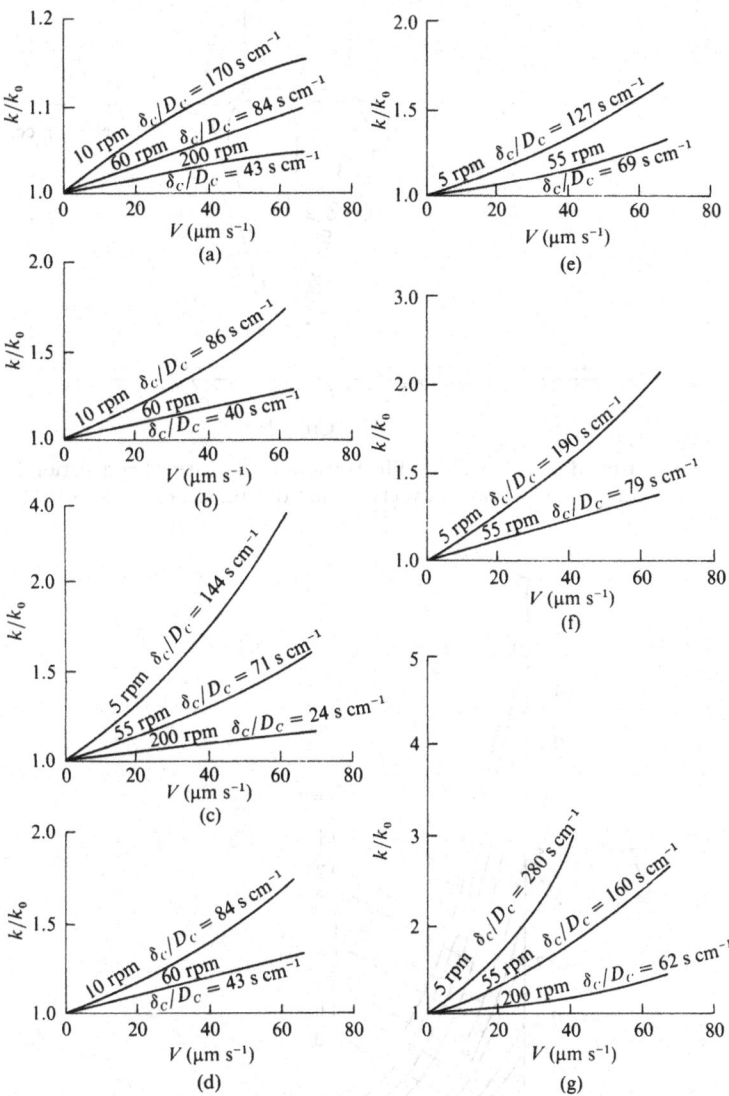

Fig. 3.10. Dependence of the effective segregation coefficient for dopants on V and ω during CZ growth of Si: (a) B, (b) Aℓ, (c) Ga, (d) In, (e) P, (f) As and (g) Sb from Kodera.[7]

because both W_∞ and R can change in this process leading to changes in a/ω needed to be used in the data analysis. In practice, both the altered matter transport equation (Eq. (3.10a)) and the altered fluid motion

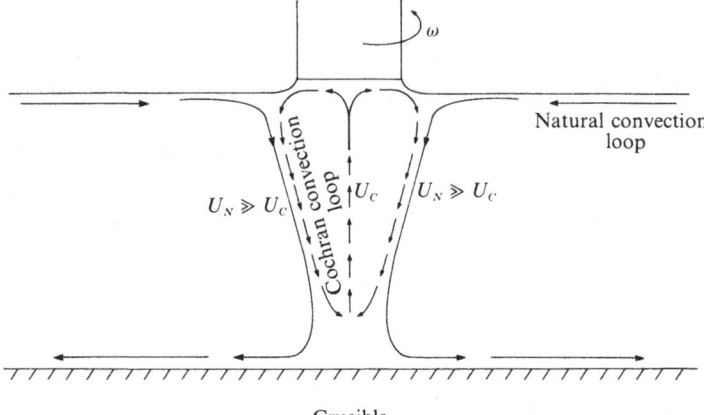

Fig. 3.11. Schematic illustration of the natural convection loop spinning the Cochran convection loop during CZ crystal growth.

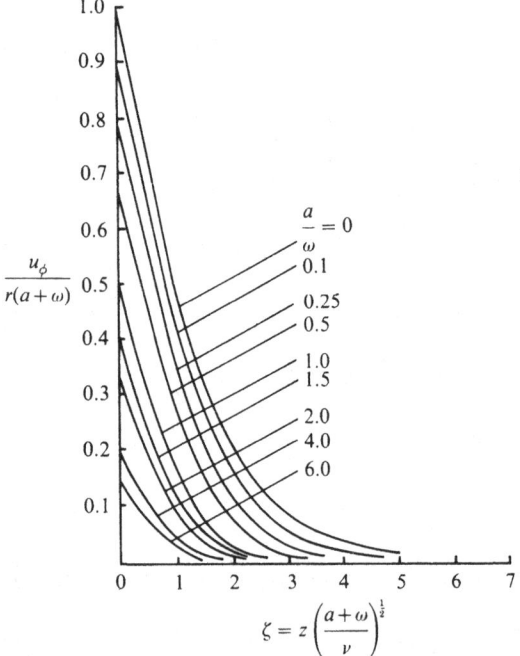

Fig. 3.12. Circumferential velocity distribution, u_ϕ, near a rotating disc in a perpendicularly impinging flow for different a/ω. ($a/\omega = 0$ is Cochran flow.)

Table 3.2. *Relationship between a and δ_m (for 95% reduction in $u_\phi/r\,(a+\omega)$).*

a/ω	$\dfrac{0.05}{(1+a/\omega)}$	$\zeta^*_{a/\omega}$	$\delta_m/\nu^{1/2}\omega^{-1/2}$
0	0.05	3.8	3.8
0.1	0.045	3.2	3.05
0.25	0.04	2.7	2.41
0.5	0.033	2.4	1.95
1.0	0.025	2.05	1.45
1.5	0.02	1.9	1.20
2.0	0.0167	1.75	1.0
4.0	0.01	1.6	0.715
6.0	0.00835	1.4	0.53

equation of Tifford and Chu (increased β'_u) need to be considered. In the final analysis, Fig. 2.31 illustrates what happens when both neutral and forced convection are involved.

Case 4 ($D_{SC} = u_\infty = \delta\tilde{\Delta}G_0 = 0$, chemical reaction in the liquid): In this case, we shall illustrate the effect of chemical reaction in the solute-rich layer of liquid at the interface and will choose the $C + O \rightleftharpoons CO$ reaction as a useful example. For this case, our differential equation becomes

$$D^j \frac{\partial^2 C_L^j}{\partial z^2} + V \frac{\partial C_L^j}{\partial z} - \beta_r^j(C_L^j - C_L^{*j}) = 0; \quad j = C, O, CO \qquad (3.13a)$$

where $\beta_r^O = \beta_r^C = -\beta_r^{CO}$ and the equilibrium constant, $\hat{K}$, is given by

$$\hat{K} = [CO]/[C][O] \qquad (3.13b)$$

In order to gain a relatively simple analytical solution to Eq. (3.13a), it is necessary to replace C_L^{*j}, which will vary with position, by a constant value $C_L^{*j} = C_\infty^j$ over the complete boundary layer. For such an approximation the solution becomes

$$C_L^j(z) = C_\infty^j + A^j \exp(-B^j z) \qquad (3.13c)$$

where

$$A^j = (1 - k_0^j)C_\infty^j / \left\{ k_0^j - \frac{1}{2}\left[1 - \left(1 + 4\beta_r^j D^j/V^2\right)^{1/2}\right]\right\} \quad (3.13d)$$

and

$$B^j = \frac{V}{2D^j}\left[1 + \left(1 + 4\beta_r^j D^j/V^2\right)^{\frac{1}{2}}\right] \quad (3.13e)$$

This solution is completely stable for C and O since $\beta_r > 0$ for these species; however, it is not stable for CO when $-(4\beta_r D/V^2)^{CO} > 1$ and oscillations should result for the CO content. Even though there is no mixing in the liquid for this case, the value of k^j is not unity but is given by

$$k^j = \frac{\left[1 + \left(1 + 4\beta_r^j D^j/V^2\right)^{1/2}\right]k_0^j}{\left\{2k_0^j\left[1 - \left(1 + 4\beta_r^j D^j/V^2\right)^{1/2}\right]\right\}} \quad (3.13f)$$

which is less than unity for C and O but greater than unity for CO. From the foregoing, we see that the effect of chemical reaction in the interface layer is to steepen the solute profile and reduce the interface concentration for the reactant species and to do the reverse for the product species. When $(4\beta_r D/V^2)^{CO} > -1$, an oscillatory solution is possible leading to periodic solute profiles for C, O and CO. If the overall concentration is high enough, this will also generate a periodic value of V.

Another important example in this category is O partitioning during Si crystal growth via the CZ technique. For three decades, competent investigators have been conducting experiments to specify the value of k_0^O. However, the collection of viable experiments has led only to $\sim 0.25 < k_0^O < \sim 5$. Something is wrong with the reasoning of these experiments and what seems to be the missing factor is that there are multiple O species in the Si melt rather than just a single species with a single value of k_0.

Most of the O in a Si melt comes from the dissolution of the SiO_2 crucible holding the melt. SiO_2 is comprised of SiO_4 tetrahedra with the O's being corner-shared between two tetrahedra. About 95% of the free energy of SiO_2 formation from Si and O_2 goes to form these tetrahedra with the remaining 5% going to form the particular crystallographic arrangement of these tetrahedra. Thus, the tetrahedra are very stable species and the dissolution of the SiO_2 into the melt may be via SiO_4 species that then dissociate to O species or via the removal of single O species or both. The important point is that multiple oxygen species probably exist in the melt with the two dominant forms being Si–O–Si and 2Si–(SiO$_4$)–2Si (let us call these simply O and SiO$_4$ species).

To make sense of the experimental data, let us make the reasonable assumption that $k_0^{SiO_4} \ll 1$ and $k_0^O > 1$ and $C_\infty^O \gg C_\infty^{SiO_4}$. Partitioning at the interface leads to a build-up of SiO$_4$ and a depletion of O in the interface boundary layer with the following reaction occurring between the two O species

$$\text{Si} + 4\text{O} \overset{\rightarrow}{\leftarrow} \text{SiO}_4; \quad K^* = \frac{[C_L^{SiO_4}]}{[C_L^O]^4}$$

Although the concentration of SiO$_4$ species entering the solid is expected to be small, their role in the formation of SiO$_2$ precipitates in the Si crystal is expected to be large!

In addition to this chemical reaction factor, melt convection is also important for this case; thus, we are really dealing with a combination of Eqs. (3.10a) and (3.13a) with $C_L^* \approx C_\infty$. A steady state solution can be gained for both C_L^O and $C_L^{SiO_4}$ by following the procedure of Eqs. (3.10) and a new expression gained for both k^O and k^{SiO_4} that illustrates the synergism between the convection and chemical reaction processes.[8]

Solving the simultaneous transport equations with associated boundary conditions requires a numerical approach because of the coupling due to the β_r^j type of coefficients. However, in the special case where one of the species has a partition coefficient much smaller than 1 (species 2, SiO$_4$) while the other species has a k_0-value near unity (species 1,O), we expect a huge concentration pile-up in the melt near the interface for species 2 compared to that for species 1 so that the net reaction is always the conversion of species 2 to species 1. In this special case, the original problem of solving two coupled differential equations becomes a two-step problem where each equation can be solved separately. First, the redistribution profile for species 2 can be solved from

$$D_2 \frac{\partial^2 C_2}{\partial z^2} + V \frac{\partial C_2}{\partial z} - \beta_u'[1 - \exp(-p_2'z)](C_2 - C_{2,\infty}^*)$$
$$- \beta_r^2(C_2 - C_{2,\infty}^*) = 0 \quad (3.14a)$$

subject to the usual boundary conditions. With a known distribution for $(C_2 - C_{2,\infty}^*$, the redistribution profile for species 1 can be obtained by solving

$$D_1 \frac{\partial^2 C_1}{\partial z^2} + V \frac{\partial C_1}{\partial z} - \beta_u'[1 - \exp(-p_1'z)](C_1 - C_{1,\infty}^*)$$
$$+ n\beta_r^2(C_2 - C_{2,\infty}^*) = 0 \quad (3.14b)$$

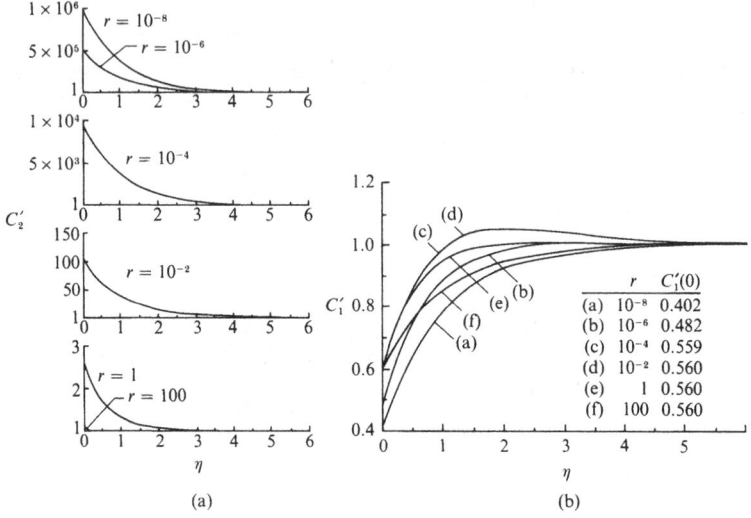

Fig. 3.13. Calculated plots of $C'_2(SiO_4)$ and $C'_1(O)$ as a function of $\eta = V_z/D_2$ or $\eta = V_z/D_1$ respectively, for a range of reaction rates, $r = D_2\beta_r^2/V^2$ and the set conditions $k_{0,1} = 2.5$, $k_{0,2} = 10^{-6}$, $\beta'_u = 0$ and $C^*_{2,\infty}/C^*_{1,\infty} = 0.1$.

subject to its boundary conditions. Here, for the decomposition of SiO_4 into O and Si, $n = 4$. Defining $r = D_2\beta_r^2/V^2$ with the special case of $k_{01} = 2.5$, $k_{02} = 10^{-6}$, $\beta'_u = 0$, $n = 4$ and $C^*_{2,\infty}/C^*_{1,\infty} = 0.01$. Fig. 3.13 provides calculated values of $C'_2 = C_2/C^*_{2,\infty}$ versus $\eta = V_z/D_2$ and $C'_1 = C_1/C^*_{1,\infty}$ versus $\eta = V_z/D_1$.[8] We see that C'_2 changes very strongly with r and that significant variations are also generated in the C'_1 profile.

3.2   Non-planar interfaces ($V = $ constant)

When the interface is not planar, initially due to slightly curved isotherms, but the macroscopic growth direction is one-dimensional (see Fig. 3.14), lateral diffusion of solute occurs so that C_i builds up at the retarded regions of the interface (for $k_0 < 1$). Since the temperature is generally lower at these retarded points, the interface position may be stabilized. This leads to lateral chemical inhomogeneity in the crystal cross-section. Generally, the greater the interface curvature, the greater the degree of chemical segregation between the crystal center and the crystal edge.

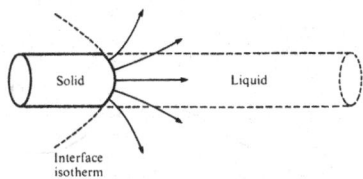

Fig. 3.14. Schematic illustration of liquid diffusion ahead of a convex-shaped interface.

Even when the interface is macroscopically planar, it often departs from smoothness on a microscopic level and presents a periodic morphology to the liquid. Prominent examples of this are to be found in (i) cellular or filamentary growth, (ii) eutectic crystallization and (iii) interface layer growth. For an interface growing at velocity V in the z-direction and periodic with period λ in only the x-direction with interface shape $Z = f(x)$ as illustrated in Fig. 3.15(a), the general steady state solution to the solute distribution ahead of the interface is given by

$$\frac{C_L}{C_\infty} = 1 + \sum_{n=0}^{\infty} [a_n \cos(\varepsilon_n x) + b_n \sin(\varepsilon_n x)] \exp(-\beta_n z) \qquad (3.15a)$$

with

$$\beta_n = \frac{V}{2D} \left\{ 1 + \left[1 + \left(\frac{2D\varepsilon_n}{V} \right)^2 \right]^{1/2} \right\} \qquad (3.15b)$$

and $\varepsilon_n = 2\pi n/\lambda$. For most examples in this category, the origin in the moving coordinate system can be chosen at a symmetry point on the surface so that $f(x)$ is an even function and all the $b_n = 0$. The a_n are determined by applying the interface conservation condition along $Z = f(x)$. For an interface that exhibits only a single protuberance as illustrated in Fig. 3.15(b), the steady state solute distribution is given by

$$\frac{C}{C_\infty} = 1 + \int_0^\infty a(\varepsilon) \cos(\varepsilon x) \exp[-\beta(\varepsilon)z] d\varepsilon \qquad (3.15c)$$

where all possible values of the separation constant ε are allowed and β is given by Eq. (3.15b).

Eqs. (3.15) can be readily altered in form when the interface shape is such that it displays rectangular, hexagonal, cylindrical or other symmetry. Of course, in all cases, the appropriate reduced coordinate systems must be selected in order to obtain such a "separation of variables" type of solution.

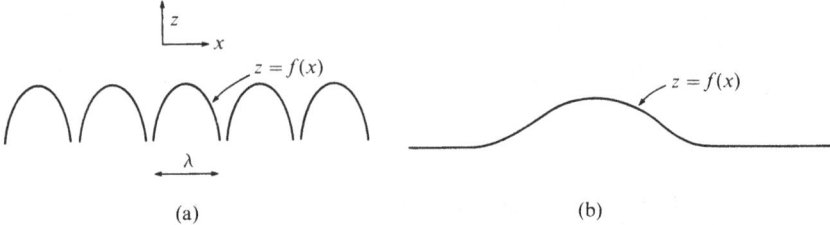

Fig. 3.15. Schematic illustration of non-planar interface shapes (a) interface periodic in the x-direction and (b) a single bump in the x-direction.

To illustrate the utility of Eqs. (3.15), let us consider the three cases illustrated in Fig. 3.16, the idealized eutectic, the idealized macroscopic interface and idealized interface layer motion.

3.2.1   Idealized eutectic

We consider the growth of a binary lamellar eutectic and our main approximation is to treat the interface as being perfectly flat. A second and less significant approximation is to approximate the spatially varying interface concentration, $C_i(x)$, by the constant eutectic concentration, C_E. Thus, the interface conservation condition becomes

$$\frac{\partial C_L}{\partial z} = -\frac{V}{D}(1 - k_\alpha)C_E \quad ; \quad 0 < x < \frac{\lambda_\alpha}{2}, z = 0 \qquad (3.16a)$$

$$= -\frac{V}{D}(1 - k_\beta)C_E \quad ; \quad \frac{\lambda_\alpha}{2} < x < \frac{\lambda_\alpha + \lambda_\beta}{2}; z = 0 \quad (3.16b)$$

where k_α and k_β are the k_0 values for the α and β phases, respectively. Taking $\partial C_L/\partial z$ in Eq. (3.15a) and setting it equal to Eqs. (3.16), the orthogonality operation yields for the a_n

$$a_0 = \frac{C_E}{C_\infty}\left[(1 - k_\alpha)\frac{\lambda_\alpha}{\lambda} + (1 - k_\beta)\frac{\lambda_\beta}{\lambda}\right] \qquad (3.16c)$$

$$a_n = \frac{\lambda V C_E}{\pi^2 n^2 D C_\infty}(k_\beta - k_\alpha)\sin\left(\frac{\pi n \lambda_\alpha}{\lambda}\right) \qquad (3.16d)$$

The difference in solute content between the centers of the lamellae, ΔC_α^β, is thus

$$\frac{\Delta C_\alpha^\beta}{C_\infty} = \sum_{n=0}^{\infty} a_{2n+1} = \frac{\lambda V C_E}{\pi^2 D C_\infty}(k_\beta - k_\alpha)\sum_{n=0}^{\infty}\frac{\sin[(2n+1)\pi\lambda_\alpha/\lambda]}{(2n+1)^2}$$

$$(3.16e)$$

Since $C_i \approx C_E$ across the interface, this approximation is a reasonable one. However, because the interface shape departs appreciably from flat-

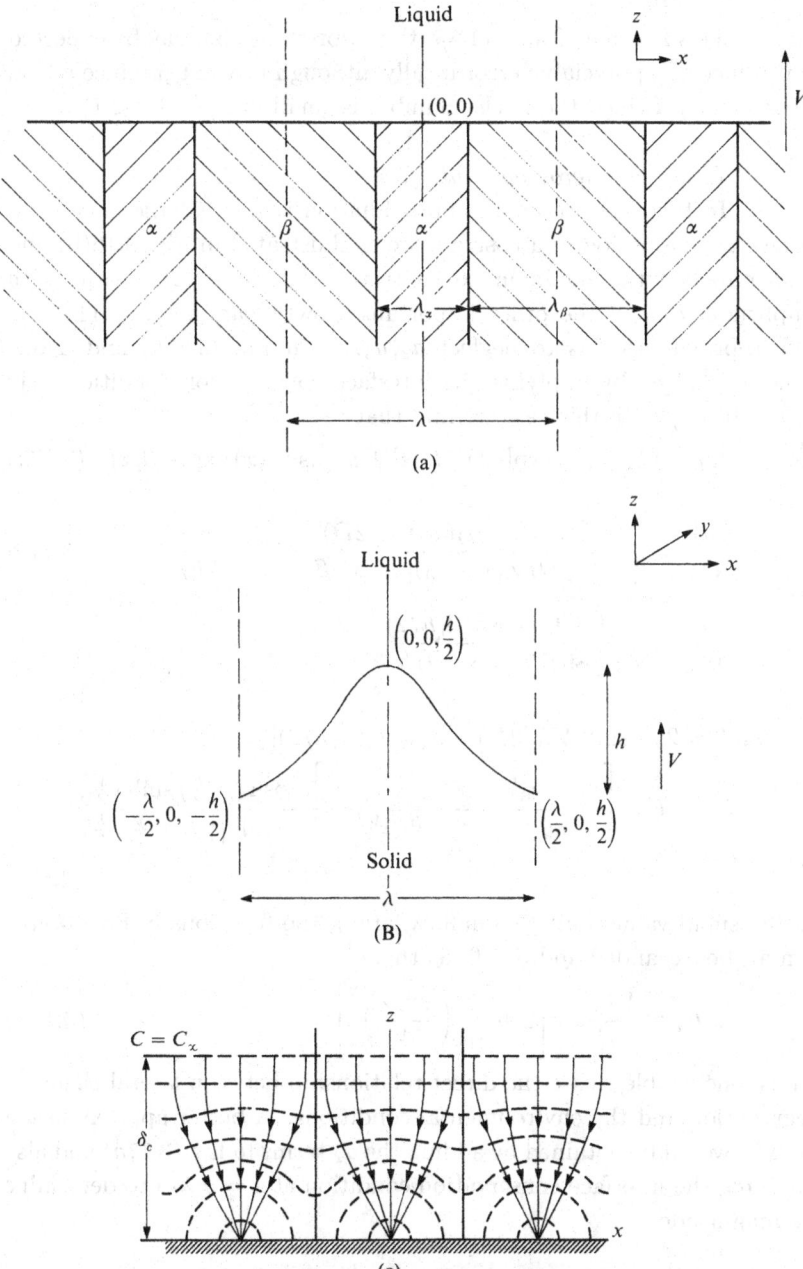

Fig. 3.16. Schematic illustration of interface condition for idealized cases of solute partitioning (a) binary eutectic lamellae with a flat interface, (b) cosine-shaped macroscopic interface for single phase continuum growth and (c) crystal growth via the TLK mechanism neglecting the movement of the ledge and the height of the ledges.

ness on the size scale of λ_α and λ_β, this approximation may be expected to produce an appreciable error locally although it won't produce a long range error provided the Péclet number is small ($V\lambda_j/2D \ll 1$).

3.2.2 Idealized macroscopic interface

In this case the main approximation is to treat the curved interface as one period of a cosine wave as illustrated in Fig. 3.16(b) and determine how the maximum concentration difference, ΔC_i, varies with amplitude, h, and the other important growth parameters. The second approximation is to neglect $a_2, a_3, \ldots$ in Eq. (3.15a) and determine a_0 and a_1 by applying the interface conservation condition only at $x = 0, \pm\lambda/2$. In this case we find that

$$C_L - C_\infty \approx a_0 \exp[-(V/D)z] + a_1\cos(\varepsilon_1 x)\exp(-\beta_1 z) \quad (3.17a)$$

where

$$\frac{a_1}{C_\infty} = \frac{\sinh(Vh/2D)}{[1 - \beta_1 D/V(1 - k_0)]\cosh[(\beta_1 - V/D)h/2]} \quad (3.17b)$$

$$\frac{a_0}{C_\infty} = \frac{(1 - k_0)\cosh(\beta_1 h/2)}{k_0\cosh[(\beta_1 - V/D)h/2]} \quad (3.17c)$$

and

$$\Delta C_i = 2[a_0\sinh(Vh/2D) - a_1\cosh(\beta_1 h/2)] \quad (3.17d)$$

$$= 2C_\infty\left[\frac{(1 - k_0)}{k_0} - \frac{1}{1 - \frac{\beta_1/(V/D)}{(1-k_0)}}\right]\frac{\cosh\left(\beta_1\frac{h}{2}\right)\sinh\left(\frac{Vh}{2D}\right)}{\cosh\left[\left(\beta_1 - \frac{V}{D}\right)\frac{h}{2}\right]}$$

$$(3.17e)$$

In the small value limit (h small, λ large), the functions in Eq. (3.17e) can all be expanded and one finds that

$$\Delta C_i \approx \frac{VhC_\infty}{D}\left[2 + \frac{1}{k_0}\left(\frac{2\pi D}{\lambda V}\right)^2\right] \quad (3.17f)$$

Thus, one is able to see the direct relationship between lateral chemical segregation and the environmental conditions. A better approximation to ΔC_i would be obtained by adding the a_2 terms to Eq. (3.17a) and also applying the interface conservation condition at $x = \pm\lambda/4$ to determine its magnitude.

3.2.3 Idealized interface layer motion

A third interface non-planarity effect that should be considered is that of an array of steps moving parallel to the interface by a long range constant flux of solute towards the interface from the adjacent nutrient phase (solution or vapor). For the case where both interface diffusion and the lateral motion of step ledges can be neglected, Chernov[9] used a complex plane analysis to map the diffusion problem into one that looks like Fig. 3.16(c). Here, the layer edges have been treated like a periodic array of point sinks on the flat plane and the bulk solute source is at a distance $z = \delta_c$ from the interface (edge of the boundary layer). The Chernov[9] treatment neglects moving interface effects and his solution for the concentration profile is

$$C_L - C_\infty = A\ell n \left(\left\{ \sin^2 \left(\frac{\pi x}{\lambda} \right) + \sinh^2 \left[\frac{\pi}{\lambda}(z + \delta_c) \right] \right\}^{1/2} \right) \quad (3.18)$$

Eqs. (3.16) also provide a suitable solution as $\delta_c \to \infty$ provided we allow $k_\beta = 0$. This solution includes the effects of interface motion in the z-direction and the finite size of the ledges but it neglects the ledge motion in the x-direction. One cannot readily modify Eqs. (3.16) to include the V_x contribution because the basic steady state differential equation is not separable in x and z when both the V_z and V_x contributions are included. However, a solution that is periodic in time is separable (see Chapter 4). Thus, although both solutions give a good representation of the long range diffusion field (small Péclet number effect), neither will give a good representation close to the interface. One would need a sum of moving line source solutions to represent adequately the near field condition to the layer edge.

3.3   Film formation

The main purpose of this section is to show that the concepts and mathematics of this chapter apply to thin film formation from the vapor phase. Although some unique features enter when one considers solute partitioning during the film forming process, the novel features are not extensive. This material follows directly from Chapter 5 of the companion book[3] where the general film growth rate is given by Eq. (5.20). This result applies to Si film formation from SiH_4 or $SiC\ell_4$ and thus involves the redistribution of at least two chemical species at layer edges and on the terraces. The inclusion of a dopant, in the form of PH_3 or B_2H_6 say, adds an additional chemical family to be considered and, for completeness, one should also consider the presence of a gettering agent like H_2O or $C\ell$ which influences the amount of P or B that is available to be incorporated in the film for a given gas partial pressure of PH_3 or B_2H_6.

Conceptually, the P doping process can be illustrated by Fig. 3.17(a) and the partitioning of at least five constituents occur at the ledges, $SiC\ell_2*$ and PH_2* diffuse to the ledges while $HC\ell$, $H*$ and $C\ell*$ are all diffusing away from the ledges with the ledge reactions being considered as

$$SiC\ell_2 ** + 2H * \overset{\rightarrow}{\leftarrow} Si + 2HC\ell + 4* \tag{3.19}$$

$$PH_2 * + * \overset{\rightarrow}{\leftarrow} P + 2H* \tag{3.20}$$

where the $*$ refers to a chemical bond with the terrace or ledge. Because there are additional $H*$ and $C\ell*$ on the terrace in the path of the moving ledge, these will also partition at the ledge with their appropriate partition coefficients (very small). There is potentially a site conservation problem in the ledge region due to too many atoms being built up at the ledge, i.e., we have the constraint

$$n_{s\ell}^{SiC\ell_2*} + n_{s\ell}^{PH_2*} + n_{s\ell}^{H*} + n_{s\ell}^{C\ell*} + n_{s\ell}^{HC\ell} + n_{s\ell}^{\hat{v}} = n_0 \tag{3.21}$$

where n_0 is the total number of available sites, $\hat{v} \equiv$ vacancy and the different species are all assumed to have some average area. As the temperature is lowered, because $n_{s\infty}^j$ increases, $n_{s\ell}^{\hat{v}}$ decreases so that D_s^j decreases and compressive stresses begin to develop in the adlayer and a significant ΔG_E contribution begins to be developed as a consequence of ledge growth. The system can resolve its difficulties in at least two ways; (1) a type of Frenkel defect formation develops in the adlayer to create an adlayer vacancy and place the adlayer atom on the top of the adlayer (chemisorbed atom $\rightarrow$ physisorbed atom plus vacancy) and (2) trapping of an adlayer species (H or $C\ell$ for example) in the ledge to reduce $n_{s\ell}^j$. This aspect of the problem will dominate at lower temperatures but, in what is to follow, let us presume that our substrate is at a high enough temperature that $n_{s\ell}^{\hat{v}}/n_0$ is sufficiently high where transport of all surface species is not a problem.

The steady state surface concentration of P will be proportional to the flux of PH_3 from the gas to the surface and this is proportional to the partial pressure P_{PH_3} in the gas phase. Thus, from the earlier part of this chapter, we can expect that the steady state concentration of P in the solid will be proportional to P_{PH_3} in the gas phase. Such a relationship is demonstrated in Fig. 3.17(b).

The steady state growth rate for the film will be proportional to $P_{SiC\ell_4}$ or P_{SiH_4} in the gas so that the maximum degree of dopant incorporation, X_{max}^D, is given by the ratio of the two fluxes; i.e.,

$$X_{max}^D = J^D/J^{SiC\ell_4} = D_g^D P^D / D_g^{SiC\ell_4} P^{SiC\ell_4} \tag{3.22}$$

Since $V_\ell \propto P^{SiC\ell_4}$, we expect to see X_{max}^D vary inversely with V. This is the domain where the effective distribution coefficient, k, is unity provided V_ℓ is large enough for conservation to obtain in the adlayer and the dopant concentration in the solid will be just $n_{s\infty}^D = J^D \tau_s^D$. However, as V_ℓ decreases because $J^{SiC\ell_4}$ decreases, other processes have time to occur in the adlayer dopant profile near the ledge so that the adlayer system behaves in a non-conservative way and $k < 1$. Such processes might be back-diffusion to the surface due to an interface field or reaction of the dopant with a foreign species to form a volatile constituent which leaves the adlayer. Both a schematic representation of expected dopant profile with growth rate and some actual data for P in Si are given in Fig. 3.17(c) while Fig. 3.18 illustrates the decrease of B doping in Si due to B_2H_6 reaction with H_2O to form BHO_2 and H_2.[10]

For a set of widely spaced ledges, the concentration profiles in Fig. 3.19 illustrate (a) the pile-up of all non-Si-bearing adsorbed intermediate species (k-species) at the ledges, (b) the movement of the ledges due to the influx of key Si-bearing species to the ledges and (c) the pile-up of dopant species at the ledges. For each of these j-species, in a reference frame moving with the ledge, we have

$$D_s^j \frac{\partial^2 n_s^j}{\partial x^2} + V_\ell \frac{\partial n_s^j}{\partial x} - \frac{(n_s^j - n_{s\infty}^j)}{\tau_s^j} = 0 \qquad (3.23a)$$

at steady state when no interface fields are present. Here, the bracketed term incorporates exchange between the adlayer and the vapor phase. In general, we should expect an interface field in the x-direction that moves with the ledge as well as one in the z-direction that moves with the macroscopic interface. Let us choose these to be of the form

$$\frac{\partial}{\partial x} \delta \widetilde{\Delta G}_{0x}^j = \frac{\partial}{\partial x} \bar{\beta}_x^j \exp(-\alpha_x^j x); \quad \frac{\partial}{\partial z} \delta \widetilde{\Delta G}_{0z}^j = \frac{\partial}{\partial z} \bar{\beta}_z^j \exp(\alpha_s^j x)$$

$$(3.23b)$$

in the adlayer ahead of the ledge and in the solid beneath the terrace, respectively. Approximating these fields by constant fields over the region $x < (\alpha_x^j)^{-1}$ and $z < -(\alpha_z^j)^{-1}$, Eq. (3.23a) for the adlayer ahead of the ledge must be replaced by

$$D_s^j \frac{\partial^2 n_s^j}{\partial x^2} + \left(V_\ell - \frac{\alpha_x^j \bar{\beta}_x^j D_s^j}{\kappa T}\right) \frac{\partial n_s^j}{\partial x} - \frac{(n_s^j - n_{s\infty}^j)}{\tau_s^j} + \frac{\alpha_z^j \bar{\beta}_z^j D_{S_-}^j}{\kappa T} n_{S_-}^j = 0$$

$$\text{for} \quad 0 < x < (\alpha_x^j)^{-1} \qquad (3.23c)$$

and

$$D_s^j \frac{\partial^2 n_s^j}{\partial x^2} + V_\ell \frac{\partial n_s^j}{\partial x} - \frac{(n_s^j - n_{s\infty}^j)}{\tau_s^j} + \frac{\alpha_z^j \bar{\beta}_z^j D_{S_-}^j}{\kappa T} n_{S_-}^j = 0$$

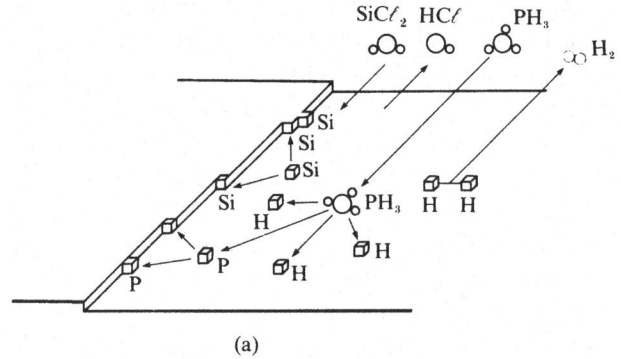

(a)

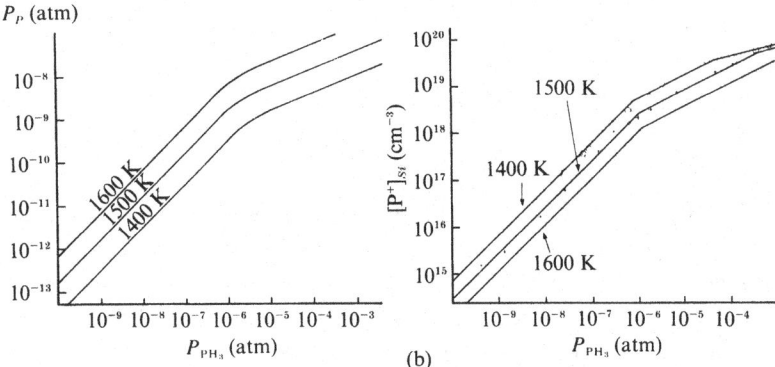

(b)

Fig. 3.17. P doping of Si films: (a) schematic representation of film growth, (b) calculated partial pressures of monatomic phosphorous, P_P, and experimental donor concentration, $[P^+]$, versus input PH_3 content. The Si growth rate was 0.45 μm min^{-1}.[10]

$$\text{for} \quad x > (\alpha_x^j)^{-1} \tag{3.23d}$$

In Eqs. (3.23c) and (3.23d), $D_{S_-}^j$ and $n_{S_-}^j$ are the diffusion coefficient and concentration of the j-species, respectively, in the solid immediately beneath the adlayer. By defining $V_\ell' = V_\ell - \alpha_x \bar{\beta}_x D_s/\kappa T$ and $n_{s\infty}' = n_{s\infty} + \tau_s \alpha_z \bar{\beta}_z D_{S_-} n_{S_-}/\kappa T$, one sees that Eqs. (3.23c) and (3.23d) can be put in the form of Eq. (3.23a) with the use of effective ledge velocities and effective equilibrium concentrations. An additional pair of equations applies for the region behind the ledge at $x < 0$.

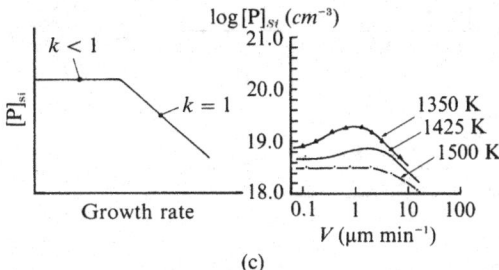

(c)

Fig. 3.17. (c) Schematic and actual representation of P doping of Si as a function of film growth rate V for $P_{PH_3} = 1.2 \times 10^{-4}$ atm.[10]

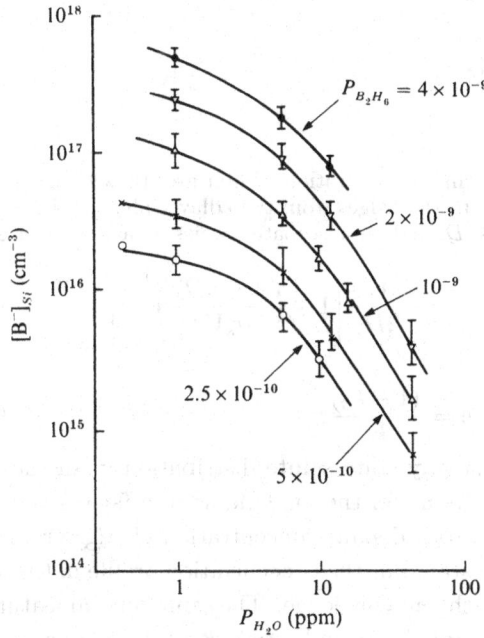

Fig. 3.18. B doping in Si at $1150\,^{\circ}$C for constant B_2H_6 input in atmospheres as a function of water vapor content of the gas phase.[10]

For the case where the ledges are far enough apart that only a single ledge need be considered and where all interface fields are absent, the solutions to Eq. (3.23a) in the adlayer are of the form

$$n_s - n_{s\infty} = A \exp(-\xi_- x) \qquad x > 0 \qquad (3.24a)$$

$$= A \exp(-\xi_+ x) \qquad x < 0 \qquad (3.24b)$$

where

128   *Steady state solute partitioning*

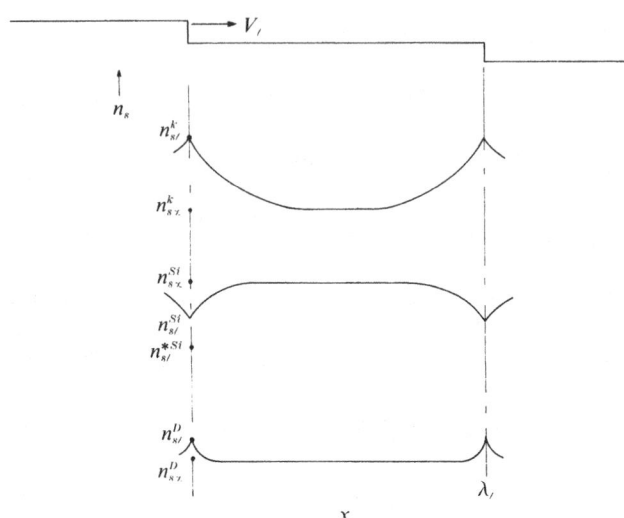

Fig. 3.19. Schematic illustration of film growth with ledge motion due to Si supply to the ledges from the adlayer plus partition of both a dopant species, D, and an adsorbate species, k, at the layer edges.

$$\xi_{\mp} = -\frac{V_\ell}{2D_s}\left[1 \mp \left(\frac{1+4D_s}{\tau_s V_\ell^2}\right)^{1/2}\right] \qquad (3.24c)$$

and

$$A = \frac{(1-k_0)}{k_0}n_{s\infty} \qquad (3.24d)$$

Thus, we see that the steady state solute distribution in the adlayer is almost the same as that found for the solid/liquid interface except for the decay term. This leads to a dopant concentration of $n_{s\infty}^j$ crystallizing into the ledge at steady state but the concentration profile in the adlayer will be asymmetrical at the moving ledge. The same general features will also obtain when we include ledge and interface field effects except that the asymmetry will be different, and, since $V_\ell \neq V_\ell'$, $k \neq k_0$. Further, if $\bar{\beta}_z < 0$, the value of $n_{s\infty}'$ may become sufficiently small that a negligible dopant profile develops in the adlayer.

When the ledges are close together, the solution must be periodic in x which is not possible with Eq. (3.23a) unless the moving reference frame term can be neglected. By converting Eq. (3.23a) to its dimensionless form, it can be shown that this is always possible for small Péclet number; i.e., Eq. (3.23a) becomes

$$\frac{\partial^2 n_s'}{\partial x'^2} + 2\hat{P}_C\frac{\partial n_s'}{\partial x'} - \frac{(n_s'-1)}{\tilde{\eta}} = 0 \qquad (3.25a)$$

with

$$2\hat{P}_C(1 - k_0)n'_{s\ell} = -\left(\frac{\partial n'_s}{\partial x'}\right)_{\ell+} + \left(\frac{\partial n'_s}{\partial x'}\right)_{\ell-} \qquad (3.25b)$$

for the ledge conservation condition with no field effects and the ledge height equals the adlayer thickness. Here, $\hat{P}_C = V_\ell\lambda/2D_s$, $\tilde{\eta} = D_s\tau_s/\lambda^2$, $x' = x/\lambda$ and $n'_s = n_s/n_{s\infty}$. Clearly, for $\hat{P}_C \ll 1$, the middle term in Eq. (3.25a) can be neglected leading to a general solution in the form of cosh and sinh functions.

To evaluate the specifics of the solution in a general case, we need to consider a ledge height, h, and an adlayer thickness, a, as illustrated in Fig. 3.20(a). Here, the flux is normal to the terraces but it could have just as readily been oriented at some specific angle. With this specific configuration, we find the perpendicular ledge flux, J'_ℓ, the perpendicular ledge velocity, V'_ℓ, and the ledge length, h', are given by

$$J'_\ell = J\sin\left(\phi_2 - \frac{\pi}{2}\right); \quad V'_\ell = V_\ell\cos\left(\phi_2 - \frac{\pi}{2}\right); \quad h' = h/\cos\left(\phi_2 - \frac{\pi}{2}\right)$$
$$(3.26)$$

We can now evaluate diffusion in the adlayer and will use x to represent distance in the adlayer as illustrated in Fig. 3.20(b). The periodic terrace regions will be labeled Region 1 while the periodic ledge regions will be labeled Region 2. We will assume that the surface lifetime, τ_s, and surface diffusion coefficient, D_s, are constant for each region, i.e., we neglect corner effects at the lower ledge and upper ledge locations. We shall see that this is the simplification allowing the sinh contribution to be dropped from the general solution when we require continuity and conservation of species at the junctions between Regions 1 and 2.

The mathematical description of the boundary-value problem is
Region 1:

$$\frac{\partial^2 n'_{s1}}{\partial x'^2} - \frac{(n'_{s1} - 1)}{\tilde{\eta}_1} = 0; \qquad 0\tilde{<}x'\tilde{<}1 \qquad (3.27a)$$

$$x' = \frac{x}{\lambda}; \quad n'_{s1} = \frac{n_{s1}}{n_{s1\infty}}; \quad n_{s1\infty} = J\tau_{s1}; \quad \tilde{\eta}_1 = D_{s1}\tau_{s1}/\lambda^2 \qquad (3.27b)$$

Region 2:

$$\frac{\partial^2 n'_{s2}}{\partial x'^2} - \frac{(n'_{s2} - 1)}{\tilde{\eta}_2} = 0; \qquad -1\tilde{<}x'\tilde{<}0 \qquad (3.27c)$$

$$x' = \frac{x}{h'}; \quad n'_{s2} = \frac{n_{s2}}{n_{s2\infty}}; \quad n_{s2\infty} = \frac{J'_\ell\tau_{s2}}{1 + k_0V'_\ell\tau_{s2}};$$

$$\tilde{\eta}_2 = \frac{D_{s2}\tau_{s2}}{h'^2(1 + k_0V'_\ell\tau_{s2})} \qquad (3.27d)$$

With this description, we see that the net flux falling on the ledges is

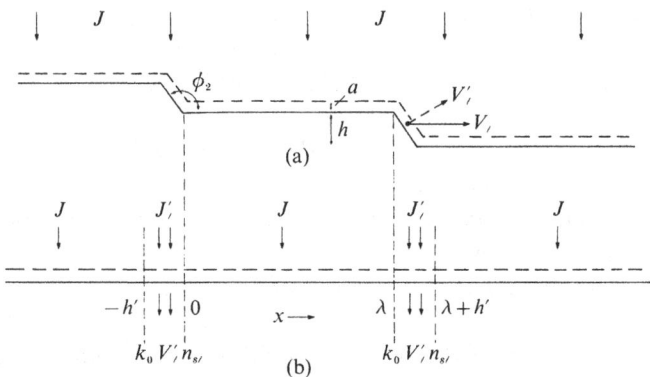

Fig. 3.20. (a) Schematic illustration of ledges of height h growing laterally due to a matter flux, J, into an adsorption layer of thickness a and (b) one-dimensional representation of fluxes to, from and within the adlayer.

just the difference between that entering the adlayer from the vapor or solution and that leaving the adlayer due to crystallization at the ledge. The general adlayer solutions are

$$n'_{s1}(x') - 1 = A_1 \cosh\left(\frac{x' - \frac{1}{2}}{\tilde{\eta}_1^{1/2}}\right) + B_1 \sinh\left(\frac{x' - \frac{1}{2}}{\tilde{\eta}_1^{1/2}}\right), \quad 0 \bar{<} x' \bar{<} 1$$

$$(3.27e)$$

and

$$n'_{s2}(x') - 1 = A_2 \cosh\left[\frac{x' - \frac{1}{2}}{\tilde{\eta}_2^{1/2}}\right] + B_2 \sinh\left[\frac{x' - \frac{1}{2}}{\tilde{\eta}_2^{1/2}}\right], \quad -1 \bar{<} x' \bar{<} 0$$

$$(3.27f)$$

where, in Region 1, we have $x' = x/\lambda$ and, in Region 2, we have $x' = x/h'$. Applying the continuity and conservation conditions at the junctions between the two regions leads to $B_1 = B_2 = 0$ and

$$A_2 = \frac{(n_{s1\infty} - n_{s2\infty})}{n_{s2\infty}\cosh(1/2\tilde{\eta}_2^{1/2}) - \alpha_{12}n_{s1\infty}\cosh(1/2\tilde{\eta}_1^{1/2})} \qquad (3.27g)$$

$$\frac{A_1}{A_2} = \alpha_{12} = -\frac{(D_{s2}/D_{s1})(n_{s2\infty}/n_{s1\infty})}{(\tilde{\eta}_2/\tilde{\eta}_1)^{1/2}} \frac{\sinh(1/2\tilde{\eta}_2^{1/2})}{\sinh(1/2\tilde{\eta}_1^{1/2})} \qquad (3.27h)$$

We thus find that the species profile in the adlayer will be of either form illustrated in Fig. 3.21. For the solvent species, $k_0 > 1$ and the flow is towards the ledge; for the majority of solute species, $k_0 < 1$ and the flow is away from the ledge.

Rather than use the coordinate system presented in Fig. 3.20, one

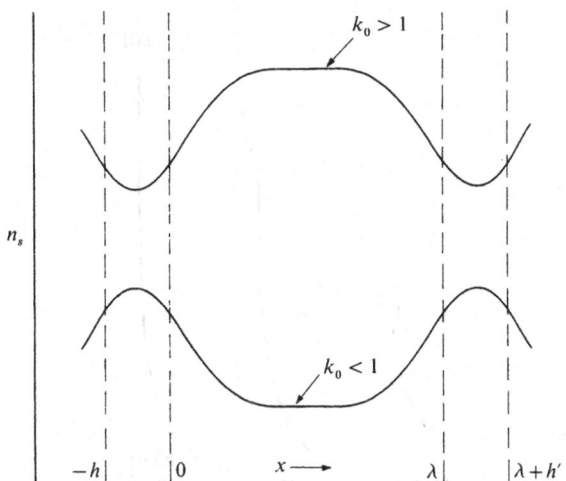

Fig. 3.21. Solute profiles in terrace and ledge adlayer for $k_0 \gtrless 1$.

often prefers to place the origin at a ledge when $h = a$ and $\phi_2 = \pi/2$ (the typical case people treat). For this case, the solution is given by

$$n'_s(x') - 1 = (n'_{s\ell} - 1)\frac{\cosh[(x' - \frac{1}{2})/\tilde{\eta}^{1/2}]}{\cosh(1/2\tilde{\eta}^{1/2})} \qquad (3.28a)$$

where

$$n'_{s\ell} = \frac{\tanh(1/2\tilde{\eta}^{1/2})}{\tanh(1/2\tilde{\eta}^{1/2}) - \tilde{\eta}^{1/2}\hat{P}_C(1 - k_0)} \qquad (3.28b)$$

and $\tilde{\eta} = D_s \tau_s / \lambda^2$. The two limiting solutions to Eq. (3.28b) are

$$n'_{s\ell} \to [1 - 2\tilde{\eta}\hat{P}_C(1 - k_0)]^{-1} \quad \text{as} \quad \tilde{\eta} \to \infty \qquad (3.29a)$$

$$n'_{s\ell} \to [1 - \tilde{\eta}^{1/2}\hat{P}_C(1 - k_0)]^{-1} \quad \text{as} \quad \tilde{\eta} \to 0 \qquad (3.29b)$$

In Fig. 3.22, $n'_{s\ell}$ is plotted versus $\tilde{\eta}^{1/2}$ for a range of values of the parameter $\hat{P}_C(1 - k_0)$ so that the whole range of behavior may be seen.

If we consider a specific case like Si growing via MBE or CVD with a dopant, then Eqs. (3.28) must hold for both the Si-bearing species and the dopant species in the adlayer. In addition, for the CVD case, we also have the adsorbate species partitioning at the ledge with $k_0 <<< 1$. For the Si-bearing species, one could use Eq. (3.28b) with $k_0 \approx \Omega^{-2/3}/n_{s\ell}$ if one wished; however, it is easier to modify Eq. (3.25b) for the case where essentially pure film is being formed by replacing $(1 - k_0)n'_{s\ell}$ by

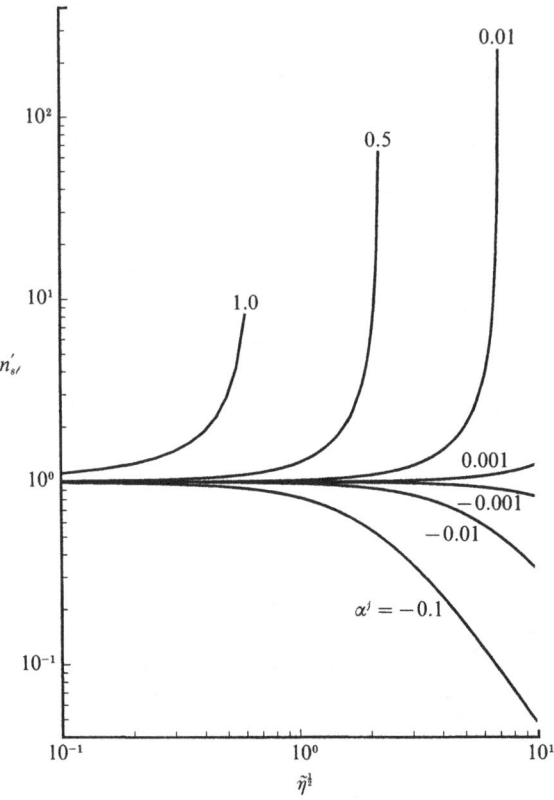

Fig. 3.22. Plot of dimensionless ledge concentration, $n'_{s\ell} = n_{s\ell}/n_{s\infty}$, as a function of the dimensionless transport parameter, $\tilde{\eta} = D_s\tau_s/\lambda^2$, for several values of the partitioning parameter $\alpha^j = [\hat{P}_C(1 - k'_0)]^j$.

$(\Omega/an_{s\infty})^{-1}$ so that Eq. (3.28b) becomes an equation for $\hat{P}_C^{Si}$; i.e.,

$$\hat{P}_C^{Si} = \left(\frac{\Omega}{a}\right)_{Si} \frac{(n_{s\infty}^{Si} - n_{s\ell}^{Si})}{\tilde{\eta}^{Si}} \tanh(1/2\tilde{\eta}^{Si}) \tag{3.30}$$

Since $\hat{P}_C^j/\hat{P}_C^{Si} = D_s^{Si}/D_s^j$, the determination of $\hat{P}_C^{Si}$ from Eq. (3.30) allows $\hat{P}_C^j$ to be known for any dopant or adsorbate of interest. In all cases, $n_{s\infty}$ is defined as $\tau_s J$ for that species where J is the flux from the gas phase or the solution phase to unit area of surface. Most people use Eq. (3.30) as a defining equation for V_ℓ and assume that equilibrium obtains for the host species at the ledge; i.e. $n_{s\ell} = n_s^*$. In Fig. 3.23, $\hat{P}_C^0/(1 - n_s^{*'})$ is plotted versus $\tilde{\eta}^{0^{1/2}}$ for several values of the important parameters $\alpha^0 = [(\Omega/a)(J\tau_s)]^0$. Here, the superscript 0 refers to the

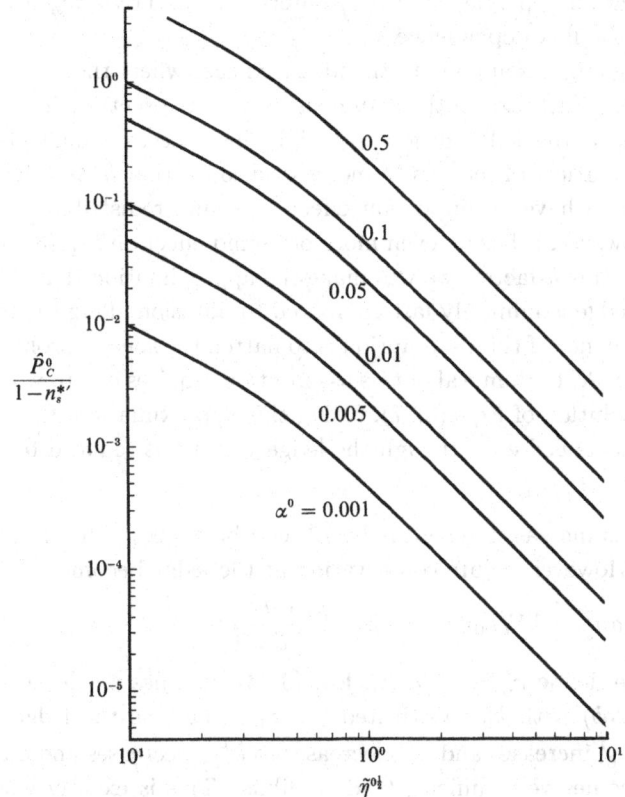

Fig. 3.23. Plot of modified host atom Péclet number, $\hat{P}_C^0/(1 - n_s^{*'})$, as a function of the dimensionless host transport parameter, $\tilde{\eta}^0 = [D_s\tau_s/\lambda^2]^0$ for several values of the adlayer area parameter, $\alpha^0 = [(\Omega(a)n_{s\infty}]^0$.

host film material while a superscript j refers to a minor constituent like a dopant.

In practice, for a given $\tilde{\eta}^0$ and α^0, Fig. 3.23 allows us to obtain $\hat{P}_C^0$ if $n_s^{*'}$ is known. If D_s^j/D_s^0 and τ_s^j/τ_s^0 are also known, then $\hat{P}_C^j$ and $\tilde{\eta}^j$ are known so that Fig. 3.22 can be used to obtain $n_{s\ell}^{'j}$ for a given k_0^j. Since the dopant concentration in the film is just given by $k_0^j n_{s\ell}^j$, an effective partition coefficient, $k_{eff}^j = k_0^j n_{s\ell}^{'j}$ can be used, where

$$\frac{k_{eff}^j}{k_0^j} = \frac{n_{s\ell}^j}{\tau_s^j J^j} \tag{3.31}$$

The orientation dependence of k_{eff}^j enters Eq. (3.31) via $n_{s\ell}^j$ as does a portion of the flux dependence.

The foregoing has applied to the idealized case where the ledge height and adlayer width are both of dimension a. In practice, for a given orientation, θ, the ledge height, h, might increase to a multiple layer height for a variety of reasons (λ increases also) so that $h \gg a$. This will reduce $\tilde{\eta}^0$ and have an important effect on solute redistribution in the adlayer; however, it has an even more profound effect on Eq. (3.25b). To understand this h-factor, we will make the approximation that diffusion along the ledge is infinitely fast compared to diffusion along the terrace. The consequence of this assumption is to flatten the species profile at the ledge in Fig. 3.21. Removal of this assumption requires one to work with the exact solution of Eqs. (3.27). Using this approximation, Eq. (3.25a) appears unchanged even though the ledge velocity is reduced to

$$V_\ell' = V_\ell/(h/a) \tag{3.32}$$

Because λ is increased accordingly, $\hat{P}_C^0$ can be expected to increase via Fig. 3.23. However, solute conservation at the ledge becomes

$$V_\ell' a n_{s\ell}^j - h V_\ell' k_0 n_{s\ell}^j = -2a D_s^j \left(\frac{\partial n_s}{\partial x} \right)_\ell \tag{3.33}$$

Thus, if we define $k_0' = (h/a)k_0$, Eq. (3.33) becomes of identical form to Eq. (3.25b) with k_0' substituted for k_0. Thus, as the ledge height increases, $k_0'^j$ increases and $n_{s\ell}^j$ decreases so k_{eff}^j decreases and one finds that it becomes very difficult to dope films. This is exactly what one finds in practice! It arises because $k_0'^j$ can be $\sim 10^2$–$10^4 k_0^j$ so that, instead of a pile-up of the j-species existing on the terrace adjacent to a moving ledge, a depletion zone can occur. When one includes the effect of an interface field in the z-direction, one must replace Eq. (3.31) by

$$\frac{k_{eff}^j}{k_0^j} = \frac{n_{s\ell}^j}{\tau_s^j J^j + (\tau_s^j \alpha_z \bar{\beta}_z D_{S-}^j n_{S-}^j / \kappa T)} \tag{3.34}$$

Depending upon the sign of $\bar{\beta}_z$, this contribution can either lower ($\bar{\beta}_z > 0$) or raise ($\bar{\beta}_z < 0$) k_{eff}^j. Of course, these interesting results will need to be modified somewhat at large h/a due to the breakdown of our initial assumption concerning the infinitely rapid ledge diffusion.

As an illustrative example, suppose we consider the following situation: $k_0^j = 0.01$, zero interface field, $D_s^j/D_s^0 = \tau_s^j/\tau_s^0 = 1$, $(1 - n_s^{*'}) = 0.8$, $(\Omega/a)n_{s\infty}^0 = 10^{-2}$, $J^j/J_0 = 10^{-4}$ and $\theta = 0.1°$. We shall assume that $D_s^0 \tau_s^0$ is such that $\tilde{\eta}_0(h/a = 1) = 10^2$ and consider the effect of h/a growing from 1 to 10^3 on $\hat{P}_C^0, n_{s\ell}^j$ and k_{eff}^j. These results can be determined by (i) using Eq. (3.30) or Fig. 3.23 to obtain $\hat{P}_C^0$, (ii) with $\hat{P}_C^0$ and $k_0^{'j}$, using Eq. (3.28b) or Fig. 3.22 to obtain $n_{s\ell}^{'j}$ and (iii) using Eq. (3.31) to determine $k_{eff}^j = k_0^j n_{s\ell}^j$. It was found that $n_{s\ell}^{'j} \approx 1$ for $h/a < 10^4$ and that $\hat{P}_C$ increased monotonically from 4×10^{-5} at $h/a = 1$ to 4×10^{-3} at $h/a = 10^2$ to 0.08 at $h/a = 10^4$.

In the foregoing, the assumption has been made that the adatom distribution on a terrace is symmetrical with respect to the ledge and this led to the "cosh" function of Eq. (3.28a). Although this may be correct for the 0-species, it is unlikely to be correct for the j-species. For solutes, even without interface fields acting at the ledge, the binding potential of a j-species will be different in the ledge adjacent to the lower terrace than it is adjacent to the upper terrace. Thus, the equilibrium adlayer concentration on the terrace adjacent to these two locations must be different. In fact, even for a pure Si film, the calculated adlayer concentration on the terrace adjacent to the lower ledge is different than that adjacent to the upper ledge. Fig. 3.24 shows the calculated results for a monolayer high $[2\bar{1}\bar{1}]$ ledge on Si(111) using a potential energy function composed of a Lennard–Jones two-body potential and an Axelrod–Teller three-body potential.[11] For solutes, when interface fields are added and surface lifetimes are adjusted one can expect some concentration difference, $\Delta n_{s\ell}^j(h/a)$, to exist across the ledge (or along the terrace) and this will be a function of h/a as indicated. For such a situation, the coefficients B_1 and B_2 in Eqs. (3.27e) and (3.27f) will not be zero but will be proportional to $\Delta n_{s\ell}'$.

Problems

1. You are growing an AB crystal from a solution where complete dissociation and complete ionization have occurred to give a far-field concentration $C_\infty^{A^+} = C_\infty^{B^-}$ at $z > \delta_C$. The electrostatic potential difference between the interface and solution is $\phi_i - \phi_\infty > 0$ for steady state one-dimensional crystal growth at velocity V so that a positive electric field $E \approx (\phi_i - \phi_\infty)/\delta_C$ resists the movement of A^+ to the interface and enhances the movement of B^-. Show that the interface concentrations, C_i^+ and C_i^-, are given by

$$\left(\frac{C_i}{C_\infty}\right)^{\pm} = \frac{V/\Omega(DC_\infty)^{\pm} - (1/\delta_C \mp qE/2kT)}{(\mp qE/2kT - 1/\delta_C)}$$

where Ω is the molecular volume of AB and q is the ion charge.

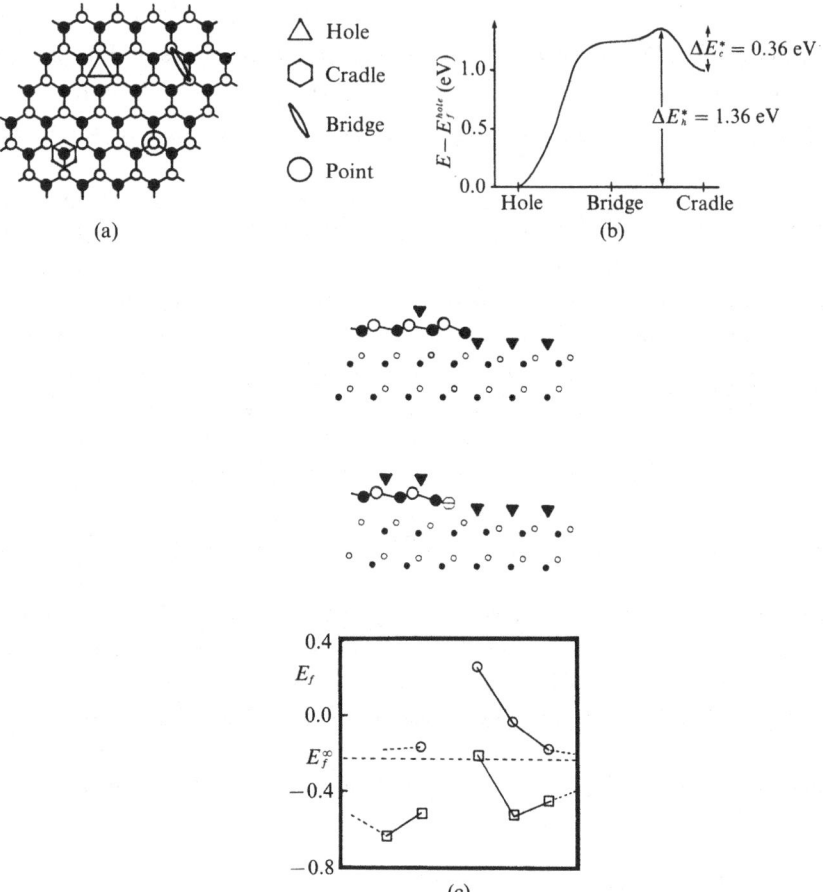

Fig. 3.24. Non-symmetric potential fields, for adatoms near ledges on the Si(111) plane, (a) different adatom sites, (b) minimum energy change for adatom diffusion between the hole and cradle site and (c) formation energies for hole adatoms near the lower (circles) and upper (squares) $[2\bar{1}\bar{1}]$ ledges where E_f^∞ is the infinite terrace value. Adatom positions are shown by the filled triangles for the two cases.[11]

2. Show that the steady state solution to Eq. (3.1c) in the moving interface reference frame can be greatly simplified by the

substitutions

$$C' = C \exp \left\{ \int \left[\frac{V-U}{D} + \frac{\partial}{\partial z} \left(\frac{\delta \tilde{\Delta} G_0}{\kappa T} \right) \right] dz \right\}$$

$$D' = D \exp \left\{ -\int \left[\frac{V-U}{D} + \frac{\partial}{\partial z} \left(\frac{\delta \tilde{\Delta} G_0}{\kappa T} \right) \right] dz \right\}$$

What form does the general solution for C' take? Give the general solution for C.

3. During steady state growth with convection and the boundary-layer approximation (BPS equation[(4)]), make a plot of the discontinuity in solute flux at $z = \delta_c$, as a function of the parameter $V\delta_c/D$ for a range of k_0-values.

4. Use the data of Fig. 3.10(a) for B in Si with Eq. (3.11a) to find the values of p and thus β_u (Eq. (3.10g)) giving the best match for each rotation rate. Using $D_L = 4 \times 10^{-5}$ cm^2 s^{-1}, determine values of β'_u and p' for each rotation rate. Why should these values depend on V?

5. Consider Eqs. (3.13) and (3.14) for the freezing of an Fe melt containing C_∞^C and the equilibrium amount of O in the melt for $P^{CO} = 10^{-2}$ atm over the melt. For $k_0^C = 0.25, k_0^O = 0.1, \log \hat{K} = 1160/T + 2.00$ and all the CO formed boils off. Plot the reduced steady state concentrations of C and O in the solid Fe, C_S^C/C_∞^C and C_S^O/C_∞^O, as a function of the parameter $\beta_r D/V^2$ ($D \sim 5 \times 10^{-5}$ cm^2 s^{-1}, $V \sim 10^{-4} - 10^{-1}$ cm s^{-1}). What is the rate of production of CO during steady state freezing?

6. For an interface of mild curvature, $h \sim \lambda/4$, use Eqs. (3.17) to plot $\Delta C_i/C_\infty$ as a function of Péclet number, $\hat{P}_C = V\lambda/2D$, for a range of k_0 values.

7. For an idealized eutectic interface, use Eqs. (3.15) and (3.17) to plot $\Delta C_\alpha^\beta/C_E$ as a function of Péclet number, $\hat{P}_C = V\lambda/2D$, for $\lambda_\alpha = \lambda_\beta = \lambda/2$ and a range of $k_\beta - k_\alpha$.

8. Prove that Eq. (3.11d) is correct by integrating Eq. (3.10a) once and evaluate the integration constant by requiring that the interface conservation condition be satisfied.

4

Macroscopic and microscopic solute redistribution

As already presented in Eq. (3.1), for a one-dimensional system the general solute conservation equation for the liquid in a constant velocity moving coordinate frame is given by

$$\frac{\partial C_L}{\partial t} = \frac{\partial}{\partial \xi}(-J) + \dot{Q} \qquad (4.1a)$$

$$= \frac{\partial}{\partial \xi}\left\{\frac{D_C \partial C_L}{\partial \xi} + \left[V + \frac{D_C}{\kappa T}\frac{\partial \delta \widetilde{\Delta G_0}}{\partial \xi} - u(\xi)\right]C_L\right\} + \dot{Q} \quad (4.1b)$$

↑	↑	↑	↑	↑
Fick's	*Moving*	*Field*	*Moving*	*Non-*
first law	*interface*	*effect*	*fluid*	*conservative*
effect	*effect*		*effect*	*system*
				effect

In Eq. (4.1), $\dot{Q}$ is the exchange rate of solute between this one-dimensional fluid and the environment while ξ is the distance measured relative to the moving coordinate frame that is generally attached to the interface. In practice, this can be the macroscopic loss rate (or gain rate) from the melt surface during CZ crystal growth or it can be associated with the lateral transport of solute from a fluid slice wherein the one-dimensional approximation is being utilized or it can be due to a chemical reaction occurring in the melt. Here, the field effect can be due to either an externally applied field (electric generally) or an internal interface field of the type discussed in Chapters 1 and 3.

To obtain the solute distribution in the crystal as a function of position $z = Vt$, we must first solve Eq. (4.1) subject to the appropriate initial and boundary conditions and obtain an expression for $C_i(t)$, the concentration in the liquid at the interface. Then, if we choose to neglect transport in the solid, we find that

$$C_S(z) \equiv \hat{k}_i C_i(z/V) \qquad (4.2a)$$

where $\hat{k}_i$ includes all interface field effects. When we choose to be completely exact, the following equation must be solved for $C_S(z,t)$

$$\frac{\partial C_S}{\partial t} = \frac{\partial}{\partial \xi} \left[D_{CS} \frac{\partial C_S}{\partial \xi} + \left(V + \frac{D_{CS}}{\kappa T} \frac{\partial}{\partial \xi} \delta \widetilde{\Delta G}_{0S} \right) C_S \right] \qquad (4.2b)$$

subject to

$$C_S(0) = k_i C_L(0,t) \qquad (4.2c)$$

at the interface end of the crystal and some other appropriate boundary condition at the seed end, i.e., either zero flux, constant concentration, etc. The solute profile in the solid is further coupled to that in the liquid via the interface conservation condition

$$VC_i - k_i VC_i = J_{Li} - J_{Si} \qquad (4.3)$$

The complete solution for the liquid also requires the specification of its far-field condition (constant concentration, zero flux, etc.) and its initial condition (generally constant concentration).

From the Chapter 1 discussion on $\hat{k}_i$, it was predicted that the interface solute distribution will approximate an equilibrium distribution in the particular phase during growth so long as $V << \bar{\beta} \alpha D_C / \kappa T$.[1] This condition will be satisfied during many crystalline and non-crystalline phase transformation processes so that this interface field effect segregation contribution can be approximated to ride just on top of the macroscopic profile. There will be changes in the equilibrium segregation portion of the profile during a transient condition, in direct proportion to the changes in $C_L(+\Delta z_L)$ with time; i.e., the interface will "see" the concentration at $z = +\Delta z_L$ as its effective far-field concentration and will seek to change its local profile as this effective far-field concentration changes.

During the thermal oxidation of P-doped Si, the above described conditions hold with $V \sim 20$ Å min^{-1}, $\bar{\beta}_{Si} \sim 0.3$ eV, $\alpha_{Si}^{-1} \sim 100$ Å, $\kappa T \sim 0.1$ eV and $D_C^P \sim 5 \times 10^{-13}$ Å^2 min^{-1} [2] so that the interface segregation in response to the unique field created by the formation of the Si/SiO$_2$ interface should be relatively independent of the oxidation velocity. For this example, Fig. 4.1 shows the evolution of the P pile-up in the early stages of oxidation. Special attention should be paid to the 10 min oxidation where one notes that much of the excess P at the

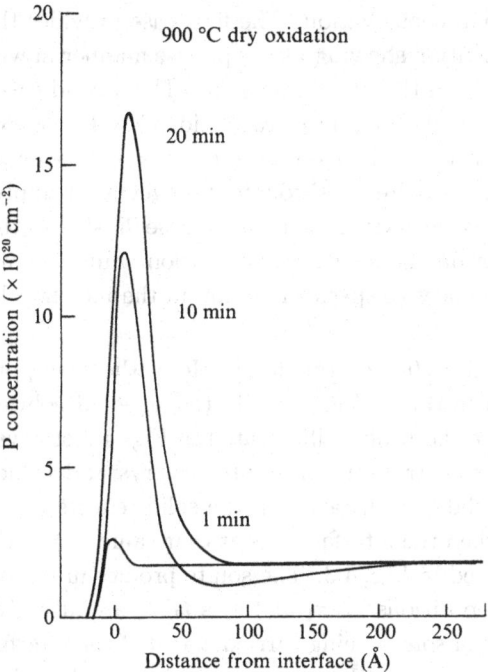

Fig. 4.1. Evolution of the P pile-up during the early stages of Si oxidation.[2]

interface reached that location by field-driven diffusion from the bulk Si rather than by interface partitioning (the depletion zone is the tell-tale indicator). Fig. 4.1 is somewhat reminiscent of Fig. 3.5.

4.1 Initial transient (V = constant, planar front, no mixing)

The initial condition for normal freezing is generally $C_L(\xi, 0) = C_\infty$, although some of the very recent studies of laser melting of ion-implanted regions require the use of $C_L(\xi, 0) = f(\xi)$. Here, we shall restrict ourselves to the simple case of $f(\xi) = C_\infty$ and shall focus our attention initially on five illustrated cases: (1) a semiinfinite sample is considered with crystallization initiated at one end at time $t = 0$, no external electric field, $\hat{E}$, is applied and diffusion in the solid is neglected ($D_{CS} = 0$), (2) the same geometry and $\hat{E} = 0$ but diffusion in the solid is considered for the special case of $D_{CS} = D_{CL}$, (3) the same geometry and $D_{CS} = 0$ but $\hat{E} = $ constant, (4) the effect of a simple interface field and (5) the effect of non-conservation. The first case provides the most generally applicable solution showing us the precise manner in which the solute content builds up in the liquid and solid. The second case allows us to evaluate the effect of diffusion in the solid (C in Fe for example) in an extreme situation while the third case teaches us how greatly the solute distribution can be altered through the agency of applying an electric field to a solidifying crystal. The fourth case illustrates the effect of interface adsorption on the solute redistribution while the fifth case teaches us the consequences of species reaction in the interface layer.

Case 1 ($D_{SC} = \delta\widetilde{\Delta G}_0 = 0$, conservative): Here, the first portion of crystal will have concentration $k_0 C_\infty$ while the rejected solute builds up in the liquid to give the profiles illustrated in Fig. 4.2 as crystallization continues. The concentration frozen into the crystal does not move because $D_{CS} = 0$, so the concentration in the solid reflects the concentration in the liquid when the interface was at that value of $z = Vt$. This relationship is illustrated in Fig. 4.3. The solute profiles in the liquid at four specific interface positions z_1–z_4 at times t_1–t_4 are also given and the increasing amount of solute being carried ahead of the interface may be noted. This solute excess in the liquid is exactly equal to the solute depletion present in the solid as illustrated by the hatched areas in Fig. 4.2.

The exact solutions for the solute distributions in the liquid and solid for the case of zero density difference may be obtained by using the Laplace transform technique with the following quantitative results[3]

$$\frac{C_L(\xi, t)}{C_\infty} = \left\{ 1 + \frac{(1 - k_0)}{2k_0} \exp\left(\frac{-V\xi}{D_{CL}}\right) \text{erfc}\left[\frac{\xi - Vt}{(4D_{CL}t)^{1/2}}\right]\right.$$

$$- \frac{1}{2}\text{erfc}\left[\frac{\xi + Vt}{(4D_{CL}t)^{1/2}}\right] - \frac{(\frac{1}{2} - k_0)}{k_0}$$

$$\times \exp\left\{ -\left[(1 - k_0)\frac{V}{D_{CL}}(\xi + k_0 Vt)\right]\right\}$$

$$\left. \times \text{erfc}\left[\frac{\xi + (2k_0 - 1)Vt}{(4D_{CL}t)^{1/2}}\right]\right\} \tag{4.4a}$$

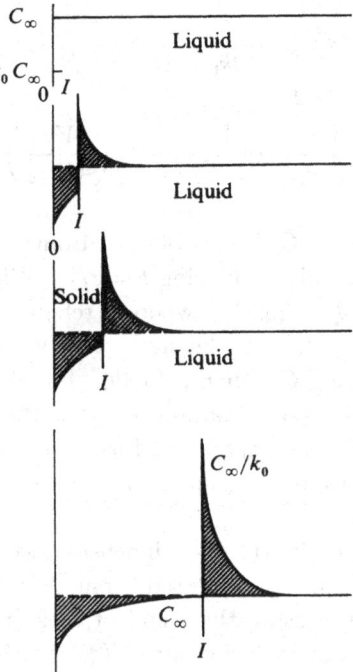

Fig. 4.2. Typical solute profile changes in the solid and liquid during initial transient crystallization at V = constant and no melt mixing.

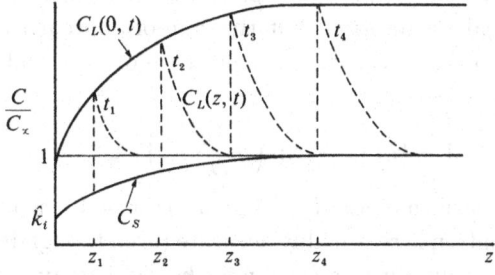

Fig. 4.3. Schematic plots of $C_L(0,t), C_S(z)$ and the liquid boundary layer profile changes for the four times t_1, t_2, t_3 and t_4 during the initial transient.

and

$$\frac{C_S(z)}{C_\infty} = \frac{1}{2}\left\{1 + \text{erf}\left[\left(\frac{Vz}{4D_{CL}}\right)^{1/2}\right] + (2k_0 - 1)\right.$$

$$\left. \times \exp\left[-k_0(1-k_0)\frac{Vz}{D_{CL}}\right]\text{erfc}\left[(2k_0-1)\left(\frac{Vz}{4D_{CL}}\right)^{1/2}\right]\right\}$$

(4.4*b*)

The solute distribution in the solid, $C_S(z)$ is obtained from $C_L(\xi, t)$ by setting $\xi = 0$, multiplying by k_0 and introducing $t = z/V$. When there is a density difference, $u/V = (d_L - d_S)/d_L$ where d refers to density, Eqs. (4.4) hold but with $V' = (d_S/d_L)V$ replacing V, $z' = V't$ replacing z and $C'_\infty = (d_S/d_L)C_\infty$ replacing C_∞ in Eq. (4.4*b*). For the case of finite interface field, the simplest approximation to use is the effective partitioning approximation and replace k_0 by $\hat{k}_i$ in Eqs. (4.4). Of course, this would not reveal the presence of the depletion zone in Fig. 4.1 at $t = 10$ min.

A natural variable in Eq. (4.4*b*) is the dimensionless quantity $V'z'/D_{CL}$ and, in Fig. 4.4, $C_S(z')/C'_\infty$ is plotted versus $V'z'/D_{CL}$ for a range of k_0 values. Before discussing the trend of these results, it is useful to consider an approximate solution for $C_S(z')$ which may be obtained by knowing that C_S rises continuously from k_0 at $z = 0$ and tends asymptotically to C_∞ with increase of z. From conservation of solute considerations, the areas between C_S and C_∞ must be equal to the area between C_L and C_∞ in Fig. 4.2 for all positions of the interface. If we make the critical assumption that the rate of approach of C_S to C_∞ is proportional to $C_\infty - C_S$ then C_S must be of exponential form and is given by[4]

$$\frac{C_S(z)}{C_\infty} = k_0 + (1-k_0)\left[1 - \exp\left(\frac{-k_0 Vz}{D_{CL}}\right)\right]$$

(4.5)

If one makes a comparison plot of Eq. (4.5) onto Fig. 4.4, one finds that this approximate equation is reasonably accurate (even though it doesn't satisfy the differential equation), especially for $k_0 \ll 1$ (maximum error of $\sim 20\%$ for $k_0 = 0.1$). Further, by considering this approximate equation, one can readily see that the "characteristic distance" of the initial transient distribution is approximately $D_{CL}/k_0 V$. This may be several centimeters in length for small k_0 and normal crystal pulling rates while, for geological systems, it can be orders of magnitude larger.

By setting $z = Vt$ in Eq. (4.5), one finds that the time required for C_S/C_∞ to rise from k_0 to unity is $\tau_{rise} \sim D_{CL}/k_0 V^2$ whereas, in the liquid boundary layer of thickness D_{CL}/V, the solute redistributes itself

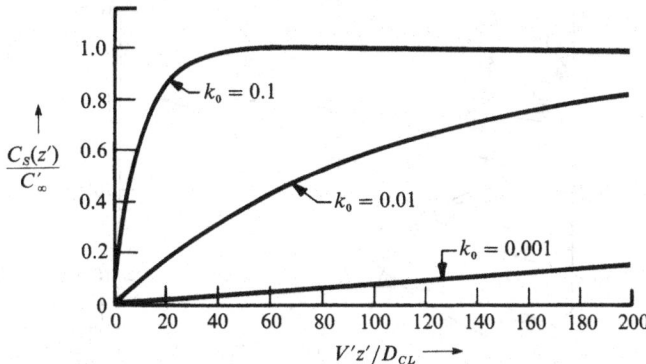

Fig. 4.4. Exact calculation (Eq. (4.4b)) of initial transient profiles in the solid for three k_0 values.[3]

in a time $\tau_{diff} \sim D_{CL}/V^2$. Since, for small k_0, we have $\tau_{diff} \ll \tau_{rise}$, diffusion in the liquid boundary layer takes place sufficiently rapidly that the solute profile in the boundary layer is nearly that for a quasi-steady state profile at each instant of time. Thus, Eq. (4.5) is called the *quasi-steady state solution* and it is little wonder that it matches Eq. (4.4b) for $k_0 \ll 1$.

This observation may be utilized to formulate an integral equation for C_L which may be solved to provide the solute distribution in the crystal when k_0 is not a constant but is a function of interface concentration, time, etc. In such a case, an approximate solution to this non-linear situation is

$$\frac{C_L}{C_\infty} \approx 1 + \left(\frac{C_i}{C_\infty} - 1\right) \exp\left(\frac{-V\xi}{D_{CL}}\right) \tag{4.6}$$

provided C_i is a slowly varying function of time (holds for $k_0 \ll 1$). Although Eq. (4.6) is incorrect in detail, it is sufficiently accurate to calculate such gross quantities as the solute excess carried ahead of the interface; i.e., $[C_i(t) - C_\infty](D_{CL}/V)$.

Case 2 ($D_{SC} = D_{LC}$, $\delta\tilde{\Delta}G_0 = 0$, conservative): For the special case of $D_{CS} = D_{CL} = D_C$, exact solutions for $C_L(\xi,t)$ and $C_S(z)$ can also be readily obtained by using the Laplace transform technique.[5] It is found that, for $k_0 < 1$, less than a 10% error is involved in the use of Eq. (4.4b); however, for the case of $k_0 > 1$, quite a large error is involved in making this approximation. In Fig. 4.5(a), C_S/C_∞ is plotted versus Vz/D as solid lines for a range of k_0. The dashed curves give the absolute

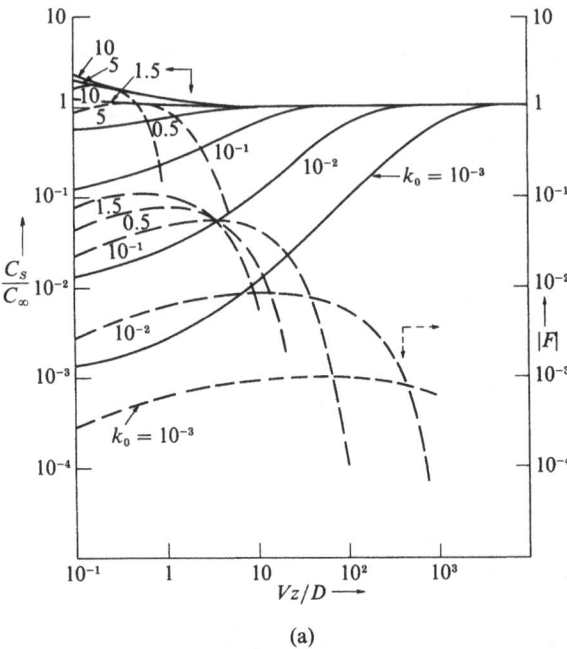

(a)

Fig. 4.5. Relative solute concentrations in the solid as a function of the dimensionless growth rate, Vz/D, for (a) the case where $D_S = D_L$ for a range of k_0 (solid lines). The dashed lines give $|F|$, the fractional error involved in assuming $D_S = 0$ instead of $D_S = D_L$.[5]

value of the fractional error $|F|$ involved in using Fig. 4.4 instead of Fig. 4.5(a). From this we can conclude that, even for this extreme case where $D_{CS} = D_{CL}$, diffusion in the solid does not significantly alter the frozen-in ($D_{CS} = 0$) solute distribution. Thus, for almost all situations, one can have a good degree of confidence that Eq. (4.4b) is a good approximation to the actual value of $C_S(z)$ in the crystal.

Case 3 ($D_{SC} = 0$, conservative, $\hat{E}$ = constant): When transport is completely neglected in the solid but a constant electric field, $\hat{E}$, is applied to the crystallizing system, an exact mathematical solution can also be obtained provided we maintain $C_L = C_\infty$ by mixing the $\xi = \infty$ liquid with an infinite bath of solute content C_∞.

By using the transformation

$$V_E = V - \hat{M}_{\hat{E}}\hat{E} \tag{4.7a}$$

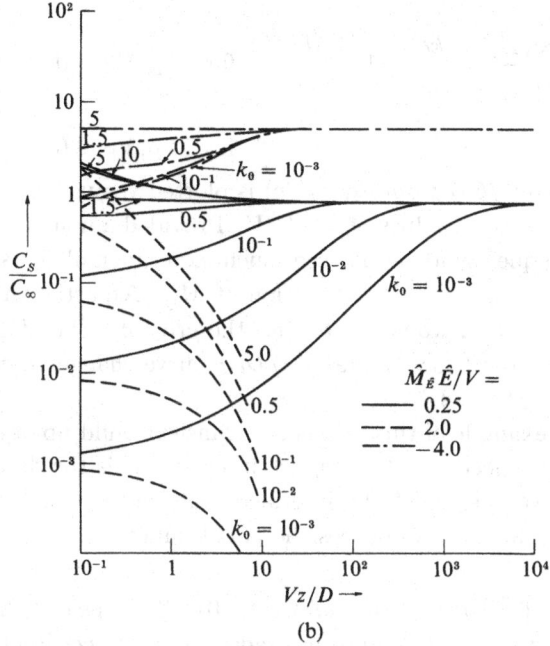

(b)

Fig. 4.5. (b) Three values of the field effect migration parameter, $\hat{M}_{\hat{E}}\hat{E}/V$, for a range of k_0.[5]

and

$$k_E = (V/V_E)k_0 \qquad (4.7b)$$

where $\hat{M}_{\hat{E}}$ is the effective ionic mobility of solute in the liquid, $C_L(\xi, t, 0)$ given by Eq. (4.4a) can be changed to $C_L(\xi, t, \hat{E})$ by replacing V with V_E and k_0 with k_E in Eq. (4.4a). By defining an additional parameter, $\theta^2 = Vz/D_{CL}$, the initial transient distribution for solute in the solid is given by[5]

$$\frac{C_S(z, \hat{E})}{C_\infty} = \frac{k_0}{2k_E} \left\{ 1 + \mathrm{erf}\left(\frac{V_E\theta}{2V}\right) \right.$$

$$\left. + (2k_E - 1)\exp\left[-k_E(1 - k_E)\left(\frac{V_E}{V}\right)^2\theta^2\right]\mathrm{erfc}\left[(2k_E - 1)\frac{V_E\theta}{2V}\right] \right\}$$

$$(4.7c)$$

From this equation it should be noted that, as $z \to \infty$, a steady state solute distribution is achieved which has the form

$$\frac{C_S(\infty, \hat{E})}{C_\infty} = \frac{k_0}{k_E} = 1 - \frac{\hat{M}_{\hat{E}}\hat{E}}{V} \quad \text{for} \quad k_E, V_E > 0 \qquad (4.7d)$$

$$= 0 \qquad \text{for} \quad k_E, V_E < 0 \qquad (4.7e)$$

In Fig. 4.5(b), $C_S(z)/C_\infty$ from Eq. (4.7c) is plotted versus Vz/D_{CL} for a range of k_0 and several values of $\hat{M}_{\hat{E}}\hat{E}/V$. Provided a non-circulatory convection technique could be devised, such experimental plots would provide a procedure for the determination of $\hat{M}_{\hat{E}}$. A particularly simple method, for $k_0 < 1$, would be to find the critical value of $\hat{E}/V$ at which the slope of the $C_S(z)$ versus Vz/D_{CL} curve changes sign. This condition gives $\hat{M}_{\hat{E}}(\hat{E}/V)_c = 1$.

An important example of this case is the transient build-up of H^+ and OH^- in the water ahead of a freezing ice interface. It is the differential partitioning of two species at the interface that leads to the interface electric field that, in turn, influences the partitioning.

Case 4 ($D_{SC} = 0$, conservative, $\delta\widetilde{\Delta}\tilde{G}_0 \neq 0$): To consider the total transient circumstances where interface adsorption is occurring at the same time as gross solute partitioning due to interface movement at $V = $ constant, we shall use a simplified form for the interface field. Instead of the decaying exponential form used earlier, we shall employ a square-well potential as illustrated in Fig. 4.6(a). At equilibrium with a static interface, equalization of chemical potential for the solute species should result in a box-like distribution of dopant near the interface as illustrated in Fig. 4.6(b). For a specific example, we shall consider P as a dopant during the thermal oxidation of Si.

The step-like structure of Fig. 4.6 allows us to treat the interface as if it were a separate chemical phase with a greater affinity for P. The square-well potential is a severe approximation to the sort of interface fields one normally expects; however, if the width is chosen to correspond to a single lattice distance, the square well may be a very good approximation for the monolayer model of interface segregation. We shall leave the width of the well, δ, as an adjustable parameter. The total normalized content of the equilibrium distribution, given by the product $k_i\delta$, is an important material parameter that can be determined from Auger sputter profiles.

To simplify the problem, we assume that the SiO_2 is impermeable to P and that the diffusion constant in the interface phase, D_i^P, is infinite.

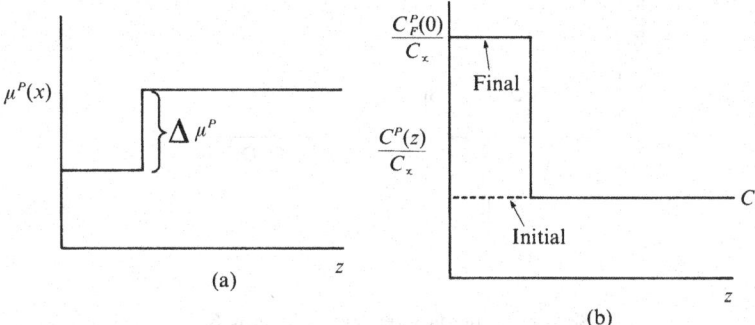

Fig. 4.6. (a) Square-well chemical potential profile, $\mu^P(z)$, at time $t = 0$ and (b) initial (dashed) and final (solid) phosphorus concentration profiles for this potential profile $(C_F^P(0)/C_\infty = \exp(\Delta\mu^P/\kappa T))$.[6]

This is analogous to the crystallization case where the Si acts as the liquid and the SiO$_2$ as the solid but with $k_0 = 0$. The initial conditions and the time-dependent boundary conditions at the interface are illustrated in Fig. 4.7. An exact mathematical solution can be obtained for this case by the use of the Laplace transform technique and, in Fig. 4.8, the calculated solution to the initial transient distribution for P in Si is given for four different times with $\delta = 40$ Å and $k_i = 12$.[6] The values for C_∞, V and D_{Si}^P correspond to published values for the oxidation of Si at 900 °C ($C_\infty = 1.3 \times 10^{20}$ cm^{-3}, $D_{Si}^P = 4.25 \times 10^3$ Å^2 min^{-1} and $V = 3$ Å min^{-1}). Fig. 4.8(a) shows the P distribution after 0.005 min and corresponds fairly well to the initial condition. Fig. 4.8(b) shows the interface beginning to fill up with P accompanied by steep concentration gradients in the Si. In Figs. 4.8(d) and (e), the final stages of the initial transient, after 50 and 200 min, are illustrated. Note that, in the time elapsed between the two profiles, the concentration gradient near the interface has changed from positive to negative values. This transition marks a reversal in the direction of P flux in the region of the interface. As the chemical potential well fills up with P, the normal redistribution process that accompanies the rejection of dopant from the SiO$_2$ begins to dominate the profile. The redistribution does not begin to cause a P pile-up in the bulk of the Si, however, until after ~ 250 Å of oxide has been grown ($\sim 10\delta$).

The effect of the potential well shape on the evolution of the dopant profile was also investigated; however, so long as the $k_i\delta$ characteristic remained constant, the relative height and width of the well had no

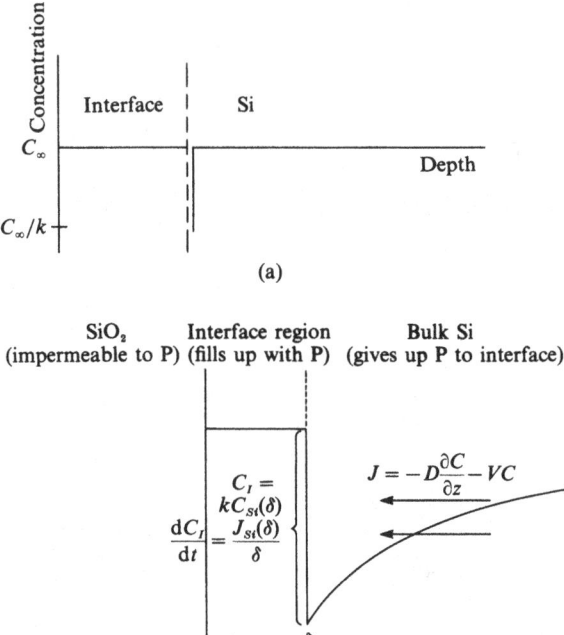

Fig. 4.7. Schematic of (a) the initial condition and (b) the time-dependent interface boundary condition to illustrate the mathematical procedure for solving the square-well potential initial transient.[6]

significant effect on the P profile evolution. One of the Auger sputter profiles is illustrated in Fig. 4.9 along with the calculated results for the diffusion analysis. Despite the crude "square-well" approximation employed in the diffusion calculation, fairly good agreement is found between the theoretical and experimental profiles. By studying the evolution of the solute distribution with time, it has been determined that the interface region will fill to within 90% of its equilibrium value after 90 min of oxidation at 900 °C and after 420 min of oxidation at 800 °C. Both of these results have been confirmed by Auger measurements. The actual width of the interface remains unknown using this technique since the experimental profiles are broadened by the effects of ion knock-on mixing. Subsequent experiments showed that the interface segregation process was a well-defined reversible adsorption process with temperature giving an activation energy for k_i of $Q = -0.26$ eV.[6]

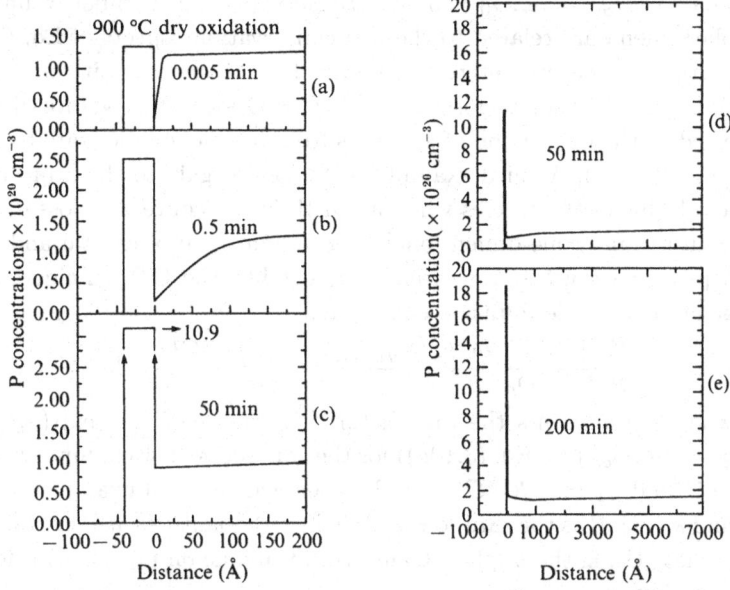

Fig. 4.8. Calculated initial transient phosphorus profiles due to a square-well interface potential of width, $\delta = 40$ Å and $k_i = 12$ for a range of times. Between 50 and 200 min, the depleted zone in the Si becomes filled with P due to rejection from the SiO_2.[6]

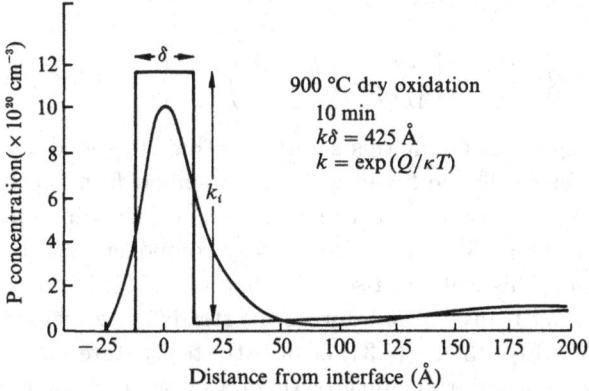

Fig. 4.9. Comparison between the calculated P profile and an experimental Auger sputter profile obtained for a Si sample oxidized for 10 min at 900 °C. For the theoretical profile, δ has been arbitrarily set at 25 Å while the parameter, $k_i\delta$ was determined from the Auger measurements.[6]

Case 5 $(D_{SC} = \delta \widetilde{\Delta} G_0 = 0$, *non-conservative):* A number of important phenomena are related to the non-conservation aspects of Eq. (4.1). In particular, we are interested in the loss of one constituent via reaction with another to form a third $(C + O \rightarrow CO)$. We are also interested in the conversion of a species from one polymeric state to another $(nO \rightleftharpoons O_n)$. A third example is the loss or gain to the atmosphere of a volatile solute species while a fourth is the convective loss or gain of solute from a one-dimensional longitudinal slice of crystal and melt in a practical crystal growth environment like the CZ technique. In this general case, the diffusion equation is

$$\frac{\partial^2 C^j}{\partial z^2} + \frac{V}{D_C} \frac{\partial C^j}{\partial z} + \frac{q_1}{D_C} C^j + \frac{q_2}{D_C} = \frac{\partial C^j}{\partial t} \tag{4.8a}$$

where q represents the rate factor; i.e., $q_1 = -\beta_u'[1 - \exp(-p'z)]$ and $q_2 = -q_1 C_\infty$ (see Eq. (3.10a)) for the convective flow example while $q_1 = \beta C^k$ and $q_2 = -\beta C^k C^{*j}$ for the chemical reaction example between j and k with β as the rate constant and the * quantities are the equilibrium values. Using the Laplace transform technique on Eq. (4.8a) leads to an equation of the form

$$\frac{d^2 \bar{C}^j}{dz^2} + \frac{V}{D_C} \frac{d\bar{C}^j}{dz} + \left(\frac{q_1 - S}{D_C}\right) \bar{C}^j + \frac{(q_2 + C_\infty)}{D_C} = 0 \tag{4.8b}$$

which has a solution, for constant q_1 and q_2, given by

$$\bar{C} = A_1 \exp(-az) + A_2 + (q_2 + C_\infty)/(q_1 - S) \tag{4.8c}$$

where

$$a = -\left\{ \frac{V}{2D_C} + \left[\frac{V^2}{4D_C} - \left(\frac{q_1 - S}{D_C}\right) \right]^{1/2} \right\} \tag{4.8d}$$

The interesting point about this solution is that, provided $V^2/4 > q_1$, the solution for C^j in the liquid will have a similar form to that given by Eq. (4.4a) but with a smaller or larger D_{eff} depending on the sign of q_1; however, if $V^2/4 < q_1$, an oscillatory component will be present in the solution. This will give rise to fluctuations of C^j in the solid both during the initial transient and during the steady state regimes.

In Chapter 3, Eqs. (3.25)–(3.34) dealt with steady state solute redistribution in the adlayer during film growth for a surface of fixed orientation, θ. Here, we are interested in the initial transient distribution for such a train of ledges moving at constant Péclet number, $\hat{P}_C$. In dimensionless form, the transient transport equation and ledge solute conservation condition are

$$\frac{\partial^2 n'_s}{\partial x'^2} + 2\hat{P}_C \frac{\partial n'_s}{\partial x'} - \frac{(n'_s - 1)}{\tilde{\eta}} = \frac{1}{\tilde{\eta}} \frac{\partial n'_s}{\partial t'} \tag{4.9a}$$

and

$$2\hat{P}_C(1 - k'_0)n'_{s\ell} = \left(\frac{\partial n_s}{\partial x'}\right)_{\ell+} + \left(\frac{\partial n'_s}{\partial x'}\right)_{\ell-} \tag{4.9b}$$

where $x' = x/\lambda$, $t' = t/\tau_s$, $\tilde{\eta} = D_s\tau_s/\lambda^2$, $\hat{P}_C = V_\ell\lambda/2D_s$, $k'_0 = (h/a)k_0$ and $n'_s = n_s/n_{s\infty}$ for ledges of height h and spacing λ. One interesting thing to notice about this pair of equations is that, even with a transient growth of h/a, $\hat{P}_C$ can remain constant, i.e., as λ grows in size proportional to t^n, V_ℓ decays in size proportional to t^{-n} so that the product stays constant. However, the Laplace transform technique cannot be used here unless $h = $ constant.

For $k'_0 = $ constant, the transformed differential equation is

$$\frac{\partial^2 \bar{n}'_s}{\partial x'^2} + 2\hat{P}_C \frac{\partial \bar{n}'_s}{\partial x'} - \frac{(\bar{n}'_s - 1)}{\tilde{\eta}} - \frac{S\bar{n}'_s}{\tilde{\eta}} + \frac{n_{s\infty}}{\tilde{\eta}} = 0 \tag{4.9c}$$

For a single ledge, this yields a transformed plane solution

$$\bar{n}'_s = \frac{(1 + n_{s\infty})}{(1 + S)} + A \exp(\beta x') \tag{4.9d}$$

where

$$\beta = -\hat{P}_C \pm [\hat{P}_C^2 + (1 + S)/\tilde{\eta}]^{1/2} \tag{4.9e}$$

This has a general form similar to Eqs. (4.8c) and (4.8d) so that a solution in the untransformed plane may be easily attained. Once again, a periodic solution for multiple ledges is not attainable by this technique except in the limit that $\hat{P}_C \ll 1$.

A periodic solution, based upon the quasi-steady state approach for $\hat{P}_C \ll 1$, can be written and this is

$$n_s(x', t) = [n_{s\ell}(t) - n_{s\infty}(t)] \cosh\left(\frac{x' - \frac{1}{2}}{\tilde{\eta}^{1/2}}\right) + n_{s\infty}(t) \tag{4.9f}$$

$$n_{s\infty}(t) = n_{s\infty}\{1 - \exp[-(k_d + k_a)t]\} \tag{4.9g}$$

where k_d and k_a are the desorption and adsorption rate constants in the adlayer[1] and

$$\frac{n_{s\ell}(t)}{n_{s\infty}(t)} = \frac{\tanh(1/2\tilde{\eta}^{1/2})}{\tanh(1/2\tilde{\eta}^{1/2}) - \tilde{\eta}^{1/2}\hat{P}_C(1 - k'_0)} \tag{4.9h}$$

In this example, $h = h(t)$, so $k'_0 = h(t)/ak_0$ and $\tilde{\eta} = D_s\tau_s/[\tan\theta/h(t)]^2$. The concentration in the thickening ledge, as a function of position, x_ℓ, along the ledge in the initial transient is

$$n_s(x_\ell) = k_0 n_{s\ell}(x_\ell) \quad \text{for} \quad x_\ell = \int_0^t V_\ell(t)dt \qquad (4.9i)$$

This assumes no diffusion in the solid.

For evaluating the initial transient in the presence of convective mixing, we use the time-dependent version of Eq. (3.10b), i.e.,[7]

$$\frac{\partial^2 C}{\partial \tilde{\eta}^2} + \frac{\partial C}{\partial \tilde{\eta}} - \beta_u[1 - \exp(-p\tilde{\eta})](C - 1) = \frac{\partial C}{\partial \tau} \qquad (4.10a)$$

subject to the boundary conditions

$$(1 - k_0)C + \frac{\partial C}{\partial \tilde{\eta}} = 0 \quad \text{at} \quad \tilde{\eta} = 0, \tau > 0 \qquad (4.10b)$$

$$C \to 1 \qquad \text{at} \quad \tilde{\eta} \to \infty, \tau > 0 \qquad (4.10c)$$

and the initial condition

$$C = 1 \qquad \text{at} \quad \tau = 0, \tilde{\eta} \geq 0 \qquad (4.10d)$$

Using the Laplace transform technique, we find that

$$
\begin{aligned}
C(\tilde{\eta}, \tau) = &1 + \frac{(1 - k_0)}{2} \exp\left(-\tilde{\eta}/2\right) \left\{ \left[1 + \frac{(p - k_0)}{\alpha_1}\right] \mathcal{L}_- \right. \\
&+ \left. \left[1 - \frac{(p - k_0)}{\alpha_1}\right] \mathcal{L}_+ \right\} \\
&- \frac{\beta_u(1 - k_0)}{2p\alpha_1} \exp\left[-\left(\frac{1}{2} + p\right)\tilde{\eta}\right](\mathcal{L}_- - \mathcal{L}_+) \quad (4.10e)
\end{aligned}
$$

where

$$\alpha_1 = [(1 - k_0)^2 + 2\beta_u]^{1/2}, \quad \alpha_2 = \left(\frac{1}{4} + \beta_u\right)^{1/2} \qquad (4.10f)$$

and

$$
\begin{aligned}
\mathcal{L}_\pm(\tilde{\eta}, \tau) = &\frac{\exp(-\alpha_2\tilde{\eta})\mathrm{erfc}(\tilde{\eta}/2\tau^{1/2} - \alpha_2\tau^{1/2})}{k_0 \pm \alpha_1 + 2\alpha_2} \\
&+ \frac{\exp(+\alpha_2\tilde{\eta})\mathrm{erfc}(\tilde{\eta}/2\tau^{1/2} + \alpha_2\tau^{1/2})}{k_0 \pm \alpha_1 - 2\alpha_2} \\
&- \left[\frac{k_0 \pm \alpha_1}{k_0 \pm \alpha_1 - 2\alpha_2}\right] \\
&\times \exp\left\{\frac{1}{2(k_0 \pm \alpha_1)[\tilde{\eta} + \frac{1}{2}(k_0 \pm \alpha_1 - 2\alpha_2)\tau]}\right\} \\
&\times \mathrm{erfc}\left[\frac{\tilde{\eta}}{2\tau^{1/2}} + \frac{1}{2}(k_0 \pm \alpha_1)\tau^{1/2}\right] \quad (4.10g)
\end{aligned}
$$

By putting $\tilde{\eta} = 0$ and replacing τ by $\tilde{\eta}_1$ in Eq. (4.10f) and then multiplying by k_0, we obtain the value of C_S at the position $\tilde{\eta}_1$. Thus, the corresponding solute concentration appearing in the crystal is given by

$$C_S(\tilde{\eta}_1) = k_0 + k_0 \frac{(1 - k_0)}{2} \left\{ \left[1 + \frac{(p - k_0)}{\alpha_1} \right] \mathcal{L}_-(0, \tilde{\eta}_1) \right.$$
$$\left. + \left[1 - \frac{(p - k_0)}{\alpha_1} \right] \mathcal{L}_+(0, \tilde{\eta}_1) \right\}$$
$$- \frac{\beta_u k_0 (1 - k_0)}{2 p \alpha_1} [\mathcal{L}_-(0, \eta_1) - \mathcal{L}_+(0, \eta_1)] \qquad (4.10h)$$

where the origin of $\tilde{\eta}$ is chosen as the onset of growth.

Fig. 4.10 shows the calculated initial transient profiles in the solid for three k_0 values and a range of β_u. All of the curves start from $C = k_0$ at $\tilde{\eta}_1 = 0$ and rise towards a given plateau determined by the value of β_u. It is noted that the characteristic distance also decreases as β_u increases. This decrease in transient time is reasonable because, the larger the lateral convective flux, the smaller is the time needed for the system to remove or add the excess or depleted solute atoms in the boundary layer and reach the final steady state profile. As expected, when $\beta_u = 0$, Eqs. (4.10) revert to Eqs. (4.4).

4.2   Intermediate transient (planar front, conservative)

At any point during the initial transient or steady state regimes, an increase or decrease in freezing velocity produces bands of higher or lower solute content in the solid, respectively, for $k_0 < 1$ (reversed for $k_0 > 1$). If V increases abruptly to $V + \Delta V$, the solute distribution in the liquid at the interface must first rise and then become steeper (k_0 may also change to $k_0 + \Delta k_0$) to carry away the larger amount of solute partitioned per unit time from the interface region. The amount of solute in the enriched layer ($k_0 < 1$) is less than before (if $\Delta k_0 = 0$), the difference having been deposited in the solid as an excess over $C_S(V)$ during the transient change in C_L. If the steady state regime had already been attained before the velocity began to fluctuate, the solute content in the solid would fluctuate above and below C_∞ with the same periodicity; i.e., k would fluctuate about the value $k = 1$. The qualitative changes in k ($k = C_S/C_\infty$) for an abrupt change from V to V_1 are illustrated in Fig. 4.11(a) for the case of $k_0 < 1$ and for a cyclic change from $V \rightarrow V_1 \rightarrow V$ in Figs. 4.11(b) and (c). To understand the quantitative character of the changes, two special situations need to be considered: (1) no mixing and (2) partial convective mixing.

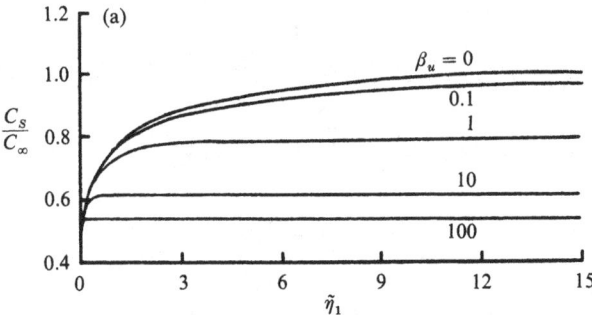

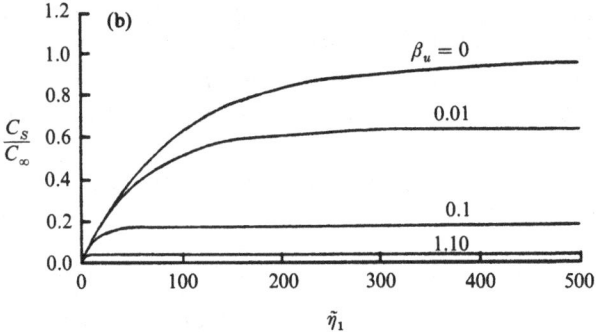

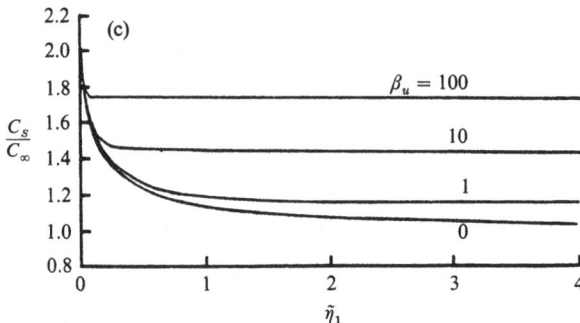

Fig. 4.10. Initial transient solute redistribution profiles in the grown solid, for the case of a convectively mixed melt, as a function of normalized length, $\tilde{\eta}_1 = Vz/D_L$, for (a) $k_0 = 0.5$, (b) $k_0 = 0.01$ and (c) $k_0 = 2.0$ and a number of β_u values.[7]

Case 1 ($D_{SC} = u_\infty = \delta\widetilde{\Delta G}_0 = 0$, $V = constant$, V_1/V impulse): For the no mixing situation, a quantitative expression for $C_S(\Delta z)$, where $\Delta z = V_1 t$, can be found for the case where a steady state distribution has been established in the liquid at velocity V and then the velocity is abruptly changed to the value V_1. Again the Laplace transform procedure is utilized and $C_S(\Delta z)$ is found to be given by[3]

$$\frac{C_S(\Delta z)}{C_\infty} = \left\{ 1 - \frac{1}{2}\mathrm{erfc}\left[\left(\frac{V_1\Delta z}{4D_{CL}}\right)^{1/2}\right]\right.$$

$$+ (1 - k_0)\left(\frac{\frac{1}{2} - V/V_1}{k_0 - V/V_1}\right)\exp\left[-\left(1 - \frac{V}{V_1}\right)\frac{V_1\Delta z}{D_{CL}}\right]$$

$$\times \mathrm{erfc}\left[\left(\frac{V}{V_1} - 1\right)\left(\frac{V_1\Delta z}{4D_{CL}}\right)^{1/2}\right] + \frac{(2k_0 - 1)}{2}\left(\frac{1 - V/V_1}{k_0 - V/V_1}\right)$$

$$\left.\times \exp\left[-k_0(1 - k_0)\frac{V\Delta z}{D_{CL}}\right]\mathrm{erfc}\left[(2k_0 - 1)\left(\frac{V_1\Delta z}{4D_{CL}}\right)^{1/2}\right]\right\}$$

$$(4.11)$$

As in the case of the initial transient distribution, this equation may be put in dimensionless form by plotting k versus $V_1\Delta z/D_{CL}$ as the independent variable. However, since Eq. (4.11) contains two parameters, k_0 and V/V_1, a family of curves must be generated. Such plots are given in Fig. 4.12 where we can see that (i) the magnitude of k increases, for fixed k_0, as V/V_1 decreases, (ii) the magnitude of k increases, for fixed V/V_1, as k_0 decreases and (iii) the magnitude of Δz for the maximum k, plus the total width of the solute excess region, increases as k_0 decreases. A solution for C_S/C_∞ can be readily obtained for the case where $\Delta k_0 \neq 0$ by following the same mathematical procedure.

One generally thinks of the velocity change as having been brought about by abruptly changing the furnace temperature, the crystal pull rate, the fluid stirring, etc; i.e., by alterations of the gross system conditions. Obviously, by such manipulations, one can produce a wide variety of C_S distributions. However, in certain cases, a natural interface process can operate which leads to the development of periodic growth rate fluctuations. This is due to the process of interface adsorption wherein special impurity atoms attach to the interface as a function of time and poison the kink sites where layer growth is occurring. This slows down the rate of interface advance while more and more of the total driving force for growth becomes consumed in the molecular attachment part of the overall process.[1] At some minimum growth velocity (which may be $V = 0$), it becomes energetically efficient for the system to store excess free energy in the form of defects (ΔG_E). At this point in the process, rapid layer growth occurs to cover over the poisoned layer and thus incorporate defects into the solid. Now, new growth occurs with no poisoned

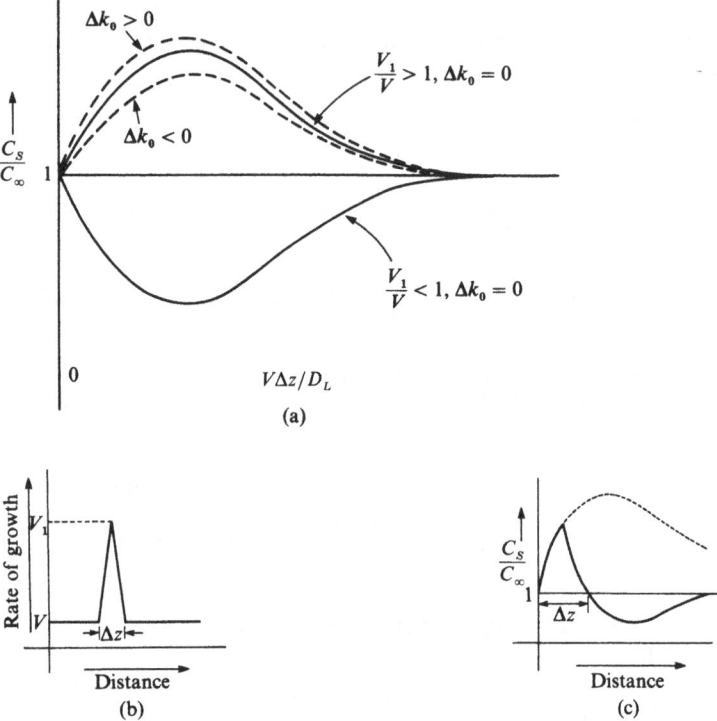

Fig. 4.11. (a) Schematic illustration of concentration fluctuations in the solid due to an abrupt growth rate change from V to V_1 as a function of dimensionless distance, $V\Delta z/D_L$ (Δk_0 is the associated change in k_0 due to the $V \rightarrow V_1$ transition), (b) proposed growth rate profile after steady state has been attained and (c) the consequent change in solute content (schematic) grown into the solid as a result of this rate change profile.

sites so the initial growth velocity is again quite high. This condition endures until a new critical concentration of site-poisoning adsorbent develops. The overall process keeps repeating in a cyclic fashion so long as the nutrient phase contains the critical adsorbing species. Many examples of this type of impurity banding can be found in geological crystals and in laboratory crystals grown from aqueous solutions.

A third possibility would arise from Eqs. (4.8) when q_1 is positive and $q_1 > V^2/4$. Since the solution in the transformed plane will contain periodic oscillations, the mathematical solution arising from the inverse

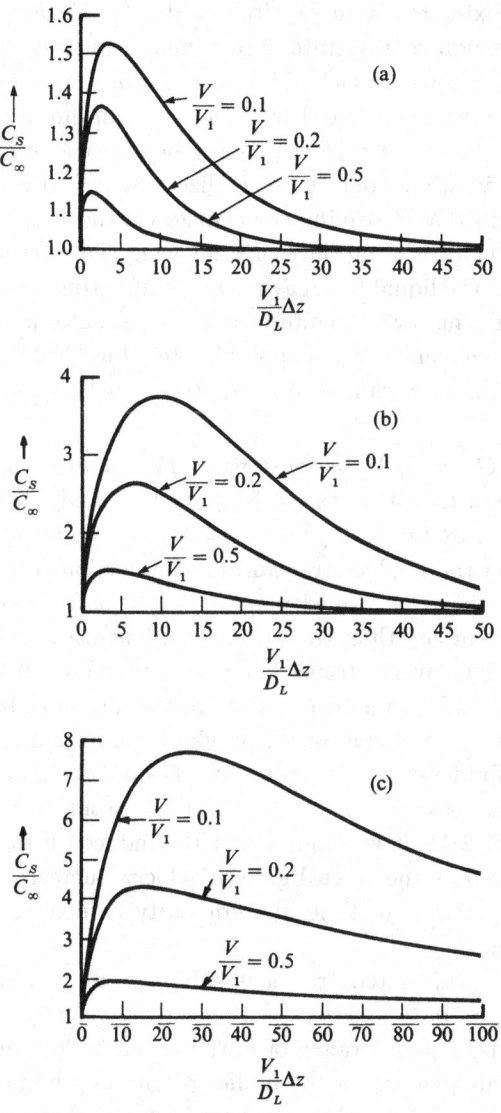

Fig. 4.12. Calculated intermediate transient solute profiles in the crystal due to an abrupt growth rate change from V to V_1, after steady state had been achieved, as a function of the dimensionless distance parameter $V_1 \Delta Z / D$ for three different k_0 values: (a) $k_0 = 0.5$, (b) $k_0 = 0.1$ and (c) $k_0 = 0.01$.[3]

transform step in the process will also yield an oscillatory transient solution.

To obtain an approximate solute distribution due to the rate change, one might use the pseudo steady state approximation just as was done for the initial transient case. In fact, the solute rise in the change from V to V_1 might be treated as an initial transient from an initial liquid of composition C_∞/k_0. However, this would only be reasonable for small Δz. A better approximation would be to realize that it is the interface conservation condition that is driving the change via the change in the concentration gradient at the interface. By allowing the interface concentration gradient in the liquid to decay exponentially from its initial to its final state and utilizing both the interface conservation condition and the overall conservation condition, an approximate solute distribution in both the liquid and the solid can be obtained (see Problem 4.3).

Case 2 ($D_{SC} = \delta \widetilde{\Delta G}_0 = 0$, $V = $ constant, V_1/V impulse, $u_\infty \neq 0$): For this case, one starts with Eqs. (4.10a), (4.10b) and (4.10c) but Eq. (4.10d) is replaced by Eq. (3.10f) with $q = Vp/V_1$ replacing p given by Eq. (3.10g). Using the Laplace transform technique, one can readily obtain a solution in the transformed plane but it is at least an order of magnitude more complex than the initial transient case.[7] Because of this complexity, the inverse transform was not found. Instead, a simple numerical method was used to calculate the desired solute concentration profiles using a couple of simplifying assumptions: (1) the change in effective fluid velocity, β_u, due to $V \to V_1$ occurs abruptly and (2) Eq. (3.10c) holds so that $\beta_{u_1}/\beta_u = (V/V_1)^2$. From the melt flow analysis of Figs. 2.29–2.32, it was found that the induced melt convection is slightly reduced as the crystal growth velocity increases so that $\beta_u \propto V^{-2+\Delta}$ where Δ is a small negative quantity (which we neglect, here, as second order).

Fig. 4.13 gives the calculated transient solute profiles in the solid for solutes with $k_0 = 0.01$ and $k_0 = 0.5$, abrupt velocity changes, $V/V_1 = 0.1$, 0.2 and 0.5 and a range of initial effective flow velocities, β_u. Figs. 4.13(a) and (d) confirm the earlier results for the no-mixing case.[3] As expected, for a given V/V_1, when the value of β_u increases, the extent of the transient response decreases such that not only does the height of the concentration excess decrease but the time required to reach the new steady state also decreases. When β_u becomes large enough, the sudden build-up of solute atoms near the interface due to the $V \to V_1$ transition can be immediately removed by the strong lateral convection. In this limit, the transient solute profile behaves much

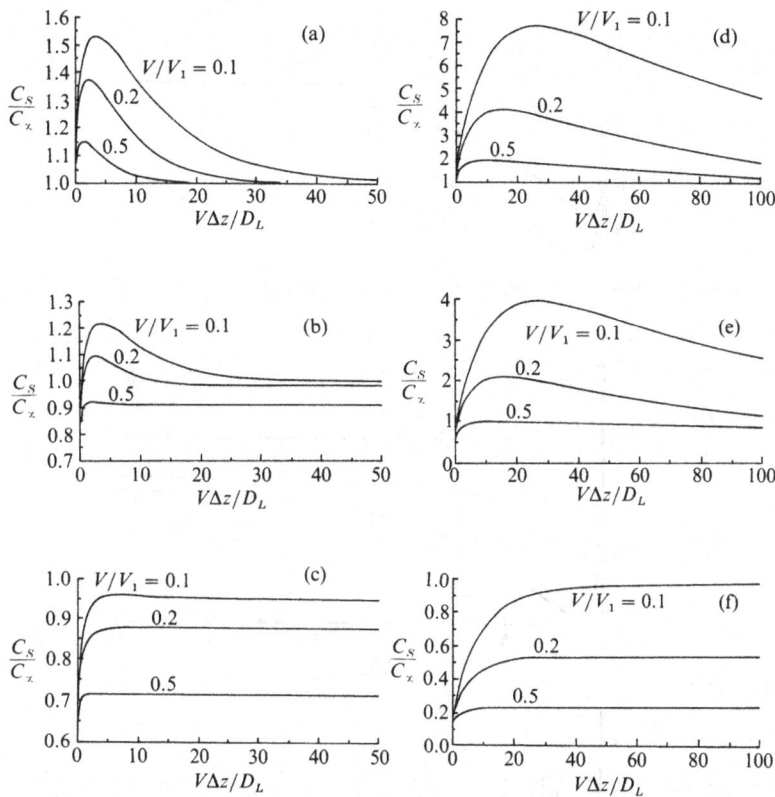

Fig. 4.13. Calculated intermediate transient profiles in the crystal, C_S/C_∞, as a function of dimensionless distance, $V\Delta z/D_L$, and a range of velocity impulses, V/V_1, (a) $k_0 = 0.5$, $\beta_u = 0$, (b) $k_0 = 0.5$, $\beta_u = 1$, (c) $k_0 = 0.5$, $\beta_u = 10$, (d) $k_0 = 0.01$, $\beta_u = 0$, (e) $k_0 = 0.01$, $\beta_u = 0.02$ and (f) $k_0 = 0.01$, $\beta_u = 0.2$.[7]

like an *initial* transient characterized by a very small rise-time. For a given β_u, the transient response becomes less severe the smaller is V_1/V; thus, the critical value of β_u for which the bump in the solid concentration starts to disappear decreases as V_1/V decreases. In Fig. 4.14, corresponding results are presented for the case of $k_0 = 2.0$.

Case 3 (V = constant, rotational striations): Single crystals pulled from alloy melts by the CZ method generally exhibit microscopic impurity patterns which are related to crystal growth rate fluctuations from a variety of sources. The general striation pattern without magnification is shown on an etched $Si + A\ell$ doped crystal in Fig. 4.15(a).[8] A magnified

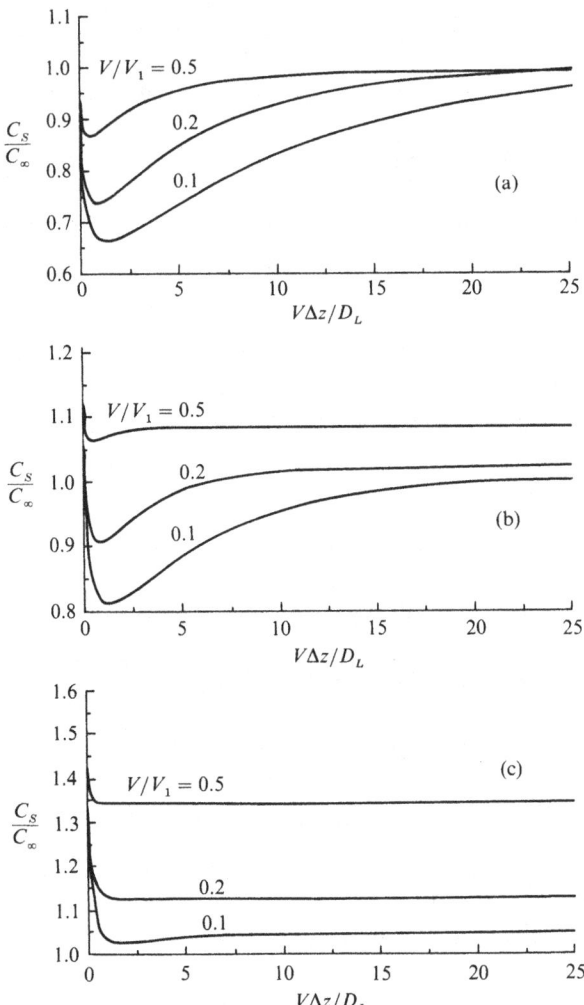

Fig. 4.14. Calculated intermediate transient profiles in the crystal C_S/C_∞, as a function of dimensionless distance, $V\Delta z/D_L$, and a range of velocity impulses V/V_1 for $k_0 = 2.0$ (a) $\beta_u = 0$, (b) $\beta_u = 2$ and (c) $\beta_u = 20$.[7]

view is given in Fig. 4.15(b) and, from these, it is seen that the striations appear as etched lines of varying lengths which are not necessarily
parallel to each other. In going from the outer to the central part of
the crystal, at a distance $\sim$ 2–3 mm from the edge, there is a rapid and
discontinuous decrease in striation frequency. The striations that extend

from the outer to the central part are of variable length and intensity. Some striations lie entirely within the central part (see Fig. 4.15(b)) and they are also of variable length and intensity. The intensity of any one striation is not constant over its entire length and, for striations that lie entirely within the central portion of the crystal, the intensity is highest in the striation's central portion and decreases towards its ends. One should equate intensity with solute excess.

The major investigators of this type of microsegregation have been Gatos, Witt and associates.[9-11] They observed that one of the major microsegregation bands had the same period as the period of rotation of the crystal but, although pulling a segment of crystal without rotation eliminated these inhomogeneities in that segment, other inhomogeneities (striations) were still there. Growing a crystal of InSb in Skylab, by melting back one-half of an earth-grown crystal and refreezing it using the unmelted part of the crystal as a seed, showed that the compositional inhomogeneities present in the earth-grown seed are totally eliminated in the crystal segment grown in the absence of gravity. We expect that this was because no form of convection was present in the melt. Crystals grown in vertical crucibles under earth conditions, where thermally or solute-driven convection is absent, should be free of these microheterogeneities as well.

The rotational striations are thought to owe their origin to the fact that the axis of crystal rotation does not exactly coincide with the thermal axis of the freezing point isotherm. Due to the thermal asymmetry in the melt, the microscopic growth rate must gradually increase from a minimum to a maximum and then decrease again to a minimum within each rotational cycle. Thus, the portion of the crystal grown under increased rate is larger than the corresponding portion grown under decreased rate in any cycle. Consequently, a greater dopant concentration (for $k_0 < 1$) is expected in the portion grown at the increased rate than in the portion grown at the decreased rate (see Fig. 4.12). Thus, a crystal grown under rotation should exhibit impurity inhomogeneities consisting of alternately broad and narrow bands. The narrow bands, whose thickness becomes very small depending on the extent of thermal asymmetry, constitute the rotational striations. Under pronounced thermal asymmetry, the microscopic growth rate assumes negative values and back-melting takes place within each rotational cycle.

Using either regular mechanical vibrations or regular electrical current pulses through the interface, it was noted[9] that regularly spaced chemical inhomogeneities were built into the crystal which did not in-

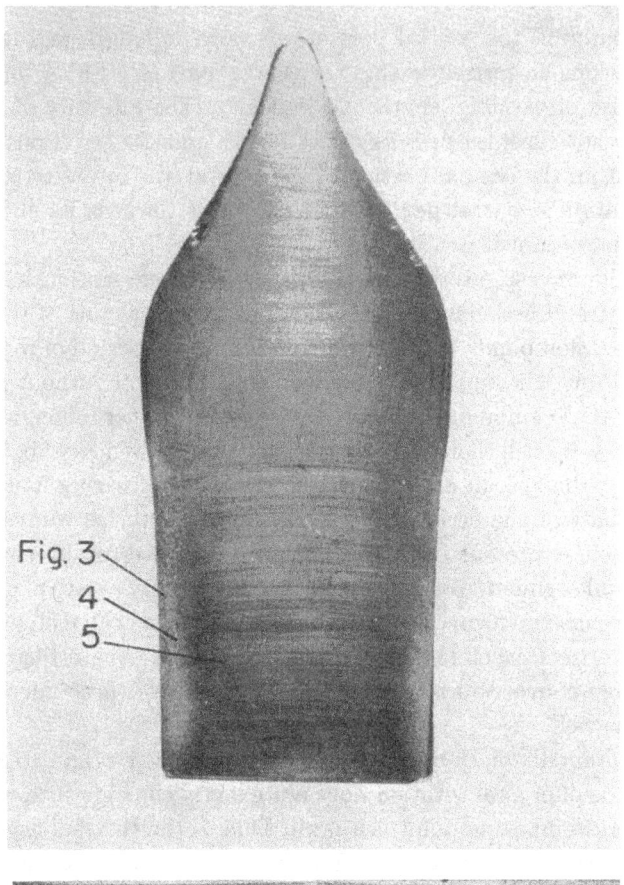

Fig. 4.15. General striation pattern on the (110) plane of a Si + Aℓ-doped crystal; (a) macroscopic view and (b) microscopic view of both the outer and central portions of the crystal.[8]

terfere significantly with the other inhomogeneities present. Thus, the intentionally generated inhomogeneities could be used as time markers to determine a microscopic growth velocity in contrast to the macroscopic growth velocity determined by the pull rate. They were able to show that the microscopic growth rate is not the same as the pulling rate and that, as shown in Figs. 4.16(a) and (b), in fact it varies significantly depending on the thermal conditions prevailing during growth. It was found that, during conditions of turbulent convection, the crystal underwent pronounced transient back-melting and the average microscopic growth rate was independent of and up to 20 times greater than the average macroscopic growth rate. For conditions of oscillatory convective instability in the melt, the crystal exhibited fluctuations in dopant concentration with a periodicity identical to that of the thermal oscillations in the melt. Under stabilizing thermal gradients (no temperature fluctuations in the melt), the microscopic and macroscopic growth rates were identical and fluctuations in dopant concentration could not be detected.

Using a spreading resistance probe with a resolution of less than 5 μm, microprofiling of these surfaces was able to correlate quantitatively the dopant fluctuations with the microscopic growth rates (see Fig. 4.17). A quantitative analysis in Si (Fig. 4.18) showed that there is good agreement between solute redistribution theory for a fluctuating growth rate and the experimental results on microsegregation. From Fig. 4.18 we see that, for a growth rate fluctuation of ~ 5 μm s^{-1} maximum, the concentration changes by $\sim 15\%$ in a distance ~ 75 μm. Using Eq. (4.5) to look at an initial transient change from the steady state condition, at velocity V, we must replace C_∞ and V in Eq. (4.5) by C_∞/k_0 and $\underline{\Delta V}$, respectively. This leads to

$$\frac{\Delta C}{C_\infty} \approx (1 - k_0)\frac{\underline{\Delta V} z_1}{D_{CL}} \approx 0.12$$

for $D_{CL} \approx 2 \times 10^{-5}$ cm^2 s^{-1}, where $\underline{\Delta V}$ is an average excess velocity (approximately 2/3 of the maximum) and ΔC is the excess solute content in the solid above the average value of C_∞. Because of the uncertainty in D_{CL}, one must consider this as excellent agreement between theory and experiment.

Case 4 ($D_{SC} = \delta \widetilde{\Delta G_0} = 0$, $V = oscillatory$): Using a time-dependent extension of the BPS model, Wilson[12] investigated the quantitative relationship between periodic variations in growth rate and the compo-

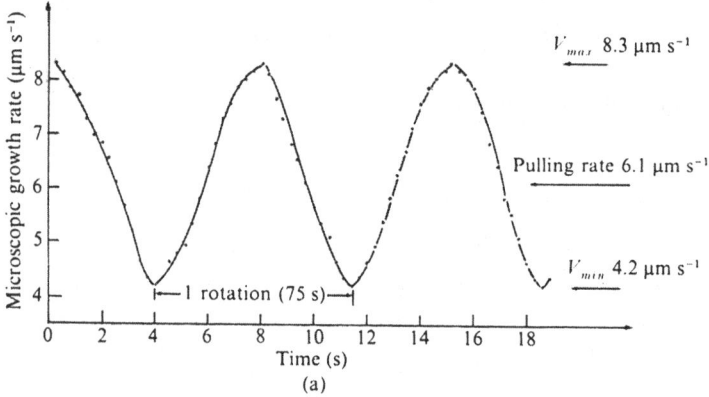

Fig. 4.16. Cross-section of a Te-doped InSb crystal grown with seed rotation and induced rate striations of constant frequency plus determined microscopic growth rates; (a) for thermal symmetry conditions which show that the variation in rate striation spacing reflects changes in the microscopic growth rate (decreased growth rate → closer spacing of striations). (Courtesy of H. Gatos.)

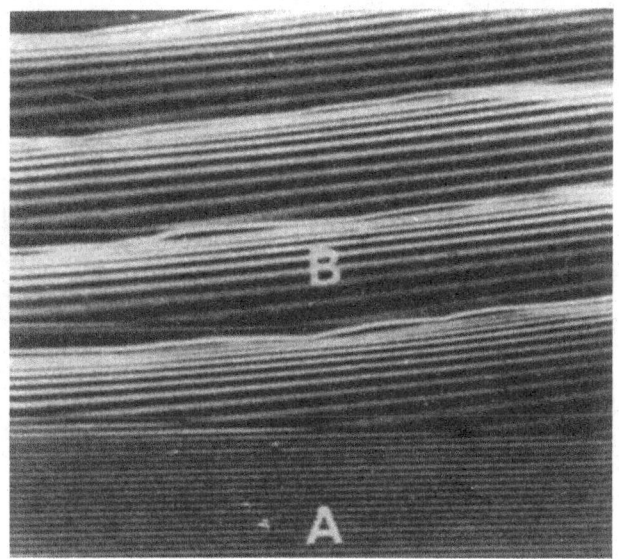

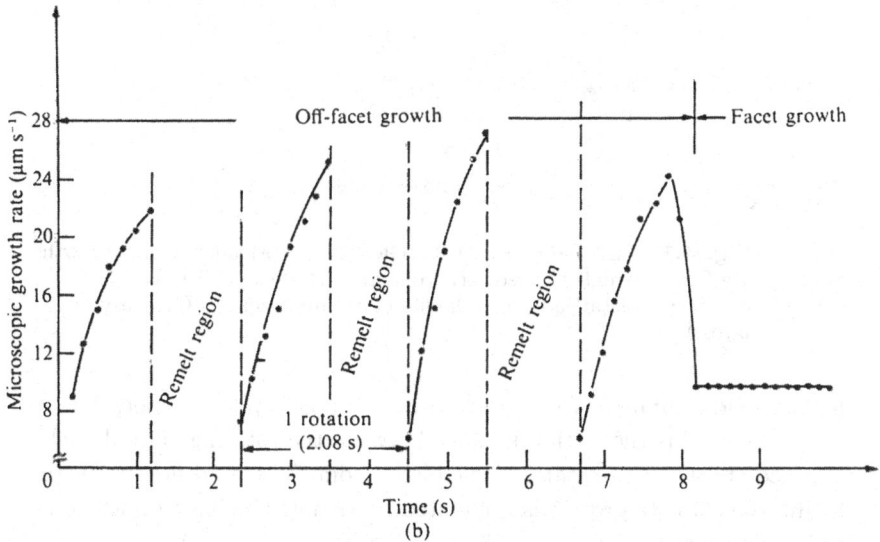

Fig. 4.16. (b) For pronounced thermal asymmetry (region B is with rotation while region A was grown without rotation). (Courtesy of H. Gatos.)

sitional inhomogeneities in crystals grown by the CZ technique. The Navier–Stokes equation, the continuity equation and the diffusion equation were all solved numerically. Two dimensionless constants appear:

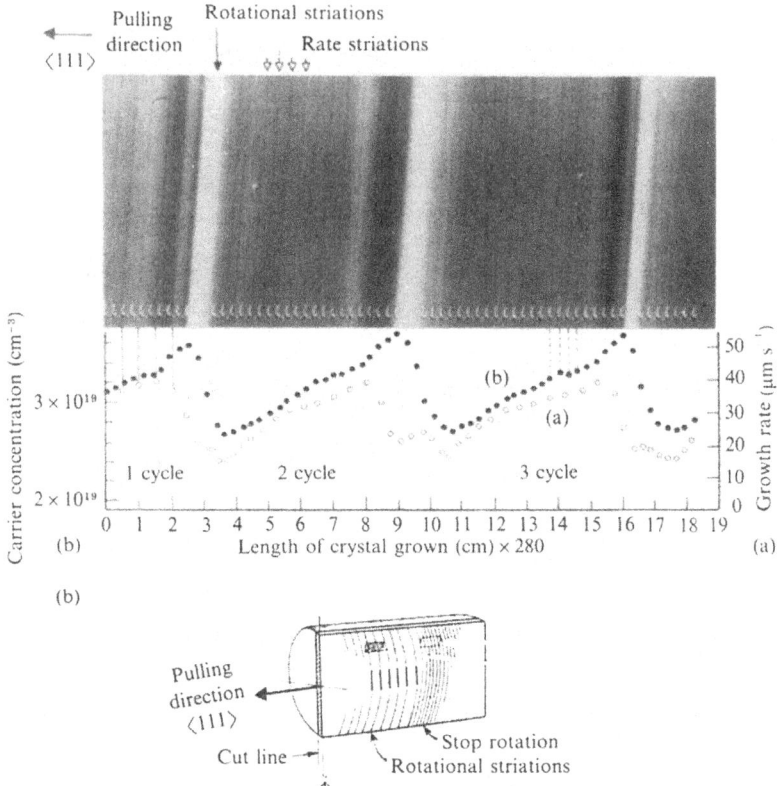

Fig. 4.17. CZ Ge crystal with Ga-doping: (a) microscopic growth rate from current-induced striations and (b) dopant distribution from spreading resistance measurements at 10 μm spacing. (Courtesy of H. Gatos.)

the Reynolds number, $Re = \omega d^2/\nu$ and the Schmidt number, $Sc = \nu/D_C$, where d is the initial distance between the rotating crystal/melt interface and the non-rotating crucible bottom. The fluid flow through the interface due to growth is accounted for by introducing a dimensionless suction parameter

$$a(t) = V(t)\omega^{-1/2}\nu^{-1/2} \tag{4.12}$$

where $V(t)$ denotes the (dimensional) growth rate. The growth rate is assumed to vary sinusoidally about the average growth rate with the same period as the crystal rotational period. Hence

$$a(t) = \bar{a}[1 + A\sin(\omega t)] \tag{4.13}$$

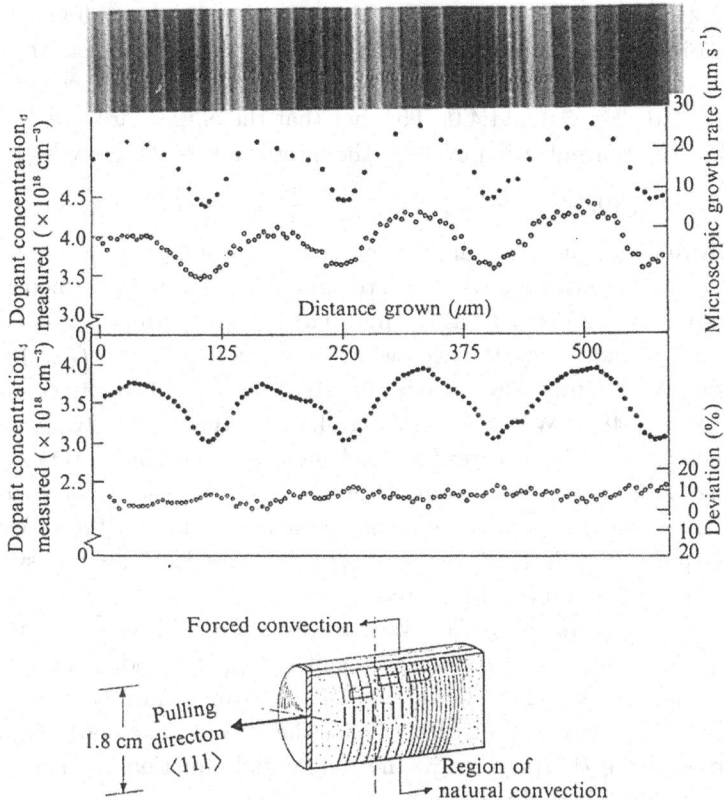

Fig. 4.18. CZ Si crystal with Sb-doping at $\omega = 5$ rpm: upper black dots give the microscopic growth rate; open dots give the corresponding dopant concentration; lower black dots give the theoretical dopant concentration (BPS) and the lower open dots give the deviation between theory and experiment. (Courtesy of H. Gatos.)

where the amplitude A may be less than or greater than unity (melt-back).

The dependence of the crystals' doping profile on Re is negligible as long as Re is sufficiently large, which it typically is in most crystal growth situations ($Re \gtrsim 10^3$). Fig. 4.19(a) indicates how the concentration profile $C(\xi, t)$ in the melt very near the interface changes with time. The dimensionless distance and time variables are $\xi = Re^{1/2}(\nu/\omega)^{1/2}z$ and $t = \omega t'$. Five individual concentration profiles, equally spaced in the time cycle, have been plotted with $n = 0, 1, \ldots, 4$ corresponding to the instantaneous growth rates $a_n = \bar{a}[1 + A\sin(2\pi n/5)]$. These results apply to the specific case of $k_0 = 0.01$, $\bar{a} = 0.03$, $A = 0.9$ and $Sc = 20$. In Fig. 4.19(b), the objects which resemble ellipses represent the concentration in the melt at the interface as a function of interface position for $A = 1.0$, 2.0, 3.0 and 4.0. The fact that the ellipses are tilted indicates that the concentration cycle at the interface lags the growth cycle; i.e.,

$$C(0,t) = \frac{1}{2}(C_{max} + C_{min}) + \frac{1}{2}(C_{max} - C_{min})\sin(t - \phi) \quad (4.14)$$

where C_{max} and C_{min} are the maximum and minimum values of the interface concentration while ϕ is the phase lag. For the parameter set corresponding to Fig. 4.19(a), we find that $C_{max} = 1.610$, $C_{min} = 1.318$ and $\phi = 0.58$ rad. Thus, the average concentration $\frac{1}{2}(C_{max}+C_{min}) = 1.464$, which is slightly lower than the steady state $(A = 0)$ condition of $C_{SS} = 1.488$. We note from Fig. 4.19(b), that $\bar{C} = \frac{1}{2}(C_{max} + C_{min})$ decreases, while ϕ increases, as A increases. For the parameter set of Fig. 4.19(a), the concentration profile in the crystal, $C(\ell)$, and the instantaneous rate profile, $\hat{a}(\ell)$, are presented in Fig. 4.19(c). Because of the phase lag effect, the concentration profile is decidedly non-symmetric about its maxima and minima.

These results are useful guidelines from a qualitative viewpoint; however, in a quantitative sense they suffer from the fundamental weakness associated with extension of the BPS treatment into the time domain (see Eqs. (3.9) and (3.10)). Thus, the effects quantitatively ascribed to only the $\partial C / \partial t$ term in the differential equation are really due to $\overset{\bullet}{q} + \partial C / \partial t$ where $\overset{\bullet}{q}$ is given by Eq. (3.9c).

Before closing this case, two features of CZ crystal growth that one generally ignores should be brought to light because they bear on the non-rotational striations. The first is that low frequency vibrations in the system can lead to surface waves on the top of the melt with wavelengths in the range $\sim 0.5 - 5$ cm for frequencies in the range $\sim 10^2 - 1$ Hz. Fluid velocity effects associated with such surface waves only penetrate ~ 1 wavelength into the fluid. Thus, a fluid motion effect from this source should be present at the solid/liquid interface, especially near the periphery of the crystal. It should modify the momentum boundary layer thickness in some non-periodic way depending on the surface wave character and thus lead to a class of chemical inhomogeneities in the outer regions of the crystal which will show up on etching as striations.

The second feature deals with the hydrodynamic analysis of fluid flow for a non-ideal, finite radius rotating disc (see Chapter 2). It has been found that the distribution of streamlines at a disc surface exhibits a special pattern, such as illustrated in Fig. 4.20. These streamlines move

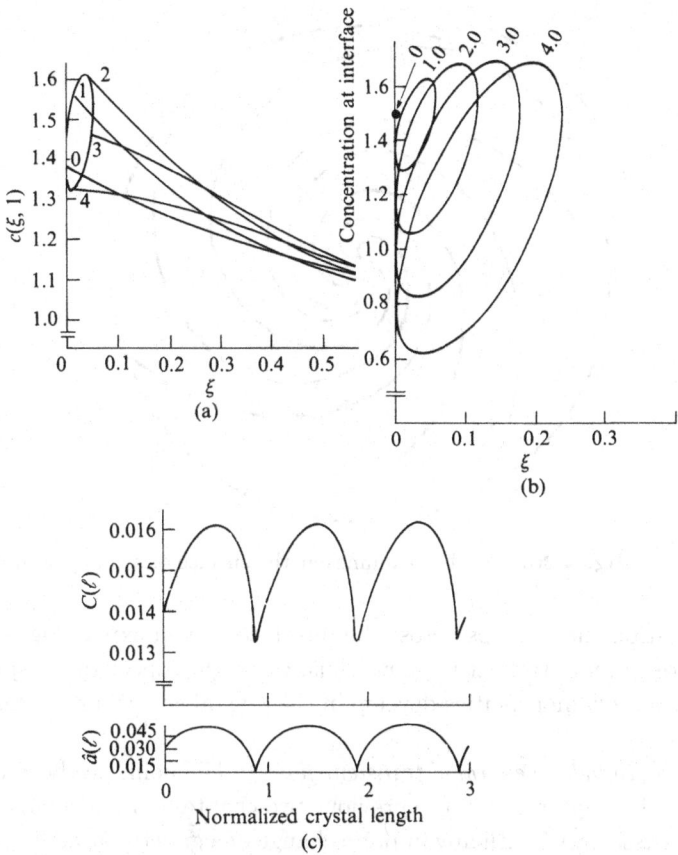

Fig. 4.19. (a) Concentration profiles in the melt near the interface as a function of the dimensionless distance, ξ, for five uniformly spaced instants of time in the growth cycle, (b) concentration in the melt at the interface as a function of interface position for $A = 0, 1.0, 2.0, 3.0$ and 4.0 and (c) the normalized concentration profile, $C(\ell)$, and the dimensionless growth rate profile, $\hat{a}(\ell)$, in the crystal as a function of the normalized crystal length, ℓ (distance grown in ℓ rotation cycles).[12]

across the surface and, for a flat untransforming disc, give a constant time-average momentum and solute distribution that is independent of r and ϕ. However, for a transforming interface, such as we have under consideration, we must consider that rotating spiral-shaped oscillations of the momentum and solute boundary layers occur to influence the local growth conditions. In addition, since the spiral arms of the solute distribution are more widely spaced as r increases, the concentration

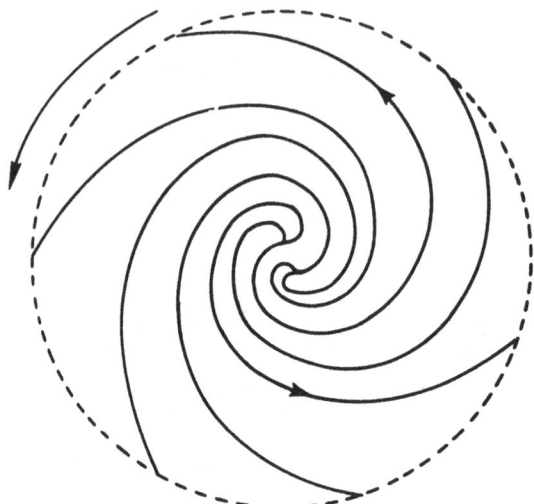

Fig. 4.20. Spiral streamlines at the surface of a rotating disc.

fluctuation that sweeps across the interface like an expanding wave has its largest amplitude at large r. Thus, we might expect to see spiral-like chemical inhomogeneities develop in the crystal as a result of this effect.

Case 5 (Impulse-generated transient flows): The point has been made in both this chapter and in the previous two chapters that a change in fluid flow is associated with any impulse change of a growth parameter such as ω or V. Further, because the normal flow situation is a mixture of forced and natural convection, we cannot theoretically know the magnitude of the convection parameter b (see Eq. (2.37)) or β'_u (see Eq. (3.10a)) and some experimental technique must be developed to reveal the magnitude of the appropriate parameter. One such technique is the "Solute Marker and Transfer Function Technique".[13] It involves using a well-known dopant like P in Si whose solute profile in the solid can be described by

$$C_S(t) = k(t)C_\infty(t) \qquad (4.15a)$$

where $k(t)$ is given from Eqs. (3.11b) and (3.10a), i.e.,

$$k(t) = \frac{k_0}{1 - (1 - k_0)\left(4/\{3 + [1 + (4D_L/V^2)\beta'_u(t)]^{1/2}\}\right)} \qquad (4.15b)$$

To deal with transient situations, it is assumed that the solute redistribution after an abrupt change of a growth parameter can still be given by Eqs. (4.15a) and (4.15b) while the response of the dopant incorporation to this change will enter $\beta'_u(t)$ through the following transfer function

$$\beta'_u(t) = \beta'_{u_i} + (\beta'_{u_f} - \beta'_{u_i})[1 - \exp(-t/\tau_R)] \qquad (4.15c)$$

where the subscripts i and f, respectively, represent the initial and final steady state values for β'_u and τ_R is the associated time constant. To illustrate the use of this technique, we shall consider two situations: (1) an abrupt change in crystal rotation rate, $\Delta\omega$, and (2) an abrupt change in crystal pull velocity, ΔV.

Fig. 4.21(a) gives a typical P longitudinal profile in a crystal of 40 mm diameter pulled at 64 mm hr^{-1} from an 8 kg melt where the rotation rate suffered the intentional abrupt transient from 25 rpm to 4 rpm and again later to 25 rpm where the length of each segment was over 60 mm. No meaningful change in the crystal diameter was observed. Here, the locations at which the step changes in ω were applied are indicated by the arrows. We note from Fig. 4.21(a) that, despite having some fluctuations due to the time-dependent melt instability, the transient response in this C_S^P profile from one constant condition growth section to another is clearly revealed and the transition time, τ_R, is very small. Fig. 4.21(b) is the corresponding bulk melt P profile, C_∞^P, calculated on the basis of mass conservation. Using Figs. 4.21(a) and (b) and Eqs. (4.15a) and (4.15b), the corresponding $\beta_u = D_L\beta'_u/V^2$ profile can be determined and it is plotted in Fig. 4.21(c). It is found that, for $\omega = 25$ rpm, the average value of β_u is about 1560 while, for growth under $\omega = 4$ rpm, the average β_u is about 110. For these values of V and D_L, the magnitude of β'_u is about a factor of 10 lower than this and natural convection is expected to be relatively strong.

Fig. 4.22(a) shows a typical longitudinal P transient profile for a crystal grown under $V = 25 \to 250 \to 25$ mm hr^{-1} abrupt pull rate transitions. The crystal diameter decreased monotonically after the abrupt increase in V and then increased again after the abrupt decrease in V. Following the usual procedure, C_∞^P was calculated and β_u was determined with the result indicated in Fig. 4.22(b). Because we are dealing with a change in V, here, β_u and β'_u are no longer simply proportional to each other so we need to plot $\beta_u V^2 = D_L\beta'_u$ to see the transient behavior in β'_u. This is presented in Fig. 4.22(c) indicating that β'_u changes from a value ~ 3–4 s^{-1} at $V = 25$ mm hr^{-1} to ~ 400 s^{-1} at $V = 250$ mm hr^{-1} ($D_L \sim 2 - 3 \times 10^{-5}$ cm^2 s^{-1}) with a long time constant, τ_R. This very large melt convection change arises because of the large temperature redistribution near the interface and indicates that natural convection is the dominant flow mechanism during this CZ growth of Si.

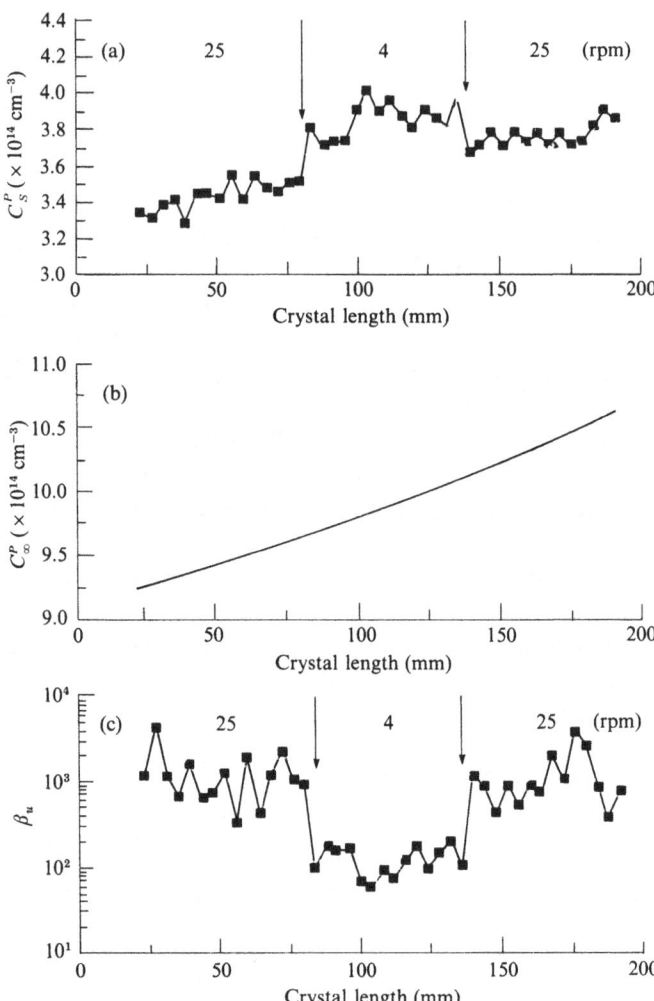

Fig. 4.21. Longitudinal (a) P solid concentration, (b) P liquid concentration and (c) the corresponding β_u profile along the axis of a crystal which was grown under $\omega = 25 \to 4 \to 25$ rpm abrupt crystal rotation transitions.[13]

This study tends to substantiate the view that the Cochran analysis should not be used to describe quantitatively fluid flow during CZ crystal growth.

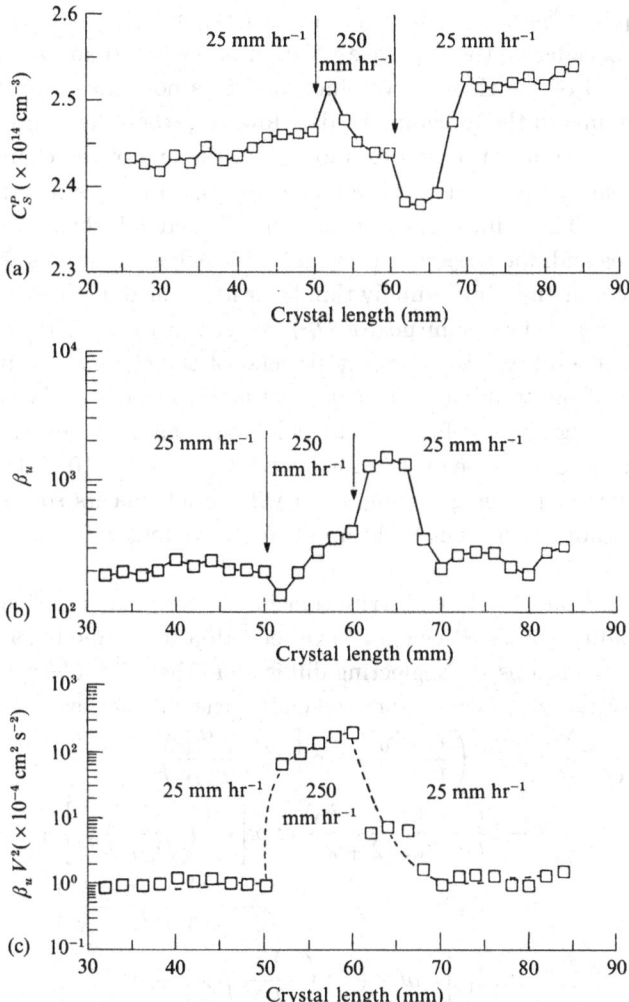

Fig. 4.22. Longitudinal (a) P solid concentration, (b) the corresponding β_u and (c) the corresponding $\beta_u V^2$ profiles along the axis of a crystal which was grown under $V = 25 \rightarrow 250 \rightarrow 25$ mm hr^{-1} abrupt crystal growth velocity transitions.[13]

4.3   Terminal transient (V = constant, planar front, no mixing)

The foregoing has dealt with the freezing of a semiinfinite channel of liquid whereas, in reality, one is always concerned with finite systems. So long as the interface position is greater than $\sim 10D_{CL}/V$ from the end of the channel, the interface does not sense the finiteness of the system and the foregoing holds. However, when the solute boundary layer in the liquid begins to impinge on the end of the channel, the average concentration in the layer must begin to rise as illustrated in Fig. 4.23(a). The solute concentration in the solid will then rise to very large values and, for $D_{CS} = 0$, it eventually reaches eutectic proportions (even if only in an infinitesimally thin layer for some dilute cases). Thus, depending upon the magnitude of D_{CL}/V, either a thin or thick solute-rich zone of solid will be found at the end of the channel. A practical utilization of the terminal transient effect is to be found in the equiaxed zone of castings or ingots or in a weld joint. There, when the different crystal interfaces begin to grow closer together ($\sim 10^{-2}$–10^{-3} cm), the presence of the neighboring grain will be felt via its solute profile and the terminal transient build-up of solute content will occur at both interfaces.

The quantitative solute distribution in the terminal transient liquid can be readily evaluated using a wave reflection technique to satisfy the boundary conditions.[3] Neglecting diffusion in the solid, $C_S(z_2)$ is given in terms of the distance z_2 from the end of the channel by

$$
\begin{aligned}
\frac{C_S(z_2)}{C_\infty} =& 1 + 3\left(\frac{1-k_0}{1+k_0}\right)\exp\left[-2\left(\frac{V}{D_{CL}}\right)z_2\right] \\
& + 5\frac{(1-k_0)(2-k_0)}{(1+k_0)(2+k_0)}\exp\left[-6\left(\frac{V}{D_{CL}}\right)z_2\right] + \cdots \\
& \cdots + (2n-1)\frac{(1-k_0)(2-k_0)\cdots(n-k_0)}{(1+k_0)(2+k_0)\cdots(n+k_0)} \\
& \times \exp\left[-n(n+1)\left(\frac{V}{D_{CL}}\right)z_2\right] + \cdots
\end{aligned}
\tag{4.16a}
$$

If $k_0 < 1$, this series diverges to infinity for $z_2 = 0$. For all other values of z_2, the series converges and does so quite rapidly for the larger values of z_2. If $k_0 > 1$, the series converges slowly even for $z_2 = 0$. As in the case of the initial transient, Eq. (4.16a) contains the dimensionless quantity Vz_2/D_{CL} as a natural variable. In Figs. 4.23(b) $C_S(z_2)/C_\infty$ is plotted versus Vz_2/D_{CL} for several values of the parameter k_0.

Since theory predicts that a second phase will always be produced

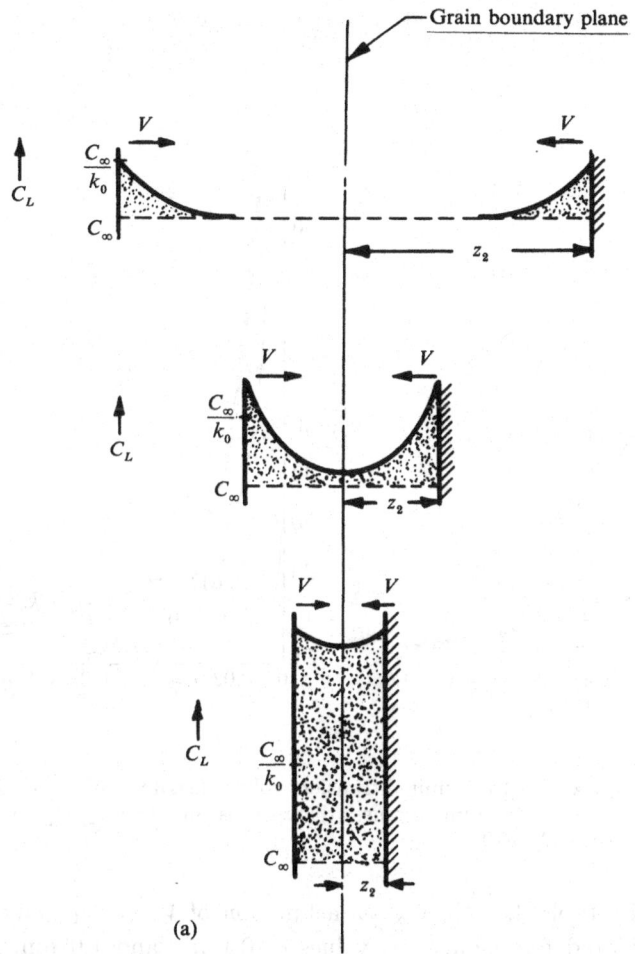

(a)

Fig. 4.23. (a) Representation of concentration build-up in the liquid as two grains closely approach each other.

provided diffusion in the solid is neglected, even if only in an infinitesimally thin layer between the grains, then it is the thickness of this layer that will be of interest to us. Below a certain thickness, the second phase, if it nucleates, will not remain stable as a continuous film but will break up into small sphere-like particles for surface energy reasons. By considering Eq. (4.16a), it can be shown that, for $V z_2 / D_{CL} \gtrsim 0.1$, the concentration of the liquid is almost constant. Therefore, if a second phase begins to form at a certain z_2, the width of the second phase will

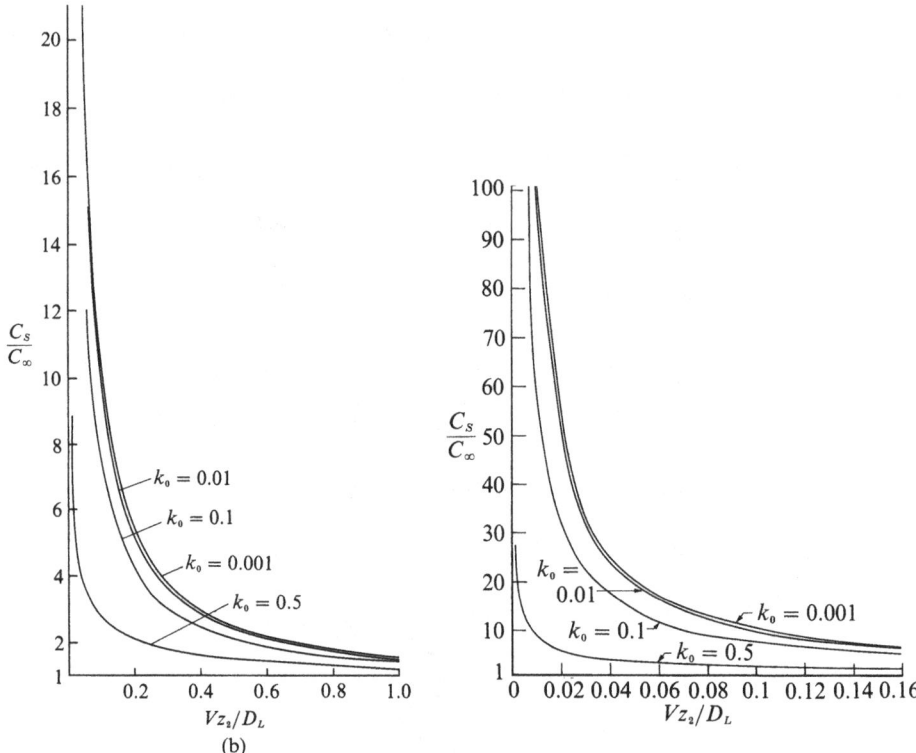

Fig. 4.23. (b) Terminal transient solute distribution in the solid as a function of the dimensionless distance parameter Vz_2/D_L for several k_0 with $D_S = 0$.

be $2z_2$. In Table 4.1, C_S/C_∞ as a function of V and k_0 values have been tabulated for nominal z_2 values of 0.1 μm and 1.0 μm. These calculations assume that the grain size is equal to or larger than that required for a steady state solute build-up before solute boundary layer impingement. This steady state size, d_{ss}, is also given in Table 4.1.

From Table 4.1, we see that C_∞ can be several orders of magnitude below the solid solubility limit and still a continuous film of second phase can form at the boundary, especially for $k_0 \ll 1$. Table 4.2 presents approximate k_0 values for a number of solutes in delta Fe, Mo and Ti. From this table, we can expect that the solidification-induced build-up of O and S at the Fe grain boundaries will be large so that stable oxides and sulfides will form even when the initial concentrations of these elements in the melt are very small.

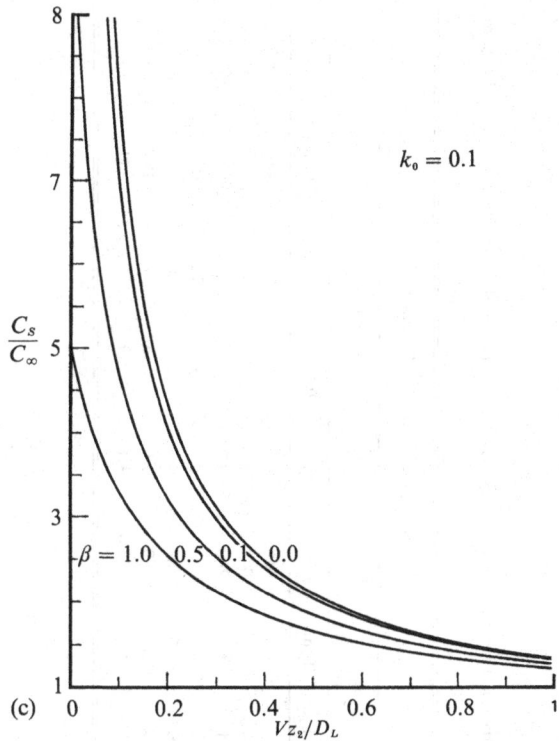

Fig. 4.23. (c) Comparison of terminal transient solute distributions for $k_0 = 0.1$ and $\beta = D_S/D_L = 0, 0.1, 0.5$ and 1.0 as a function of $\xi_2 = V z_2/D_L$.

The formation of FeS at the grain boundaries in Fe usually occurs as a continuous film which is deleterious to the mechanical properties of the Fe. Such a grain boundary film may be eliminated by the addition to the melt of an appropriate element that produces a more stable sulfide than FeS; i.e., the elements Mn, Ti or Zr. It has been found experimentally that Mn additions are much more effective in eliminating the FeS grain boundary films than equal concentration additions of Ti or Zr even though Ti and Zr form thermodynamically more stable sulfides than does Mn. The reason for this is that, for the same C_∞, more Mn is partitioned to the boundary region than either Ti or Zr because it has a much smaller k_0 value (see Table 4.2). From Table 4.1, $V = 1$ mm min^{-1} and $z_2 = 1$ μm gives $C_S(z_2)/C_\infty = 400$ and 16 for $k_0 = 0.1$ and 0.5, respectively. Thus, for the same C_∞, in the melt, about 20 times more Mn than either Ti or Zr will be segregated to the as-solidified grain

Table 4.1. *Grain size required for steady-state growth d_{ss} and relative solute segregation. $C_S(z_2)/C_\infty$ at 0.1 μm and 1 μm from the grain boundary as a function of V and k_0 ($D = 5 \times 10^{-5}$ cm s^{-1})*

V (mm min^{-1})	d_{ss}, cm				$C_S(z_2)/C_\infty$ at $z_2 = 0.1$ μm and 1 μm							
	$k_0 = 0.5$	$k_0 = 0.1$	$k_0 = 0.01$	$k_0 = 0.001$	$k_0 = 0.5$		$k_0 = 0.1$		$k_0 = 0.01$		$k_0 = 0.001$	
					0.1 μm	1 μm	0.1 μm	1 μm	0.1 μm	1 μm	0.1 μm	1 μm
1	6×10^{-2}	3×10^{-1}	3	30	50	16	> 1000	400	> 1000	1000	> 1000	> 1000
6	10^{-2}	5×10^{-2}	5×10^{-1}	5	18.5	6	1000	32	> 1000	45	> 1000	55
12	5×10^{-3}	2.5×10^{-2}	2.5×10^{-1}	2.5	14	4	300	17.5	800	23	1000	25
24	2.5×10^{-3}	1.2×10^{-2}	1.2×10^{-2}	1.2	10	3	150	9	200	12	300	13
48	1.2×10^{-3}	6×10^{-3}	6×10^{-3}	6×10^{-1}	6.5	2.5	38	4.5	66	6	72	6.5
240	2.5×10^{-4}	1.2×10^{-2}	1.2×10^{-1}	1.2×10^{-1}	3	1.3	9	1.6	12.5	1.7	13.5	1.8
1200	5×10^{-5}	2.5×10^{-4}	2.5×10^{-3}	2.5×10^{-2}	1.6	1	2.5	1.1	2.9	1.2	3.0	1.3

Table 4.2. *Distribution coefficient k_0 for various solutes in Fe, Mo, and Ti.*

Solute	k_0 in Fe	k_0 in Mo	k_0 in Ti
P	0.2		
Mn	0.15		0.3
Mo	0.7		
C	0.25	0.1	~ 3
O	0.1		~ 6
S	0.002		
Aℓ	0.6	0.3	0.4
Si	0.7	0.3	
Be		0.01	
B		0.02	
W		1.1	2
Fe			
Ti	0.6		
Zr	0.5		0.4

boundaries. Of course, in a real ingot, the interface will not be planar but will probably be of a dendritic morphology yielding C_i (tip) $< C_i$ (planar) (see Fig. 4.34(b)) so that somewhat less solute will be deposited at the grain boundary for $V =$ constant (for $V = \alpha t^{-1/2}$, which applies here, there is a compensating effect).

A general rule concerning the replacement of a sulfide film or any other compound at a grain boundary by a thermodynamically more stable sulfide or compound may be formulated. The addition element that will be most effective is that compound-forming element having the smallest value of k_0 in the particular solvent under consideration (e.g., Mg in Ni for sulfide formation). This rule can be broadened and applied to the replacement of any deleterous grain boundary phase by a more desirable phase.

Since the rate of solute build-up is so large for the terminal transient, transport in the solid may be expected to alter the distribution in some cases. Thus, it is useful to evaluate the effect of solid state diffusion on $C_S(z_2)$. This can be done by including a $D_{CS}\, \partial C_S / \partial z_2$ contribution to

Table 4.3. *Comparison "half-area" values for solute distributions.*

Distribution coefficient	Steady state $(V/D)z'$	Initial transient $(V/D)z$	Terminal transient $(V/D)z_2$
0.5	0.693	1.4	0.8
0.1	0.693	7.6	< 0.01
0.01	0.693	67.0	< 0.01
0.001	0.693	670.0	< 0.01

the interface conservation condition leading to the solution

$$\frac{C_S(z_2)}{C_\infty} = 1 + \sum_{n=1}^{\infty} (2n+1) \frac{\prod_{m=0}^{n}[m - k_0 - m(m+1)(d_s D_{CS}/d_L D_{CL})k_0]}{\prod_{m=0}^{n}[m + k_0 + m(m+1)(d_s D_{CS}/d_L D_{CL})k_0]}$$
$$\times \exp[-n(n+1)V z_2/D_{CL}] \qquad (4.16b)$$

The effect of D_{CS} is shown in Fig. 4.23(c) for $k_0 = 0.1$ with $(d_s D_{CS}/d_L D_{CL}) = 1$, 0.5, 0.1 and 0. A simpler way to determine if solute transport in the solid is important is to think of the solute in the solid as having an effective drift mobility, $\hat{M}_{eff}$, due to the concentration gradient where $\hat{M}_{eff}$ is determined by equating fluxes so that $\hat{M}_{eff} \sim (D_{CS}/C_S)(\partial C_S/\partial z)$.[1] When $\hat{M}_{eff} << V$, the solid diffusion effect can be neglected but, when $\hat{M}_{eff} \tilde{>} V$, the effect is significant and Eq. (4.16b) rather than Eq. (4.16a) must be utilized.

It is frequently desirable to produce, by solidification from the melt, a specimen that is of uniform concentration. If this is to be accomplished, all transient regions must be avoided or subsequently removed from the solid sample. Consequently, it is of real interest to know the approximate dimensions of the various transient zones. As a basis for comparison of the various transient characteristic lengths, the distance measured from the beginning to the point at which half of the transient solute change has occurred may be used. For example, in the case of the initial transient, the area between the $C_S(z_1)$ curve and the $C_S = C_\infty$ level represents the solute deficiency in the transient region. The value of $V z_1/D_{CL}$ corresponding to one-half of this area will be used as a measure of the transient extent. A listing of these "half-area" values is given in Table 4.3 for the three important transients.

4.4   Transient velocity (planar front, no mixing)

Consider an initially superheated melt of constant temperature in the form of a long cylinder with planar ends and place a heat sink at one end, $z = 0$ at time $t = 0$. If the temperature at $z = 0$ remains constant for all subsequent times, the liquid will cool and crystallize from this end such that the massive solid/liquid interface at $z = Z_I$ will propagate down the cylinder according to the law $Z_I = 2\alpha t^{1/2}$ where α has values in the range 10^{-2}–0.5 cm s$^{-1/2}$ for metallurgical systems but in the range 10^{-6}–10^{-3} cm s$^{-1/2}$ for geological systems. This simple relationship holds provided that (a) the interface state variables (see Chapter 1) stay relatively constant and (b) $\Delta G_K << \Delta G_{sv}$. When interface attachment kinetics and changing interface state variables are taken into account, the crystallization velocity will be of the general form

$$V = \alpha t^n \tag{4.17}$$

where n may take on different values during the evolution of the crystal. It is generally at large times that the transport-limited growth occurs and $n \to -1/2$ in Eq. (4.17).

In metallurgical systems, just as in geological systems, the heat extraction conditions at $z = 0$ of the very long cylinder, or the heat input to the large mass of liquid from which our crystal is growing, can change and often does so rather abruptly. For the most general situation, one must be prepared to consider cases with $n \lessgtr 0$ in Eq. (4.17) and V may even be cyclical in character.

If we begin with $n = -\frac{1}{2}$, the mathematical solution to the solute redistribution equation yields the qualitative results shown in Fig. 4.24 for C_S/C_∞ and C_L/C_∞ as a function of interface position, Z_I. In the liquid, the interface concentration is constant for all time and, as Z_I increases, more and more solute is partitioned at the interface and carried in the solute boundary layer. The effective thickness of the layer is $\sim D_{CL}/V$ and increases from ~ 2 μm to ~ 200 μm as Z_I increases from ~ 0.1 cm to ~ 10 cm for $\alpha \approx 10^{-1}$ cm s$^{-1/2}$ and $D_{CL} \approx 10^{-5}$ cm^2 s^{-1} (metallurgical values) and from ~ 2 μm–2 cm to 200 μm–200 cm for similar values of Z_I when $\alpha \approx 3 \times 10^{-5}$ cm s$^{-1/2}$ and $D_{CL} \approx 10^{-8}$–10^{-12} cm^2 s^{-1} (geological values). Because the interface concentration in the liquid is constant with Z_I, that frozen into the crystal will be constant along the length of the crystal but reduced in magnitude by the factor k_0. In the limit of very small α, the concentration in the solid will approach $k_0 C_\infty$ whereas, at very high α, C_S will approach C_∞.

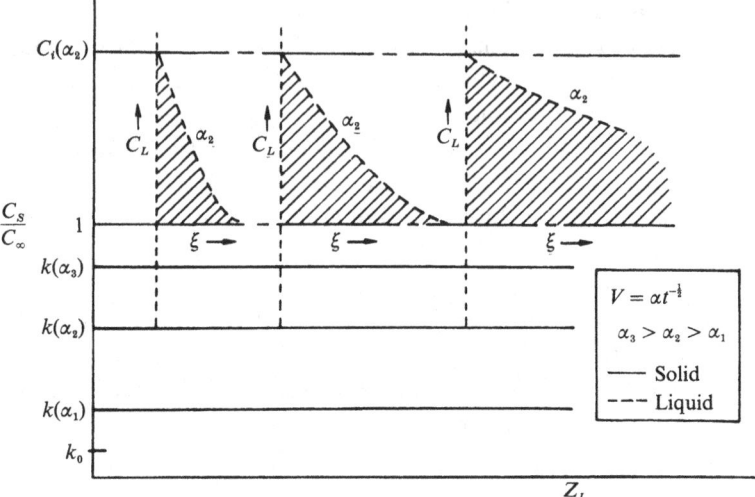

Fig. 4.24. Plot of relative solute distribution in the solid, C_S/C_∞, for three values of α plus the solute distribution in the liquid, C_L, with $\alpha = \alpha_2$ at three different interface positions, Z_I, for the $n = -\frac{1}{2}$ case.

The reason for this remarkable solute redistribution behavior is that normal diffusion scaling occurs via the relationship $z^2 = D_{CL}t$ so that solute isoconcentrates in the liquid can transport at the same rate as the interface motion (Eq. (4.17) with $n = -1/2$) once a certain interface concentration has been set. The larger is α, the steeper must this interface gradient be for stationary state diffusion so the larger C_i must be ($\hat{M}_{eff} \equiv (D_{CL}/C_i)(\partial C_L/\partial z)_0 \equiv V$).

The solute distribution in the liquid for $n = -1/2$ is given by

$$\frac{C_L}{C_\infty} = \left(\frac{1 - (1 - k_0)\left(\pi\alpha^2/D_{CL}\right)\exp\left(\alpha^2/D_{CL}\right)}{1 - (1 - k_0)\left(\pi\alpha^2/D_{CL}\right)^{1/2}\exp\left(\alpha^2/D_{CL}\right)\operatorname{erfc}\left(\alpha^2/D_{CL}\right)} \right.$$

$$\left. \times \left\{ \operatorname{erfc}\left[\left(\alpha^2/D_{CL}\right)^{1/2} z/Z_I\right] - \operatorname{erfc}\left(\alpha^2/D_{CL}\right) \right\} \right) \qquad (4.18a)$$

where $z/Z_I \geq 1$. From Eq. (4.18a), C_i is given by

$$\frac{C_i}{C_\infty} = \left[\frac{1}{1 - (1 - k_0)\left(\pi\alpha^2/D_{CL}\right)^{1/2}\exp\left(\alpha^2/D_{CL}\right)\operatorname{erf}\left(\alpha^2/D_{CL}\right)} \right]$$

$$\qquad (4.18b)$$

while $C_S/C_\infty = k_0 C_i/C_\infty$ is plotted in Fig. 4.25 as a function of the parameter $(\alpha^2/D_{CL})^{1/2}$ for several k_0.

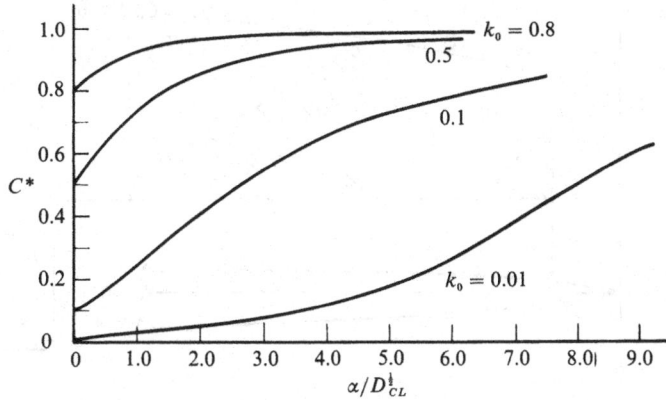

Fig. 4.25. Plot of the dimensionless parameter, $C^* = C_i/(C_\infty/k_0)$, versus the parameter $(\alpha/D_{CL}^{1/2})$ for several values of k_0.

Let us now extend our consideration to the range $n = -\frac{1}{2} + \Delta$ (let Δ be small) with all other conditions unchanged. If $\Delta < 0$, then the solute diffusion in the liquid outstrips the interface motion and C_i must therefore decay as Z_I increases even though the total amount of solute carried in the boundary layer increases. If $\Delta > 0$, then the solute diffusion cannot keep pace with the interface motion and C_i must therefore continually increase towards C_∞/k_0 at large Z_I. Since $C_S = k_0 C_i$, C_S must correspondingly decrease or increase with Z_I for $\Delta < 0$ or $\Delta > 0$, respectively. These variations in C_S are illustrated in Fig. 4.26.

Just as in the $n = 0$ case, if an abrupt rate change occurs at time t_1 for the $n = -1/2$ case, V is changed from $\alpha_1 t_1^{-1/2}$ to $\alpha_2(t - t_1)^{-1/2}$ and Z_I becomes $Z_{I_1} + 2\alpha_2(t - t_1)^{1/2}$. If we choose $\alpha_2 > \alpha_1$, solute is being partitioned at the interface at a rate faster than it can diffuse away during the interface advance so that C_i will increase towards its new plateau as illustrated in Fig. 4.27(b). Unlike the $n = 0$ case, it probably does not overshoot the new plateau. For $\alpha_2 < \alpha_1$, the inverse behavior to Fig. 4.27(b) is expected.

Examples of solute distribution changes like those shown in Fig. 4.27 are to be readily found in the geological literature on "zoning". In Fig. 4.28, a plagioclase example, produced in the laboratory at 6.1 kbar by cooling in steps of 50 °C from 1050 °C, provides a striking record of these rate change steps.[14] The clear parallel demarcation lines associated with the rate change steps outline the successive positions of the interface to give a built-in record of its growth history. Of course, it is important

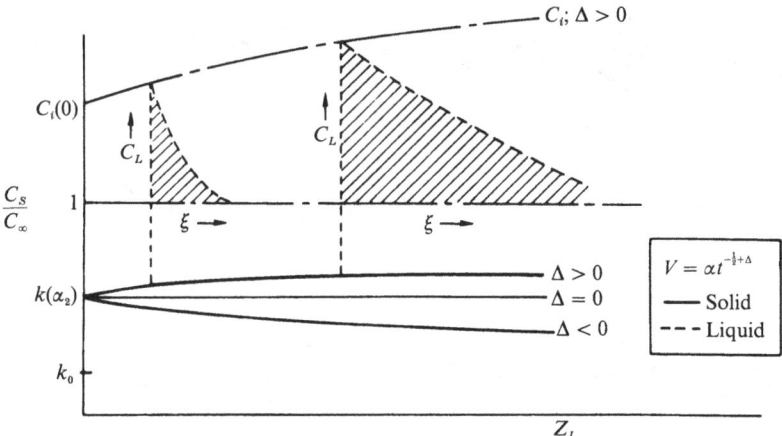

Fig. 4.26. Relative solute distribution in the solid, C_S/C_∞, and in the liquid at the interface, C_i/C_∞, as a function of interface position, Z_I, for $n = -\frac{1}{2} + \Delta$ with $\Delta \gtrless 0$.

to recognize that, in systems such as these, chemical adsorption effects or pressure fluctuations at the interface may also have contributed to the detailed solute distribution changes. In Fig. 4.28, it appears as if V was decreasing during the subsequent steps ($\alpha_3 < \alpha_2 < \alpha_1$) which is consistent with the decreasing degree of color detected in the growth layers associated with the later steps ($C_{S3} < C_{S2} < C_{S1}$).

4.4.1   LPE approximation

During the growth of crystal layers via the liquid phase epitaxial (LPE) technique, the layer growth rate is directly proportional to the flux of the major solute to the interface from solution. The growth of a GaAs layer on a GaAs substrate is usually carried out from a Ga-rich melt via one of three cooling procedures: (i) equilibrium cooling, (ii) step cooling or (iii) supercooling.[15] Starting with a melt concentration whose equilibrium temperature is T_L, the melt is cooled at some rate $\dot{T}$. The arrows in Fig. 4.29 indicate the times at which the melt is placed in contact with the substrate. For equilibrium cooling, one assumes interface equilibrium from time $t = 0$ as the cooling proceeds; for step cooling, one quenches the substrate an amount ΔT_0 at $t = 0$ and assumes interface equilibrium at all subsequent times while the substrate temperature remains constant; for supercooling, an initial quench

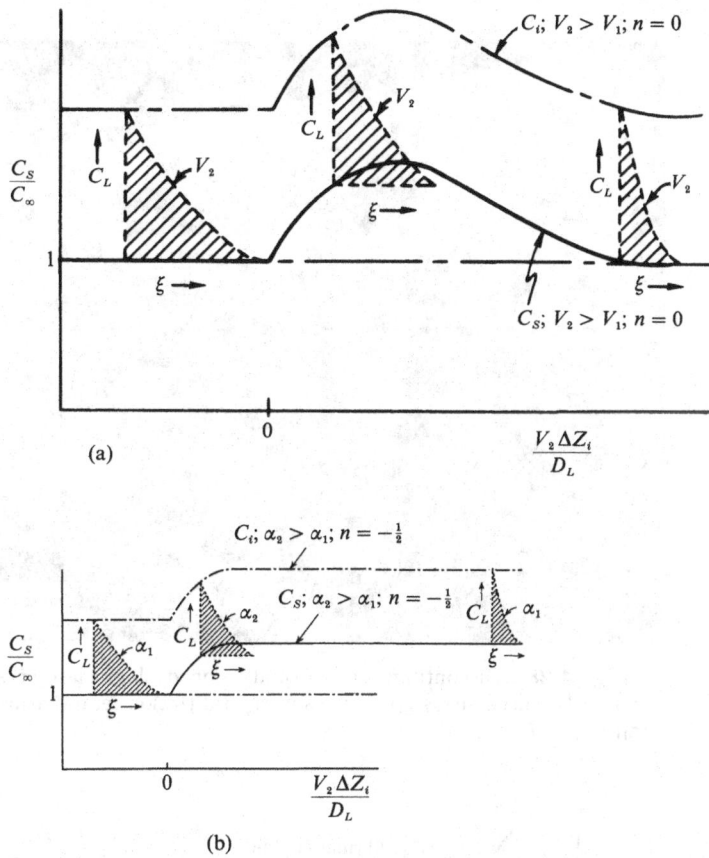

Fig. 4.27. Relative solute distribution in the solid, C_S/C_∞, and in the liquid at the interface, C_i/C_∞, as a function of $V_2\Delta Z_I/D_L$; as a result of a unit pulse rate change (a) from V_1 to $V_2 > V_1$ for the $n = 0$ case and (b) from α_1 to $\alpha_2 > \alpha_1$ for the $n = -\frac{1}{2}$ case.

step occurs but then the melt and substrate temperatures decrease at $\dot{T}$ while interface equilibrium is assumed at all times.

For the mathematical analysis of these three cases, one recognizes that the small Péclet number condition obtains; i.e., $VZ_I/2D_{CL} \ll 1$, so the movement of the interface can be neglected in the solution of the diffusion equation. This greatly simplifies the analysis because the interface can be considered to be at $Z_I = 0$ for purposes of determining the matter flux. For a semiinfinite melt and constant liquidus slope, m,

Fig. 4.28. Discontinuously, normally zoned plagioclase crystal produced by successive step changes in crystallization temperature ($\Delta T = 50\,^{\circ}$C).[14]

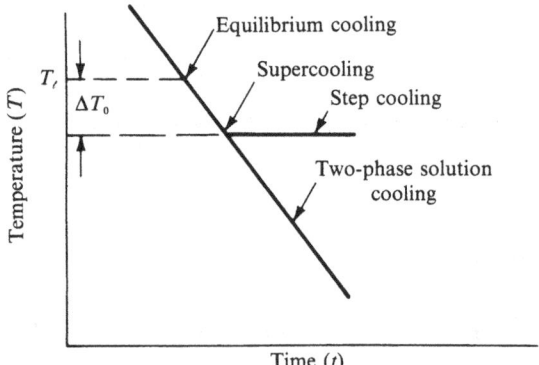

Fig. 4.29. Solution cooling procedure for four different LPE growth techniques. The arrows indicate the initial contact time between substrate and solution.[15]

$Z_I(t)$ for these three cases is given by

   (i) equilibrium cooling,

$$Z_I = (2/C_S m)(D_C/\pi)^{1/2}(2/3)\dot{T}t^{3/2} \qquad (4.19a)$$

   (ii) step cooling,

$$Z_I = (2/C_S m)(D_C/\pi)^{1/2}\Delta T_0 t^{1/2} \qquad (4.19b)$$

   (iii) supercooling

$$Z_I = (2/C_S m)(D_C/\pi)^{1/2}(\Delta T_0 t^{1/2} + (2/3)\dot{T}t^{3/2}) \qquad (4.19c)$$

Here, C_S is the concentration of solute in the epi layer material. The units for C_S are atoms of solute per unit volume of grown solid while the units for m are degrees per atom of solute per unit volume of solution. Agreement between these theoretical predictions and the experimental results are quite good as can be seen from Fig. 4.30.[15]

For a melt of finite thickness, as layer growth proceeds, the solute is depleted from the melt so a continually rising V cannot be sustained during equilibrium cooling. For this case, we expect V to rise to a maximum value and then decay as cooling continues. With a more realistic representation of the liquidus curve

$$C_L = K \exp[-(\Delta H/\mathcal{R}T_L)] \qquad (4.20a)$$

where K and ΔH are constants, the solute concentration at the growth interface will be given by

$$C_i(t) = C_0 \exp(-t/\tau) \qquad (4.20b)$$

where C_0 is C_L corresponding to T_0 and $\tau = T_0^2 \dot{T}^{-1}(\Delta H/\mathcal{R})^{-1}$. The interface position, Z_I, for growth from a solution of thickness ℓ is given by

$$Z_I = \frac{8\Delta C_I \ell}{\pi^2 C_S} \sum_n \frac{1 - \exp[-(\lambda_n^2 t/\tau)]}{(2n+1)^2}$$

$$+ \frac{C_0 \ell}{C_S} \left\{ \frac{\tan\delta}{\delta}[1 - \exp(t/\tau)] - \frac{2}{\delta^2} \sum_n \frac{1 - \exp[-(\lambda_n^2 t/\tau)]}{\lambda_n^2(\lambda_n^2 - 1)} \right\}$$

$$(4.20c)$$

where $\Delta C_I = C_I - C_0$, $\delta = \ell/(D_C\tau)^{1/2}$, $\lambda_n = \frac{1}{2}(2n+1)\pi/\delta$ and C_I is the initial concentration at the start of growth. For $(D_C t)^{1/2} \ll \ell$, the case for a semiinfinite solution, Eq. (4.20c) reduces to Eq. (4.19c) with $1/m$ replaced by dC_L/dT_L. For $(D_C t)^{1/2} \gg \ell$, the thin solution case, essentially all of the excess solute in the melt for a certain time interval

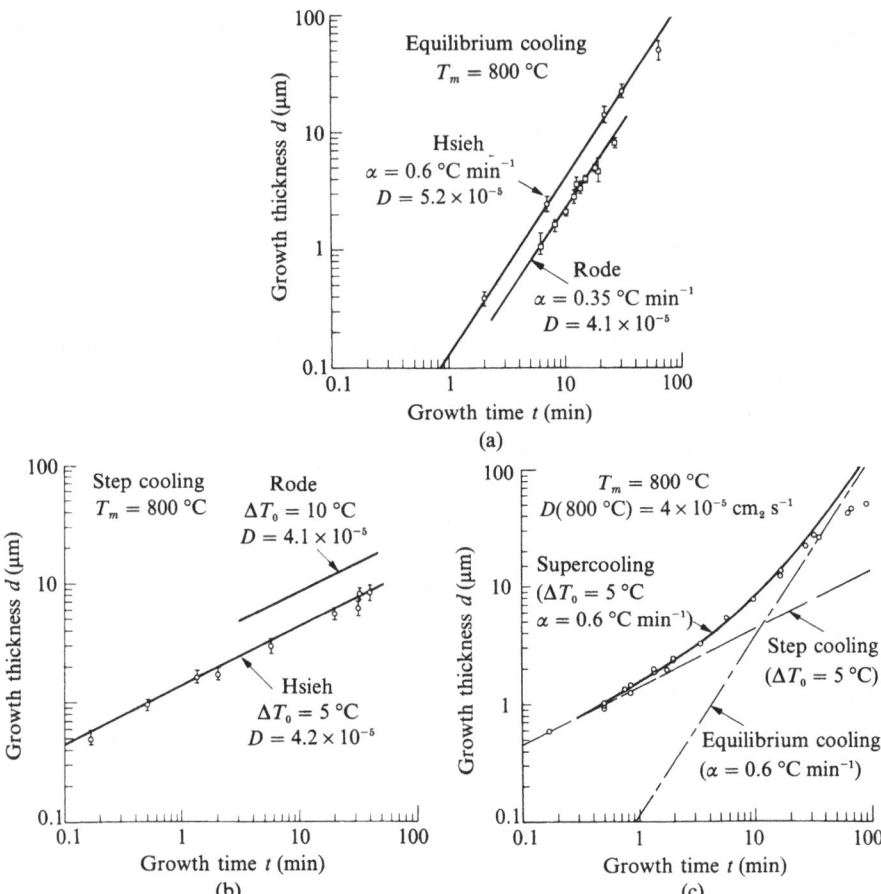

Fig. 4.30. Thickness, d, as a function of growth time t, for GaAs layers grown by (a) the equilibrium cooling technique from two different investigators, (b) the step cooling technique and (c) the supercooling technique.[15]

Δt becomes incorporated into the epi layer. Thus, with constant $\dot{T}$ as t increases, because of the liquidus curvature, less solute excess is available in the interval Δt so V must decrease.

In this general area of slowly varying interface velocity, it is useful to also consider the Laplace approximation; i.e., neglect $\partial C/\partial t$ in the diffusion equation and let the coefficients of the resulting mathematical solution be time-dependent. In Fig. 4.31, the rigorous and Laplace approximation solutions for equilibrium cooling are compared for a number

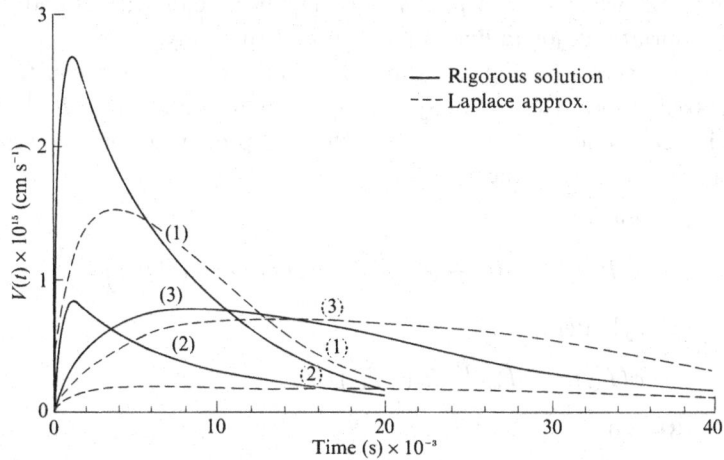

Fig. 4.31. Comparative LPE film growth velocity, V, as a function of growth time, t, for the rigorous solution and the Laplace solution with $T_0 = 850\,°C$ and (1) $\dot{T} = 200\,°C\ hr^{-1}$, $D_L/\ell = 3 \times 10^{-5}$ cm s^{-1}; (2) $\dot{T} = 200\,°C\ hr^{-1}$, $D_L/\ell = 3 \times 10^{-6}$ cm s^{-1}; (3) $\dot{T} = 20\,°C\ hr^{-1}$, $D_L/\ell = 3 \times 10^{-5}$ cm s^{-1}.

of different cooling rates ($\dot{T}$) and melt thickness, ℓ. We see that, for slow rates, the approximate solution is fairly good.

4.5   Three-dimensional distributions (no mixing)

If one chooses to neglect surface energy and to assume infinitely rapid interface attachment kinetics, there are two classes of crystal shapes for which the interface is an isoconcentrate independent of time provided the crystal expands in size in accordance with a particular time law. One class of crystals grow with a velocity proportional to $t^{-1/2}$ and are *spheroidal* in shape. The second class exhibits a prominent growth axis and the growth velocity is constant with time; i.e., they are *elliptical paraboloids*. By solving the transport equation in the appropriate coordinate system the solute distribution in the liquid can be readily determined. Since our primary interest here is in the determination of k for these shapes, we can write for all shapes

$$\frac{k(\hat{P}_C)}{k_0} = \frac{1}{1 - (1 - k_0)\phi(\hat{P}_C)} = \frac{C_i}{C_\infty} \qquad (4.21a)$$

where $\phi(\hat{P}_C)$ is different for each shape and $\hat{P}_C = VR/2D_C$ for the spheroids while $\hat{P}_C = V\rho/2D_C$ for the paraboloids with R and ρ as the appropriate major radius of curvature of the body.

The crystals having an ellipsoidal shape grow with a constant ratio of axes. Here, the semimajor axis, R, increases as $R = 2\alpha t^{1/2}$ where α is determined by undercooling, thermal properties, etc. For the most familiar shapes we have

(1) sphere:

$$\phi(\hat{P}_C) = 2\left(\hat{P}_C - \pi^{1/2}\hat{P}_C^{3/2}\exp(\hat{P}_C)\mathrm{Erfc}(\hat{P}_C^{1/2})\right) \tag{4.21b}$$

(2) cylinder:

$$\phi(\hat{P}_C) = -\hat{P}_C e^{\hat{P}_C} E_i(-\hat{P}_C) \tag{4.21c}$$

(3) slab:

$$\phi(\hat{P}_C) = (\pi\hat{P}_C)^{1/2}\exp(\hat{P}_C)\mathrm{Erfc}(\hat{P}_C^{1/2}) \tag{4.21d}$$

where E_i is the exponential integral function.

The second class of crystal shapes, that of the elliptical paraboloids, increase in length proportional to the first power of time; i.e., $R = \alpha' t$ where α' is a constant. For the familiar shapes, the circular paraboloid (needle) and the parabolic cylinder (platelet), we have

(4)   circular paraboloid:

$$\phi(\hat{P}_C) = -\hat{P}_C\exp(\hat{P}_C)E_i(-\hat{P}_C) \tag{4.21e}$$

(5) parabolic cylinder:

$$\phi(\hat{P}_C) = (\pi\hat{P}_C)^{1/2}\exp(\hat{P}_C)\mathrm{Erfc}(\hat{P}_C^{1/2}) \tag{4.21f}$$

By comparing Eqs. (4.21c) with (4.21e) and (4.21d) with (4.21f), it may be seen that, for the same value of $\hat{P}_C$, (a) the cylinder of class I and the circular paraboloid of class II are equivalent and (b) the slab of class I and the parabolic cylinder of class II are equivalent. This implies that it is the lateral growth of the class II bodies that dominates their solute distributions and makes them equivalent to their class I counterparts. This can be seen quantitatively by considering the equation of a paraboloid of revolution, $Z \approx Z_0 + bR^2$, and taking its time derivative, $dZ/dt = 2bR\,dR/dt$. If R is growing laterally with $V_\ell = \alpha' t^{-1/2} = dR/dt$, then $dZ/dt = 4b\alpha'^2 = $ constant, which is the correct axial velocity relationship for this shape.

In Fig. 4.32, k versus $\hat{P}_C^{1/2}$ has been plotted for $k_0 = 0.1$ and these major particle shapes. From this figure we note that, for the same value of $\hat{P}_C$, the sphere partitions solute more effectively than any other shape (three-dimensional diffusion). The slab is the least effective partitioner of solute as might be expected (one-dimensional diffusion). If the growth of the crystal from solution is dominated by its matter transport, the comparative solute profiles as a function of distance and of growth velocity as a function of time will be as presented in Figs. 4.33(a) and (b) respectively. The sphere is the dominant form in the ellipsoid class while the parabolic cylinder is the dominant form in the elliptical paraboloid class. At small time, the preferred shape will be the sphere while at long time it will be the needle. In Fig. 4.34(a), k (sphere) has been plotted as a function of $\hat{P}_C^{1/2}$ for a range of k_0 values. It exhibits the expected behavior; i.e., $k \to k_0$ as $\hat{P}_C \to 0$ ($V \to 0$ or $R \to 0$) and $k \to 1$ as $\hat{P}_C \to \infty$ ($V \to \infty$ or $R \to \infty$). In Fig. 4.34(b), k/k_0 has been plotted for the parabolic needle shape as a function of $\hat{P}_C$ for a range of k_0. Again, the expected asymptotic limits may be noted. If we are dealing with a situation in which many particles are growing, k will begin to change with interface position when the solute fields of the different particles begin to overlap. We are thus interested in the extent of the solute field at the surface of the growing particle. In Fig. 4.35(a), the relative solute profile, $(C - C_\infty)/(C_i - C_\infty)$, for the sphere has been plotted as a function of reduced radius for several values of the growth rate parameter α^2/D_C. For values of $\alpha^2/D_C > 1$, which should often be the case for dilute alloys, the solute distribution is confined to a tight spherical shell around the particle.

Using $X = (\alpha^2/D_C)^{1/2}(r/R)$, the solute distributions in the liquid for the simple particle shapes are

(1) sphere:

$$\frac{C - C_\infty}{C_\infty} = \frac{\left(\frac{\sqrt{2}}{X}\right)\exp(-X^2) - \left(\frac{\pi}{2}\right)^{1/2}\mathrm{Erfc}(X)}{\left(\frac{\pi}{2}\right)^{1/2}\mathrm{Erfc}\left[\left(\frac{\alpha^2}{D_C}\right)^{1/2}\right] - \left[\frac{1}{(1-k_0)}\left(\frac{D_C}{2\alpha^2}\right)^{3/2} - \left(\frac{D_C}{2\alpha^2}\right)^{1/2}\right]\exp\left(\frac{-\alpha^2}{D_C}\right)}$$

(4.22a)

(2) cylinder:

$$\frac{C - C_\infty}{C_\infty} = \frac{E_i(-X^2)}{\frac{1}{(1-k_0)}\left(\frac{D_C}{\alpha^2}\right)\exp\left(-\frac{\alpha^2}{D_C}\right) - E_i\left(-\frac{\alpha^2}{D_C}\right)}$$

(4.22b)

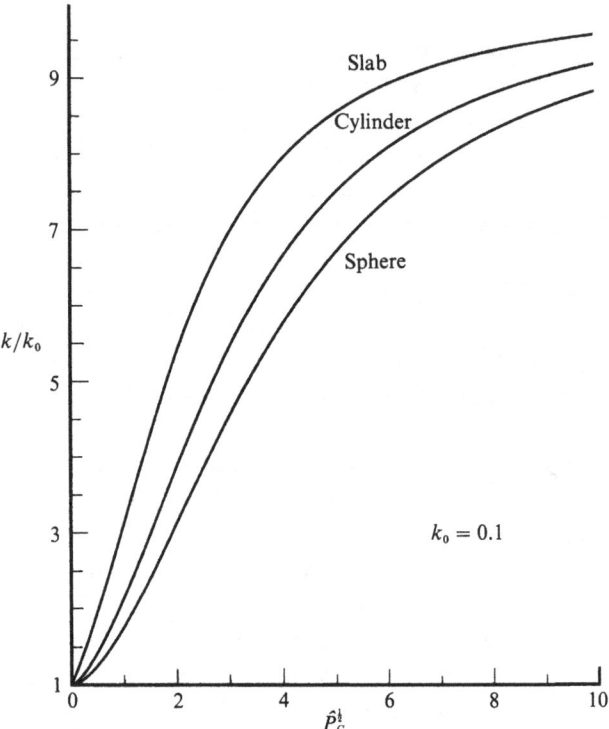

Fig. 4.32. Effective solute distribution coefficient, k, versus the Péclet number, $\hat{P}_C^{1/2}$, for the three major shapes with $k_0 = 0.1$.

(3) slab:

$$\frac{C - C_\infty}{C_\infty} = \frac{(\pi/2)\mathrm{Erfc}(X)}{\left(\frac{1}{1-k_0}\right)\left(\frac{D_C}{2\alpha^2}\right)^{1/2}\exp\left(\frac{\alpha^2}{D_C}\right) - \left(\frac{\pi}{2}\right)^{1/2}\mathrm{Erfc}\left[\left(\frac{\alpha^2}{D_C}\right)^{1/2}\right]}$$

(4.22c)

(4) circular paraboloid:

$$\frac{C - C_\infty}{C_i - C_\infty} = \frac{E_i(-\hat{P}_C g)}{E_i(-\hat{P}_C)}$$

(4.22d)

where $g = z/\rho + [(r/\rho)^2 + (z/\rho)^2]^{1/2}$ and $g = 1$ refers to the interface.

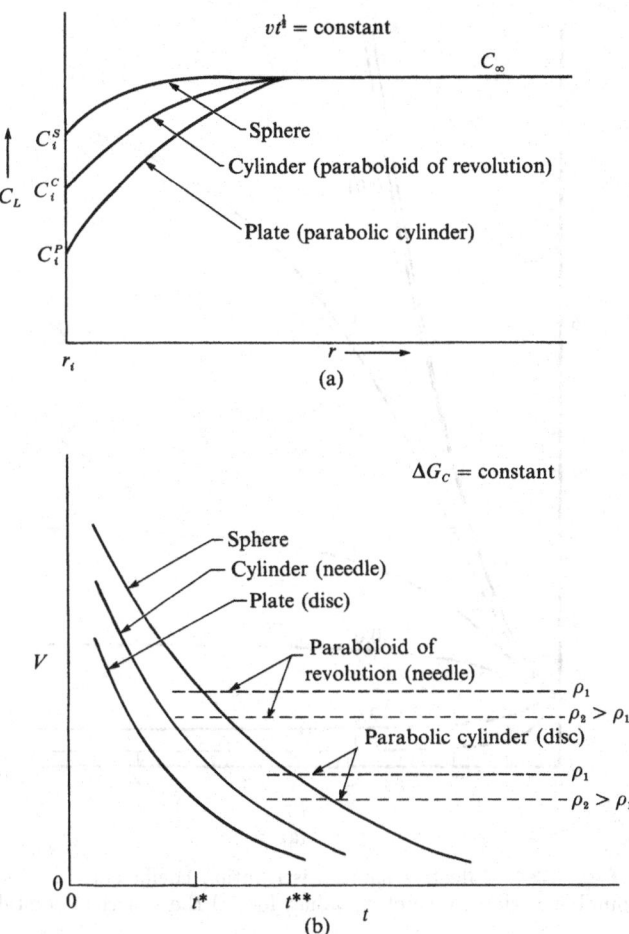

Fig. 4.33. (a) Liquid concentration, C_L, during solute crystal growth as a function of distance from the interface, r, for the important characteristic shapes and (b) relative solute crystal growth velocity as a function of time for these five characteristic shapes at constant ΔG_C. (ρ is the radius of curvature for the filament tip while t^*, t^{**} are the critical shape transition times from sphere to needle or sphere to disc, respectively.)

(5) parabolic cylinder:

$$\frac{C - C_\infty}{C_i - C_\infty} = \frac{\mathrm{Erfc}\left[(\hat{P}_C g)^{1/2}\right]}{\mathrm{Erfc}(\hat{P}_C^{1/2})} \qquad (4.22e)$$

where $g = z/\rho + [(x/\rho)^2 + (z/\rho)^2]^{1/2}$.

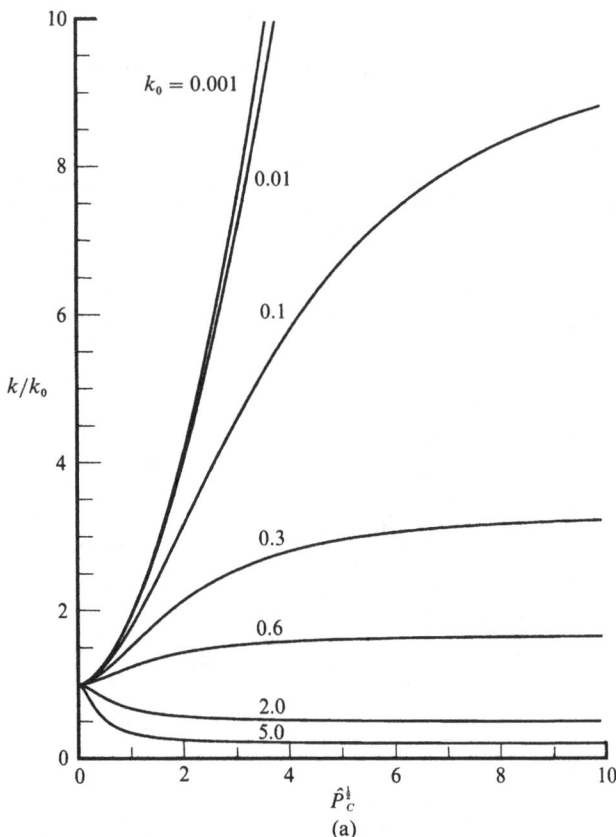

Fig. 4.34. Effective solute distribution coefficient, k, versus Péclet number with a range of k_0 values for (a) the spherical crystal shape.

In Fig. 4.35(b), the relative solute profile around the parabolic needle crystal has been plotted as a function of g for the cases of large $\hat{P}_C$ and small $\hat{P}_C$. The mathematical development of the solute profiles for these limiting shapes and for in-between shapes has been contributed to by several authors.[16–18]

4.5.1   The Laplace approximation
    By setting $\phi = C/C_\infty$, $x' = x/R$, $y' = y/R$, $z' = z/R$ and $t' = D_C t/R^2$, Eq. (4.1) becomes (for constant D_C, V, and $u = \dot{Q} =$ field $= 0$) in three dimensions

$$\nabla'^2\phi + 2\hat{P}_C\,\nabla'\,\phi = \partial\phi/\partial t' \qquad (4.23)$$

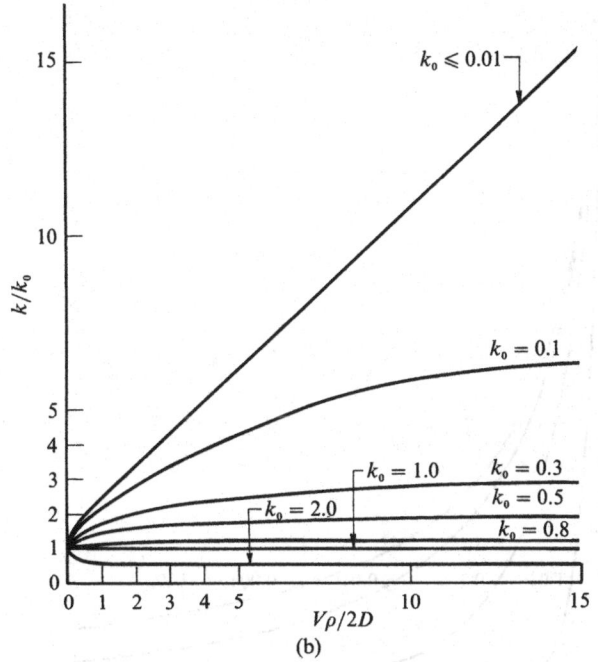

Fig. 4.34. (b) The paraboloid of revolution crystal shape.

For $\hat{P}_C << 1$, $\hat{P}_C \nabla' \phi << \nabla'^2 \phi$ and can be neglected, so the transport equation is reduced to the simple diffusion equation. In many situations it is also useful to simplify the equation further to the Laplace equation by neglecting $\partial \phi / \partial t'$ and allowing the coefficients of the solution to be slowly varying functions of time. It is the slow time dependence that allows this approximation to be used.

If we consider the growth of a spherical crystal in this approximation, the solute distribution in the liquid is given by

$$C = A + B/r \qquad (4.24a)$$

where the solution to Laplace's equation requires only that A and B be independent of r; they can be functions of time. If the crystal is increasing according to the time law $R = R(t)$, the interface condition can be used to determine B and the far-field condition to determine A. Thus, we find

$$A = C_\infty; \quad B = R\dot{R}(1 - k_o)C_\infty / \left[D_C/R - \dot{R}(1 - k_0) \right] \qquad (4.24b)$$

When $R = 2\alpha t^{1/2}$, Eq. (4.24b) gives the same result as Eq. (4.22a) in

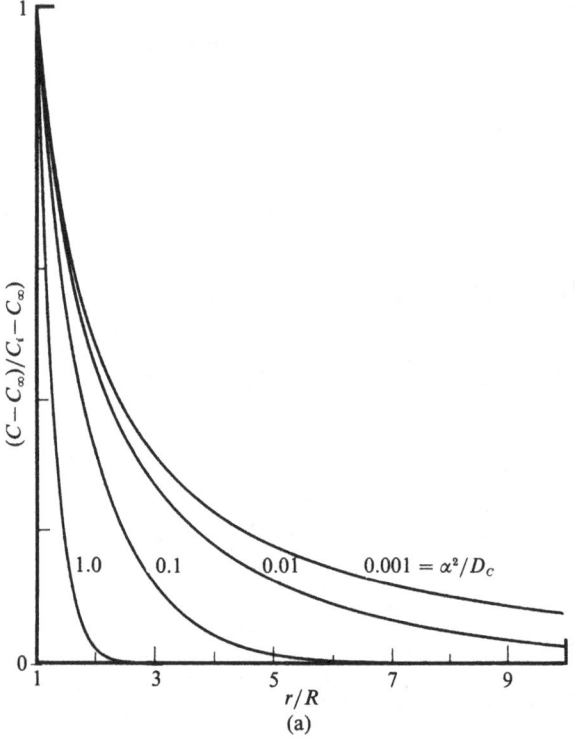

Fig. 4.35. Relative solute profiles, $(C - C_\infty)/(C_i - C_\infty)$, adjacent to growing crystals; (a) for a sphere as a function of reduced radius, r/R, with several values of the growth parameter, α^2/D_C, and $k_0 = 0.1$.

the limit of small $\hat{P}_C$. From Eqs. (4.24), k is given by

$$k = \frac{k_0}{1 - \frac{R\dot{R}}{D_C}(1 - k_0)} \qquad (4.25)$$

which tells us that k is a constant with time for the case of $R = 2\alpha t^{1/2}$ in accordance with the more exact treatment. It is clear that this approximate formula cannot hold for $\hat{P}_C \tilde{>} 0.5$ since k becomes negative in this range.

In the domain where $\hat{P}_C \ll 1$, the transport is so fast compared to the motion of the source that, as one moves away from the source, the solute field rapidly becomes independent of the source shape so that variations in source shape and source size have a negligible effect on the prime character of the solute field. Thus, the total influx of solute

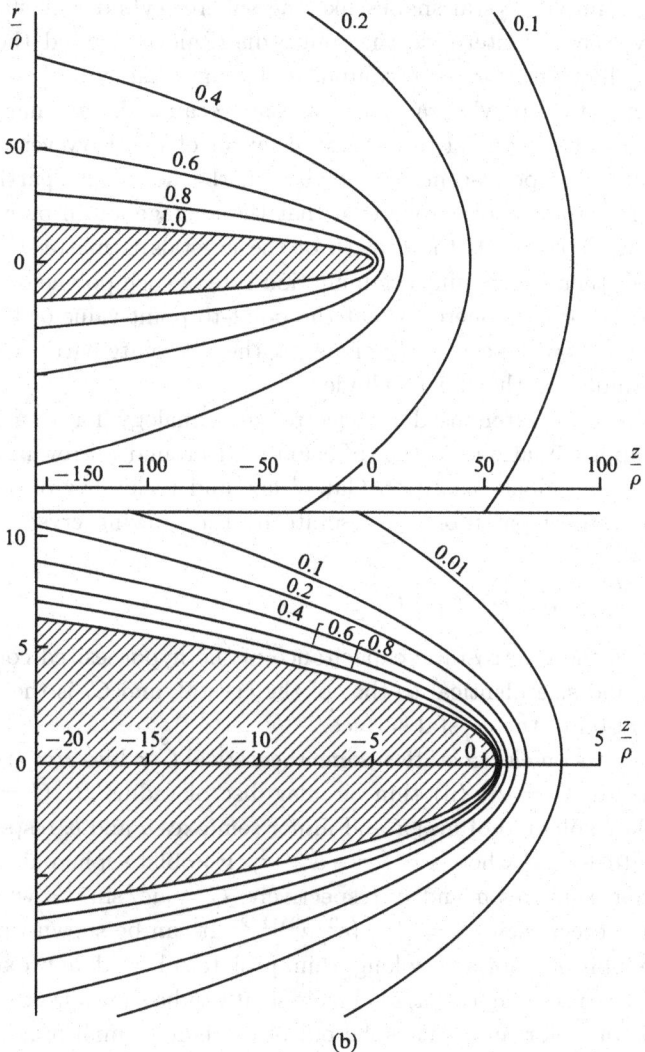

Fig. 4.35. (b) For a needle-like crystal with small $\hat{P}_C$ (top) and large $\hat{P}_C$ (bottom). The abscissas are z/ρ and the ordinates r/ρ, for $\rho =$ radius of curvature of the tip.[18]

to the crystal ($k_0 > 1$) or the total efflux of solute from the crystal ($k_0 < 1$) is independent of the shape of the crystal; i.e., we are dealing with a relatively steady type of situation so that the integral flux of solute across any isoconcentrate surface must be a constant.

For the simple crystal shapes like the sphere, cylinder or slab, it is relatively easy to determine the solute distribution around the growing body; however, for more complicated shapes like cubes, hexagonal prisms and other polyhedra, it is not easy because the angular dependences must be included in the specification of C. However, we can expect that, for point-centered polyhedra, the net solute partitioning will be close to that of a sphere so that the average k will be given by that for the sphere. If the growth of the crystals is matter-transport controlled, then the volume accumulation rate of such polyhedra will be the same as for the sphere so that the point-to-point value of V can be determined. For line-centered polyhedra, the net solute partitioning will be very similar to that for a cylinder.

Workers have often used an electrostatic analogy for treating the present approximations to the diffusion field around a growing crystal and, from arguments based on Gauss' law and Fick's law of diffusion, the total rate of partitioning of solute by the growing crystal can be written

$$\frac{dC}{dt} = 4\pi\tilde{C}D_C(\bar{C}_i - C_\infty) \qquad (4.26)$$

where $\tilde{C}$ is the electrostatic capacitance of the hypothesized conductor of shape and size identical to that of the crystal, and $\bar{C}_i$ is the average concentration of the crystal surface.

From works on electrostatics, some theoretical $\tilde{C}$ values are: (i) sphere of radius R, $\tilde{C} = R$; (ii) thin circular disc of radius R, $\tilde{C} = 2R/\pi$; (iii) prolate spheroid of major and minor semiaxes a and b, respectively, $\tilde{C} = \varepsilon\ln[(a+\varepsilon)/b]$ where $\varepsilon = (a^2 - b^2)^{1/2}$; (iv) oblate spheroid of major and minor semiaxes a and c, respectively, $\tilde{C} = ae/\sin^{-1}e$ where e is the profile eccentricity, $e = [1 - (c^2/a^2)]^{1/2}$. It can be shown from (iii), that the limit of $\tilde{C}$ for a very long, thin, prolate spheroid, approximating needle form, is $\tilde{C} = a/\ln(2a/b)$. Likewise, it can be shown from (iv) that the limit of $\tilde{C}$ for an oblate spheroid of extremely small minor axis is $\tilde{C} = 2a/\pi$, i.e., the same as given by (ii) for the circular disk. Of course, (iii) and (iv) go over into (i) in the limit of zero shape eccentricity.

Using brass patterns, $\tilde{C}$ has been measured for a variety of non-simple shapes by using a Faraday cage.[19] The absolute accuracy is limited by the additive effect of a parallel capacitance unavoidably introduced by the insertion of the lead wire required for the $\tilde{C}$ measurements. Table 4.4 illustrates (a) the effect of lead wire field short out on $\tilde{C}$ for test discs of various radii (thickness 0.33 mm), (b) the values of $\tilde{C}$ for thin hexagonal plates, (c) the effect of thickness on $\tilde{C}$ for hexagonal plates with hexagonal side length of 3.81 cm, (d) $\tilde{C}$ for needle forms and (e) $\tilde{C}$ for hexagonal prisms. In Fig. 4.36, relative values of $\tilde{C}$ for hexagonal dendritic forms with $\ell = 3.81$ cm are presented. Such data are certainly of great interest to the theorist and experimentalist wishing to analyze crystal growth data. Although the experimental accuracy is not as high as one might like, it does provide for more accurate values for $\tilde{C}$ than one might have guessed. Computer solutions for these shapes would provide more accurate results.

The foregoing capacitance technique is useful for determining the average concentration distribution around a body and, more specifically, its crystallization velocity. However, one knows that the solute concentration will not be uniform over a given face and, in many cases, what one wishes to know is the point-to-point variation along a face to determine either the segregation inhomogeneity in the crystal or the maximum supersaturation on a face. Considerable supersaturation may occur on parts of the faces, especially near corners and vertices. It is presumably this excess supersaturation at crystal corners that, in certain circumstances, causes dendritic rather than polyhedral growth and it is therefore desirable to be able to estimate how this supersaturation excess depends on crystal geometry.

Defining the supersaturation number, $\bar{\sigma}$, as the difference between the highest and the lowest concentration at the crystal surface divided by the integral of the normal solute gradient, q, taken over the trace of the crystal surface, $\bar{\sigma}$ for a regular polygonal plate of n sides of length a is given in Table 4.5 as $\bar{\sigma}_1$. In Table 4.5, $\bar{\sigma}_2 = \pi/4n^2$ which is the leading term of a Taylor's series expansion for $\bar{\sigma}$ in powers of n^{-1}. This gives values of $\bar{\sigma}$ that are too large, the correction term being of order n^{-3}. Another upper bound on $\bar{\sigma}$ is $\bar{\sigma} = \bar{\sigma}_3$ in Table 4.5 which is found by assuming the concentration field to be associated with a line sink of unit strength. This approximation gives $\bar{\sigma}_3 = -(2\pi)^{-1}\ell n[\cos(\pi/n)]$ and, we note from Table 4.5 that $\bar{\sigma}_1, \bar{\sigma}_2$ and $\bar{\sigma}_3$ approach each other as n increases.

A useful formula for the solute variation at the coordinate u on the face measured relative to a face corner ($u = a/2$) is given by

$$\frac{C(u) - C(a/2)}{qa} = \left[2n\bar{\sigma} - \frac{1}{4\cos(\pi/n)}\right]\left(\frac{2u}{a}\right)^2$$
$$+ \left[-n\bar{\sigma} + \frac{1}{4\cos(\pi/n)}\right]\left(\frac{2u}{a}\right)^4 + \cdots \quad (4.27)$$

Table 4.4. *Capacitance,* $\tilde{C}$, *studies using brass patterns.*[19]

(a) *Effect of lead wire short out on test discs.*

Disc radius (cm)	5.08	4.45	3.82	3.18	2.54	1.91	1.27
Measured $\tilde{C}$ ($\mu\mu$F)	2.56	2.18	1.79	1.43	1.08	0.75	0.43
Theoretical $\tilde{C}$ ($\mu\mu$F)	3.60	3.15	2.70	2.25	1.80	1.36	0.90
Ratio (measured/ theoretical)	0.71	0.69	0.66	0.64	0.60	0.55	0.48

(b) *Comparison of* $\tilde{C}$ *for thin hexagonal plates and thin circular discs of equal area.*

Disk radius (cm)	3.82	3.18	2.54	1.91	1.27
Measured $\tilde{C}$ ($\mu\mu$F)	1.79	1.43	1.08	0.75	0.43
Hexagonal edge $\tilde{\ell}$ (cm)	4.18	3.49	2.79	2.09	1.40
Measured $\tilde{C}$	1.81	1.42	1.10	0.72	0.47
% excess disc over hexagon	−1.1	0.7	−1.8	4.2	−8.5

(c) *Thickness effect on* $\tilde{C}$ *for hexagonal plates* ($\tilde{\ell} = 3.81$ cm).

Thickness (cm)	0.033	0.084	0.127	0.228	0.482
Measured $\tilde{C}$ ($\mu\mu$F)	1.71	1.70	1.75	1.81	1.93
Relative value	1.00	0.99	1.01	1.04	1.12

(d) *Thickness effect on cylindrical rods* (*length* $= 15.2$ cm).

Radius b (cm)	0.0318	0.0418	0.0635	0.155	0.765
Measured $\tilde{C}$ ($\mu\mu$F)	0.89	0.96	1.08	1.26	1.81
Theoretical $\tilde{C}$ ($\mu\mu$F)	1.23	1.29	1.38	1.61	2.31
Ratio (measured/ theoretical)	0.73	0.74	0.78	0.78	0.78

(e) *Length effect on hexagonal prisms* (2.54 cm *cross-section, flat to flat*).

Length (cm)	25.4	20.3	15.2	10.2	5.08
Measured $\tilde{C}$	4.32	3.57	2.95	2.17	1.40

Table 4.5. *Supersaturation number, $\bar{\sigma}$, for an n-sided polygonal plate crystal.*

n	2	3	4	6	8	10	20
$\bar{\sigma}_1$	0.111	0.0570	0.0349	0.0171	0.0101	0.00674	0.00177
$\bar{\sigma}_2$	∞	0.1103	0.0551	0.0229	0.0126	0.00798	0.00198
$\bar{\sigma}_3$	0.196	0.0822	0.0492	0.0218	0.0123	0.00788	0.00196

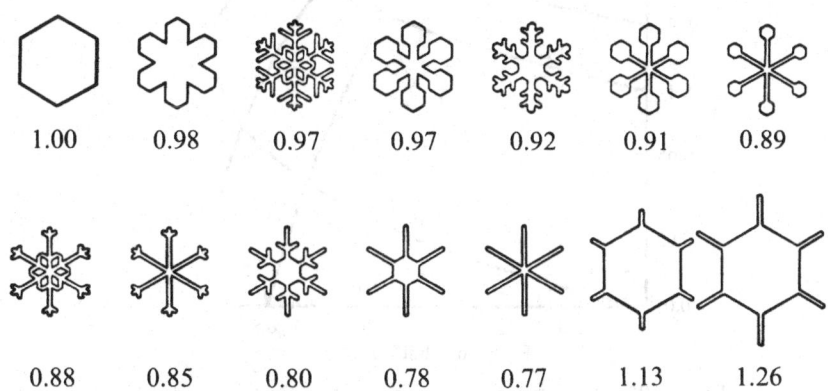

1.00	0.98	0.97	0.97	0.92	0.91	0.89

0.88	0.85	0.80	0.78	0.77	1.13	1.26

Fig. 4.36. Relative values of $\tilde{C}$ from brass pattern studies for thin hexagonal dendritic forms (hexagonal side length = 3.81 cm).[19]

4.6 Fluid mixing effects (normal freezing)

The case of partial mixing is very important in practice and three situations are worthy of note: (1) k = constant, conservative system, (2) k = variable, conservative system, and (3) k = constant, non-conservative system.

Case 1 (k = constant, conservative system): For the case of $k_0 < 1$, the solid that freezes from the liquid will be purer than the liquid so that rejection of solute into the liquid will occur at the solid/liquid interface. Stirring of the bulk liquid limits the solute-rich interface layer to a thickness δ_C. If δ_C and V are held constant, the appropriate value of k is soon developed across the layer with the interface liquid concentration becoming $C_i = (k/k_0)C_\infty$. If k remains constant throughout the freezing process, the distribution of solute in the solid, C_S, is found to be given as a function of the fraction of liquid solidified, g, by

$$C_S = kC_0(1 - g)^{k-1} \tag{4.28}$$

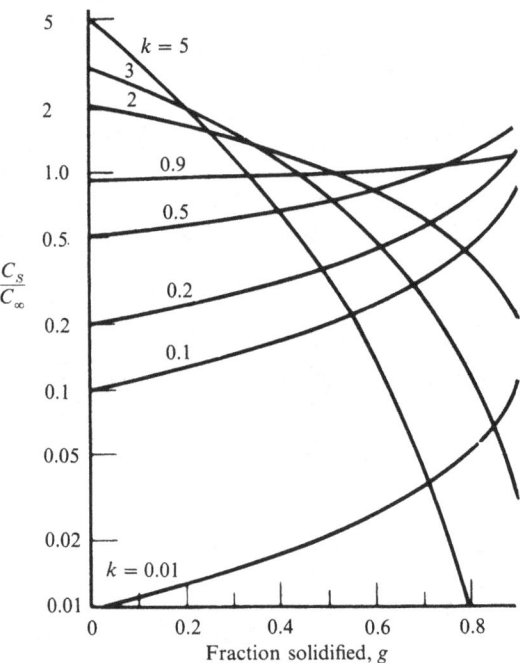

Fig. 4.37. Relative solute concentration in the solid, C_S/C_∞, as a function of fraction solidified, g, along the solid for normal freezing of a bar of unit length for a range of k-values.[20]

where C_0 represents the initial concentration of solute in the liquid. Curves of C_S/C_o as a function of g for various k are shown plotted in Fig. 4.37. The curves are concave upward for $k < 1$ so the last part to freeze is rich in solute while they are concave downward for $k > 1$ leading to solute depletion in the last portion to freeze.[20]

Case 2 (k = variable, conservative system): If k is allowed to vary with g and the total solute content is held constant, various curves result that differ from those of Fig. 4.37. One can vary $k(g)$ through either $\delta_C(g)$ or $V(g)$ and, if one assumes that the solute distribution in the boundary layer adjusts immediately to the condition needed for the new k, the value of $k(g)$ needed to produce a particular $C_S(g)$ is given by

$$k(g) = \frac{C_S(g)}{C_0 + \frac{1}{(1-g)} \int_0^g [C_0 - C_S(g)] dg} \qquad (4.29)$$

All the constants entering Eq. (4.29) must be chosen such that $k(g)$ is

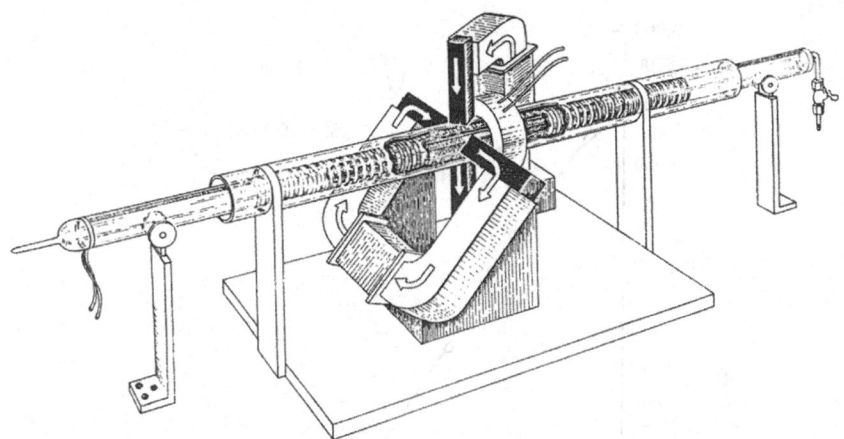

Fig. 4.38. Simple electromagnetic stirring device for normal freezing treatment.[21]

never required to fall outside the limits $k_0 \lesssim k \lesssim 1$. To demonstrate the validity of the method, a horizontal Bridgeman technique was used with an electromagnetic stirring device as shown in Fig. 4.38.[21] The crossed field coils were fixed in space and provided a rotating magnetic field perpendicular to the axis of the boat that was slowly drawn through the stirring field. With this type of field the variation of δ_C is given by

$$\frac{\delta_C}{D_C} = \left(\frac{\delta_C}{D_C}\right)_0 \exp(-m\hat{H}^2) \tag{4.30}$$

for a field strength of $\hat{H}$ Oe. From Fig. 4.39, we see that $(\delta_C/D_C)_0 = 7300$ sec cm^{-1} and $m = 3.52 \times 10^{-5}$ Oe^{-2}. Some results of experiments on Pb–Sn alloys designed to give either a linear distribution or a cosine distribution are shown in Fig. 4.40. Since $k_0 \approx 0.5$ for Sn in Pb, these distributions are strikingly different from the $k = $ constant profiles of Fig. 4.37.

The situation is very different for variable k when the variability arises from a concentration dependence of k_0 rather than from $V\delta_C/D$ control. If the phase diagram is such that $k_0 = k_0(C_L)$, the analysis procedure is to take the k_0 versus C_L plot and convert it into a type of histogram such as illustrated in Fig. 4.41. Thus, $k_0 = k_{0m}$ for $C_{L_{m-1}} < C_L < C_{L_m}$ will be utilized for $1 \lesssim m \lesssim p$. Starting with $g = 0$, we have, for $D_S^j = 0$,

$$C_L/C_0 = (1-g)^{k_1-1}, \qquad g \lesssim g_1 \tag{4.31a}$$

where

$$C_{L_1}/C_0 = (1-g_1)^{k_1-1} \tag{4.31b}$$

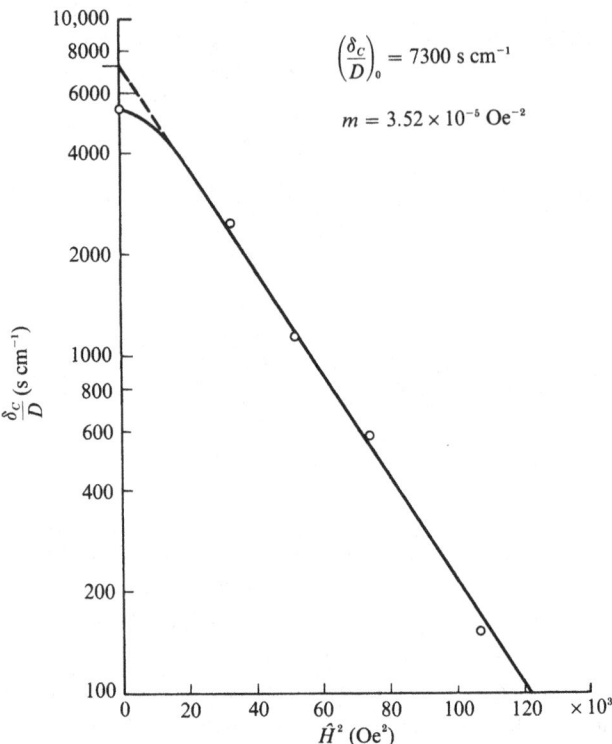

Fig. 4.39. Plot of solute boundary layer thickness, δ_C / D_C, as a function of stirring field strength, $\hat{H}^2$, using the device of Fig. 4.38.[21]

Continuing, we obtain
$$C_L / C_{L1} = (1 - g_1 - g)^{k_2 - 1}, \qquad g \leq (g_2 - g_1) \qquad (4.31c)$$
where
$$C_{L_2} / C_{L_1} = (1 - g_2)^{k_2 - 1} \qquad (4.31d)$$
Ultimately, we obtain
$$C_{L_m} / C_0 = \prod_{m=1}^{n} (1 - g_m)^{k_m - 1} \qquad (4.31e)$$
where $g_m - g_{m-1}$ is the fraction of initial liquid solidified during the mth stage.

*Case 3 (**k** = constant, non-conservative system):* Suppose the solute is volatile so that there is an exchange of solute between the liquid and gaseous phases. Since we are dealing with a non-conservative system, to

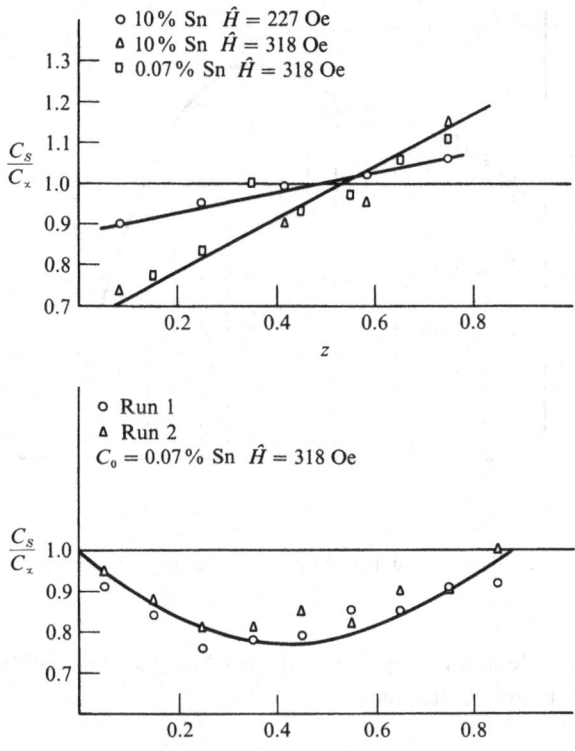

Fig. 4.40. Experimental test of programmed linear and cosine shaped solute distributions in the solid, C_S/C_∞, as a function of distance z for a bar of unit length using Pb–Sn alloys and different magnetic field strengths, $\hat{H}$.[21]

calculate $C_S(g)$ for constant k, the change of liquid concentration with g due to this exchange must be considered. Let C_L^* be the concentration of liquid in equilibrium with the gas phase so the rate of exchange with the gas phase is given by

$$\frac{\mathrm{d}C_L(g)}{\mathrm{d}t} = \frac{\hat{K}_R}{h}[C_L^* - C_L(g)] \qquad (4.32)$$

where $\hat{K}_R$ is the reaction constant and h is the ratio of melt volume to contact surface area with the gas. By calculating the change in total concentration in the melt (when the interface is at g), $C_S(g)$ can be found as a solution to the equation

$$\frac{\mathrm{d}C_S}{\mathrm{d}g} + \left(\frac{k-1}{1-g} + \frac{\hat{K}_R}{hV}\right)C_S = \frac{\hat{K}_R}{hV}C_L^* \qquad (4.33)$$

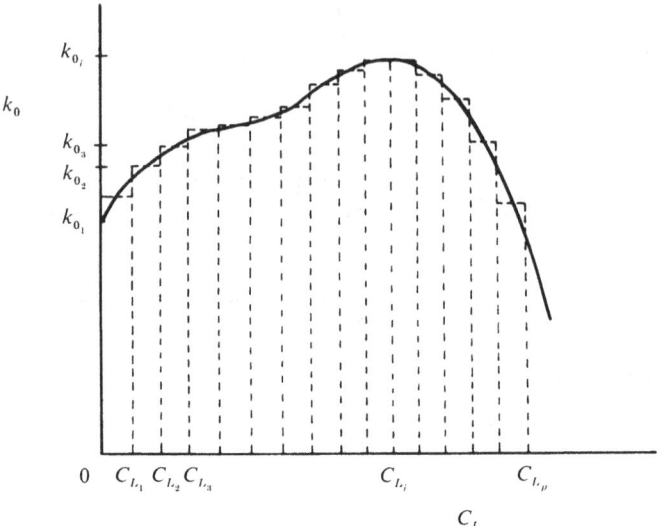

Fig. 4.41. Histogram procedure for dealing with a concentration-dependent k_0.

The solution for C_S is an integral equation which can be readily evaluated when $\hat{K}_R$, V and h are known.

4.6.1   O pick-up in Si melts

An important example in the non-conservative system category is that of O in Si when the crystals are grown via the CZ technique. O is added to the melt by chemical reaction of liquid Si with the SiO_2 crucible and lost from the melt by either incorporation into the crystal or by SiO release from the exposed upper surface. This O pick-up process is illustrated in Fig. 4.42 where (a) presents the major O fluxes overview while (b) defines the various boundary layer thicknesses involved. It has been assumed that strong convective mixing exists to give a uniform O content in the bulk melt. In addition, the transport of SiO(g) from the melt surface to the ambient gas is considered to be very rapid.

We shall initially neglect the population of SiO_x species in the melt and presume that only O is present. Defining C_L^{*O} and C_W^O as the equilibrium melt concentration of O in contact with SiO_2 and the O concentration at the crucible wall, respectively, and referring to Fig. 4.42(a), we have[7]

$$F_1 = A_W \hat{K}_W (C_L^{*O} - C_W^O) \qquad (4.34a)$$

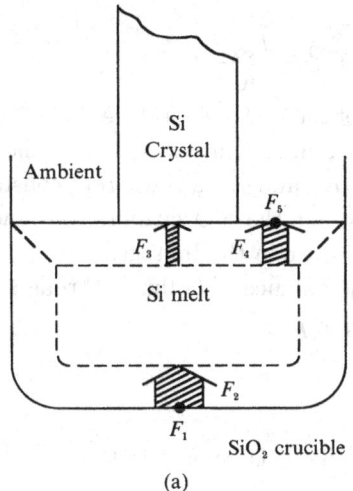

(a)

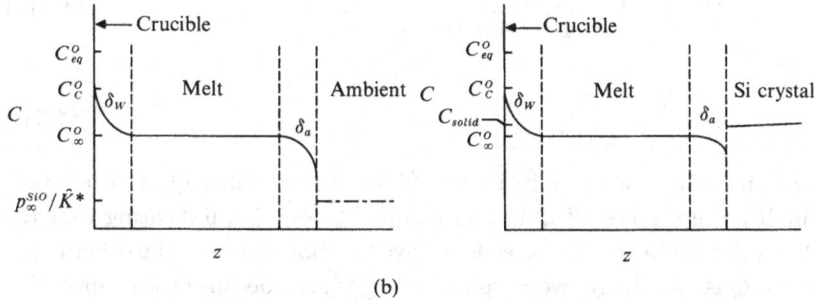

(b)

Fig. 4.42. (a) Key O fluxes during CZ Si crystal growth and (b) O concentration profiles from crucible to ambient and crystal.[7]

where $\hat{K}_W$ is the rate constant for this dissolution reaction and A_W is the area of crucible in contact with the melt. Further, from Fig. 4.42(b), we have

$$F_2 = A_W D_L^O \frac{(C_W^O - C_\infty^O)}{\delta_W} \qquad (4.34b)$$

where δ_W is the effective diffusion boundary layer thickness. In addition, we have

$$F_3 = \pi R_c^2 V k C_\infty^O \qquad (4.34c)$$

$$F_4 = A_a D_L^O \frac{(C_\infty^O - C_a^O)}{\delta_a} \qquad (4.34d)$$

and

$$F_5 = A_a \hat{K}_a \left(C_a^O - \frac{P_\infty^{SiO}}{\hat{K}^*} \right) \tag{4.34e}$$

where A_a is the area of the free melt surface, δ_a is the diffusion boundary thickness at the melt/ambient interface, C_a^O is the O concentration at the free melt surface, $\hat{K}_a$ and $\hat{K}^*$ are the rate constant and equilibrium constant, respectively, for the SiO evaporation reaction while P_∞^{SiO} is the SiO(g) partial pressure in the bulk gas.

At steady state, the balance of O fluxes through the system gives

$$F_1 = F_2 = F_3 + F_4 \tag{4.34f}$$

and

$$F_4 = F_5 \tag{4.34g}$$

Since $F_3 \ll F_4$, generally, it is ignored in this first order analysis which, when solved, leads to

$$C_\infty^O = \frac{C_L^{*O} + (A_a/A_W)QP_\infty^{SiO}/\hat{K}^*}{1 + (A_a/A_W)Q} \tag{4.34h}$$

where

$$Q = \frac{\delta_W + D_L^O/\hat{K}_W}{\delta_a + D_L^O/\hat{K}_a} \tag{4.34i}$$

The limiting cases of C_∞^O in Eq. (4.34h) arise when Q is either very small or very large. If Q is very small, $C_\infty^O \to C_L^{*O}$, indicating that the dissolution rate greatly exceeds the evaporation rate. On the other hand, when Q is very large, we have $C_\infty^O \to P_\infty^{SiO}/\hat{K}^*$ because the evaporation process is much more efficient than the dissolution process.

We shall use $C_L^{*O} = 7.5 \times 10^{18}$ atoms cm^{-3} at 1460 °C, $P^{*SiO} = 8.6 \times 10^{-3}$ atm at T_M^{Si} and $\hat{K}^* = 2.46 \times 10^{-21}$ atm cm^3 atom^{-1} at T_M^{Si}, the Si melting temperature.[7] For one typical commercial Si CZ growth system, the crucible diameter is 20 cm, the initial melt depth is 12.5 cm and the crystal is about 7.5 cm in diameter, thus, we have A_q/A_W ranging from about 0.25 at the onset of growth to about 0.86 when the melt is almost gone. Further, one uses $P_\infty^{SiO} \sim 10^{-2}P^{*SiO}$–$10^{-1}P^{*SiO}$.

Treating Q as a unified growth-condition parameter, Fig. 4.43(a) shows C_∞^O as a function of solidified melt fraction g (or A_a/A_W) for $R_c = 7.5$ cm, $P_\infty^{SiO} = 8.6 \times 10^{-4}$ atm and a range of Q. We see that the melt concentration is concave downwards and decreases as g or A_a/A_W increases, just as observed experimentally. Fig. 4.43(b) shows the effect

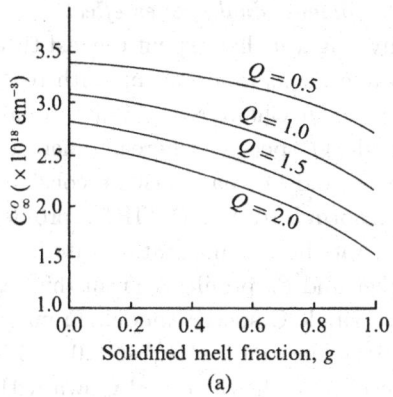

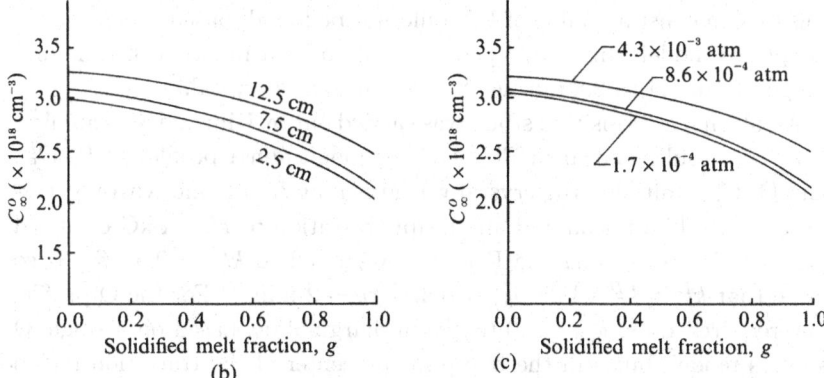

Fig. 4.43. Calculated O concentration in the bulk Si melt, C_∞^O, as a function of melt fraction solidified, g, or A_a/A_W for (a) $R_c = 7.5$ cm, $P_\infty^{SiO} = 8.6 \times 10^{-4}$ atm and a range of Q, (b) $Q = 1.0$, $P_\infty^{SiO} = 8.6 \times 10^{-4}$ atm and a range of R_c and (c) $R_c = 7.5$ cm, $Q = 1.0$ and a range of P_∞^{SiO}.[7]

of crystal diameter variation while Fig. 4.43(c) shows the effect of P_∞^{SiO} variation.

Recently, the use of a strong magnetic field to stabilize unwanted thermal convection during CZ crystal growth of Si has been invoked. Experiments show that the O content in the grown crystals is substantially reduced and achieves a better longitudinal uniformity. This is consistent with expectations wherein an increase in $\hat{H}$ will increase Q via a substantial increase in δ_W.

4.6.2   O partition coefficient, dual species effect

Competent investigators have spent the last three decades trying to obtain a reliable value for k_0^O and end up with results ranging from 0.25 to 5.[7] As discussed in Chapter 3, this is probably because they considered only a single O species, whereas, there are probably two interacting species, SiO_4 and O, that must be considered. Fig. 4.44(a) shows the Fourier Transform Infra Red (FTIR) O profile results for the Si crystal grown with abrupt changes in rotation rate (see Fig. 4.21 for the corresponding P marker and β_u profiles). From this result, one expects $k_0^O \sim 1.6$ based upon a single O species model. From other crystals, the same type of experiment led to $1.2 < k_0^O < 2.0$.[7] Fig. 4.44(b) shows the FTIR O profile results for the Si crystal grown with abrupt changes in V (see Fig. 4.22 for the corresponding C_S^P and $\beta_u V^2$ profiles). For this case, a constant value of k_0^O could not be found, based upon a single O species model. Instead, k_0^O was found to be a function of time and ranged from 1.4 to 2.8 during the growth of this crystal.[7]

An abrupt $\hat{H}$ transition study was carried out by Kim and Smitama[22] with the results shown in Fig. 4.45 for longitudinal profiles of B, C_S^B, and O, C_S^O. Initially, the crystal was grown at $\hat{H} = 0$ and with no crystal or crucible rotation and an abrupt transition to $\hat{H} = 4$ kG occurred at $z \approx 59$ mm. Using the B as a marker led to $k_0^B \sim 0.6$, β_{u_i} is so large that $k^B \approx k_0^B$ while $\beta_{u_f} \sim 0$ and $\tau_R \sim 2$ min.[7] For the O profile, one requires $4.8 < k_0^O < 7.0$ to give a marginal fit, based on a single O species model, but still the bump in the center of the transition region in Fig. 4.45(b) cannot be accounted for.

Using an O two-species model, one would expect the effective partition coefficient to vary during such transients and that a transfer function should be defined for k_0^O; i.e.,

$$k_0^O(t) = k_{0i}^O + (k_{0f}^O - k_{0i}^O)[1 - \exp{-(t/\tau_k)}] \qquad (4.35)$$

If $k_{0i}^O \approx 1.6$ in the $\hat{H} = 0$ region, then $k_{0f}^O \approx 0.7$ in the steady state $\hat{H} = 4$ kG growth regime, $\tau_k \approx 6$ min and the bump in Fig. 4.45(b) is thought to be due to the interaction of the β_u and k_0^O transfer functions. The fact that we find $k_{0f}^O < 1$ in the $\hat{H} = 4$ kG growth regime is consistent with an additional piece of information reported by Kim and Smitana.[22] They found that, a moment before the growth was terminated by a rapid quench, $C_S^O \approx 4.9 \times 10^{17}$ atoms cm^{-3} at the solid side of the interface while $C_\infty^O \sim 5.8 \times 10^{17}$ atoms cm^{-3} for the quenched liquid so that $k_0^O \sim 0.85$.

Figs. 4.46(a) and (b) are the proposed schematic profiles for the C^O

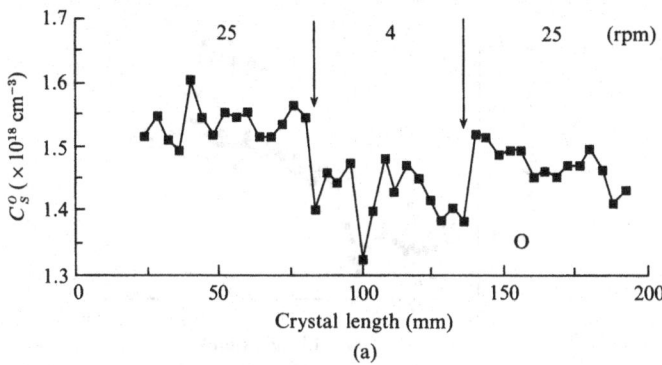

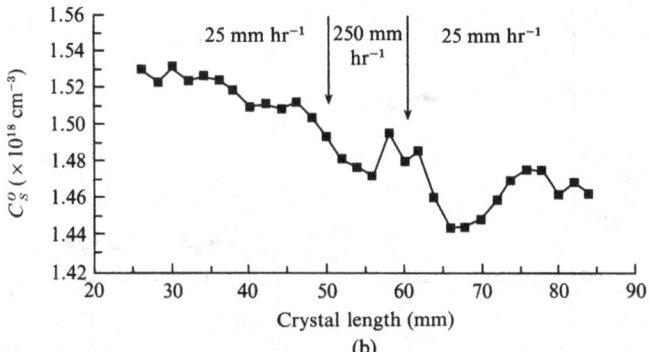

Fig. 4.44. Longitudinal O_I concentration profile in Si crystals grown with abrupt transitions. (a) $\omega = 25 \rightarrow 4 \rightarrow 25$ rpm at V = constant and (b) $V = 25 \rightarrow 250 \rightarrow 25$ mm hr^{-1} at ω = constant.[13]

and C^{SiO_4} species in (a) the steady state $\hat{H} = 0$ growth regime and (b) the steady state $\hat{H}$ growth regime. Here, the solid lines represent the real concentration profiles while the dashed lines are the profiles if no chemical reaction had been allowed to take place in the solute boundary layer. Without applying a magnetic field to the system, δ_C is relatively small because of the strong melt convection so the extent of SiO$_4$ pile-up and O depletion within δ_C are small; i.e., k^O and k^{SiO_4} are close to k_0^O and $k_0^{SiO_4}$, respectively. Because of the chemical reaction effect, one expects C^O (reaction) $> C^O$ (no reaction) and C^{SiO_4} (reaction) $< C^{SiO_4}$ (no reaction) within δ_C.

After a sudden increase of $\hat{H}$, at least three distinct effects occur simultaneously: (1) δ_C increases in magnitude due to the slow down of the melt flow velocity so C_L^O decreases in the boundary layer ($k_0^O > 1$);

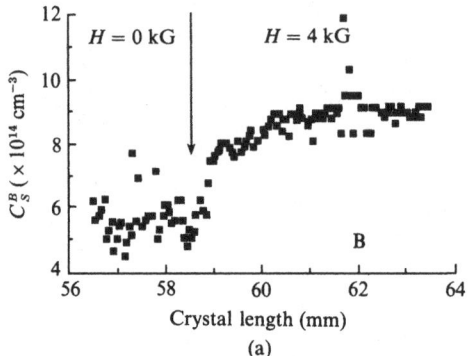

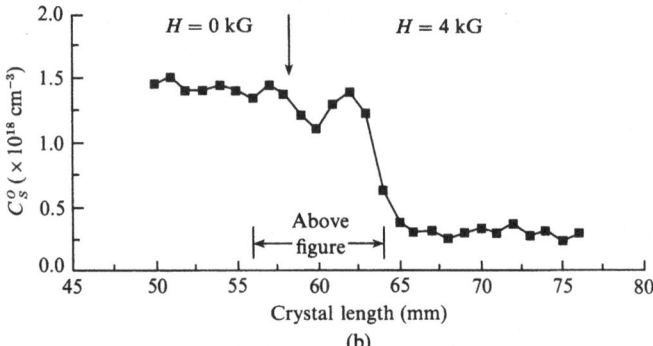

Fig. 4.45. Longitudinal concentration profiles in a Si crystal grown at constant V but an abrupt transition from ($\omega = 0$, $\hat{H} = 0$) to ($\omega = 0$, $\hat{H} = 4$ kG) for (a) B and (b) O interstitial.[22]

(2) because $k_0^{SiO_4} \ll 1$, $C_L^{SiO_4}$ increases greatly as δ_C increases so chemical reaction occurs to reduce this profile by conversion to the O species causing C_L^O to increase and (3) after $\hat{H} \to \hat{H} = 4$ kG, the melt temperature was observed to decrease by $\sim 12\,^\circ\mathrm{C}$[22] which would favor conversion of O to the SiO_4 species resulting in a decrease of C_L^O. Probably (1) has the shortest time constant while (3) has the longest time constant.

In Fig. 4.46(b), we note that $C_S^O < C_\infty^O$ ($k^O < 1$) under the strong magnetic field influence despite the fact that $k_0^O > 1$. Thus, if one didn't allow the possibility of a second interacting O species, one would conclude that $k_0^O < 1$ because of the observed $C_S^O < C_\infty^O$. We can now see that $C_S^O < C_\infty^O$ ($k < 1$) and $C_S^O > C_\infty^O$ ($k > 1$) results are possible

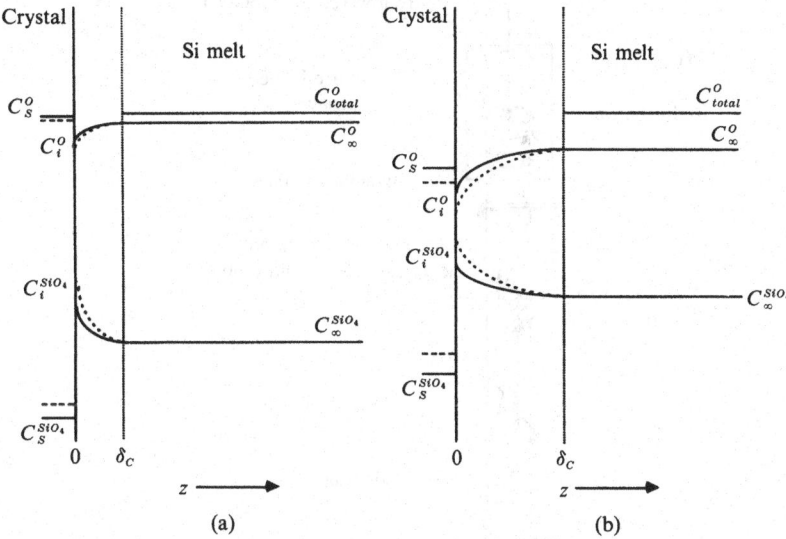

Fig. 4.46. Schematic diagrams for the C^O and C^{SiO_4} concentration profiles (a) in the steady state $\hat{H} = 0$ kG growth regime and (b) in the steady state $\hat{H} = 4$ kG growth regime. The dotted lines represent the profiles if no chemical reaction were allowed to take place between O and SiO_4.[13]

depending upon the specific set of growth conditions present during CZ Si growth and that this signals the presence of multiple interacting O species.

4.6.3 Phase diagram analysis

The foregoing concepts can be utilized to provide important information concerning two-phase equilibrium in a polycomponent system where it is necessary to know either of the liquidus or solidus surfaces and the tie-line, $\bar{K}_0$, i.e., $C_S = \bar{K}_0 C_L$ where $C_L = [C^0, C^1, \ldots, C^n]$ denotes the concentration of the $n + 1$ constituents in the liquid and $\bar{K}_0 = [k_0^0, k_0^1, \ldots, k_0^n]$ denotes the gross partition coefficient for elements between the two phases. To obtain the essential $\bar{K}_0$ components, place the polycomponent alloy in a sealed container such as illustrated in Fig. 4.47. Melt and stir the alloy until it is homogeneous and then allow freezing to begin slowly at one end under conditions of complete mixing.

After the charge has been solidified, it is sectioned at various positions along its length and analyzed chemically. Suppose Fig. 4.48(a) represents the distribution of the ith constituent along the bar so that $C_S^i(z')$ is

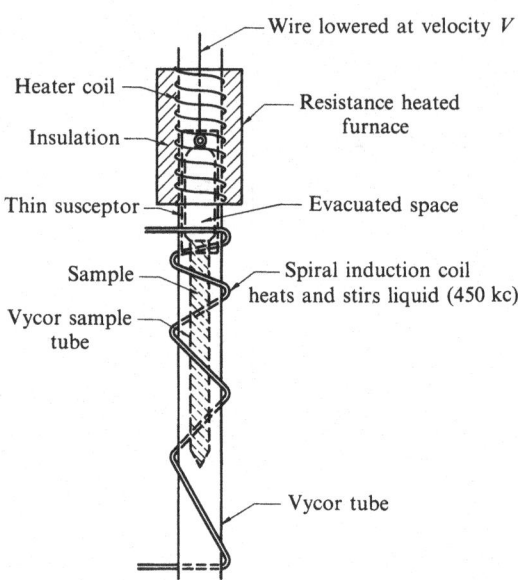

Fig. 4.47. Controlled solidification apparatus for phase diagram studies.

the measured concentration at the position z'. Thus, when the interface was at z', the volume from z' to L was liquid and $C_S^i(z')$ was frozen into the solid. If we denote the concentration of this liquid by $C_L^i(z')$, then $k_0^i(z') = C_S^i(z')/C_L^i(z')$. Because we have a closed system and the charge is of constant cross-section, we have, for $D_S^j = 0$,

$$k_0^i(z') = \frac{C_S^i(z')}{(L - z') \int_{z'}^{L} C_S^i(z)\mathrm{d}z} \tag{4.36}$$

Similarly $k_0^0(z'), k_0^1(z'), \ldots, k_0^n(z')$ may be determined for this liquid of composition $(C_L^0(z'), C_L^1(z'), \ldots, C_L^n(z'))$. Since the concentration of the liquid varies with z', the tie-line, $\bar{K}_0$, may be determined over a range of liquid concentrations from a single charge. By starting with charges of distinctly different composition, $\bar{K}_0$ can be obtained for the entire composition field in a small number of experiments.

Various types of phase boundaries in the system may also be determined by considering the variations of k_0^i as a function of position along the bar as illustrated in Fig. 4.48(b). The cases to be considered are (i) complete solid solubility where $k_0^i(z)$ is a continuous function of z from $z = 0$ to $z = L$, (ii) eutectic phase boundary where $k_o^i(z)$ suffers a discontinuity at z' and becomes equal to unity from z' to L. The order

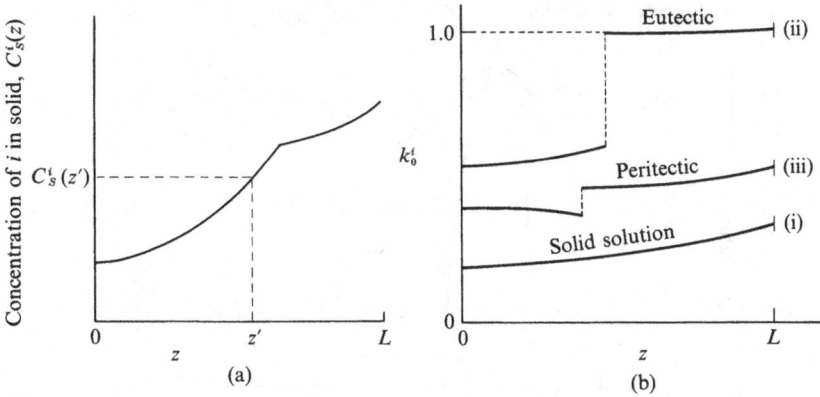

Fig. 4.48. (a) Possible distribution, $C_S^i(z)$, for the jth constituent along the length of the solidified bar and (b) variation of k_0^j versus z along the bar used for determining the types of phase boundaries present.

of the eutectic is the number of the $n + 1$ constituents that suffer this type of discontinuity at $z = z'$ and (iii) peritectic phase boundary where $k_0^i(z)$ suffers a discontinuity at z' and increases to a value less than unity and is continuous again from z' to L. The order of the peritectic is the number of the $n + 1$ constituents that suffer this type of discontinuity at the value of $z = z'$.

It is often difficult to determine the liquidus maximum position relative to the stoichiometric compound as illustrated in Fig. 4.49(a) or to discriminate between certain eutectic and peritectic reactions as illustrated in Fig. 4.49(b). The controlled freezing technique can be used to sort out these dilemmas by comparing the concentration of the last part to freeze relative to the average concentration of the bar. For Fig. 4.49(a), if the initial concentration is to the right of the liquidus maximum, the last part to freeze will be rich in B. If the initial concentration is to the left of the maximum, the last part to freeze will be deficient in B. Thus, the position of the maximum can be determined to within the accuracy of weighing the initial samples. Following the same procedure for Fig. 4.49(b) with initial samples of concentration a, b and c, the last part to freeze will always be richer in component B for the peritectic diagram but, for the eutectic diagram, the last part to freeze will be deficient in B.

The only serious limitation of the method depends on the ability to

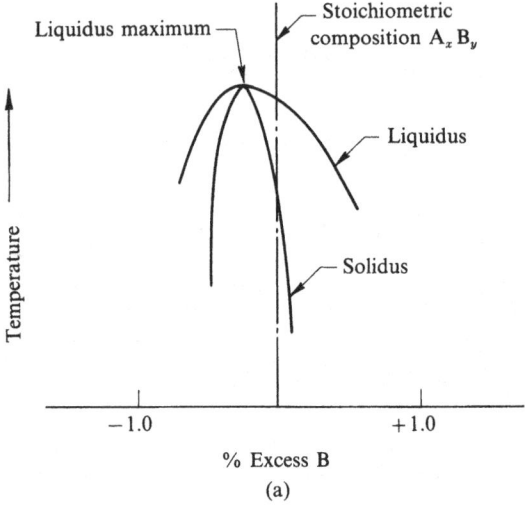

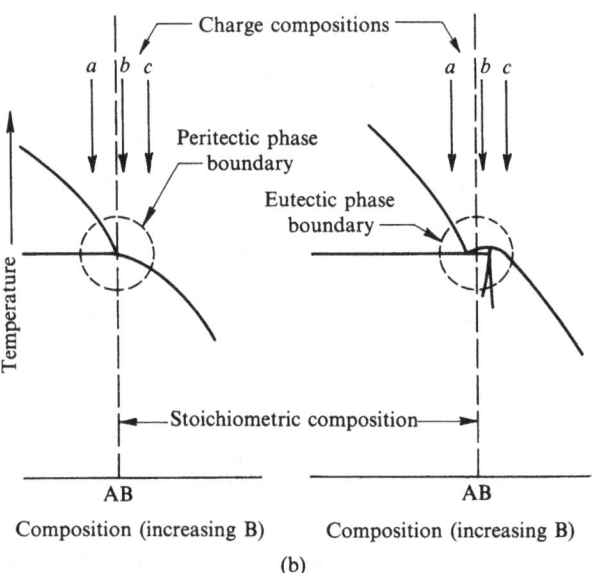

Fig. 4.49. (a) Illustration of liquidus and solidus curves near a compound composition and (b) starting compositions a, b and c used to discriminate the presence of a peritectic versus a eutectic reaction.

mix the liquid completely because, in the experiment, one measures $\bar{K}$, rather than $\bar{K}_0$. By considering the BPS equation (see Eq. (1.11a)), it can be readily seen that $k_0 < k < 1.01k_0$ provided $(1 - k_0)(V\delta_C/D_C) < 10^{-2}$. Thus, for strong stirring $(\delta_C/D_C \sim 10^2)$, one requires $V \lesssim 10^{-4}$ cm s^{-1} to achieve this level of accuracy.

It is of interest to note that, during the casting of polycomponent alloys, the interface between solid and liquid is always dendritic in nature. The liquid concentration at the tips of the dendrites is only slightly in excess of the liquidus concentration under most conditions and, if there is no strong fluid motion in the liquid, solute rejected at the tips of the dendrites diffuses laterally to the liquid channel regions between the dendrites. The lateral freezing of the channel walls is generally so slow $(V_\ell \sim 10^{-3}\text{--}10^{-4}$ cm s$^{-1})$ and the interdendritic channels so narrow $(\Delta y \sim 10^{-3}$ cm) that diffusional mixing is essentially complete in the 1–10 s it takes for these channels to close $(D_C t > \Delta y^2)$. Thus, the concentration locus along the liquidus surface taken by these channels, and the phase boundaries intersected as the walls freeze together, can be directly revealed and easily displayed by the above suggested normal freezing experiments on the same initial alloy composition. From a practical point of view, it is possible to do casting alloy design via such laboratory controlled solidification experiments. By starting with different charge compositions, one can find those starting compositions that avoid deleterious second phases and thus provide stronger and more sound castings. Fig. 4.50 shows a section of the $A\ell$–Cu–Mn system with various concentration loci during freezing from different initial concentrations.[23]

4.7   Zone melting

The technique of Fig. 3.1(b), invented by Pfann[17] leads to many special cases of interest. We will restrict our attention to only single and multiple passes under constant conditions.

4.7.1   *Single pass (constant conditions)*

If a molten zone of length ℓ is passed through a charge of constant cross-section, constant concentration C_0 and constant k as in Fig. 3.1(b), then the resulting solute distribution in the solid $C_S(z/\ell)$ is given by

$$C_S(z/\ell)/C_0 = \{1 + (1 - k)\exp[-(k\rho_s/\rho_L)(z_1/\ell)]\} \qquad (4.37)$$

where ρ_S/ρ_L is the solid/liquid density ratio. In Fig. 4.51, Eq. (4.37) has been plotted for $C_S(z/\ell)/C_0$ as a function of the number of zone lengths, z/ℓ, for several k, for a charge ten zone lengths long and for $\rho_S/\rho_L = 1$.

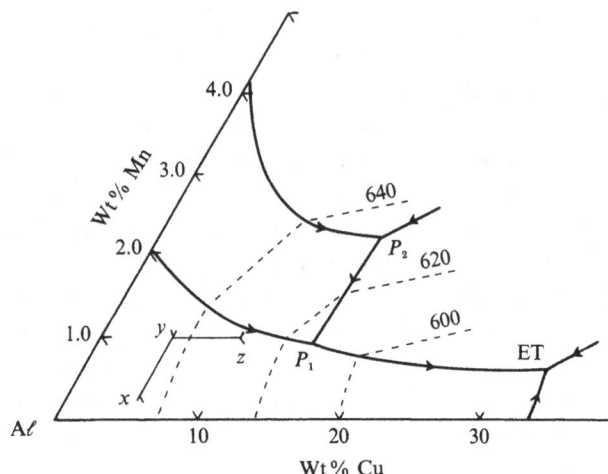

Fig. 4.50. Section of the Aℓ–Cu–Mn phase diagram showing the direction of liquid concentration change during freezing for different starting compositions. Phase boundaries are intersected at P_1 and P_2 while ET is a ternary eutectic.[23]

From Fig. 4.51, one notes that zone melting of a uniform rod results in an initial and final transient where $C_S(z_1)/C_0$ is considerably different from unity. An improvement in the uniformity can be made by passing the zone in the reverse direction. Even if k is very small and unity is not attained in the forward pass, it is found that a substantially uniform concentration exists in all but the last zone after the reverse pass.

4.7.2   Multiple pass (constant conditions) [20]

Consider the passage of a second zone through the first pass solute distributions of Fig. 4.51. The zone accumulates solute as it passes through the initial region and leaves behind a reduced and somewhat longer initial transient region. When the front of the molten zone reaches the edge of the last zone length, the slope of the solute distribution will begin to rise sharply. Thus, the pile-up at the end of the ingot progresses backwards one zone length during the second pass and one additional zone length for each succeeding pass. Multiple passes, therefore, lower the initial transient, raise the final transient and decrease the length of the intermediate region. In Fig. 4.52, $C_S(z/\ell)/C_0$ is plotted from the analytical result obtained from treating a semiinfinite charge versus the number of zone lengths, z/ℓ, for the first eight passes with $k = 0.25$ and

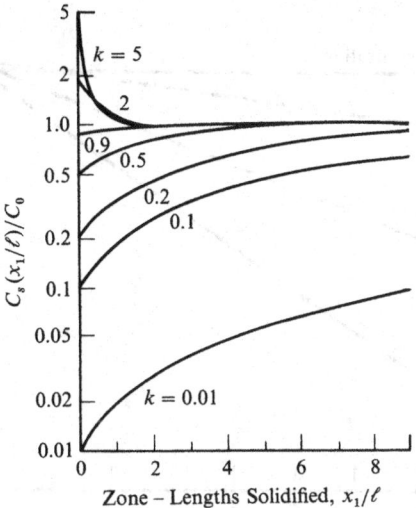

Fig. 4.51. Relative concentrations in the solid, C_S/C_∞, after single pass zone melting as a function of distance in zone lengths, z/ℓ, solidified for several values of k.[20]

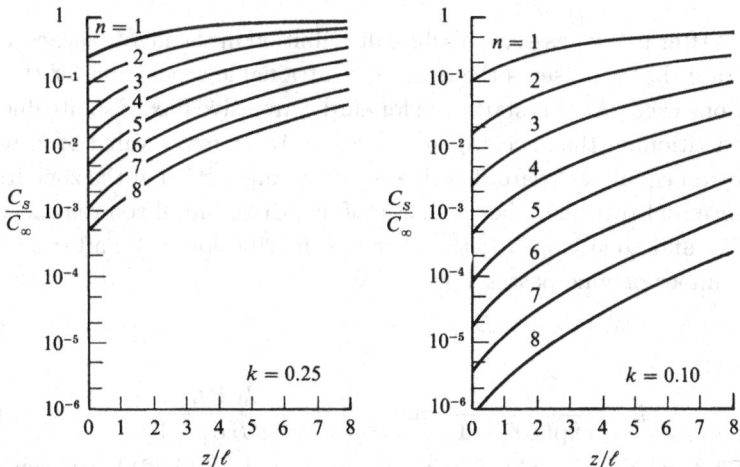

Fig. 4.52. Relative solute distribution, C_S/C_∞, in a semiinfinite solid as a function of distance in zone lengths, z/ℓ, along the bar for $k = 0.1$ and 0.25 after n zone passes.[20]

$k = 0.1$. For a finite length charge that is $(z/\ell) + 8$ zone lengths long, the results are identical to those given in Fig. 4.52.

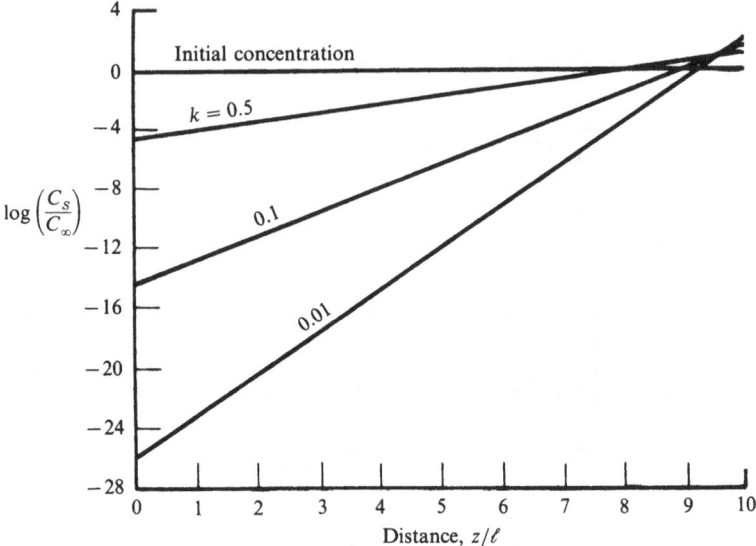

Fig. 4.53. Ultimate relative solute distribution, C_S/C_∞, in a semi-infinite solid as a function of distance in zone lengths, z/ℓ, along the bar for several values of k.[20]

After many passes, the solute distribution in the ingot reaches a steady state that represents the maximum attainable separation for the conditions used. At this stage, the forward convective flux of solute due to the partitioning that accompanies freezing is, at every value of z, opposed by an equal backward flux due to the mixing action of the zone from the terminal transient. For a charge of length L, initial solute concentation C_0, and zone length ℓ, the ultimate distribution, C_S^{**}, after an infinite number of zone passes is given by

$$C_S^{**} = A \exp(Bz) \tag{4.38a}$$

with

$$k = \frac{B}{\exp(B\ell) - 1} \quad \text{and} \quad A = \frac{C_0\, BL}{\exp(BL) - 1} \tag{4.38b}$$

The ultimate distributions calculated from Eqs. (4.38) have been plotted in Fig. 4.53. For $\ell = 1$ and $L = 10$, it may be seen that these profiles are very steep for small k. For example, if $k = 0.1$, $L = 10$ and $\ell = 1$, C_S^{**} at $z_1 = 0$ is approximately $10^{-14}C_0$.[20]

The foregoing theoretical result is not realistic in a practical situation because it is based upon the assumption of a conservative system. However, in nature, as we make a material more pure, the system can lower its free energy by dissolving into itself some of its contacting environment to provide entropy. Even the most inert crucible at normal purity levels will lose its inertness as the charge becomes more pure so that a proper theoretical analysis of the ultimate distribution must include the non-conservation flux of impurity from the dissolving crucible. The greater the enthalpy of solution of the crucible material in the melt, the smaller will be the rate of crucible dissolution.

One is always interested in minimizing the time and expense required to obtain a desired yield of material of specified purity using the zone refining technique. In practice, the design problem consists of optimizing the parameters that affect the degree and time of separation, the selection of a container plus heating and stirring means as well as the zone travel mechanism. The relevant parameters for a given crucible material appear to be: (1) number of passes, n, (2) zone length, ℓ, (3) interzone spacing, i, (4) travel rate, V, (5) diffusion boundary layer thickness, δ_C, (6) heater power per zone, P_h, and (7) stirring power per zone, P_s. In general, small ℓ is desired because that produces a better separation, at least for a large n. Small ℓ and i make for small time per pass. Large V also makes for small time per pass; however, large V leads to $k \to 1$ where no separation is possible. One can circumvent this problem by stirring to reduce δ_C and thus k. One tends to find experimentally that, if $V\delta_C/D_C \approx 1$, the optimum zone refining efficiency is achieved; i.e., $P_h + P_s$ is minimized for a given degree of purification.

The foregoing formulae apply only when the solid/liquid interface is smooth. When the interface becomes cellular or dendritic, k changes abruptly to a value close to unity as illustrated in Fig. 4.54 where the purification per pass, P_p, is plotted as a function of $V\delta_C/D_C$. The non-flat nature of a cellular interface produces lateral diffusion of solute to the liquid grooves between the cell walls so that excess solute is incorporated into the solid and P_p decreases. This lateral solute transport is over such small distances ($\lambda \sim 50 \ \mu$m) that it is relatively uninfluenced by stirring of the bulk liquid ($\delta_C > 50 \ \mu$m). Thus, in all purification applications, the first device design criterion must be that the crystallization conditions never intrude upon that domain where the cellular interface morphology is stable (see Chapter 5).

4.7.3 Temperature gradient zone melting

This zone movement technique is illustrated in Fig. 4.55 and is useful for preparing single crystals of high melting point materials or for preparing p/n junctions in semiconducting materials.[24] The starting

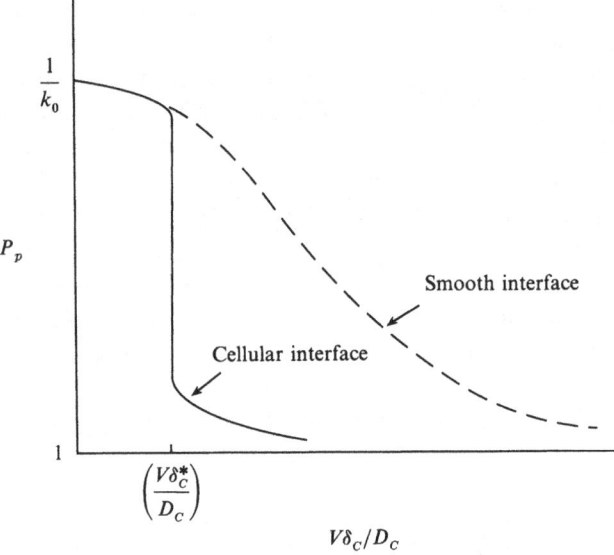

Fig. 4.54. Variation in degree of purification, P_p, for a single zone pass as a function of the stirring parameter, $V\delta_C/D_C$, for the case of a smooth interface and the case of a cellular interface.

procedure is to place a thin zone of high alloy content material (Si + C for example) between a single crystal seed (SiC) and a charge of source material (powdered or sintered SiC). This combination is placed in a temperature gradient, $\mathcal{G}$, with the highest temperature, T_{max}, far below the melting point of the source material, T_A, and where the lowest temperature, T_{min}, is below the liquidus temperature of the thin zone of high alloy content material. Thus, only the zone will melt and it will begin to migrate slowly ($V \sim 10^{-4}$ cm s^{-1} $- 10^{-6}$ cm s^{-1}) towards the higher temperature region.

The zone migration occurs because the solvent concentration, C_2, at the high temperature side of the zone (T_2) is above that, C_1, at the low temperature side of the zone (T_1) so diffusion of solvent occurs across the zone. Such diffusion undersaturates the high temperature interface causing dissolution of solvent (source material) and supersaturates the low temperature interface causing deposition of solvent (single crystal SiC).

Such zone propagation in a temperature gradient can occur for a variety of zone geometries; i.e., slab, cylinder or sphere. Such fluid zones readily occur as inclusions during crystal growth, casting or ingot for-

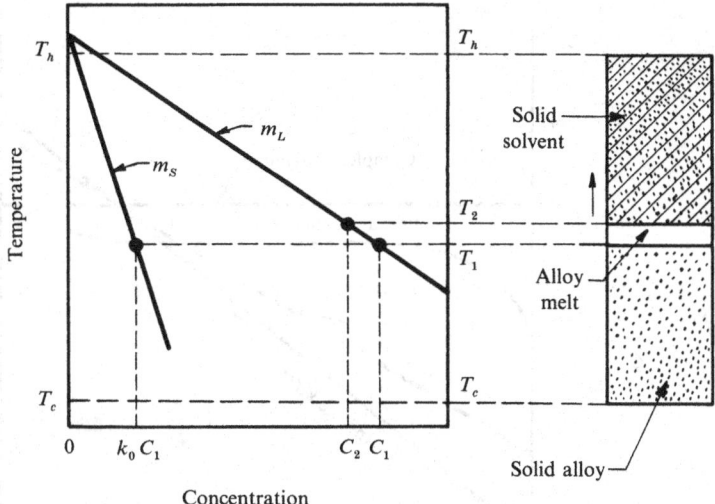

Fig. 4.55. Partial phase diagram plus the solid and liquid charge used in temperature gradient zone melting.

mation where the migration rates are limited not just by diffusion across the zone but also by the interface reaction for either dissolution or deposition. Defining $\theta = 0$ and $\theta = \pi$ as the dissolution and deposition interfaces, respectively, C_0^* and C_π^* are the equilibrium solvent concentrations. Thus, the maximum flux of solvent across the zone when there is no convection present is $D_{CL}(C_0^* - C_\pi^*)/\ell$ when the interface reactions are infinitely rapid. The concentration profile for this case is illustrated in Fig. 4.56. When the interface reaction is sluggish, driving force must be consumed by this process and the overall zone migration rate is reduced. Fig. 4.56 gives the concentration profiles across the zone in this instance for the cases of no mixing, partial mixing and complete mixing. For the no mixing case, the zone velocity is given by

$$V_\pi = -[\varepsilon\mathcal{G}/m]/\{1 - (\varepsilon/\ell m)[(1/\beta_\pi) - (1/\beta_0')]\} \tag{4.39}$$

for the uniform attachment mechanism controlling the interface reaction. Here, $\mathcal{G} = (T_0 - T_\pi)/\ell$, $\varepsilon = -D_C/(1 - k_0)C_\pi$ and $\beta_0' = \beta_0 V_\pi/V_0$.

This zone movement process can be driven by *any* electrochemical potential difference, $\Delta\eta$, across the zone not just a temperature potential difference, ΔT. Electrostatic potential difference, $\Delta\phi$, via electric field application and gravitational potential difference, Δg, via centrifugal force application[24] are both expected to produce ready zone movement. The zone movement process also occurs for vapor zones and even for

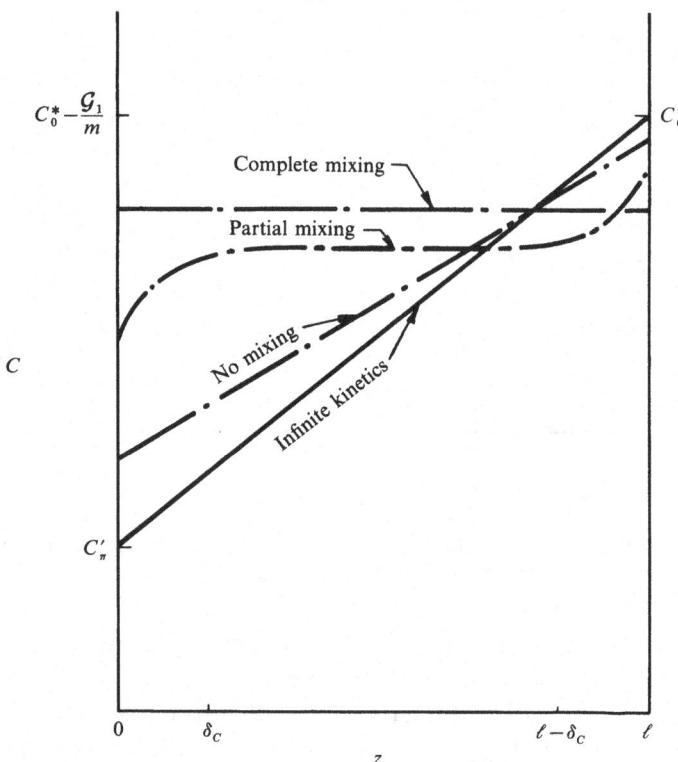

Fig. 4.56. Solvent concentration profile, C, in the molten zone as a function of distance, z, through the zone for the conditions of (i) infinite atomic kinetics with no mixing and (ii) finite atomic kinetics with no mixing, partial mixing and complete mixing.

solid zones albeit the latter zones move at extremely slow rates through their host solid where matter transport is generally limited to interfacial diffusion rather than to volume diffusion.

Problems

1. A liquid solution contains two solute constituents A and B at concentrations C_∞^A and C_∞^B that can react to form an A_xB_y compound via the reaction

$$xA + yB \overrightarrow{\longleftarrow} A_xB_y$$

During steady state planar front freezing of this solution, calculate the maximum value of C_∞^A allowable for no A_xB_y compound formation at the crystallizing interface for the cases of (a) no interface fields present, (b) interface fields present.

2. If the crystal in problem 1 is a semiconductor, A is n-type and B is p-type, use the approximate formulation for the initial transient distributions in the crystal to develop a relationship between the ks, C_∞s and V/D_L describing the formation of a p/n junction at exactly half the characteristic distance for the n-type solute.

3. (a) Use the approximate solute distribution in an unmixed liquid
$$\frac{C_L}{C_\infty} = 1 + \left(\frac{C_i}{C_\infty} - 1 \right) \exp \left(-\frac{V}{D_L} \xi \right)$$
where C_i is a slowly varying function of time, and the solute conservation condition between liquid and crystal to develop a differential equation describing the change of C_i with time.

(b) Evaluate the solute distribution in the solid for the general case (i) $V = $ constant and $k_0 = f(C_i)$ and for the specific case (ii) $V = $ constant and $k_0 = \alpha C_i^{-1/2}$.
(c) Evaluate the solute distribution in the solid for the case where $k_0 = $ constant and $V = V_0 + \beta t$.

4. You are growing a semiconductor crystal from a melt containing n-type and p-type solutes at C_∞^n and C_∞^p, respectively. Give an expression for the applied field $\hat{E}$ needed with steady state growth to produce an exactly charge-compensated crystal.

5. Develop an expression for the approximate intermediate transient solute distribution in the solid as the freezing velocity is changed from V_1 to $V(t) = V_2 - (V_2 - V_1) \exp(-\beta t)$. Use the Laplace approximation for the solute distribution in the liquid and use solute conservation between liquid and solid. Neglect interface field effects.

6. Develop an approximate terminal transient solute distribution in the solid by assuming a complete mixing approximation in the liquid after the interface has reached a distance D/V from the end of the sample. Assume that no excess solute above C_∞ had deposited in the solid until a distance $z_2 = D/V$ from the end of the sample had been reached. Show how this result is modified by including diffusion in the solid in this terminal transient region.

7. (a) Using an approximate mathematical procedure, provide an equation describing the change in the interface concentration

in the liquid, $C_i(t)$, with time for the general case where the interface is translating at a velocity $V = \beta t^n$.

(b) Using the result of (a), compare the $n = 0$ case with the initial transient solution given earlier for $V = \text{constant} = \beta$ and $k_0 = 0.1$, $D_L = 5 \times 10^{-5}$ cm^2 s^{-1}. Plot the two results for $\beta = 10^{-4}$ and 10^{-1} cm s^{-1}.

8. In the RAMP cooling technique for LPE (liquid phase epitaxy) of a solution having a liquidus line given by $C_L = K \exp(-\Delta H/ \mathcal{R}T)$ and a constant cooling rate $\dot{T}$ from an initially saturated condition, develop an approximate expression for the velocity of the crystal layer as a function of time, $V(t)$ for the two cases: (i) the solution layer is infinitely thick and (ii) the solution layer has a finite thickness ℓ (assume $\Delta G_K = \Delta G_E = 0$).

9. During crystal growth from a stirred liquid of solute boundary layer thickness δ_C, it is usually assumed that the solute carried in the interface layer is negligible and doesn't influence the C_S/C_∞ profiles. For a constant V and constant δ_C, calculate an expression for the initial transient build-up of this interface solute profile and compare its magnitude to that for the unstirred case.

10. You have a stirred liquid solution at saturation and quench it an amount ΔT^*. As a result of the quench, N spherical crystallites nucleate per cubic centimeter and begin to grow. Around each spherical crystal a constant solute boundary layer thickness of magnitude δ_C forms. The initial solution concentration is C_∞ and the concentration of the solid forming is $C_S = \text{constant}$.

(a) Derive an approximate expression for the velocity, $V(t)$, of the crystal as a function of time using a constant liquidus slope and $\Delta G_K = \Delta G_E = 0$ approximation.

(b) Solve for the no stirring case ($\delta_C = \infty$) by using the Laplace approximation for the solute distribution

$$C_L(r,t) \approx \frac{A(t)}{r/R} + B(t)$$

where R is the crystal radius.

(c) In case (a), what minimum delay time τ and maximum cooling rate $\dot{T}$ can be tolerated without nucleating any new crystals. Assume that no crystals will nucleate if the supercooling at any point in the solution does not exceed ΔT_c (which is less than ΔT^*).

11. You have a large three-component melt (85% Ga, 15% As+ dopant) which is supersaturated by an amount ΔC_∞. You seed it in the center of the melt and a spherical crystal grows at a velocity $\dot{R} = \beta t^{-1/2}$.

(a) Make schematic diagrams for the As and dopant concentration profiles in the melt for several values of the crystal radius R.

(b) On a diagram, at some value of R, draw all the key concentrations involved in the partitioning of the total supersaturation, C_∞/C_i, into all the subprocesses contributing to the growth of the crystal.

(c) Give approximate quantitative expressions for the radial distribution of the dopant in this crystal and for the amount of dopant being carried ahead of the interface as a function of R.

(d) Given that the atomic density of Ga is 50% greater than the atomic density of As, at what value of R does convection in the liquid initiate via Rayleigh instability assuming an isothermal solution?

$$\Delta T_{stable} < \frac{R_c \alpha_L \nu}{g\rho^{-1}(\mathrm{d}\rho/\mathrm{d}T)d^3}$$

12. For the GaAs crystal growth of question 4 with large β, suppose you have both n and p-type dopants in the melt such that $C_\infty^n = C_\infty^p$ and $k_0^n = \frac{1}{2}k_0^p$. Neglecting any vacancy effects and, assuming that $D_L^n = D_L^p$ and $D_S^n \gg D_S^p$, for a large negative electrostatic potential difference between the interface and the bulk solid ($\Delta\phi_i < 0$), qualitatively plot the distribution of n and p-type dopant in the crystal as a function of R.

13. You have some liquid inclusion drops of radius R in a single crystal and wish to move them through the crystal by application of a temperature gradient and an electric field. Assuming that the dissolution and deposition kinetics are infinitely rapid, calculate the migration velocity, V, as a function of temperature gradient and electric potential gradient in the liquid droplet, $\mathcal{G}_L$ and $\hat{E}_L$, respectively for constant value of liquidus slope, m_L, surface tension, γ, relative ionic mobility $\hat{M}_s$ and liquid concentration C_L. Calculate V for $\hat{E}_L = 0$ and then for $\hat{E}_L \neq 0$.

14. Use the transformation given by Eqs. (4.7a) and (4.7b) to show that the BPS equation for the effective k-value (Eq. (1.11a)) still

holds for ionic systems but with V_E and k_E replacing V and k_0, respectively, provided E/V is a constant.

5

Morphological stability of interfaces

This chapter and the next chapter is where all the earlier chapters come together to determine specific crystal shape effects. This is perhaps best illustrated by recalling the "Coupling Equation" from Chapter 1 with its temperature, concentration and pressure analogues; i.e.,

$$\Delta G_\infty = \Delta G_{sv} + \Delta G_E + \Delta G_K \qquad (5.1a)$$

$$\Delta T_\infty = \Delta T_C + \Delta T_T + \Delta T_E + \Delta T_K \qquad (5.1b)$$

$$\Delta C_\infty = \Delta C_C + \Delta C_T + \Delta C_E + \Delta C_K \qquad (5.1c)$$

and

$$\Delta P_\infty = \Delta P_C + \Delta P_T + \Delta P_E + \Delta P_K \qquad (5.1d)$$

Eqs. (5.1) are a major thermodynamic constraint on the growing crystal system combining the interactive aspects of temperature distribution, solute distribution, interface reaction plus surface creation and excess energy storage (as well as fluid flow and electrical potential fields in some cases). For constrained or unconstrained growth, the average crystal front velocity is generally known or can be obtained as a function of time by the macroscopic solution to the heat or matter transport equation. The microscopic shape of the interface is not unique but may select one of several possible morphologies that satisfy all the heat and mass transport constraints and also satisfy Eqs. (5.1). Thus, they can all be considered as metastable dynamic interface states. Under a given set of crystal growth conditions, one of these morphologies is the most kinetically stable and that is the crystal morphology that should be observed experimentally (provided there is infinite growth time available under constant experimental conditions). This chapter deals with the concepts and techniques for predicting the dominant crystal morphology in a specific situation while Chapter 6 provides abundant experimental examples to compare with the theory.

Perhaps the simplest example that illustrates the non-uniqueness of the interface morphology is the constrained growth case of the planar/cellular interface transition. This well-developed cellular interface morphology is illustrated in Fig. 5.1 both for the case of a Pb–Ag alloy where the liquid is rapidly decanted from the growing solid and for the case of ice/water where the transparency allows us to see the filamentary nature of the interface. The main experimental observations are that, with constant V, C_∞, $\mathcal{G}_L$ and k_0, and the extent of growth such that steady state conditions prevail as the temperature gradient in the liquid, $\mathcal{G}_L$, is lowered by steps, the macroscopic solid/liquid interface is smooth and planar for all $\mathcal{G}_L$ above some critical value, $\mathcal{G}_{Lc}$. However, as $\mathcal{G}_L$ decreases below $\mathcal{G}_{Lc}$, the apparently steady state interface morphology begins to develop first small indentations followed by irregular grooves which are, in turn, followed by a regular array of grooves across one coordinate direction of the interface. Finally, as $\mathcal{G}_L$ is further lowered, a regular array of grooves develops and these change in size as $\mathcal{G}_L$ continues to be lowered. Observations with transparent materials revealed the transition from a smooth interface to a filamentary interface as $\mathcal{G}_L$ is lowered below $\mathcal{G}_{Lc}$. This is illustrated in Figs. 5.1 and 5.2.

The initial explanation for this phenomenon was that it was associated with the onset of constitutional supercooling in the liquid ahead of the interface (see Fig. 5.3). For a steady state solute distribution with $k_0 < 1$ as in Fig. 5.3(a), the equilibrium liquidus temperature distribution for this interface layer of liquid is shown in Figs. 5.3(c) and 5.3(d). If the actual temperature distribution is as in Fig. 5.3(d), then a zone of liquid exists whose temperature is below the liquidus temperature. This zone of liquid is said to be constitutionally supercooled. Since the interface morphology development progressed as $\mathcal{G}_L$ was lowered below $\mathcal{G}_{Lc}$, it advanced as the degree of constitutional supercooling increased and it was thus presumed that the onset of constitutional supercooling led to the instability of the smooth interface (see Fig. 5.3(c)). The development of a small bump on the interface as in Fig. 5.4, allowed the tip to be in a region of increased supercooling when the conditions of Fig. 5.3(d) hold; thus, except for the restraining factor of capillarity, the bump should grow even faster and destabilize the smooth interface. In Fig. 5.5, the

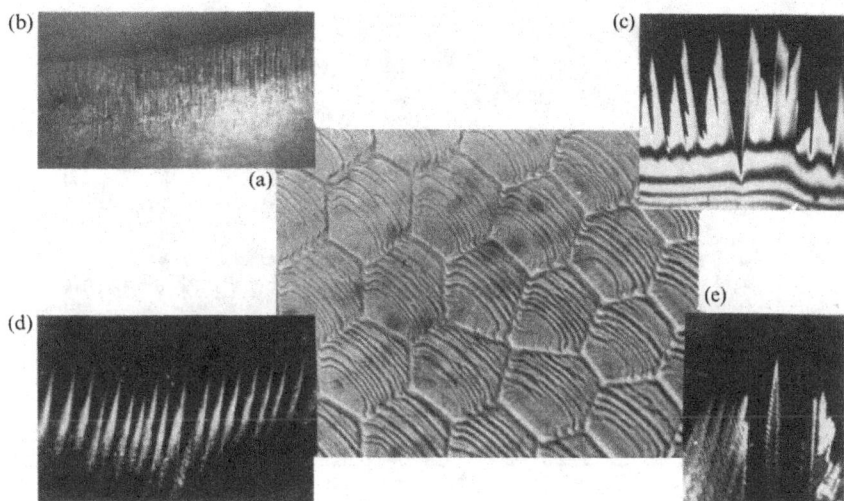

Fig. 5.1. Filament and dendrite formation during constrained growth: (a) photomicrograph of a decanted Pb crystal solidifying from a Pb + 5×10^{-4} wt% Ag melt; (b) direct observation of cell formation in ice; (c) ice bicrystal grown from very dilute HCℓ solution plus a surfactant (polarized light); (d) ice filament growth at an interface from a K_2CrO_4 solution; (e) onset of side branch formation on ice filaments (2%NaCℓ solution).

progressive development of the interface transition to regular hexagonal cells is illustrated as $\mathcal{G}_L$ is lowered below $\mathcal{G}_{Lc}$ based on the foregoing concept. We note that the area of cell boundary, into which excess solute can be stored, increases as the degree of constitutional supercooling *ahead of a smooth interface* would have increased.

The point of the foregoing example is that there appears to be a family of solutions to the solute and heat transport problems involving different detailed shapes for the interface and, under a given set of growth conditions, nature selects one of these solutions as the optimum. Without the definition of a stability condition that operates in this case, the system appears to provide n constraints but $n + 1$ variables so it is not completely specified without the specification of this stability condition.

5.1 Stability criteria for interface shape

Several different viewpoints have been voiced concerning the extra condition needed for selecting the stable interface shape. One is essentially thermodynamic in origin and recognizes that the morphologi-

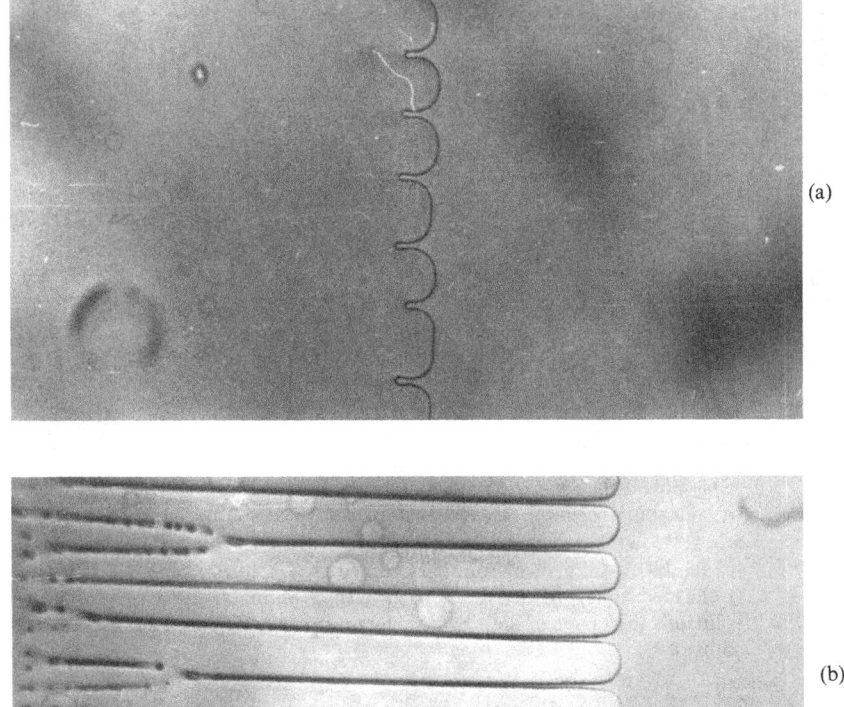

(a)

(b)

Fig. 5.2. Cellular interface in a transparent organic at 72 × Mag (succinonitrile–5.5 wt% acetone – courtesy of R. Trivedi).

cal transition of a crystal front, like a phase transition itself, is essentially a cooperative phenomenon. Kirkaldy[1] has been the leading proponent of this viewpoint and has suggested that the variational principles of irreversible thermodynamics require that a special condition be placed upon the rate of entropy production in order to specify the problem completely. This is a global viewpoint.

The second viewpoint is a more localized one, focusing its attention on an increment of interface and upon the immediately surrounding volume.

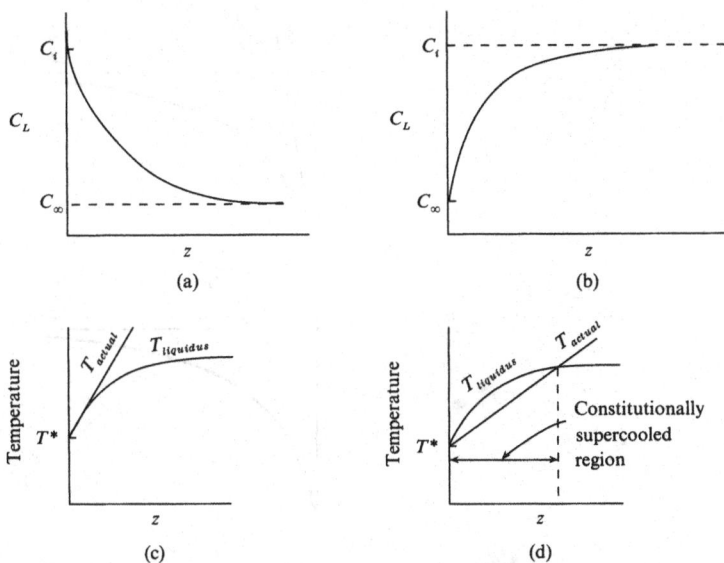

Fig. 5.3. Definition of constitutional supercooling (CS) in alloy solidification (a) solute-enriched layer in liquid at the interface for $k_0 < 1$, (b) same for $k_0 > 1$, (c) condition of no CS and (d) condition of CS.

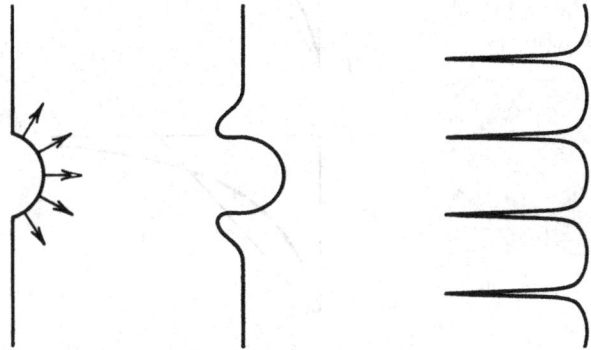

Fig. 5.4. Transition from a single interface bump to a cellular interface.

Following this viewpoint, the approach to some assessment of shape stability has been largely via two distinct paths. The stationary state development of bodies of simple shape has been considered (ellipsoids, elliptical paraboloids, etc.) and attempts have been made to determine (a) the conditions under which perturbation theory would indicate that

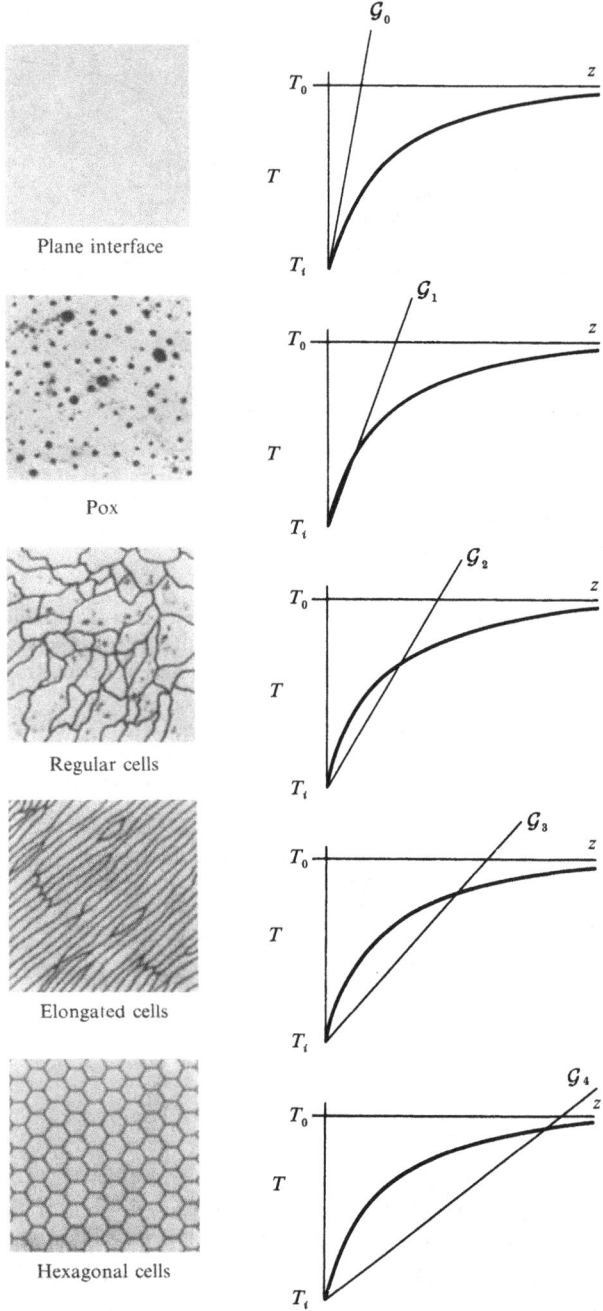

Fig. 5.5. Schematic illustration of various morphologies observed on decanted interfaces of Sn and Pb as a function of the planar interface degree of constitutional supercooling.

a particular shape is stable or unstable to small amplitude distortions and (b) which shape grows in its characteristic directions with maximum velocity (unconstrained growth) or maximum intrusion into the nutrient phase (constrained growth) for a given set of conditions and a given span of time. The first method is called the "perturbation method," with Mullins and Sekerka[2] as its leading proponents, while the second method is called the "extremum method" with Zener[3] and this author[4] as major proponents. In principle, the Perturbation Method (PM) is used to indicate when a particular shape is unstable whereas the Extremum Method (EM) is used to indicate which shape, of a group of shapes under comparison, is the most stable against perturbations (although they all may be unstable to various degrees).

As we shall see, the two methods of this second case are closely related, they just adopt different viewpoints and, in fact, the EM is a consequence of the PM. Thus, one can say that the extra needed condition for quantitatively resolving the crystal growth picture involves the response of the interface to perturbations. The perturbation wavelength that gives the fastest response for an unstable surface moves the surface most quickly to the shape that is just marginally stable. At this point, all perturbation wavelengths decay in amplitude except one which does not change in amplitude and is, in fact, the characteristic shape of the marginally stable surface. This surface is then said to be at the extremum condition. From this, one might say that *the EM bears the same relationship to the kinetics of crystallization processes that the minimum free energy postulate does to the equilibrium state in thermodynamics.*

5.1.1 Procedural philosophy for the PM

In nature, the atomic scale motions always produce microfluctuations of the thermodynamic state variables (T, C, P, ϕ) at any location in a thermodynamic system. Thus, any crystal/nutrient interface is subject to a spatially fluctuating driving force, $\bar{\mu}_i + \delta\mu_i$, which constitutes a perturbation of the system that can manifest itself as an infinitesimal bump like that illustrated in Fig. 5.4. Any bump can be thought of as the superposition of waves belonging to an orthonormal set of waves. Thus, a shape perturbation can be thought of as any member of this wave set and one is interested to see if its amplitude grows or decays in time. If it grows, the shape of the crystal front is unstable; if it decays, the shape is stable relative to that wavelength.

Suppose, as an example, one considers the transition from a somewhat spherical particle shape to a spiked or predendritic shape as illustrated

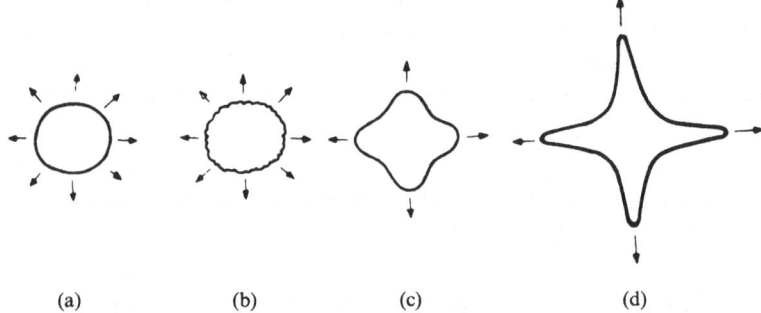

(a)   (b)   (c)   (d)

Fig. 5.6. Growth of shape perturbations on a sphere (a) as it transforms to the filamentary morphology (d).

schematically in Fig. 5.6. The appropriate orthonormal set for such
surface waves would be the spherical harmonics. For a single wave of
amplitude δ and frequency ω, growth in amplitude relative to the main
crystal surface requires the creation of surface in addition to increased
curvature at the tips of the wave so that, in the coupling equation (see
Eqs. (5.1)), both $\Delta T_E(\Delta\mu_E)$ and $\Delta T_K(\Delta\mu_K)$ must increase in magnitude (V increases also). Since ΔT_∞ is considered to be fixed, one sees
that $\Delta T_C + \Delta T_T(\Delta\mu_{sv})$ must decrease at the tips of the wave in order
that the amplitude, δ, grows in magnitude. It can be readily shown that
the magnitudes of ΔT_C and ΔT_T are controlled by the same type of
transport laws and that ΔT_C varies in the same manner as ΔT_T for this
particular example. Thus, one can conclude that a necessary condition
for dendritic growth is that the transport of heat or matter plays an
important role in the overall growth process and is more efficient for the
dendritic shape than for other particle shapes.

For the mathematical analysis of shape instability for a crystal of simple shape (ellipsoid, elliptical parabaloid, etc.) growing at velocity V,
one must first select a complete orthonormal set of functions that can
be used to describe conveniently both shape distortions as well as temperature and solute variations in the vicinity of the surface. Next, one
considers an arbitrarily shaped distortion with infinitesimal amplitude
applied to the crystal surface. This distortion can be Fourier analyzed
into the superposition of contributions from members of the selected orthonormal set of functions with amplitudes $\delta_{\ell m}$ corresponding to the ℓ
and m indices of the particular harmonic under consideration (frequency
ω_ℓ in the x-direction and ω_m in the y-direction). The aim of the subse-

quent analysis is to determine $\dot{\delta}_{\ell m}$, the time variation of the component amplitude. In most cases, to first order in $\delta_{\ell m}$, all the equations to be used are either linear or can be readily linearized, so one need treat only one harmonic at a time with the velocity of the protuberances being given by $V + \dot{\delta}_{\ell m}$. By working with Eqs. (5.1), the coupling equation, one solves for $\Delta T_C, \Delta T_T, \Delta T_K$ and ΔT_E in terms of the same set of functions. Then, Eqs. (5.1), fully expressed in terms of these functions, can be used to yield both $V(t)$ and $\dot{\delta}_{\ell m}$ by the use of the orthogonality properties, provided terms higher than first order in $\delta_{\ell m}$ can be neglected. Following this procedure, one finds for $\dot{\delta}_{\ell m}$ versus ω_ℓ or ω_m results like those illustrated schematically in Fig. 5.7.

In Fig. 5.7(a), for curve (i), all possible harmonics decrease in amplitude so the basis surface is stable. For curve (iii), $\dot{\delta}_{\ell m}/\delta_{\ell m} > 0$ for a range of frequencies so the surface is unstable. If the surface is unstable for any particular value of ω_ℓ or ω_m, that component of the distortion spectrum will grow and change the shape of the basis surface. At large (ℓ, m), $\dot{\delta}_{\ell m}/\delta_{\ell m}$ decreases because of the dominance of the ΔT_E contribution at small wavelengths. At small (ℓ, m), $\dot{\delta}_{\ell m}/\delta_{\ell m}$ also decreases because the effective boundary layer thickness of the transport fields on the undisturbed basis surface is much smaller than the perturbation wavelength. This makes lateral diffusion ineffective for the redistribution of heat or solute; i.e., the point effect of diffusion is no longer significant. Curve (ii) in Fig. 5.7(a) represents the condition of marginal stability for the basis surface. For this condition, the basis surface is the extremum surface, balancing between instability and overstability. This basis surface cannot grow any faster without becoming unstable and it is thus the surface to be selected by proper application of the EM.

The second important feature to note about the PM is that it provides a description of the time evolution of a perturbed interface so that a time and distance scale of phenomena can be ascertained. Such a description is valid so long as the perturbations and their effects remain sufficiently small to be described by linear theory. Considering a surface distortion that is only one-dimensional, if the initial surface shape gives a Fourier spectrum $\bar{\phi}(\omega_\ell)$, where ω_ℓ is the frequency of the individual harmonic, the first order PM predicts that the interface shape evolves in time like

$$\phi(x,t) = \frac{1}{2\pi} \int_{-\infty}^{\infty} \exp(i\omega_\ell x)\bar{\phi}_o(\omega_\ell) \exp\{t[f(\omega_\ell)]\} d\omega_\ell \qquad (5.2)$$

where $f(\omega_\ell) = \dot{\delta}_\ell/\delta_\ell$ as given in Fig. 5.7(a). One thus sees that, for that range of ω_ℓ where $\dot{\delta}_\ell/\delta_\ell > 0$, the shape grows exponentially with time.

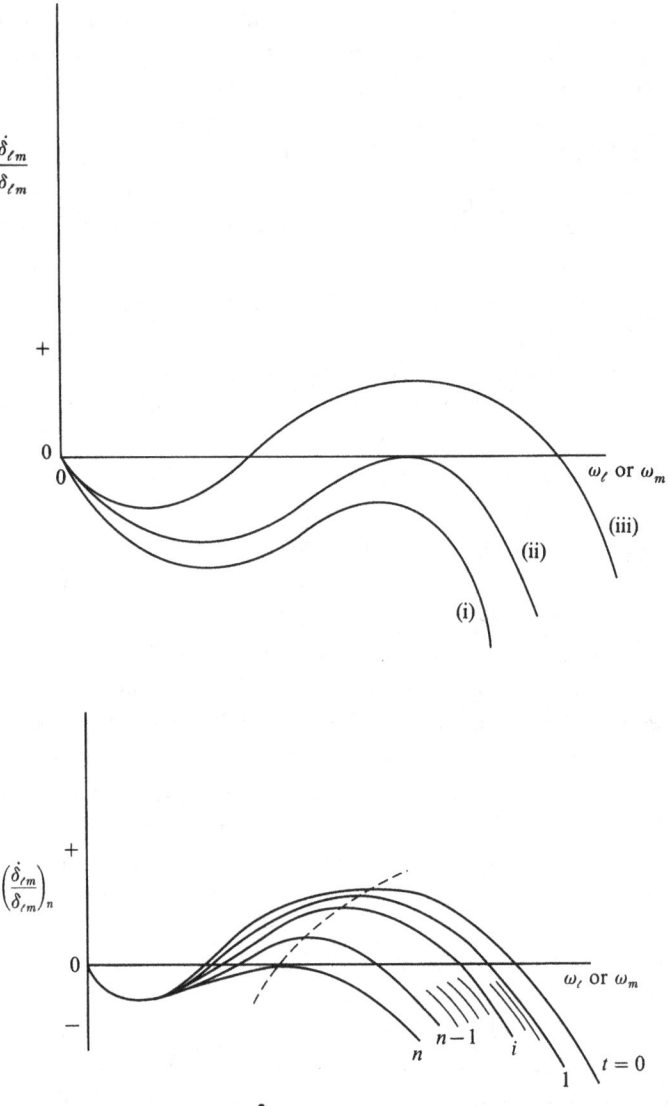

Fig. 5.7. (a) Plot of $\dot{\delta}_{\ell m}/\delta_{\ell m}$ versus either ω_ℓ or ω_m. Curve (i) is representative of a completely stable interface whereas curve (iii) is unstable for a certain frequency range and curve (ii) represents the marginally stable interface. (b) Changes in $\dot{\delta}_{\ell m}/\delta_{\ell m}$ as the basis surface changes with time from its initially very unstable shape at $t = 0$ to its final marginally stable shape at $t = n$.

Likewise, those components in the range of ω_ℓ, where $\dot{\delta}_\ell / \delta_\ell < 0$, decay exponentially with time.

5.1.2   EM versus PM

Whereas the point of view of the PM is to focus attention on the values of ω_ℓ and ω_m which yield values of $\dot{\delta}_{\ell m} / \delta_{\ell m} \lessgtr 0$ in Fig. 5.7(a), the viewpoint of the EM is to focus attention on the values of ℓ or m which yield $\dot{\delta}_{\ell m} / \delta_{\ell m}$ equal to a maximum. Thus, the most kinetically dominant momentary shape in any time increment is that shape which yields the maximum velocity of the protruding surface elements. The evolution of the basis surface from its initial state through a sequence of intermediate states to some final state will be determined mainly by surface waves of frequency given by the maximum $\dot{\delta}_{\ell m} / \delta_{\ell m}$ in Fig. 5.7(a) for the particular basis surface. In terms of a sequence $0, 1, 2, 3, \ldots, n$ of basis surfaces separated in time by small increments of Δt, the PM method would show $\dot{\delta}_{\ell m} / \delta_{\ell m}$ response spectrum shift for the ith basis surface like that illustrated in Fig. 5.7(b). The basis surface shape changes from its initial state (0) largely via the extremum path to its final state (n) which becomes the marginally stable shape for that set of growth conditions. This final shape is the extremum shape without becoming unstable.

Based on this line of reasoning, postulates can be phrased which control the crystal shape for both constrained and unconstrained crystal growth. Such postulates may only be applied to those situations where, after all the simultaneous constraints have been applied, the mathematical solution is still not completely specified. In practice, investigators often use the continuum surface approximation and do not also require that the surface creation condition (see Chapter 1) be satisfied; thus, there often appears to be one or more extra variables undetermined by the phrasing of the problem and these are resolved by application of an extremum condition. In such cases, some numerical error will be involved. The following are the general postulates:

(A) Unconstrained growth: A kinetically stable crystal shape, at any particular time during its growth from a nutrient phase, is that marginally stable shape which exhibits an extremum in the velocity of protruding surface elements with respect to small deviations in shape about this optimum shape for the particular surface element. This extremum is to have a negative second derivative with respect to such shape changes and must satisfy all the constraints implied by the coupling equation at every point of the interface.

(B) Constrained growth: A kinetically stable crystal shape, at any particular time during its growth from a nutrient phase, is that marginally stable shape which exhibits an extremum in the position of protruding surface elements with respect to small deviations in shape about this optimum shape. This extremum is to have a negative second derivative with respect to such shape changes and must satisfy all the constraints implied by the coupling equation at every point of the interface.

From these extremum postulates, one is led immediately to the following corollary deductions for crystallization from an infinite nutrient phase:

(1) Individual shape distortions on a crystal interface will develop in those crystallographic directions which lead to an extremum in either the velocity of growth or the intrusion distance of the distortions; i.e., dendrites will select a particular crystallographic direction for optimal growth.

(2) Multiple shape distortions on a macroscopic crystal interface will adopt that spacing which leads to either an extremum in the growth velocity or the intrusion distance of the array of distortions, i.e., a parallel array of filaments or a lamellar eutectic will have some optimum spacing for a particular set of growth conditions.

(3) The chemical constitution of a particular interface element of crystal, growing from an alloy nutrient, will eventually adjust to that value which leads to an extremum in the growth velocity or the intrusion distance of the surface element subject to all the system constraints; i.e., for a particular alloy concentration in the nutrient at the interface, the crystal formation that occurs need not be at the equilibrium concentration for this nutrient.

(4) The crystallographic structure (phase constitution) of a perturbed interface element of crystal growing from a nutrient phase will adjust to that class which leads to an extremum in the growth velocity or the intrusion distance for the surface element; i.e., if a metastable phase has a realistically finite probability of formation on a crystal surface and it grows at a greater velocity or intrudes further than the original phase, the metastable phase will be kinetically stabilized and will dominate the transformation.

(5) The imperfection type and content of a perturbed interface element will adjust to that which leads to an extremum in the velocity of growth or the intrusion distance of the interface element; i.e., if the defect (twin, dislocation) has a realistically finite probability of forming on the crystal and its formation allows the growth velocity or intrusion distance of that defect surface element to be increased over that of the defect-free surface, the defect will be kinetically stabilized and will propagate in the crystal interface.

For each of the above deductions, every other interface in the particular class under consideration is unstable with respect to the optimum member selected by the extremum condition and, eventually, the optimum member will appear and dominate the growth front during this infinite time period of crystallization. Thus, beyond some minimum driving force regime, even dislocation-free Si crystals would eventually develop dislocations or twin lamellae because these would allow more rapid interface movement and thereby be kinetically stabilized. However, like the condition of minimum free energy, this condition of marginal stability is an end point that is only approached asymptotically and may not be achieved in a particular experiment. In the thermodynamic domain, large activation barriers often exist between thermodynamically metastable phases and the stable phase and one finds a metastable phase like diamond exhibiting stable-like characteristics at NTP. We have had enough experience with such cases that we don't deny the correctness of the minimum free energy condition because of their presence in nature. Likewise, one should expect to find cases of kinetically metastable growth forms and should not deny the correctness of the extremum condition because of their presence. Experience will eventually ease our uncertainties in this regard.

In the sections to follow, we are required to return to the critical issue mentioned in Chapter 1 and discussed more fully in the companion book[5]; i.e., the continuum approximation to an interface versus the TLK (terrace-ledge-kink) version of an interface. Although there is no direct experimental evidence to support the continuum picture of interfaces, it is a convenient mathematical approximation and many theoretical analyses have been based upon it. In addition, *no* theoretical analysis of interface morphology made to date has properly accounted for surface creation during interface shape changes so the conclusions based upon such approximate theoretical treatments must be considered as only tentative at best.

5.1.3   *Constraints on pattern formation in crystals*

Much interest is being expressed these days, by physicists and chemists, in the pattern formation characteristics of crystals growing far from equilibrium with respect to their nutrient phase. Of course, in this regime of supersaturation, the growth figures are largely matter and heat transport dominated and, thus, bear some relationship to fractal-like figures. We know that all crystal growth is required to conform to the free energy conservation equation (see Eqs. (5.1)) wherein the $\Delta G_C + \Delta G_T$ terms favor filamentary growth morphologies via the point effect of diffusion while the $\Delta G_E + \Delta G_K$ terms favor compact crystal morphologies via bond strength effects. Thus, one should expect to be able to present a "map" of crystal shapes in a coordinate system of dimensionless driving force and dimensionless bond strength. Such a "map" has been developed by Xiao, Alexander and Rosenberger[6] using a nearest neighbor (NN)-only bond model yielding the results shown in Fig. 5.8. In Region I, crystals are constrained to acquire compact polyhedral forms because the cost of forming surface, ε_1, is quite high relative to the available driving force, $\Delta\mu$. In Region III, the crystals can become side-branched filamentary forms because the cost of forming the surface is low and the available driving force is high. As an exercise, it is useful to understand the appropriate ranges of $(\Delta\mu/\kappa T, \varepsilon_1/\kappa T)$ leading to these forms for vapor growth, melt growth and solution growth. Let us consider only the two-dimensional hexagonal filamentary form in Region II of the figure. For metals, the cohesive energy is $\sim$ 3–5 eV and $\sim$ 10 NN bonds are present so each bond, ϕ_1, has an energy $\sim$ 0.3–0.5 eV. For the basal plane $\sim$ 60% of the bonds per atom lie in the plane while $\sim$ 40% lie out of the plane; thus $\varepsilon_1 \sim$ 1 eV for such a simple bond model where ε_1 represents the surface energy per atom. In the limit that $\Delta G_\infty \to \Delta G_E$, all the driving force is stored as surface energy and none is needed for transport. Thus, in this limit, we can find the minimum value of $\Delta\mu/\kappa T$ needed, for a given $\varepsilon_1/\kappa T$ or $\phi_1/\kappa T$, to form this hexagonal filamentary shape.

Let each of the rods of this shape be of square cross-section with a length/width aspect ratio of ℓ and contain n growth units across the width. Thus, the volume of the crystal is 6 ℓn^3 while the surface area is 24 ℓn^2 giving an excess surface energy of $\Delta G_E = 24 \ell n^2 \varepsilon_1$ and an available driving force of $6\ell n^3 \Delta\mu$. We conclude that, in this limit, we

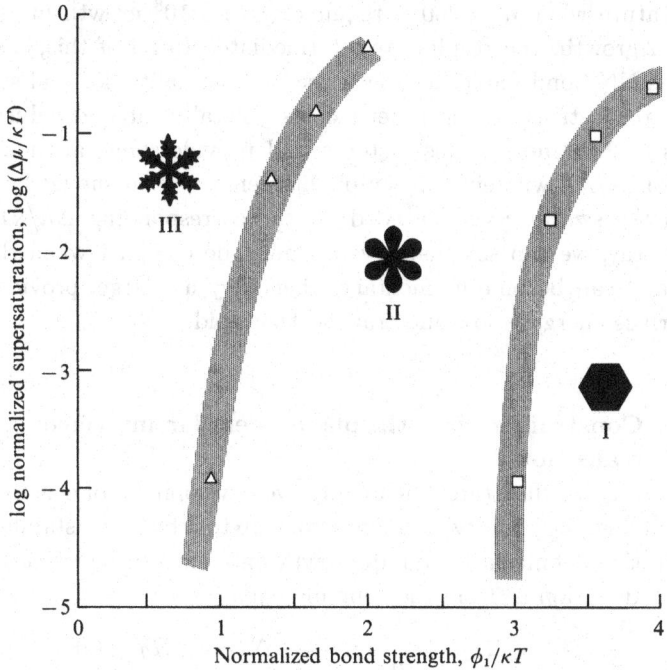

Fig. 5.8. Two-dimensional crystal growth morphology stability map in terms of the supersaturation and bond strength parameters. Region I requires compact, faceted forms. Region II yields branch-less hexagonal needles. Region III yields dendritic side-branching.[6]

require

$$\varepsilon_1 < \frac{n}{4}\Delta\mu$$

For an aspect ratio of $\ell \sim 10$ and a total number of growth units in the crystal $\sim 3 \times 10^3$, we find $n \sim 4$ so that $\varepsilon_1/\kappa T < \phi_1/\kappa T$ which requires that $\Delta\mu > 1$ eV when $\varepsilon_1 \sim 1$ eV for vapor growth. As the crystal enlarges a hundredfold (3×10^5 growth units) with the same ℓ, we find $n \sim 20$ so that $\Delta\mu > 0.2$ eV when $\varepsilon_1 \sim 1$ eV or $\phi_1 \sim 0.5$ eV. Thus, as the crystal becomes larger, the filamentary form is stabilized at a smaller driving force. For growth from the vapor, the center of Region II in Fig. 5.8 requires a growth temperature of $\sim 2000\,°C$ ($\phi_1/\kappa T \sim 2.5$) and $\Delta\mu/\kappa T \sim 10^{-2}$ so that $\Delta\mu > 0.002$ eV which requires that $n \sim 2000$ for $\ell = 10$ ($\sim 3 \times 10^{11}$ growth units).

For growth from the melt, $\Delta H_F \sim 0.1 \, \Delta H_V$, so that ε_1 is reduced by the same factor ($\phi_1 \sim 0.1$ eV) and $\phi_1/\kappa T \sim 2.5$ occurs around room temperature while $n \sim 200$ is required ($\sim 3 \times 10^8$ growth units). For solution growth, the results are intermediate. In all of this, a strictly rigorous NN bond model has been used. In actuality, for real surfaces, surface reconstruction and other factors generally enter to reduce ε_1 by factors of 2 or more. This factor would provide some latitude in the range of $\phi_1/\kappa T$ wherein this simple filamentary form should be found, and in the size of crystal needed for the corresponding $\Delta\mu/\kappa T$ used. In this way, we can see that crystals must be compact when they are small and can become filamentary when they are large, provided that the surface energy is low and transport is rapid.

5.2   Constrained growth, planar/cellular interface transition

Fig. 5.3 illustrates the qualitative development of constitutional supercooling. Let us now see its relationship to interface instability. The quantitative definition for bump growth on a smooth interface growing in the z-direction is, from the coupling equation,

$$\frac{\mathrm{d}}{\mathrm{d}z}(\Delta T_K + \Delta T_E) = \frac{\mathrm{d}}{\mathrm{d}z}(\Delta T_\infty - \Delta T_C - \Delta T_T) > 0 \qquad (5.3a)$$

For this condition, the driving force for molecular attachment and excess energy incorporation increases with the distance into the liquid while, from Fig. 5.3, this is given by

$$\Delta T_K(z) + \Delta T_E(z) = T_L(z) - T_A(z) \qquad (5.3b)$$

where the subscripts L and A refer to liquidus and actual, respectively, so that, for small values of z, Eq. (5.3b) becomes

$$\Delta T_K(z) + \Delta T_E(z) = \left\{ T_M + m_L \left[C_i + \left(\frac{dC_L}{dz} \right)_i z \right] \right\} - (T_i + \mathcal{G}_L z)$$

$$(5.3c)$$

In Eq. (5.3c), T_M is the melting temperature of pure solvent while T_i is the actual temperature at the interface. Inserting Eq. (5.3c) into Eq. (5.3a) leads to

$$\frac{\mathrm{d}}{\mathrm{d}z}(\Delta T_K + \Delta T_E) = m_L \left(\frac{dC_L}{dz} \right)_i - \mathcal{G}_L \qquad (5.3d)$$

But the gradient of solute concentration at the interface is given from the interface conservation condition (Eq. (3.2a)) so that, neglecting diffusion in the solid and interface field effects,

$$\frac{\mathrm{d}}{\mathrm{d}z}(\Delta T_K + \Delta T_E) = \frac{-m_L V}{D_L} C_i(1 - k_0) - \mathcal{G}_L \qquad (5.3e)$$

and the onset of constitutional supercooling, from Eq. (5.3a), becomes

$$\frac{\mathcal{G}_L}{V} < \frac{-m_L(1 - k_0)C_i}{D_L} \tag{5.4}$$

In Eq. (5.4), $C_i = C_S/k_0 = (k/k_0)C_\infty$, so that convection effects as well as constitution and diffusion effects are included in this "Constitutional Supercooling Criterion" (CSC).

Although the onset of CSC was initially used as the onset of planar interface instability, such a relationship holds exactly only for a highly ideal system where $\gamma = 0$, $K_S = K_L$ and $\Delta H_F = 0$. Early experiments to test Eq. (5.4) as an effective instability criterion utilized the metallic systems Pb and Sn, and used the liquid decantation technique to view the condition of the interface under a variety of growth conditions, $\mathcal{G}_L, V$ and C_∞. Crystals were grown in the horizontal Bridgeman fashion until steady state interface conditions had been reached and then the decantation of the liquid was triggered. One set of data (Sn as solute in Pb) from such an experiment is presented in Fig. 5.9 where we see that, to the available experimental accuracy, a linear relationship between C_∞ and $\mathcal{G}_L/V$ was found as predicted by Eq. (5.4).[7] The small amount of scatter is associated with an interface orientation effect, i.e., different orientations break down under slightly different conditions. The presence of "pox" or small interface indentations, as illustrated in Fig. 5.5, was taken as the signature of smooth interface instability. By matching Eq. (5.4) to the data of Fig. 5.9 under the assumption of negligible convection involvement, D_L/k was calculated to be 2.3×10^{-5} cm^2 s^{-1}, which is in the expected range for solute diffusion in liquids if we set $k \approx 1$.

As we saw from Figs. 5.1 and 5.2, definite instability leads to the formation of steep-walled grooves. For such grooves to be stable and not shrink, an analysis similar to Eq. (5.3) leads to groove stability when

$$\frac{\mathcal{G}_S}{V} < \frac{-m_L(1 - k_0)C_i}{D_L} \tag{5.5}$$

holds, where $\mathcal{G}_S$ is the temperature gradient in the solid in the vicinity of the grooves. We thus see that this is identical to the CSC equation except for the replacement of $\mathcal{G}_L$ by $\mathcal{G}_S$. If we applied the same procedure to a developing surface wave with $\gamma = 0$, we would again find the same general type of instability equation as Eq. (5.4) or Eq. (5.5) but with $\mathcal{G}_L$ replaced by $\mathcal{G}_i$ where $\mathcal{G}_i$ is some composite of $\mathcal{G}_L$ and $\mathcal{G}_S$.

For a planar interface, conservation of heat requires that $K_S\mathcal{G}_S = K_L\mathcal{G}_L + V\Delta H_F$ and, for metals, $K_S \approx 2K_L$ while, for semiconduc-

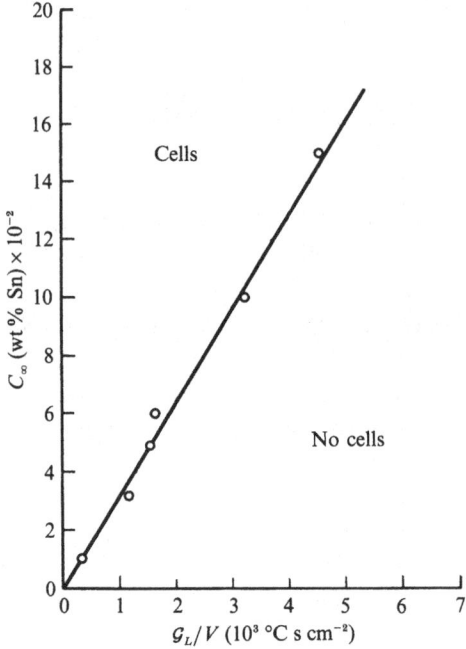

Fig. 5.9. Experimental interface morphology transition conditions for Sn as solute in Pb.[7]

tors, $K_S \approx (1/3)K_L$. Thus, $\mathcal{G}_S < \mathcal{G}_L$ for metals (at small $V\Delta H_F$) and $\mathcal{G}_S > \mathcal{G}_L$ for semiconductors (at small $V\Delta H_F$). This means that steep-walled grooves are stable at the interface for metals *before* CSC develops while, for semiconductors, steep-walled grooves are stable only *after* CSC develops. The implication here is that, for metallic systems, one should experimentally observe interface breakdown being initiated either at grain boundaries on the interface because a groove is already present there or at the groove associated with the outer edge of the crystal where it contacts the crucible wall (Bridgeman) or the liquid meniscus (CZ). In Fig. 5.10, the grain-boundary groove initiation of interface instability is shown for the Aℓ–Cu system.[8] Fig. 5.11 gives a schematic representation of the same phenomenon from the experimental findings of Schaefer and Glicksman[9] on the succinonitrile ($CNCH_2CH_2CN$) system.

For real systems, $\gamma \neq 0$ and the molecular attachment kinetics are not infinitely rapid. Thus, the critical interface velocity for the onset of instability, V_c, will, in general, not be exactly that given by V^{CSC} for

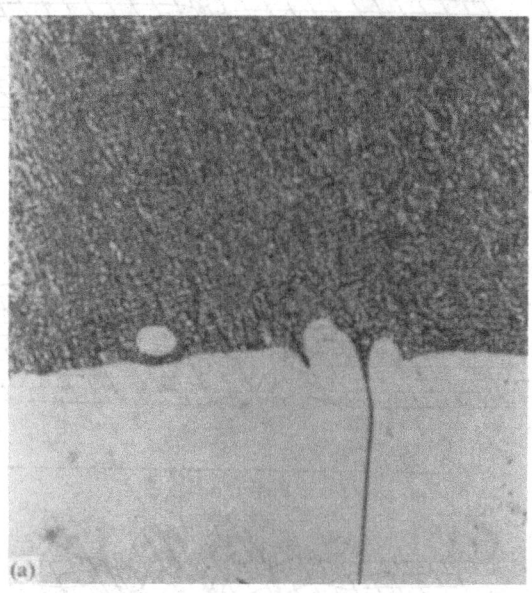

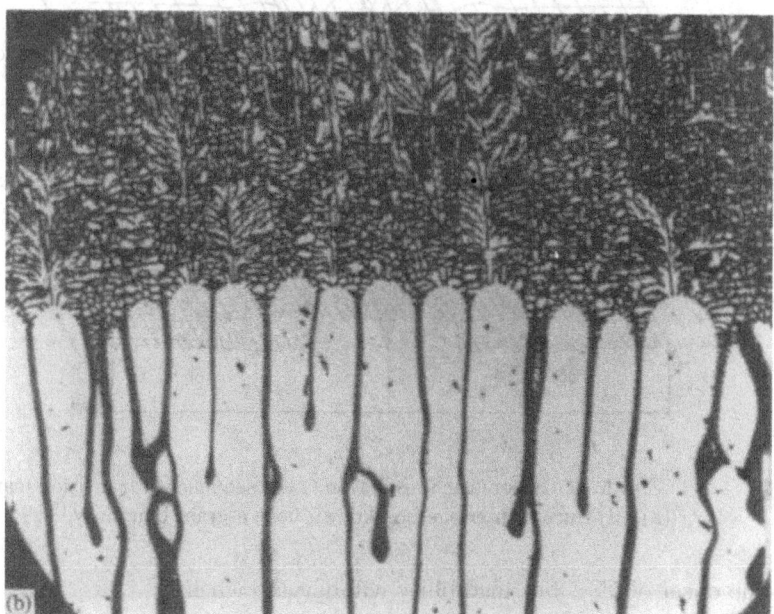

Fig. 5.10. Interface breakdown in an Aℓ–Cu alloy, (a) initial bump growth occurring at a grain boundary groove and (b) fully developed cells.[8]

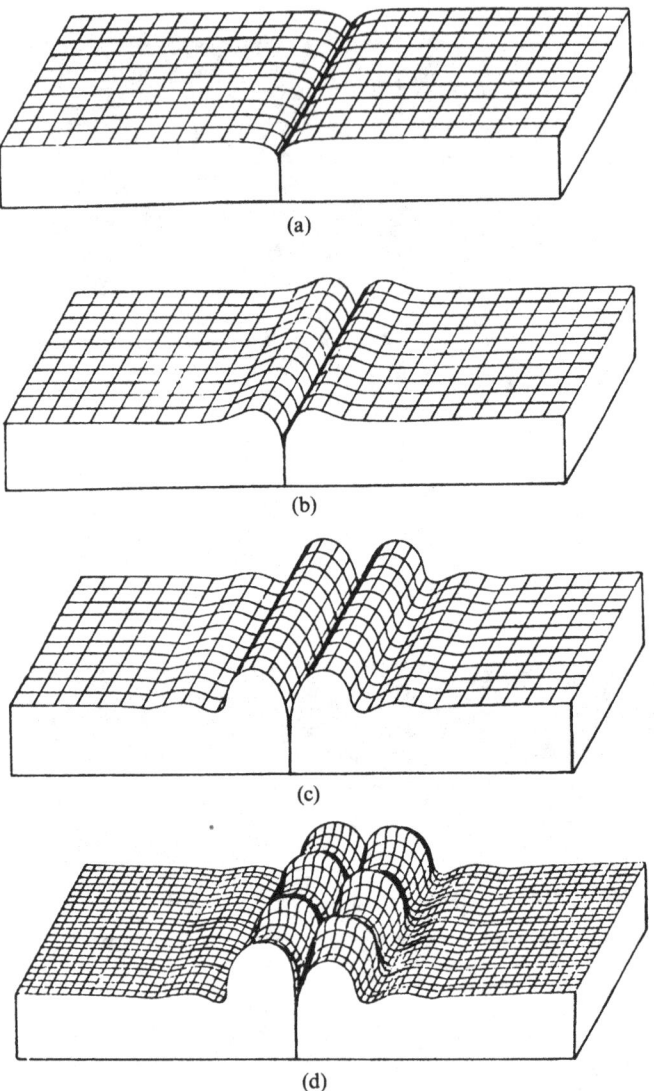

(a)

(b)

(c)

(d)

Fig. 5.11. Schematic illustration of the sequence of transition features (a)–(d) during interface breakdown near a grain boundary.[9]

the onset of CSC but instability will develop when

$$V > V_c = V^{CSC} + \Delta V_T + \Delta V_E + \Delta V_K + \Delta V_{IF} \tag{5.6}$$

Here, ΔV_T is the velocity change associated with the $K_S \neq K_L$ and

$\Delta H_F \neq 0$ effect already discussed so it is negative for metals and positive for semiconductors; ΔV_E is the velocity change associated with $\gamma \neq 0$ and surface creation so it is always positive; ΔV_K is the velocity change associated with the finite kinetic nature of the layer generation and layer motion mechanisms so it is also always positive. For a strongly faceted interface, this latter contribution can be extremely large as we shall see later. The last contribution, ΔV_{IF} is due to any interface field effect and can be positive or negative depending upon the sign of the field as illustrated later in Fig. 5.17.

For metals, the $\Delta V_E + \Delta V_K$ effects tend to counterbalance the ΔV_T effect so that

$$\text{metals}: \qquad V_c \approx V^{CSC} \qquad (5.7a)$$

For semiconductors, oxides, etc., all of the additional terms are of the same sign so that

$$\text{semiconductors}: \qquad V^{CSC} < V_c \tilde{<} \beta V^{CSC} \qquad (5.7b)$$

where $\beta \sim 2$–10 for most systems. However, for those systems where the interface orientation is at a deep cusp in the γ-plot, β can be much larger. It is this large ΔV_K factor that stabilizes large polyhedral crystal growth during the unconstrained mode of growth.

For metallic systems, one can gain a feeling for the magnitude of the effects by considering two extreme examples:

(1) A case where $C_\infty = 10^{-3}$ wt%, $m_L = -50\,°\text{C (wt%)}^{-1}$, $k_0 = 10^{-3}$, $k = 1$ and $D_L = 5 \times 10^{-5}$ cm^2 s^{-1} which leads to $\mathcal{G}_L/V_c = 10^6\,°\text{C s cm}^{-1}$. Thus, if $\mathcal{G}_L = 50\,°\text{C cm}^{-1}$ (reasonably large), $V_c = 5 \times 10^{-5}$ cm s^{-1} which is very slow indeed so the growing crystal probably has a cellular interface.

(2) A case where $C_\infty = 10$ wt%, $m_L = -10\,°\text{C (wt%)}^{-1}$, $k_0 = 10^{-2}$, $k < 1$ and $D_L = 10^{-4}$ cm^2 s^{-1} which leads to $\mathcal{G}_L/V_c = 10^8 k\,°\text{C s cm}^{-1}$. Thus, if $\mathcal{G}_L$ is again $50\,°\text{C cm}^{-1}$ and $V = 5 \times 10^{-5}$ cm s^{-1}(~ 2 cm day), strong stirring is needed to stabilize a smooth interface because $k \approx k_0$ is needed.

When one applies the PM techniques to a planar interface and neglects both $\Delta \mu_K$ and surface creation effects, the interface instability condition is found to be given by[2]

$$\frac{\mathcal{G}_L}{V} + \frac{\Delta H_F}{2K_L} < -\frac{m_L(1-k_0)C_i}{D_L}\bar{K}\mathcal{S}^* \qquad (5.8a)$$

where $\bar{K}$ is a dimensionless average thermal conductivity ($\bar{K} = (K_S + K_L)/2K_L$) and $\mathcal{S}$ is called the "stability function" which, at the $\mathcal{S} = \mathcal{S}^*$

condition, is typically between 0.7 and 0.9 for metallic systems.[10] The difference of $\mathcal{S}^*$ from unity represents the expected tendency of γ to stabilize the smooth interface so that, as γ increases, $\mathcal{S}^*$ decreases. For metallic systems, the term $\Delta H_F / 2K_L$ can generally be neglected with respect to $\mathcal{G}_L / V$ except at high V; however, for a system like ice and water, it cannot be neglected and it tends to stabilize the smooth interface. In Fig. 5.12(a), the domain of smooth interface stability is plotted. The limiting slope at large $\mathcal{G}_L / V$ is $\bar{K}(-m_L / D_L)(1 - k_0)C_i / C_\infty$ and the bending of the curve from linearity near the origin is largely due to the $\gamma > 0$ effect. Inclusion of interface attachment kinetic effects and surface creation effects would shift this solid curve downwards.

If the interface is comprised of layer edges with the layers extending in a direction making an angle ϕ to the macroscopic interface where ϕ is large, one can simply add a contribution to ΔT_K of magnitude $(\dot{\delta} / \beta_2 \Delta S_F) \sin(\omega x)$ to account for this resistance to perturbation development. As expected, this $(1/\beta_2 \Delta S_F)$ contribution enters the final result for $\dot{\delta} / \delta$ in such a way that does not change Eq. (5.8a) but only changes the rate of development of the surface undulations. However, if the interface is comprised of layer edges moving in the direction $\phi = 0$, the situation is very different because each protuberance must now have its own two-dimensional nucleation source even for infinitesimal amplitudes. The magnitude of the retardation on perturbation development will thus depend strongly on the two-dimensional nucleation frequency for the system under consideration.[5]

If we define a critical temperature gradient, $\mathcal{G}_{Lc} = -VC_\infty(1 - k_0)/ k_0 D_{CL}$, then Eq. (5.4) is defined as $m_L \mathcal{G}_{Lc} = \mathcal{G}_L$ for the no-stirring condition. From the details of $\mathcal{S}^*$, one can define a criterion for "absolute interface stability" which is

$$\hat{A}_0 = -\frac{k_0 \gamma (V/D_{CL})^2}{m_L \Delta S_F \mathcal{G}_{Lc}} > 1 \tag{5.8b}$$

A plot of $\mathcal{S}^*$ versus $\hat{A}_0$ is given in Fig. 5.12(b). Then, for a given alloy system, like Al plus Cu as solute, one can readily define the critical concentration of Cu above which interface stability occurs as a function of interface velocity, V, for a specific temperature gradient. This form of stability curve is illustrated in Fig. 5.12(c) for the case where $\mathcal{G}_L = 200\,°\text{C cm}^{-1}$. We note that the PM predicts that the CSC is a good approximation to the onset of instability. However, short wavelength perturbations at high interface velocities are stabilized because of capillarity forces.

5.2.1   *Minimal mathematics approach to planar interface instability*

For planar interface instability, use of the PM is the most mathematics intensive topic in this book, and most of the details are left to the problem section of this chapter. However, here, a simplified approach will be utilized to consider the interface stability of a planar basis surface as illustrated in Fig. 5.13(a) using the harmonic test function $\delta Y_{\ell m}$ (e.g., $Y_{\ell m} = \sin(\omega_\ell x)\sin(\omega_m y)$). The general approach is to apply the coupling equation to each surface and then evaluate the difference, i.e., for the perturbed surface, P,

$$\Delta T_\infty = (\Delta T_C + \Delta T_T + \Delta T_E + \Delta T_K)_P \tag{5.9a}$$

while, for the basis surface, B,

$$\Delta T_\infty = (\Delta T_C + \Delta T_T + \Delta T_E + \Delta T_K)_B \tag{5.9b}$$

Taking the difference yields

$$0 = (\Delta T_C^P - \Delta T_C^B) + (\Delta T_T^P - \Delta T_T^B) + (\Delta T_E^P - \Delta T_E^B) + (\Delta T_K^P - \Delta T_K^B) \tag{5.9c}$$

When the basis surface is planar, these contributions can be readily evaluated from Chapters 2, 3 and 4. Since we know that $\Delta T_C^B = -m_L(C_i^B - C_\infty)$, where $C_i^B = C_\infty/k_0$ at steady state, by analogy $\Delta T_C^P = -m_L(C_i^P - C_\infty)$. The first order steady state solution for C_L^P is given by

$$C_L(x,y,z) = C_\infty + \frac{C_\infty(1-k_0)}{k_0}\exp\left(-\frac{V}{D_L}z\right) + A\delta Y_{\ell m}\exp(-\omega_c z) \tag{5.9d}$$

where

$$\omega_c = \frac{V}{2D_L}\left\{1 + \left[1 + \left(\frac{2D_L}{V}\right)^2(\omega_\ell^2 + \omega_m^2)\right]^{1/2}\right\} \tag{5.9e}$$

Thus, it contains the basis surface solution plus a non-plane wave solution with constant A given from the solute conservation condition by

$$A = \frac{C_\infty(1-k_0)\left[\dot{\delta}/\delta V + k_0 V/D_L\right]}{k_0\left[D_L\omega_c/V - (1-k_0)\right]} \tag{5.9f}$$

Thus,

$$\Delta T_C^P = C_\infty\frac{(1-k_0)}{k_0}\left(1 - \frac{V}{D_L}\delta Y_{\ell m}\right) + A\delta Y_{\ell m} \tag{5.9g}$$

From values of ΔT_C and ΔT_T etc., Eq. (5.9c) yields for layer motion-limited attachment kinetics,

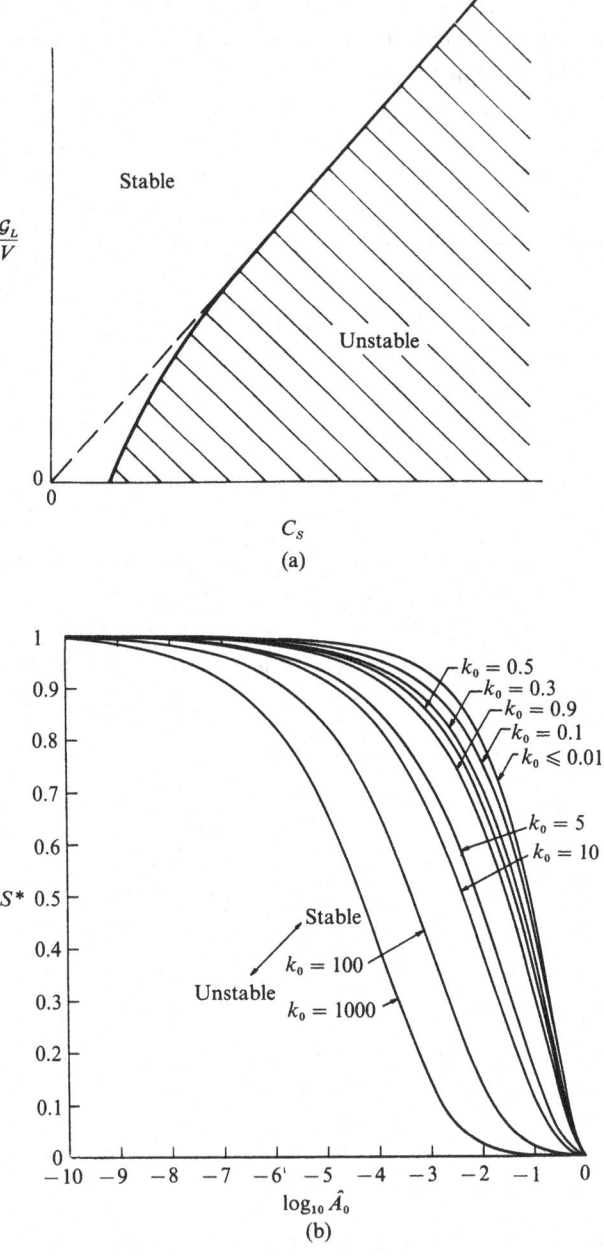

Fig. 5.12. (a) Plot of stable and unstable domains for a smooth interface during constrained growth, (b) plot of the Sekerka stability function S^* versus $\hat{A}_0$ (Eq. (5.8b)).[10]

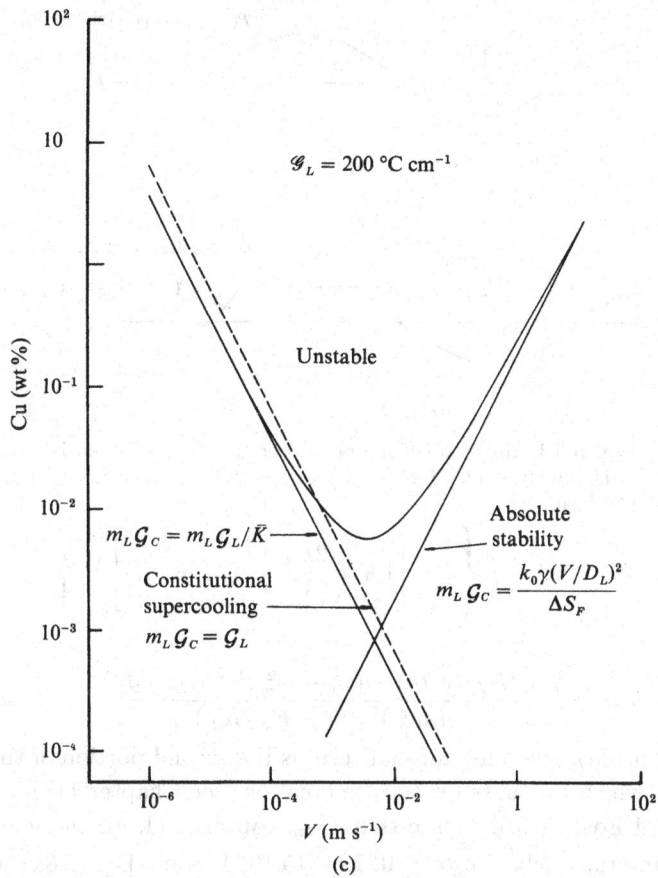

Fig. 5.12. (c) Interface stability map for unidirectionally solidified
Aℓ–Cu alloys with $\mathcal{G}_L = 200\,°\mathrm{C\,cm}^{-1}$.

$$0 = - m_L B \delta Y_{\ell m} + (b + \mathcal{G}_S)\delta Y_{\ell m}$$
$$+ \left\{ \frac{\gamma}{\Delta S_F}(\omega_\ell^2 + \omega_m^2) + \left(\frac{\partial \gamma_\ell}{\partial |\theta|}\right)_\phi \frac{\omega_\ell \omega_m (\dot{\delta}/\delta)}{\Delta S_F V_\phi} \right\} \delta Y_{\ell m} + \frac{(\dot{\delta}/\delta)}{\beta_2 \Delta S_F} \delta Y_{\ell m}$$

$$(5.10a)$$

where

$$B = \frac{V C_i (1 - k_0)}{D_L} \frac{\left[\dfrac{\dot{\delta}}{\delta}\left(\dfrac{D_L}{V^2}\right) - (\beta - 1)\right]}{[\beta - (1 - k_0)]}$$

with

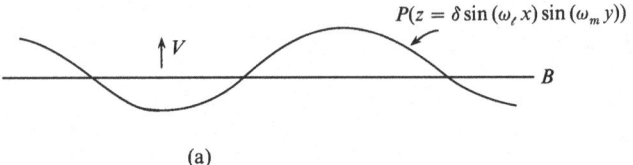

(a)

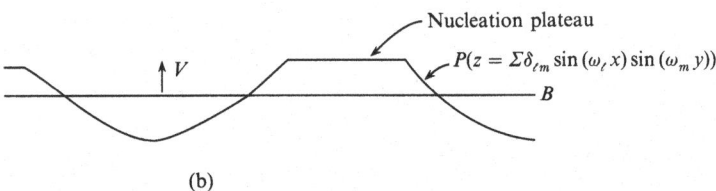

(b)

Fig. 5.13. Illustration of a perturbed interface (P) relative to a planar basis interface (B) for (a) a non-faceted interface and (b) a faceted interface.

$$\beta = \frac{1}{2}\left\{1 + \left[1 + \left(\frac{2D_L}{V}\right)^2 (\omega_\ell^2 + \omega_m^2)\right]^{1/2}\right\} \qquad (5.10b)$$

and

$$b = \frac{(\Delta H_F/K_L)(\dot{\delta}/\delta) + (\omega_\ell^2 + \omega_m^2)^{1/2}(\mathcal{G}_L - \mathcal{G}_S)}{(\omega_\ell^2 + \omega_m^2)^{1/2}(1 + K_S/K_L)} \qquad (5.10c)$$

In Eq. (5.10a), the only unusual term is the second portion of the curly bracket which accounts for surface creation (see Chapter 1). By examination of Eqs. (5.10), we see that $\delta Y_{\ell m}$ cancels and, for the marginally stable interface where $\dot{\delta}/\delta = 0$, Eqs. (5.10) become Eq. (5.8a) with $\mathcal{S}^*$ given by

$$\mathcal{S}^* = \frac{1}{2}\left[\frac{\beta^* - 1}{\beta^* - (1 - k_0)}\right] + \frac{\gamma(D_L/V)(\omega_\ell^2 + \omega_m^2)^*}{2\Delta S_F m_L C_i(1 - k_0)} \qquad (5.11)$$

Eqs. (5.10) with $\dot{\delta}/\delta = 0$ are really equations for determining the optimal frequencies ω_ℓ^* and ω_m^* yielding the marginal stability condition. Thus, β^* is the value of β with $\omega_\ell^2 + \omega_m^2 = (\omega_\ell^2 + \omega_m^2)^*$. We note that, for this marginally stable condition, both surface creation and this attachment kinetic contribution vanish. These contributions only serve to limit the rate of growth or decay of the perturbations in the unstable or stable interface domains, respectively. Thus, their main role is in the time rate of change of the interface morphology.

Under conditions where instability is encountered, the Fourier component corresponding to the maximum value of $\dot{\delta}/\delta$ (see Fig. 5.7) will grow

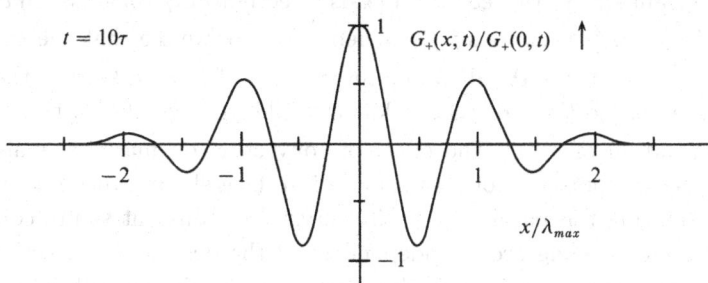

Fig. 5.14. Plot of the normalized Green's function with normalized distance at time $t = 10\tau$ after perturbation initiation.[10]

at the fastest rate. One might anticipate, therefore, that the interface might eventually evolve into a sinusoid with wavelength $\lambda_{opt} = 2\pi/\omega_{max}$. Actually, this has been tested only for the case where the surface creation term has been neglected in Eq. (5.10a). In such a case the time-dependent interface shape can be approximately characterized by the Green's function[10]

$$G_+(x,t) = \exp{(t/\tau)}\cos{(\omega_{max}x)}\exp(-x^2/4\mathcal{D}t) \tag{5.12}$$

where τ is a characteristic time and $\mathcal{D}$ is an effective diffusivity which describes the lateral spreading of the perturbation. In Fig. 5.14, a plot of $G_+(x,t)/G_+(0,t)$ is shown for the case of $t = 10\tau$. For a metal with $k_0 << 1$ and $V = 10^{-3}$ cm s^{-1}, typical values are $\tau = 0.33$ s, $\lambda_{max} = 2\pi/\omega_{max} = 5.9$ μm and $\mathcal{D} = 5.5 \times 10^{-8}$ cm^2 s^{-1}. Adding the very important surface creation term is expected to alter these numerical features appreciably but not the overall form of Eq. (5.12). Of course, the linearity approximation of the theory breaks down when the perturbation amplitudes are no longer small in comparison with their wavelengths.

From Eqs. (5.10), the general expression for $\dot{\delta}/\delta$ is given from

$$\left(\frac{\dot{\delta}}{\delta}\right)\left\{\frac{-\xi(D_L/V)}{\beta - (1-k_0)} + \frac{(\Delta H_F/K_L)}{(\omega_\ell^2 + \omega_m^2)^{1/2}(1 + K_S/K_L)}\right.$$

$$\left. + \frac{1}{\beta_2 \Delta S_F \sin\phi} + \frac{\omega_\ell \omega_m}{\Delta S_F V_\phi}(\partial\gamma_\ell/\partial|\theta|)_\phi\right\}$$

$$= \left\{-\xi V\left[\frac{\beta - 1}{\beta - (1 - k_0)}\right] + \left[\frac{(K_S/K_L)\mathcal{G}_S - \mathcal{G}_L}{1 + (K_S/K_L)}\right] - \frac{\gamma}{\Delta S_F}(\omega_\ell^2 + \omega_m^2)\right\} \tag{5.13a}$$

where

$$\xi = m_L C_i/D_L \tag{5.13b}$$

The right side of this equation leads directly to Eq. (5.8a) when $\dot{\delta}/\delta = 0$ and $\omega_\ell^2 + \omega_m^2 = (\omega_\ell^2 + \omega_m^2)^*$. When $\dot{\delta}/\delta > 0$, we see that the smaller is β_2, the greater is the dynamic retardation of $\dot{\delta}/\delta$. Likewise, the larger is the torque term of the γ-plot, $\partial\gamma_\ell/\partial|\theta|$, the greater is the dynamic retardation of $\dot{\delta}/\delta$. The effect of convection on marginal stability or on the development of an unstable interface shape can be assessed by considering the parameter ξ. If a crystal of constant solute content in the solid is being grown independent of the degree of convection, then $\xi = $ constant since $C_i = C_S/k_0$. However, if it is the bulk liquid solute content, C_∞, that is held constant during convection, then ξ decreases because C_i decreases and this tends to stabilize the planar basis surface. However, the heat transfer aspects are also changed and, if $\mathcal{G}_S$ and V are held fixed during convection changes, $\mathcal{G}_L$ must decrease which tends to destabilize the planar interface. Thus, whether increased convection stabilizes or destabilizes a smooth planar interface depends upon the details of the specific situation.

Neglecting the surface creation and the attachment kinetic terms in Eq. (5.13a), Hurle[11] found that the dominant perturbation wavelength, λ_{opt}^* is given by

$$\lambda_{opt}^* \sim (\mathcal{G}_L V)^{-n} \tag{5.14}$$

where $n = 1$ for strong convection and $n \sim 1/3$ for weak convection. We shall see in Chapter 6 that the cell size during cellular solidification obeys an equation of similar form.

5.2.2   Instability of a faceted interface

If, for simplicity, we consider a dislocation-free crystal, then the source of layers comes from two-dimensional nucleation events. The basis surface has an undercooling ΔT_K^B and a face area $A >> A_0$, the area of surface with a probability of 1 that a new nucleation event will occur within A_0 by the time the previous layer has grown out to cover A_0. The molecular attachment kinetics of this interface are thus governed by Eqs. (1.5).[5] In the small driving force limit, $\Delta G_i/\mathcal{R}T << 1$, we find that

$$V \approx \Delta T_K V_0 A_0 \exp(-\bar{\alpha}/\Delta T_K) \tag{5.15a}$$

where

$$\bar{\alpha} = \pi h \gamma_\ell^2 / 3\Delta S_F \mathcal{R}T \tag{5.15b}$$

Here, h is the layer height of edge energy γ_ℓ and V_0 is approximately a

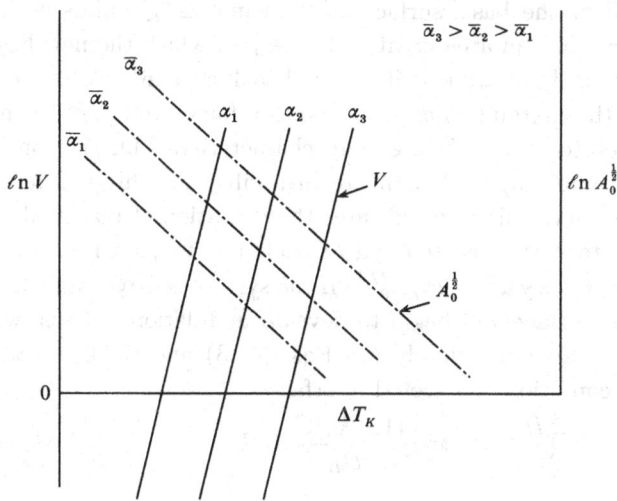

Fig. 5.15. Master plot connecting $A_0^{1/2}$(cm) and V(cm s^{-1}) with ΔT_K(°C) for the expected range of the material parameter $\bar{\alpha}$(°C).

constant ($V_0 \sim 10^{27}$). In addition, we have

$$A_0 = \pi R_0^2 = \pi \left(\beta_2 \Delta S_F \Delta T_K \frac{h}{V} \right)^2 \tag{5.16a}$$

leading to

$$A_0 = \pi^{1/3} \left(\frac{\beta_2 \Delta S_F h}{V_0} \right)^{2/3} \exp\left(\frac{2\bar{\alpha}}{3\Delta T_K} \right) \tag{5.16b}$$

Incorporating Eq. (5.16b) into Eq. (5.15a) leads to

$$V \approx \Delta T_K V_0^{1/3} (\beta_2 \Delta S_F h)^{2/3} \exp(-\bar{\alpha}/3\Delta T_K) \tag{5.17}$$

so that we now have expressions for V and A_0 that vary only with ΔT_K. These functions are schematically plotted in Fig. 5.15 for several $\bar{\alpha}$ since this is the dominant material parameter of concern to us here. Thus, as V increases, ΔT_K must increase so A_0 decreases as does its effective wavelength $A_0^{1/2}$.

For an unstable faceted interface as illustrated in Fig. 5.13(b) with nucleation plateau area $A_{\ell m}$, the mathematical analysis problem is definitely a non-linear one so many complexities would arise in trying to follow the evolution of the perturbed interface shape. However, the condition of marginal stability for the faceted surface can be determined by recognizing that $2\pi/A_o^{1/2} = \omega_c = (2\pi^{1/2}/h\beta_2 \Delta S_F \Delta T_K)V$ is a critical frequency for the perturbed system. The perturbed surface cannot grow ahead of the basis surface at the same ΔT_K unless each nucleation plateau has an area greater than A_0 on which the new layers can form. A typical situation is illustrated in Fig. 5.16(a) where curve (i) represents the marginal stability condition for an interface of the same thermal, solute and surface energy characteristics but is non-faceted. Curves (ii) and (iii) both indicate instability for this non-faceted interface but curve (iii) also indicates the condition of marginal stability when the interface is faceted and restricted to frequencies smaller than a critical frequency $\omega_c = 2\pi/A_0^{1/2}$. If the system is driven a little harder, this faceted surface will begin to develop undulations. Thus, with this insight, one can immediately use Eqs. (5.13) and (5.10) to write the instability condition for faceted interfaces:

$$\frac{\mathcal{G}_L}{V} + \frac{\Delta H_F}{2K_L} < -m_L\frac{(1-k_0)C_i}{D_L}\bar{K}\mathcal{S}_c \qquad (5.18a)$$

where

$$\mathcal{S}_c = \frac{1}{2}\left[\frac{\beta_c - 1}{\beta_c - (1-k_0)}\right] + \frac{\gamma(D_L/V)\,\omega_c^2}{2\Delta S_F m_L C_i(1-k_0)} \qquad (5.18b)$$

Since $\omega_c < \omega^*$ (see Fig. 5.16(a)) and $\beta_c < \beta^*, \mathcal{S}_c < \mathcal{S}^*$. In the vicinity of $\omega \approx \omega^*, \mathcal{S}^* \sim 1$ with both terms being $\sim \frac{1}{2}\mathcal{S}^*$. Since ω_c may be much less than ω^*, the second term in Eq. (5.18b) is easily seen to become negligible for this condition. The first term in Eq. (5.18b) becomes $(D_L\omega_c/V)^2/2[k_0 + (D_L\omega_c/V)^2]$ which will be very small when $k_0 >> (D_L\omega_c/V)^2$. Thus, we see that, under the proper conditions, $\mathcal{S}_c$ can be reduced below $\mathcal{S}^*$ by one to several orders of magnitude. From Eq. (5.18a), this allows $\mathcal{G}_L/V$ to be reduced by several orders of magnitude before instability of the faceted interface sets in. Thus, the facet stabilizes the planar interface in the presence of a large degree of CSC.

Even in metallic systems, where $\bar{\alpha}$ is small and two-dimensional nucleation is relatively easy, $\omega_c < \omega^*$. Thus, by considering the orientation dependence of interface instability, careful experiments should reveal information concerning the two-dimensional nucleation kinetics of the facet plane. Certainly, in this author's early experiments with interface instability on Pb crystals, the {111} facet plane was qualitatively observed to resist breakdown to an unstable interface state at a lower value of $\mathcal{G}_L$ (for fixed V, C_∞, etc.) than all other orientations.[7] Certainly, if Pb does not have a spherical γ-plot, the {111} is the lowest energy orientation and should break down to cells first rather than last if it were an atomically rough interface than a TLK interface.

To understand Eqs. (5.18) relative to Fig. 5.16(a) more fully, consider the case where you are growing an oxide tricrystal of orientations θ_1, θ_2

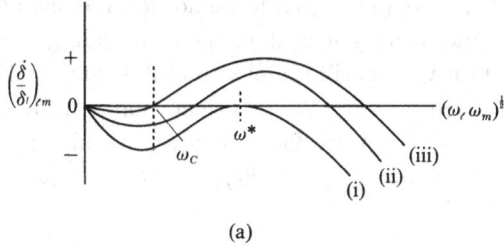

(a)

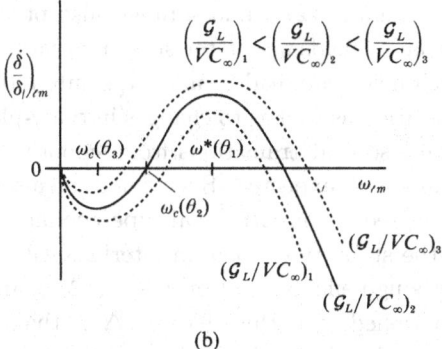

(b)

Fig. 5.16. (a) Three perturbation response curves for a non-faceted planar interface under conditions of increasing driving force from (i) to (iii). Curve (iii) is the marginally stable condition if this surface were faceted and V such that $A_0^{1/2} = 2\pi/w_c$. (b) Perturbation response curves for three orientations θ_1, θ_2 and θ_3 at different values of the parameter $\mathcal{G}_L/VC_\infty$.

and θ_3 under constrained conditions from an alloy melt (C_∞) at velocity V and temperature gradient $\mathcal{G}_L$. Under these growth conditions, you determine that the θ_1 interface has broken down into a well-developed cellular structure, the θ_2 interface exhibits small pox-like indentations while the θ_3 interface exhibits no sign of instability. You also know that the γ-values for the three interfaces are only slightly different. For this set of growth conditions, the $(\mathcal{G}_L/VC_\infty)_2$ curve of Fig. 5.16(b) illustrates the perturbation response for the θ_1 interface. On this same plot the w_c values for the θ_2 and θ_3 orientations have been indicated. Thus, at $\mathcal{G}_L/VC_\infty = (\mathcal{G}_L/VC_\infty)_2$, the lateral diffusion effect more than offsets the curvature effect for $w_{\ell m}$ just to the right of $w_C(\theta_2)$ so that $(\dot{\delta}/\delta)_{\ell m} \widetilde{>} 0$ and this orientation can begin to become unstable. If the growth conditions were shifted to $\mathcal{G}_L/VC_\infty = (\mathcal{G}_L/VC_\infty)_1$, the θ_2 orientation would become absolutely stable, however, if the growth conditions were shifted to $\mathcal{G}_L/VC_\infty = (\mathcal{G}_L/VC_\infty)_3$, it would become unstable and well-developed cells would form. For both of these shifts in growth conditions, the θ_3 orientation would still remain absolutely stable. Because of the foregoing, we anticipate that the θ_1 orientation is a vicinal face while the θ_2 and θ_3 orientations are facet planes. We also anticipate that, although $\gamma(\theta_1) \approx \gamma(\theta_2) \approx \gamma(\theta_3)$, $\gamma_\ell(\theta_1) > \gamma_\ell(\theta_2) > \gamma_\ell(\theta_3)$ so that $\mathcal{S}_C(\theta_3) < \mathcal{S}_C(\theta_2) < \mathcal{S}^*$.

5.2.3   Applied fields and interface field effects

As discussed in Chapters 1, 3 and 4, long range adsorption forces give rise to local thermodynamic fields near interfaces or surfaces. It was also shown earlier that such fields can lead to marked solute redistribution effects during crystallization as well as to changes in the γ-plot from a rounded to a cusped shape so that transitions from smooth interface growth to faceted growth might be expected. Such fields can also lead to significant changes in the degree of constitutional supercooling ahead of a planar interface and in the stability of such an interface with respect to cellular oscillations. Although effects of adsorption on ΔT_E and ΔT_K are present as already mentioned, it is the effect on ΔT_C that is novel and that we will discuss here. In Fig. 5.17, the four possible cases of interface field acting on a perturbed interface are illustrated. To evaluate the qualitative effect of the field on perturbation growth or decay, it is only necessary to ascertain the direction of the field component parallel to the interface. If this component points in the direction of the peaks of the surface wave, it impedes the point effect of diffusion and stabilizes the planar front. If it points in the direction of the troughs of the surface wave, it enhances the point effect of diffusion and stabilizes the filamentary front. In case (a) of Fig. 5.17, where the interface field is such as to attract the solute to the interface from both solid and liquid, the enhanced point effects are opposing. For the liquid side, the point effect diminishes while, for the solid side, the point effect is enhanced. Fig. 5.17, case (b), gives the maximum stability enhancement while case (c) gives the maximum destabilizing effect to the smooth front.

To ascertain the net point effect of diffusion for cases (a)–(d) in Fig. 5.17, we need only look at the relative lateral flux contribution, J'_x, for the two phases due to the field effect. Although one expects the thermodynamic potential variation, $\delta\widetilde{\Delta G}_0$, normal to the surface to decay in somewhat of an exponential form, let us consider only its linear ap-

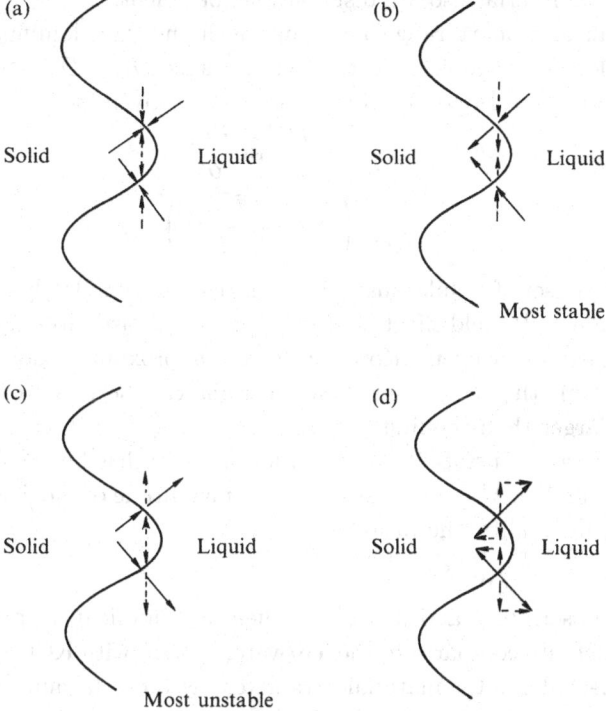

Fig. 5.17. Schematic illustration of the direction of the field-driven component to the total diffusion flux in the solid and liquid at the interface for all possible types of interface fields (dashed lines give the x-component). For case (b), the planar interface is the most stable while, for case (c), it is the most unstable.

proximation for simplicity; i.e.,

$$\delta\Delta\tilde{G}_0 = \bar{\beta}(1 + \tilde{\alpha}r_n) ; \qquad r_n \overset{<}{\sim} -\tilde{\alpha}^{-1} \tag{5.19}$$

where r_n is the distance measured along the interface normal direction and $\bar{\beta}$ is the magnitude of the field at the interface ($r_n = 0$); $\bar{\beta} = \tilde{\beta}_S$ or $\tilde{\beta}_L$ and $\tilde{\alpha} = \alpha_S$ or $-\alpha_L$ for solid or liquid, respectively. Using Eq. (5.19), one finds

$$J'_x = \frac{\hat{M}\partial\delta\Delta\tilde{G}_0}{\partial x} = -\frac{DC}{\kappa T}\bar{\beta}\tilde{\alpha}\frac{\partial r_n}{\partial x} \tag{5.20a}$$

where $\hat{M}$ is the mobility and κ is Boltzmann's constant. Thus, at any interface location

$$J'_{xL}/J'_{xS} = -D_L\tilde{\beta}_L\alpha_L/k_iD_S\tilde{\beta}_S\alpha_S \tag{5.20b}$$

where k_i is the interface solute distribution coefficient.

Even weak and short range fields in the liquid can dominate over strong and longer range fields in the solid because $D_L/k_iD_S \gg 1$ generally. Since the total lateral diffusion flux, J_x, is given by

$$J_x = -D\frac{\partial C}{\partial x} - \frac{DC}{\kappa T}\tilde{\beta}\tilde{\alpha}\frac{\partial r_n}{\partial x} \tag{5.21a}$$

$$\sim -\frac{DC}{\lambda^*}\left(10^{-2} + \frac{\tilde{\beta}\tilde{\alpha}}{\kappa T}\lambda^*\right) \tag{5.21b}$$

very near the onset of cellular instability, one cannot completely evaluate the magnitude of the field effect relative to the $\bar{\beta} = 0$ case without solving the total transport problem. However, for the approximate choice given by Eq. (5.21b), the second term on the right can be many orders of magnitude larger than the first.

The quantitative instability problem for this case has been solved by the author and Ahn[12] and, to evaluate the magnitude of the field effect on interface instability, the ratio

$$\Phi = (\dot{\delta}/\delta)_f/(\dot{\delta}/\delta)_{nf} \tag{5.22}$$

where the subscripts f and nf refer to field and no field, respectively. To make a definite comparison, the ice/water system with KCℓ as solute has been selected and the material parameters used for the numerical assessment are listed in Table 5.1. In Fig. 5.18, Φ has been plotted versus V (in cm s^{-1}) for a range of $\tilde{\beta}_S/\kappa T$ and $\tilde{\beta}_L/\kappa T$ with $\alpha_S = \alpha_L = 10^6$. From Fig. 5.18(a) one notes that, even for $\tilde{\beta}_L/\kappa T$ as small as 10^{-4}, quite significant differences in instability arise compared to the zero field case. One notes also that, as $\tilde{\beta}_S$ goes from strongly positive to strongly negative values, the interface instability effect increases in accordance with expectations from Fig. 5.17. From Figs. 5.18(b) and (c) one notes that, as $\tilde{\beta}_L$ goes from strongly positive to strongly negative values, Φ goes from strongly negative to strongly positive values. Thus, relative to the zero field (nf) predictions, the system goes from a strongly destabilized condition to a strongly stabilized condition in complete accord with Fig. 5.17.

Although the quantitative computations reveal a marked change in the degree of CSC for this range of interface field parameters, the major portion of the interface instability effect comes from the enhanced lateral diffusion at the interface. One can see that very small fields in the liquid are capable of producing large alterations in the onset of instability. This leads one to conclude that any match between theory

Table 5.1. *Essential parameter data for the ice/water–KCℓ system.*

$\Delta H_F = 80$ cal g^{-1}	$\mathcal{G}_L = 25$ K cm^{-1}
$T_M = 273.0$ K	$K_L = 1.33 \times 10^{-3}$ cal cm^{-1} s^{-1} K^{-1}
$m_L = 1.7$ K mol^{-1} ℓ$^{-1}$	$K_S = 5.3 \times 10^{-3}$ cal cm^{-1} s^{-1} K^{-1}
$k_0 = 5 \times 10^{-4}$	$\alpha_S = \alpha_L = 10^6$ cm^{-1}
$C_0 = 10^{-2}$ mol ℓ$^{-1}$	$\omega = 10^4$ cm^{-1}

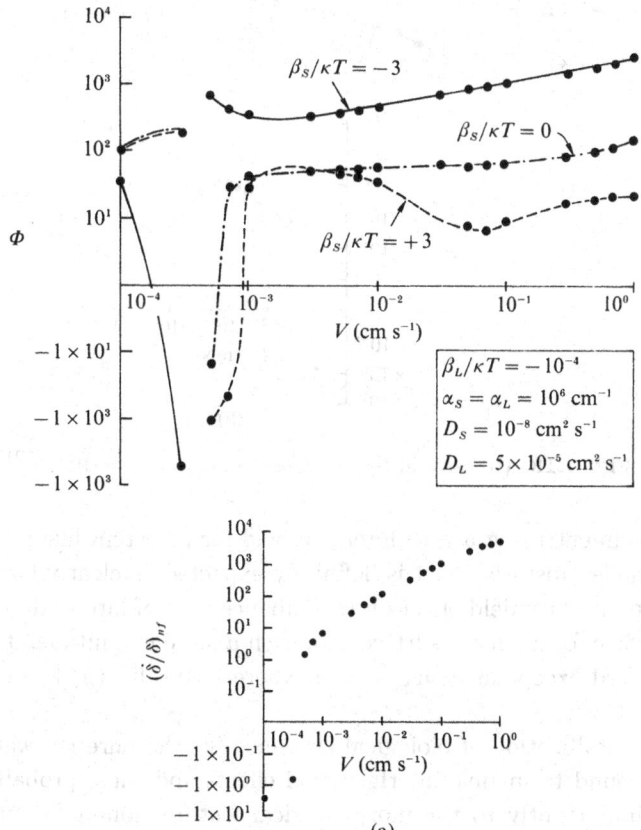

Fig. 5.18. Plot of $\Phi = (\dot{\delta}/\delta)_f/(\dot{\delta}/\delta)_{nf}$ and $(\dot{\delta}/\delta)_{nf}$ versus V in the ice/water system with KCℓ as solute for (a) several values of β_S ($\beta_L/\kappa T = 10^{-4}$, $\alpha_L = \alpha_S = 10^6$).[12]

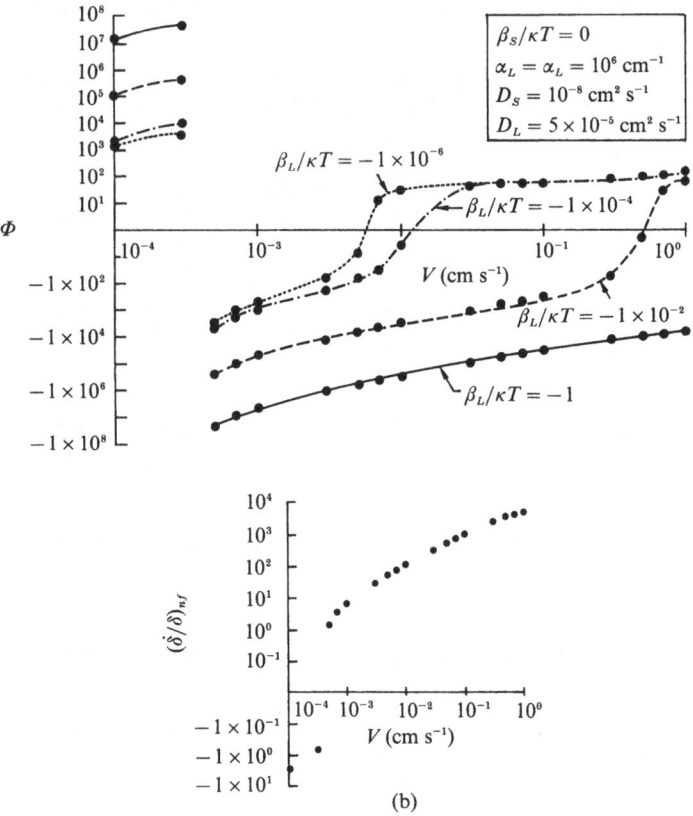

Fig. 5.18. (b) Several $\beta_L > 0$ ($\beta_S = 0$, $\alpha_L = \alpha_S = 10^6$).[12]

and experiment using a zero interface field for a system like ice/water, where an electrostatic effect is definitely expected, is clearly fortuitous. Since the interface field effect on the enhancement of lateral diffusion at the interface for a curved surface is of such a large magnitude, it cannot be neglected except in clearcut cases where a zero field is known to be present.

The crystallization of biological systems, like the pure ice/water system, is bound to involve interface field effects and these probably contribute importantly to the morphological features found for such systems. For systems with a zero interface field but a strongly enhanced interface solute diffusion coefficient compared to the bulk nutrient solute diffusion coefficient, instability will be enhanced just as if an interface field of the type given in Fig. 5.17(c) were present.

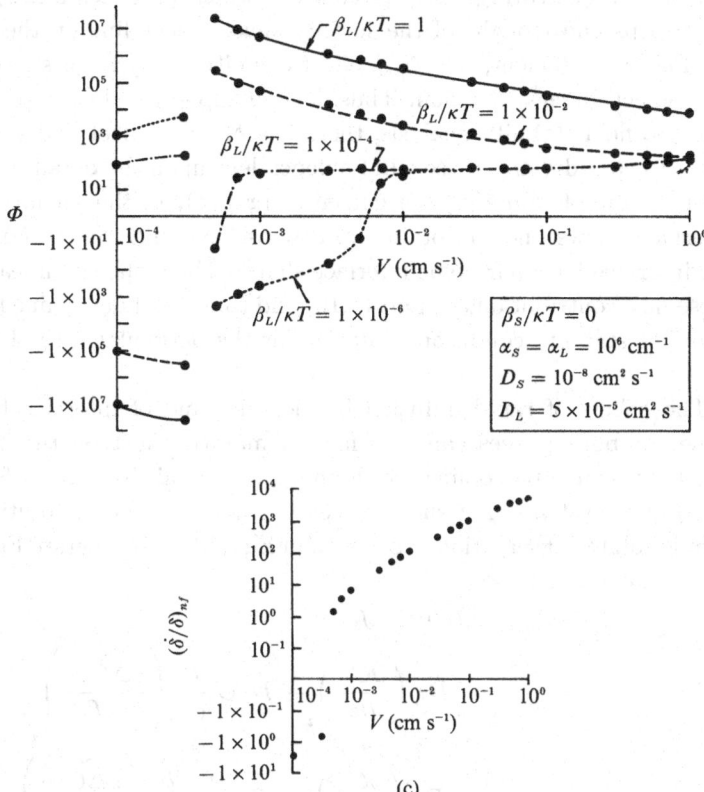

Fig. 5.18. (c) Several $\beta_L < 0$ ($\beta_S = 0$, $\alpha_L = \alpha_S = 10^6$).[12]

The applied field case gives a uniaxial field (generally) far from the interface but a three-dimensional field close to a perturbed interface because of the different material constants of the crystal and nutrient phases. This can best be illustrated for the electric field case. The applied field produces enhanced transport of particular solutes in a particular direction (either in the field direction or against it) in the bulk phases. In the interface region, if the solid is less conductive than the nutrient, the current lines will concentrate in the regions of the troughs for the surface waves shown in Fig. 5.17. If the key species has enhanced diffusion in the direction of the current flow, then lateral diffusion from the tips is enhanced and the presence of the field generates instability with this sign of current flow. For oppositely directed current, the field enhances the smooth interface stability. Of course, there is also an interface Peltier heating effect to consider which produces a ΔT_T effect at the tips and troughs of the surface wave. Depending on the sign of the Peltier coefficient, this ΔT_T effect can either oppose or support the ΔT_C effect under discussion. Thus, for the applied field case, relative to the zero field stability analysis, there is a $\delta(\Delta T_C)$ effect and a $\delta(\Delta T_T)$ effect to consider and these will be dependent upon the detailed current distribution. Evaluating the current, $I(x, y, z)$, in the vicinity of the interface allows one to obtain expressions for $\delta(\Delta T_C)$ and $\delta(\Delta T_T)$ on the interface for a harmonic interface shape. Then, one can linearly add these new contributions to Eqs. (5.10) and (5.13) and determine how the interface stability condition is altered for the particular case of applied field.

The effects of rapid diffusion in the solid and of interface fields in either or both phases can also have a marked effect on the onset of constitutional supercooling. Such an experimental situation is found in TiO_2 alloy systems. For such a case, it is not satisfactory to utilize the simple solute conservation expression in Eq. (5.3e) to generate Eq. (5.4). Instead, one must use

$$V(1 - k_i)C_i = J_L(0) - J_S(0)$$

$$= -D_L \left(\frac{\partial C_L}{\partial z}\right)_i - D_L C_i \frac{\partial}{\partial z} \left(\frac{\delta \widetilde{\Delta G}_{0L}}{\kappa T}\right)_i$$

$$+ D_S \left(\frac{\partial C_S}{\partial z}\right)_i + D_S k_i C_i \frac{\partial}{\partial z} \left(\frac{\delta \widetilde{\Delta G}_{0S}}{\kappa T}\right)_i \quad (5.23)$$

to obtain our expression for $(\partial C_L/\partial z)_i$. Solving Eq. (5.23) for $(\partial C_L/\partial z)_i$ and inserting it into Eq. (5.3d) yields

$$\frac{\mathcal{G}_L}{V} < -m_L \frac{(1 - k_i)C_i}{D_L} + \frac{m_L D_S}{V D_L} \left\{ \left(\frac{\partial C_S}{\partial z}\right)_i + k_i C_i \left[\frac{\partial}{\partial z} \left(\frac{\delta \widetilde{\Delta G}_{0S}}{\kappa T}\right)\right]_i \right.$$

$$\left. - \frac{D_L}{D_S} C_i \left[\frac{\partial}{\partial z} \left(\frac{\delta \widetilde{\Delta G}_{0L}}{\kappa T}\right)\right]_i \right\} \quad (5.24)$$

For case (b) of Fig. 5.17, the effect of the bracketed term on the right is to pump solute through the interface and deeper into the solid. This tends to give up-hill diffusion in the solid, reduce the concentration in the solid at the interface and reduce C_i. In the TiO_2 alloy case, the field effect in the solid is so large that interface instability does not develop even when $C_i \sim 25$ wt% in a natural convection system.

5.2.4 *Generalization of the constitutional supercooling criterion*

As a guide to impending interface instability in crystal growth from the vapor, solution, electrodeposition, etc., it is useful to express the appropriate analogues to the constitutional supercooling criterion under the more general heading of constitutional supersaturation (CSS). The onset of CSS occurs when the actual concentration gradient in the nutrient phase at the interface is greater than the equilibrium concentration gradient at the interface; i.e., the supersaturation increases with distance into the nutrient phase

$$\left(\frac{\partial C}{\partial z}\right)_i > \left(\frac{\partial C^*}{\partial z}\right)_i \tag{5.25a}$$

Neglecting diffusion in the solid, interface conservation of species requires that

$$\left(\frac{\partial C}{\partial z}\right)_i = C_i \left\{ \frac{V}{D}(k_0 - 1) - \left[\frac{\partial(\delta\widetilde{\Delta}G_0/\kappa T)}{\partial z}\right]_i \right\} \tag{5.25b}$$

where $\delta\widetilde{\Delta}G_0$ is the additional potential term in the nutrient phase due to the presence of a field ($\delta\widetilde{\Delta}G_0 = \hat{z}e\phi$ for an electrostatic field); i.e.,

$$\widetilde{\eta} = \mu_0 + \kappa T\ell\mathrm{n}(\hat{\gamma}C) + \delta\widetilde{\Delta}G_0 = \mu_0 + \kappa T\ell\mathrm{n}(\hat{\gamma}C\exp(\delta\widetilde{\Delta}G_0/\kappa T))$$

From the above expression, one can readily deduce the dependence of the equilibrium concentration on the presence of the additional potential $\delta\widetilde{\Delta}G_o$ so that

$$\left(\frac{\partial C^*}{\partial z}\right)_i = \left(\frac{\hat{\gamma}_0}{\hat{\gamma}}\right)\exp\left(\frac{-\delta\widetilde{\Delta}G_0}{\kappa T}\right)\left\{\left(\frac{\partial C_0^*}{\partial z}\right)_i - C_0^*\left[\frac{\partial(\delta\widetilde{\Delta}G_0/\kappa T)}{\partial z}\right]\right\} \tag{5.25c}$$

where C_0^* and $\hat{\gamma}_0$ hold when $\delta\widetilde{\Delta}G_0 = 0$.

For both the vapor and solution growth cases, the equilibrium nutrient concentration, C_0^*, is given by the form

$$C_0^* = A_n\exp(-\Delta H_n/\kappa T) \tag{5.26}$$

where the subscript n refers to the nutrient phase ($n = V, \tilde{S}$, etc.) and, for the vapor case, $A_V \propto (\kappa T)^{-1}$. Using $dC/dz = (dC/dT)dT/dz$ and Eq. (5.26), but neglecting the temperature dependence of A_n, Eq. (5.25c) becomes

$$\frac{\partial C^*}{\partial z} = \left(\frac{\hat{\gamma}_0}{\hat{\gamma}}\right)\exp\left(\frac{-\delta\widetilde{\Delta}G_0}{\kappa T}\right)C_0^*\left\{\frac{\Delta H_n}{\kappa T_i^2}\mathcal{G}_i - \left[\frac{\partial(\delta\widetilde{\Delta}G_0/\kappa T)}{\partial z}\right]_i\right\} \tag{5.25d}$$

where $\mathcal{G}_i$ is the temperature gradient in the nutrient phase at the interface. Placing Eqs. (5.25d) and (5.25b) into Eq. (5.25a) and rearranging leads to the onset of CSS when

$$\frac{1}{V}\left\{\mathcal{G}_i - \frac{\kappa T_i^2}{\Delta H_n}\left[\frac{\partial(\delta\widetilde{\Delta G}_0/\kappa T)}{\partial z}\right]_i\right\}$$

$$< \frac{\hat{\gamma}}{\hat{\gamma}_0}\frac{\kappa T_i^2}{\Delta H_n}\frac{C_i}{C^*}\left\{\frac{(k_0 - 1)}{D} - \frac{1}{V}\left[\frac{\partial(\delta\widetilde{\Delta G}_0/\kappa T)}{\partial z}\right]_i\right\} \quad (5.27)$$

This reduces to Eq. (5.4) when $\delta\widetilde{\Delta G}_0 = 0$, $\hat{\gamma}/\hat{\gamma}_0 = 1$ and $m_L^{-1} = C^*\Delta H_n/\kappa T_i^2$ as expected. Thus, not only can one consider any type of nutrient with this result, but one can see how the CSS is altered by altering the sign and magnitude of the applied field.

5.3   Morphological stability for other major shapes

Our main discussion of interface stability has related to a planar interface shape. Let us now extend these considerations to other shapes such as the sphere, cylinder, etc. The steady state solute profiles for these unperturbed shapes have been given by Eqs. (4.22) so that ΔT_C can be readily obtained. In Chapter 6, it will be shown that the temperature distribution has the same mathematical form as the solute distribution for a particular shape so that ΔT_T can also be readily obtained. It remains to describe the infinitesimal shape distortions and the perturbed transport fields for the different shapes and then combine all the effects in the appropriate coupling equation of Eq. (5.1) and use the orthogonality condition to separate the perturbed solution $(\dot{\delta}/\delta)$ from the unperturbed solution (V). Let us consider the results of this operation for the different shapes.

Case 1 (Sphere from a melt): An infinitesimal deviation of the particle from sphericity is imposed by an expansion of the surface into spherical harmonics so that the surface profile for the infinitesimally distorted sphere is

$$r(\theta, \phi, t) = R(t) + \delta(t)Y_{\ell m}(\theta, \phi) \quad (5.28a)$$

and the curvature in the small distortion limit is

$$\mathcal{K}(\theta, \phi) \approx -\nabla^2 r \quad (5.28b)$$

$$= \frac{2}{R} - \frac{2\delta Y_\ell m}{R^2} - \frac{\delta}{R^2} \left[\frac{1}{\sin\theta} \frac{\partial}{\partial\theta} \left(\sin\theta \frac{\partial}{\partial\theta} \right) + \frac{1}{\sin^2\theta} \frac{\partial^2}{\partial\theta^2} \right] Y_{\ell m}$$

$$(5.28c)$$

by using the Laplacian operator in spherical coordinates so that, neglecting surface creation,

$$\Delta T_E = \frac{\gamma}{\Delta S_F} \left[\frac{2}{R} + \frac{(\ell+2)(\ell-1)\delta Y_{\ell m}}{R^2} \right] \qquad (5.28d)$$

Once again, the effect of surface creation associated with $\dot{\delta} > 0$ will enter as a term proportional to $\dot{\delta}/\dot{R}$ and can be neglected at the onset of instability $(\dot{\delta}/\delta = 0)$.

The ΔT_K term depends upon the ledge generation mechanism which is likely to be two-dimensional nucleation on the facet planes. For a cubic system, this leads to eight flats on the sphere for layer creation and eight nipples where layer edges from four sources annihilate. Because of this symmetry, the $Y_{4n,4n}$ $(n = 1, 2, 3, \ldots)$ harmonics are the most likely ones to develop. Since surface bumps are most likely to develop at these nipple locations controlled by the attachment kinetics at layer edges, we expect to find

$$\Delta T_K = \frac{\dot{R} + \dot{\delta} Y_{4n,4n}}{\Delta S_F \beta_K(\theta, \phi)} \qquad (5.29)$$

For non-cubic crystal structures, the appropriate ℓ and m to choose will depend upon the symmetry of the layer sources for the particular crystallographic system.

For the matter transport, if we approximate the solutions by using Laplace's equation,[2] we have

$$C_L(r, \theta, \phi) - C_\infty = \frac{A}{r} + \frac{B Y_{\ell m}}{r^{\ell+1}} \qquad (5.30a)$$

with a similar form of solution holding for the temperature. Thus, we have

$$\Delta T_C = -m_L \left[\frac{A}{R} \left(1 - \frac{\delta Y_{\ell m}}{R} \right) + \frac{B \delta Y_{\ell m}}{R^{\ell+1}} \right] \qquad (5.30b)$$

$$\Delta T_T = \frac{a_L}{R} \left(1 - \frac{\delta Y_{\ell m}}{R} \right) + \frac{b_L \delta Y_{\ell m}}{R^{\ell+1}} \qquad (5.30c)$$

where A and B are determined by the interface conservation condition while a_L and b_L are determined by the heat boundary conditions at the interface. Combining these factors, we obtain for $\dot{\delta}/\delta$ the complex result

$$\frac{\dot{\delta}}{\delta} = Q\left\{\Delta T_\infty + \left(\left[\frac{(-2m_L)C_\infty \hat{P}_C(1-k_0)}{1-2\hat{P}_C(1-k_0)}\right]\left[\frac{\ell(K_S/K_L)+2\hat{P}_C(1-k_0)}{(\ell+1)-2\hat{P}_C(1-k_0)}\right]\right)\right.$$
$$\left. -\frac{\gamma}{R\Delta S_F}\left[(\ell+2)(\ell+1)+2+\ell(\ell+2)\frac{K_S}{K_L}+\frac{2\Delta S_F D_C \hat{P}_C}{\gamma\beta_K}\right]\right\} \quad (5.31a)$$

where

$$Q = \frac{\ell-1}{R^2\left\{\dfrac{\left[(\ell+1)+\ell\dfrac{K_S}{K_L}\right]}{\beta_K} + \dfrac{(\Delta H_F/c)}{D_T} + \dfrac{(-m_L)C_\infty(1-k_0)\left[(\ell+1)+\ell\dfrac{K_S}{K_L}\right]}{D_C[1-2\hat{P}_C(1-k_0)][(\ell+1)-2\hat{P}_C(1-k_0)]}\right\}}$$

$$(5.31b)$$

and $\hat{P}_C = R\dot{R}/2D_C$ is the Péclet number.

The first two terms in Eq. (5.31a) effectively represent the transport terms in the problem and favor growth of the harmonic. The third term, which is negative, represents the capillarity and attachment kinetic effects favoring decay of the harmonic. The question of stability reduces to the study of which effect dominates. All harmonics for which the curly bracket of Eq. (5.31a) is positive must grow. One finds that a critical instability radius, $R_C(\ell)$, exists for every harmonic such that the larger is ℓ, the larger must $R_C(\ell)$ be for sphere instability. To first order in δ, the first harmonic merely translates the sphere a distance δ, as reflected by the factor $\ell - 1$ in Q. If $C_\infty = 0$ and $\beta_K = \infty$, the critical value of R for which at least the second harmonic grows is $R_C(2) \sim 15R^*$ for metals where $K_S/K_L \approx 2$ and $R^* = 2\gamma/\Delta S_F\Delta T_\infty =$ critical homogeneous nucleation radius for the sphere. Of course, $Y_{4,4}$ is the lowest harmonic capable of producing instability in a cubic system growing under the TLK regime so that one expects $R_C(4,4) \sim 80R^*$. The dependence on C_∞ and β_K is not so clear cut quantitatively because each contribution enters both the numerator and the denominator of the expression for $\dot{\delta}/\delta$; however, qualitatively, we know that (i) as β_K decreases, R_C increases while (ii) as C_∞ increases, R_C decreases.

Case 2 (Cylinder from a melt):[13] The next simplest case to the sphere is the right circular cylinder which approximates to the dendrite and whisker forms often met in nature. We shall consider perturbations of both the circular shape and the radius as illustrated in Fig. 5.19 so that

$$\delta Y_{\ell,m/R}(\phi, z) = \delta \exp(i\ell\phi)\exp[i(m/R)z] \quad (5.32a)$$

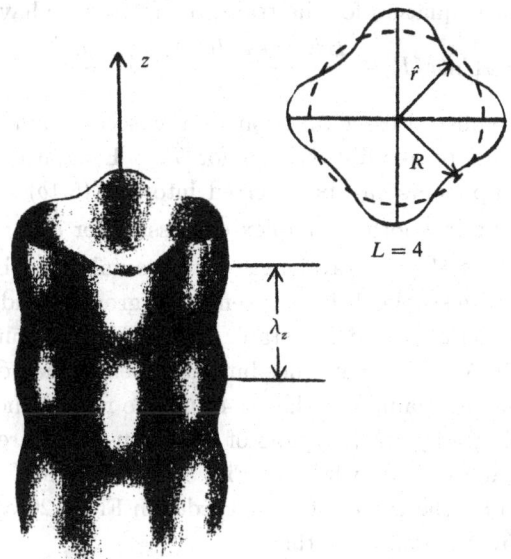

Fig. 5.19. Model chosen to describe perturbations on the surface of an infinite circular cylinder.[13]

For the case of certain long wavelength perturbations in the z-direction ($\ell = 0, m < 1$), the surface to volume ratio of the continuum model cylinder (for fixed volume) decreases so that such a cylinder is always unstable with respect to this type of perturbation.

Just as for the sphere, the equation of the perturbed cylinder is

$$r(\phi, z) = R + \delta Y_{\ell, m/R}(\phi, z) \tag{5.32b}$$

Although ℓ is restricted to integral values, m is not and it can have any real positive value. Following the small distortion approach, the curvature is

$$\mathcal{K}(\phi, z) = \frac{1}{R} + \frac{\delta}{R^2}(\ell^2 + m^2 - 1)Y_{\ell, m/R} \tag{5.32c}$$

so that ΔT_E is readily given. If we are considering a $\langle 100 \rangle$ cylinder axis in a cubic system with the TLK mechanism operating, then

$$\Delta T_K = \frac{\dot{R} + \dot{\delta}\,Y_{4n, m/R}}{\Delta S_F \beta_K(\phi, z)} \tag{5.33}$$

Again, the symmetry of the layer sources and the median ridges is fourfold.

If we use the Laplace equation for the transport factors we have

$$\frac{C_L(r,\phi,z)}{C_\infty} = A\ell\mathrm{n}(r/R) + \frac{BK_\ell(m\,r/R)\delta Y_{\ell,m/R}(\phi,z)}{K_\ell(m)} \qquad (5.34)$$

where $K_\ell(r)$ is a modified Bessel function of the second kind. An analogous equation can be readily written for T_L. Evaluating ΔT_C and ΔT_T, all the components can be inserted into Eq. (5.1b) and orthogonality used to obtain another complex expression for $\dot{\delta}/\delta$. When the ratio $(\dot{\delta}/\delta)/(\dot{R}/R) > 1$, the wavelength of the axial perturbation, $\lambda_z = 2\pi/(m/R)$, must be stable. If for a given set of growth conditions, there exist many wavelengths that are stable, the extremum condition may be used to obtain λ_{z-opt}. This procedure has been carried out using pure Ni as a practical example with $\ell = 4$ and the results shown in Figs. 5.20(a) and (b). In Fig. 5.20(a), plots of R_C versus ΔT_∞ are given for several possible values of β_K while, in Fig. 5.20(b), plots of λ_{z-opt} versus ΔT_∞ are given for the values of R_C recorded in Fig. 5.20(a). The main results of this overall study are that:

(i) both R_C and λ_{z-opt} decrease with increasing ΔT_∞ and increasing C_∞ and

(ii) both R_C and λ_{z-opt} increase with decreasing β_K and increasing γ.

Case 3 (Disc-shaped nucleus on a substrate):[14] Two cases are of interest here: (i) a two-dimensional nucleus growing on a substrate having a slight supersaturation of adsorbed atoms in a conservative condition wherein no mass is added or removed from the system and (ii) a two-dimensional nucleus growing on a substrate under mild supersaturation and in the presence of a vapor source (a non-conservative condition). For case (i), the transport equation and boundary conditions are identical to those used for an infinite right circular cylinder with a continuum interface so the morphological stability results may be expected to be identical. It is found that the $\ell = 2$ perturbation always grows slower than the mean radius R of the nucleus and will eventually be overtaken by the growing nucleus. Thus, absolute instability can only set in for harmonics larger than $\ell = 4$. As expected, once unstable, the amplitude of the perturbation increases monotonically with time and, if unchecked, would soon evolve into skeletal and dendritic morphologies.

In case (ii), the non-conservative system, the matter transport equation is slightly altered due to the presence of an annihilation/creation term (see Eq. (4.1)). In Fig. 5.21(a), the geometry of the growing nucleus is given. In Fig. 5.21(b), the values of $\dot{\delta}_\ell/\delta_\ell$ (in units of $\Omega n_{eq}(P/P^* - 1)/h\tau_s$) are given for several ℓ as a function of $x = R/X_s$ for $x^* = R^*/X_s$ where $X_s = D_s\tau_s$. Here, $\beta_K = \infty$ has been used and heat transport neglected. We note that a given perturbation is unstable and grows only when x lies between two limiting values $x_{C_1} = R_{C_1}/X_s$ and $x_{C_2} = R_{C_2}/X_s$. The instability interval of the nucleus is entirely determined by the critical nucleus size $x^* = R^*/X_s$ which is the threshold value for nucleation imposed on the system by its supersaturation. In Fig. 5.21(c), values of $\dot{\delta}_\ell/\delta_\ell$ versus x are given for different values of x^* for a specific ℓ. As x^* increases, the perturbation grows more slowly until it disappears completely at a certain value of $x^* = x_C^*$. Beyond this value, the nucleus is always stable and each harmonic determines its own x_C^*; i.e., for $\ell = 2\text{--}8$, one finds that $x_C^*(2) \approx 4 \times 10^{-2}$, $x_C^*(3) \approx 2.9 \times 10^{-2}$, $x_C^*(4) \approx 2.1 \times 10^{-2}$, $x_C^*(5) \approx 1.7 \times 10^{-2}$, $x_C^*(6) \approx 1.4 \times 10^{-2}$, $x_C^*(7) \approx 1.2 \times 10^{-2}$ and $x_C^*(8) \approx 0.99 \times 10^{-2}$. Since $x_C^*(\ell)$ decreases as ℓ increases, if $x^* > x_C^*(2)$, no perturbation will grow and the nucleus is entirely stable.

Unlike the conservative system, for this case there is no limiting nucleus radius beyond which a perturbation will always grow. The key difference between the conservative and the non-conservative cases is that, in the latter unlike the former, the growth front is fed from both sides; i.e., from the internal area of the nucleus as well as from the outer area so that the total flux input at the wave crest is not greatly different from that at the trough. A large external flux at a point of positive curvature is aided by a correspondingly small internal flux because, from this direction, one sees a negative curvature at the same point and vice versa. Since, in the non-conservative system, $x^* = R^*/X_s$ determines instability (favored by a small x^*), a low substrate temperature ($X_s \to \infty$ as $T \to 0$) and a high supersaturation ($R^* \to 1$ as $\Delta G_\infty \to \infty$) favors instability.

For a conservative system, the nucleus is unstable for all harmonics $\ell > 2$ as soon as the nucleus size, R, exceeds a critical radius, R_C. Thereafter, unbounded growth of the perturbation occurs and complete instability develops. On the other hand, the nucleus in the non-conservative system case is found to be much more stable. Instead of one limiting radius, R_C, there are now two radii, R_{C1} and R_{C2} (or x_{C1} and x_{C2}) which are, respectively, thresholds for the onset and "turning off" of a given perturbation. Amplification of the perturbation and shape instability are possible only between these two limits. A comparative behavior of the two systems is given in Fig. 5.22. From a practical point of view,

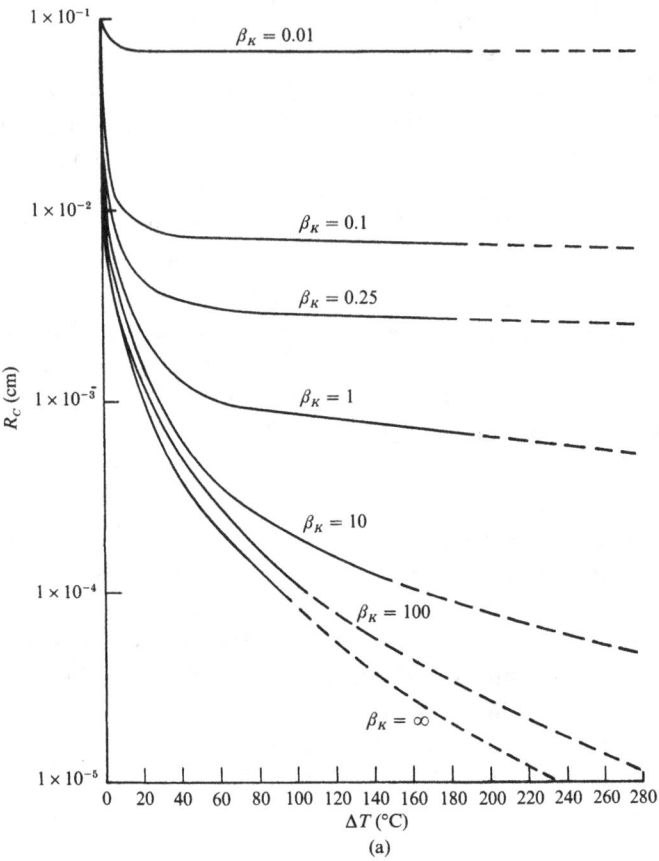

Fig. 5.20. Key instability sizes versus melt undercooling, ΔT_∞, for an infinite circular cylinder of pure Ni at several molecular attachment coefficients, β_K: (a) R_C = radius at which $[(\dot{\delta}/\delta)/(\dot{R}/R)]_{max} = 1.$[13]

defects are most likely to develop in the nucleus only during the period of growth wherein $R_{C1} < R < R_{C2}$ even though the shape may stabilize for $R > R_{C2}$.

Introducing the requirement of crystallographic features in these disc growth considerations reveals that a stable polyhedral expanding shape requires $(\dot{\delta}_\ell/\delta_\ell)/(\dot{R}/R) = 1$ for certain specific ℓ. For real instability of such polyhedral shapes with n-fold symmetry, one requires at least that $\dot{\delta}_{2n}/\delta_{2n} > \dot{\delta}_n/\delta_n$ and $(\dot{\delta}_{2n}/\delta_{2n})/(\dot{R}/R) > 1$. Such an analysis would go far beyond infinitesimal values of δ_ℓ so that higher order perturbation theory would be needed.

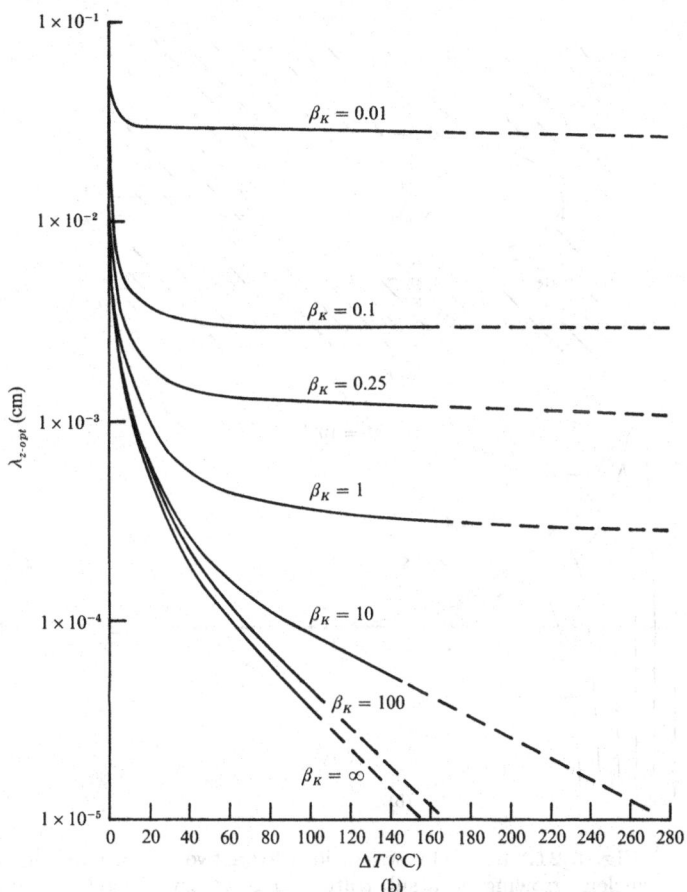

Fig. 5.20. (b) λ_{z-opt} = optimum side-branch spacing at $R = R_C$ (units of β_K are cm s^{-1} °C^{-1}).[13]

Case 4 (Stress generated film instability): In 1972, Asaro and Tiller[15] studied the effect of an in-plane stress on the amplitude development of a harmonic undulation of amplitude δ and frequency ω as illustrated in Fig. 5.23. Without the stress σ_0, the amplitude of all waves would naturally decay by surface and volume diffusion. However, at some stress level, $\dot{\delta}$ becomes positive and grows strongly with σ_0 to create a crack in the material. Here, we consider a film growing at rate V, while the epitaxial constraints to the substrate create the stress σ_0, and we wish to calculate $\dot{\delta}(\omega)$ via surface diffusion to see under what conditions the film

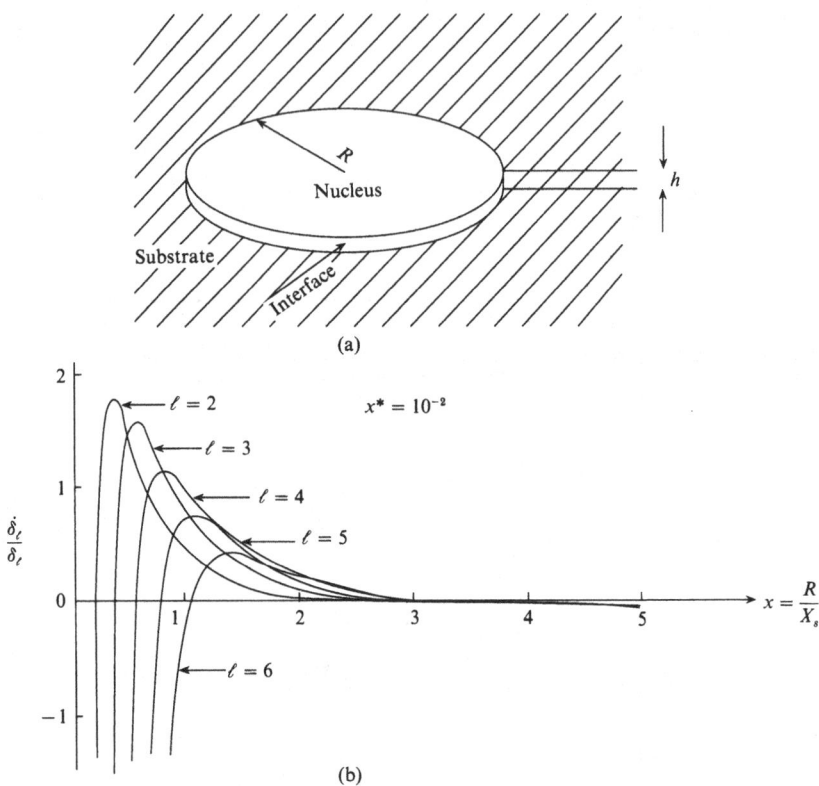

Fig. 5.21. Instability conditions for a two-dimensional disc-shaped nucleus growing on a substrate: (a) geometry, (b) relative amplification rate $\dot{\delta}_\ell/\delta_\ell$ of the different circular harmonics as a function of the relative radius, $x = R/X_s$, for a fixed relative value of the nucleation radius, $x^* = R^*/X_s$.[14]

would be island-like and under what conditions it would be relatively flat.

If we ignore the effect of film thickness, the local stress tensor is given by [15]

$$\sigma_{xx} = \sigma_0 - \delta\sigma_0(\omega^2 y - 2\omega)\exp(-\omega y)\cos(\omega x) \qquad (5.35a)$$

$$\sigma_{yy} = -\omega^2 \delta\sigma_o(1 - \omega y)\exp(-\omega y)\sin(\omega x) \qquad (5.35b)$$

and

$$\tau_{xy} = -\omega\delta\sigma_0(1 - \omega y)\exp(-\omega y)\sin(\omega x) \qquad (5.35c)$$

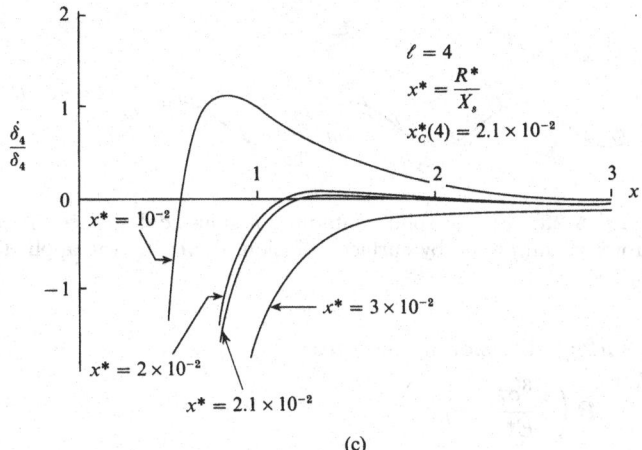

(c)

Fig. 5.21. (c) $\dot{\delta}_4/\delta_4$ as a function of x for varying values of x^*.[14]

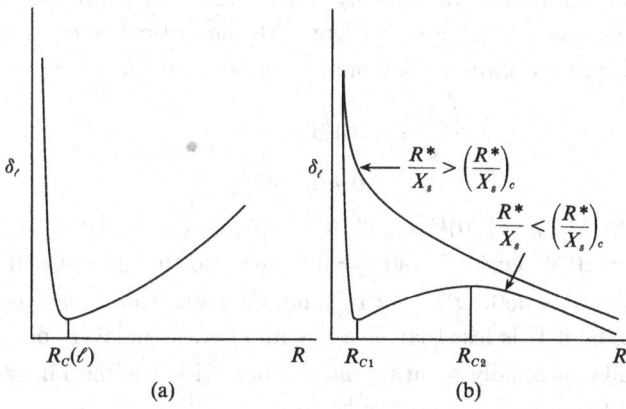

Fig. 5.22. Relative comparison of perturbation amplitude, δ_ℓ, as a function of disc radius, R, between (a) a conservative and (b) a non-conservative system.[14]

Thus, σ_{xx} is reduced as δ increases so there is a thermodynamic driving force to increase δ provided the cost of forming the new surface is not too great. The wave amplitude can grow by surface diffusion out of the troughs and towards the peaks as indicated in Fig. 5.23. Of course, volume diffusion in the gas phase of a somewhat volatile species is also a possibility. Simple perturbation analysis (neglecting surface creation)[15] shows that, provided $\sigma_0 > (E^*\gamma\omega)^{1/2}$, the wave amplitude

Crystal half space

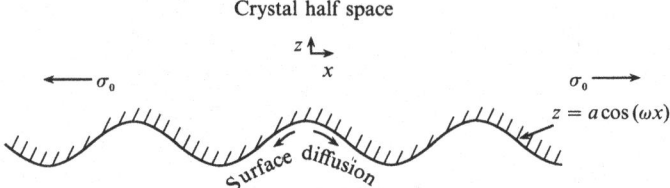

Fig. 5.23. Elastic solid containing a sinusoidal surface profile that grows in amplitude by surface diffusion, driven by the applied stress, σ_0.

grows via surface diffusion at the rate

$$\frac{\dot{\delta}}{\delta} = B\left(\frac{\omega^3\sigma_0^2}{E^*} - \gamma\omega^4\right) \qquad (5.36a)$$

where

$$B = \Omega^2 n_s D_s/\kappa T \qquad (5.36b)$$

where Ω is the atomic volume, n_s is the surface adatom density and D_s is the surface diffusion coefficient. At the critical frequency, $\omega^* = (3/4)\sigma_0^2/E^*\gamma$, we obtain a maximum value for $\dot{\delta}$ given by

$$\frac{\dot{\delta}_{max}}{\delta} = 0.1B\sigma_0^8/\gamma^3 E^{*4} \qquad (5.37a)$$

$$= 0.1B\tilde{\varepsilon}^4 E^{*4}/\gamma^3 \qquad (5.37b)$$

For $\Omega = 20$ Å^3, $n_s = 10^{14}$ cm^{-2}, $\gamma = 10^3$ erg cm^{-2}, $E^* = 10^{12}$ dyne cm^{-2}, $D_s = 10^{-12}$ cm^2 s^{-1} and $\tilde{\varepsilon} = 0.05$, we find that $B = 4 \times 10^{-44}/\kappa T$ so that $\dot{\delta}_{max}/\delta \sim 350$, $\omega^* \sim 2 \times 10^6$ and $\lambda^* \sim 300$ Å at a displacement $\tilde{\varepsilon}$ of 0.05. Thus, if V is less than $\sim$ a few hundred monolayers per second, the film will continually separate into islands. If $V \overset{\sim}{>}$ several microns per second, a relatively flat film may be formed.

An exact solution for this case would require that the stress tensor be given in both the growing film as well as the substrate. For such a situation, one might find that the Stranski–Krastanov model of film growth still obtains.

Problems

1. For the constrained steady state growth, planar interface case at V = constant, determine $C_L(z)$, $T_L(z)$ and $T_S(z)$ for both the basis surface and the perturbed surface with an infinitesimal sinusoidal perturbation, $Z_i = \delta\cos(\omega x)$. Combine these results

in the coupling equation and prove that Eqs. (5.10a) and (5.8a) are both correct.

2. You are considering the planar interface breakdown of an idealized material of given thermal and constitutional properties via the variation of only the growth velocity, V. You have determined that the marginal stability point for an orientation of non-faceted character occurs at ω^* for the velocity V^*. At other orientations of increasingly steeper cusps in the γ-plot for this material, the marginal stability condition is found to be at ω_c with growth velocity V_c. Show that $\omega_c/\omega^* = V_c/V^*$.

3. For the unconstrained growth of a spherical crystal from a pure melt at supercooling ΔT_∞, evaluate the temperature distribution for both the spherical basis surface and for a sphere with an infinitesimal perturbation of the form $R_i = R + \delta Y_{\ell m}$ where Y is a spherical harmonic. Insert this information into the coupling equation and calculate the crystal radius, R_c, at which instability occurs for the $(4,4)$ harmonic.

4. Consider the example of Problem 3 but from an alloy melt of composition C_∞. Prove that Eqs. (5.31) are correct.

5. For the unconstrained growth of an infinitely long cylindrical crystal from a pure melt at supercooling ΔT_∞, evaluate the temperature distribution for both the cylindrical basis surface and for a cylinder with an infinitesimal perturbation of the form $r = R + \delta Y_{\ell,m/R}$ where Y is a cylindrical harmonic. Insert this information into the coupling equation and calculate the crystal radius, R_c, at which instability occurs for the $(4, m/R)$ harmonic.

6. For growth of a single crystal in the normal freeze mode at constant $V\mathcal{G}C_\infty$ and no stirring, determine the critical initial transient length, z_C at which the onset of constitutional supercooling occurs. For constant stirring conditions (fixed δ_C/D), show how z_C varies with δ_C/D. Make a plot of z_C vs. δ_C/D for fixed V, $\mathcal{G}$ C_∞ and initial liquid length, L.

7. For an infinitely long, laterally insulated tube of binary alloy melt of concentration C_∞ and superheat $\hat{\Delta T}$, a perfect heat sink of temperature $T_H - \Delta T'_\infty$ is placed in intimate contact with the melt at $z = 0$ and time $t = 0$ so that the solid/liquid interface moves down the tube at $V = \alpha t^{-1/2}$. For the case of no convective mixing, calculate the value of $\hat{\Delta T} = \hat{\Delta T}_C$ needed to avoid the onset of constitutional supercooling as a function

of C_∞, the usual solute parameters and the thermal parameters of the problem. For the case of convective mixing, such that $V\delta_C/D$ is a constant in time, calculate how $\hat{\Delta T}_C$ varies with δ_C/D.

6

Dynamic interface morphologies

In this chapter, we utilize the basic concepts and some of the earlier quantitative results to predict and understand many of the morphological feature developments that are found in crystals and films. We begin with macroledge development in both films and crystals.

6.1   Layer flow instabilities

For vicinal surface orientations, although the growth conditions are not such as to produce the massive instability characterized by Eq. (5.13a), a range of less severe growth conditions exist wherein instabilities can develop in the layer front flowing across the crystal surface at $V_\ell >> V$. Thus, instead of the layer front being straight and smooth or faceted, it is rumpled with edge waves that are growing in amplitude as the front proceeds. This is just the two-dimensional analog of the instabilities already discussed (see Fig. 5.21). Like its three-dimensional counterpart, microsegregation events and microdefects are built into the growing crystal when such layer flow instabilities develop and the quantitative assessment of the onset of these instabilities is important to the growth of high quality crystals.

There are two types of perturbation consequences that interest us here: (1) that giving rise to a ledge density instability leading to ledge bunching and the growth of h/a as indicated in Fig. 6.1(b) and (2) that giving rise to ledge front instability of the lateral kind as indicated in Fig. 6.1(c). These could just be considered as ω_ℓ and ω_m effects for planar

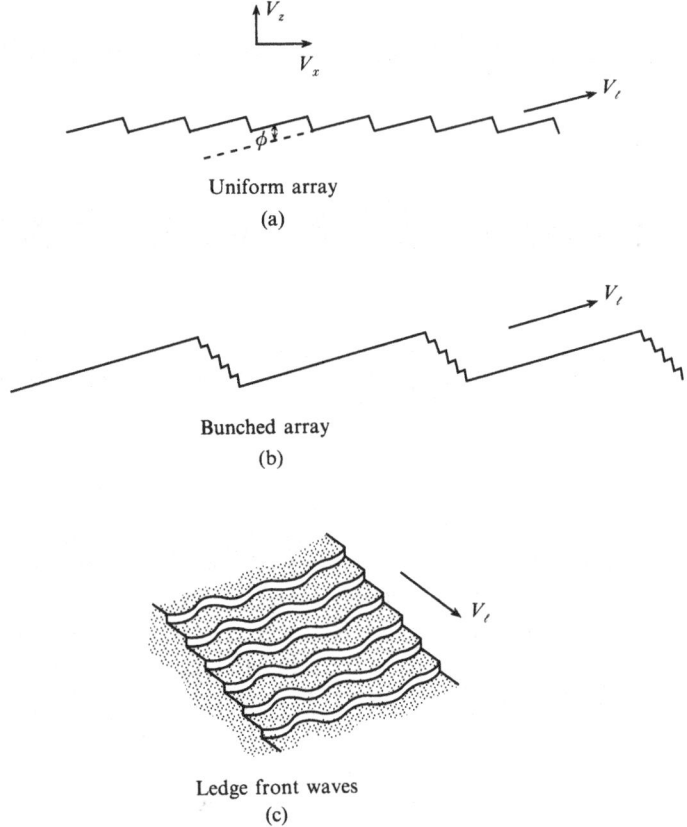

Fig. 6.1. Layer flow configurations: (a) monatomic ledges making an angle ϕ with the interface isotherm, (b) bunching of the monatomic ledges to form multistep height ledges moving at angle ϕ and (c) ledge front instability showing undulation development.

front instability. Since the type (2), or ω_m, instability has already been discussed in the previous chapter, we shall focus our attention here on the type (1) instability, layer bunching or the evaluation of h/a for fixed crystallographic angle ϕ of the interface.

From Figs. 3.19–3.23, one is able to gain an understanding of how $n_{s\ell}$ changes as h/a increases. From the qualitative solution (Fig. 3.21), one can also see that $\Delta n_{s\ell}$ will increase as h/a increases and this could be calculated for a specific system. In a similar vein, the steady state heat evolution due to condensation at the ledges will lead to a variation of local temperature across the ledge and this is also expected to increase as h/a increases. As the temperature along a ledge decreases, τ_s must increase leading to an increased value of $n_{s\infty}$. Thus, a given $\Delta T_\ell(h/a)$ leads to a given $\Delta n_{s\infty}(h/a)$ and when this equals $\Delta n_{s\ell}(h/a)$ calculated from the solute partitioning solution, for the same h/a, the ledge has reached its equilibrium height for the particular set of growth conditions involved. This is the explanation for a given ledge height, h, in the bunched arrays of Fig. 6.1(b). The transient development of h may be quite long in some cases with the process involving the dynamic capture of faster moving monatomic ledges by the step at the upper terrace location and the dynamic release of monatomic ledges by the step at the lower terrace. Thus, any given step has both a velocity component along the terrace direction plus a component perpendicular to the terrace. As we shall see at the end of Chapter 7, this gives rise to a unique set of chemical striations in the crystal.

Using the LPE growth technique and GaAs(100) or Si(111) wafers, spherically shaped by lapping, Bauser and Strunk[1] have provided the schematic representation of their interface morphologies in Fig. 6.2(a) for GaAs LPE layers. Fig. 6.2(b) shows the central facet surrounded by a terraced area while Fig. 6.2(c) shows the transition from a terraced surface to a terrace-free surface (unresolved ledges). This terraced region probably corresponds with the cusp region of the γ-plot so that the ΔT_E contribution to ledge stability is likely to be significant.

When an obstacle is present on the interface, the layer-flow behavior is quite different for the on-facet versus off-facet regimes. This is illustrated in Fig. 6.3, and in a more expanded fashion in Fig. 6.4, for a circular-shaped depression type of obstacle. With layer flow from left to right in Fig. 6.4, we note (i) terraces of average height 0.2 μm on a 40 μm thick LPE layer interfacing with the obstacles, (ii) terraces of average height ~ 2.0 μm on a 45 μm LPE layer due to enhanced bunching near the obstacle and (iii) a semicircular terrace ~ 9.5 μm high surrounding the obstacle on a 65 μm thick layer with average terrace height ~ 2.5 μm. These microscopic layer-flow features can lead to important microsegregation defects in materials as discussed in Chapter 7.

It is interesting to note that this type of layer-flow behavior is also observed on decanted interfaces of high purity Pb crystals (see Fig. 6.5) which lends support to the proposal that the TLK model also operates on metal interfaces during crystallization from the melt.[2]

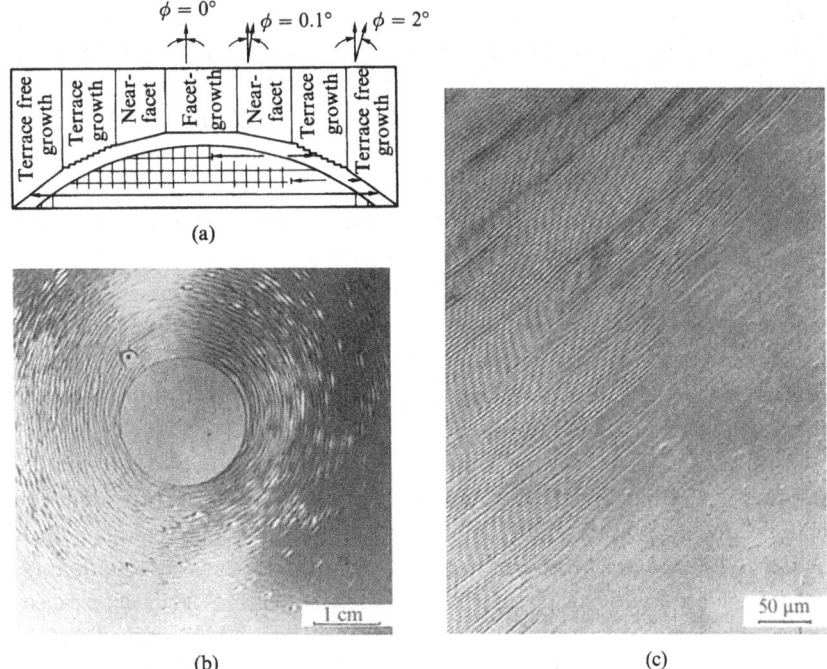

Fig. 6.2. (a) Schematic representation of diverse ledge character due to different orientations of the growth on a spherically shaped substrate during LPE growth. (b) Surface of such a GaAs LPE layer (radius of curvature = 0.5 m). The central facet is surrounded by a terrace which makes a transition to a terrace-free surface in (c).[1]

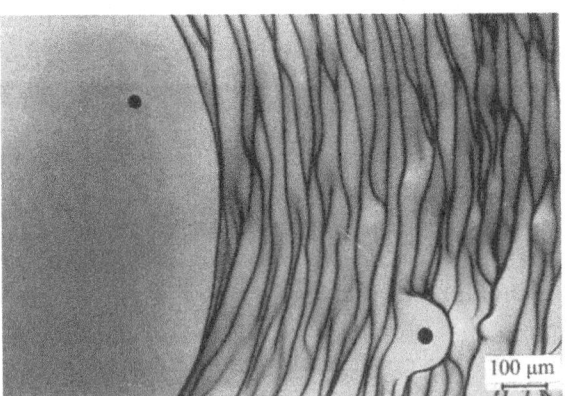

Fig. 6.3. Small obstacles (particles) located in a facet area and in a terraced area demonstrate the different stability characteristics of these two regions (GaAs LPE).[1]

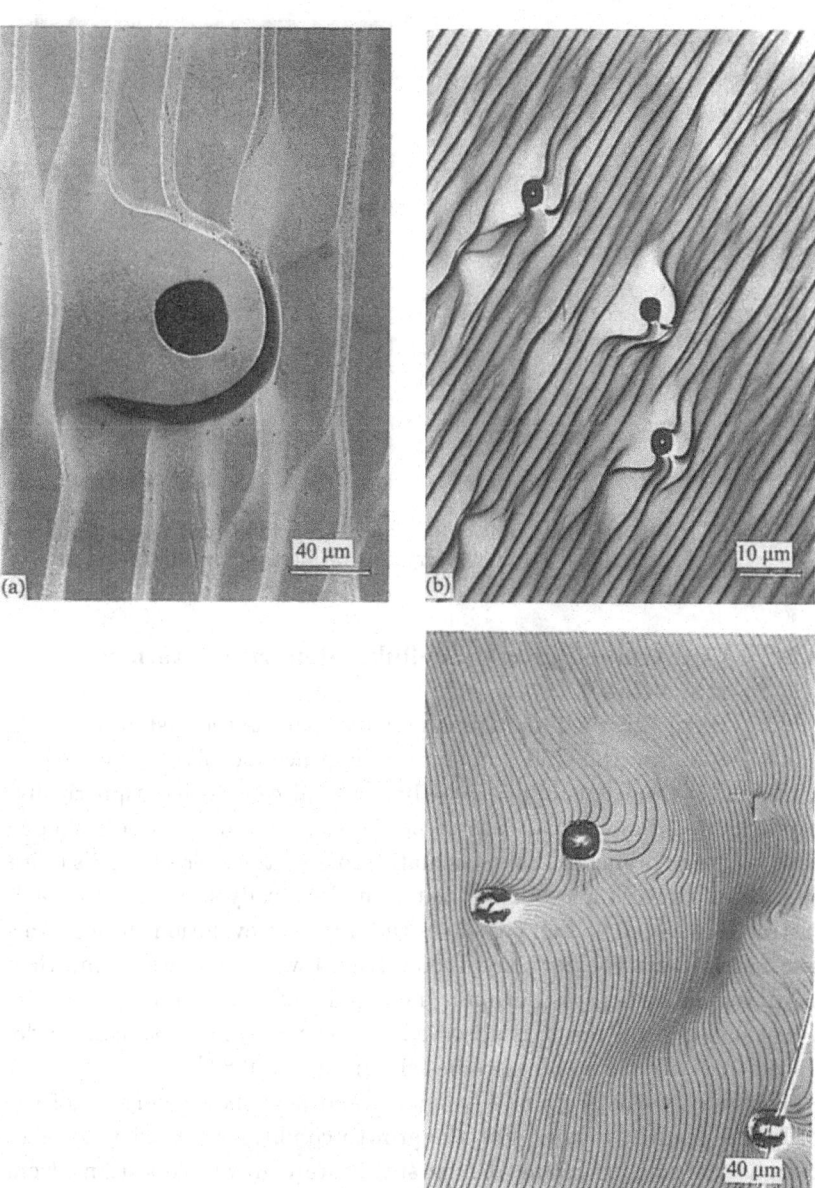

Fig. 6.4. (a) Plan view of a depression in a terrace region showing enhanced step bunching close to the depression, (b) depressions in an earlier stage of development and (c) smaller ledge height terraces interfacing with depressions showing that complete bunching as in (a) does not occur for small ledge heights (GaAs LPE).[1]

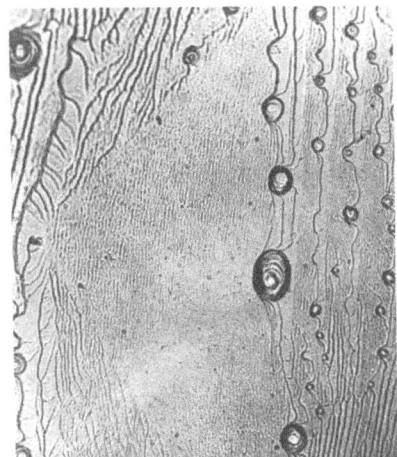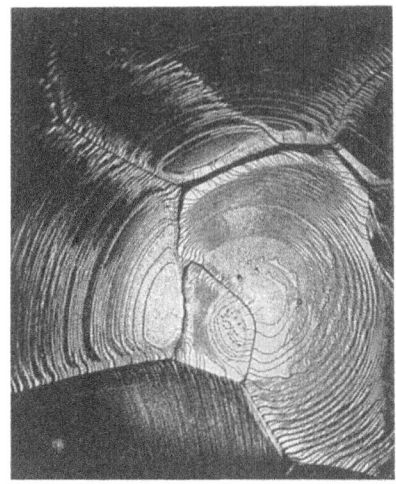

Fig. 6.5. Representation of typical ledge and terrace stuctures observed on rapidly decanted crystals of high purity Pb during growth from the melt.[2]

6.2   Constrained growth, cellular/dendritic interface transition

From Figs. 5.2 and 5.5, one notes that, as the system is driven more deeply into the CSS domain, the interface morphology undergoes a transition from flat-topped two-dimensional cells to hemispherically-capped three-dimensional cells to needle-shaped cells to needle-shaped cells with side branches. The schematic cross-sections for such cells (with the same spacing λ and cell wall gap δ) under steady state growth are illustrated in Fig. 6.6. Fig. 6.7 shows the transient evolution through this set of shape patterns, for the succinonitrile–4 wt% acetone system, when the interface velocity is abruptly increased from zero to 3.4 μm s^{-1}.[3] The same general result is observed for faceted systems as can be deduced from the Ge etching studies shown in Fig. 6.8.[4]

The goal of this section is to understand how the spacing, λ, of the cells and dendrites varies with the growth conditions and what leads to the onset of side branching. Under steady state growth conditions, from Fig. 6.6, one sees that the radius of curvature at the tips of the cells, ρ_t, can vary from being much greater than λ, close to the planar interface breakdown condition, to being much less than λ, close to the onset of the side-branching condition. The length of the liquid channels surrounding the filament can vary for the $\rho_t/\lambda \gg 1$ domain, but, at best at the

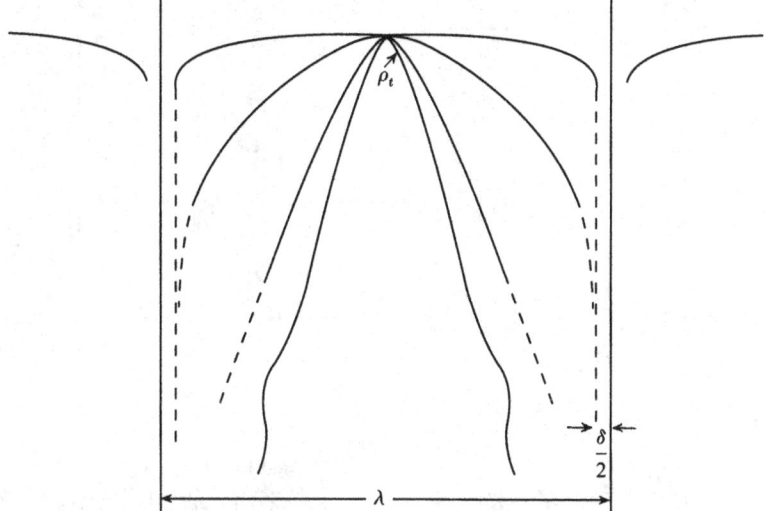

Fig. 6.6. Schematic illustration of cell profile development as a result of increased $VC_\infty/\mathcal{G}_L$ during metal crystal growth. The radius of curvature of the cell tip, ρ_t, is observed to sharpen.

junctions with two or more other cells, eutectic is generally formed. For regular cells with $\rho_t/\lambda \sim 1$, a thin film of eutectic is generally observed at the cell walls so it is approximately true to say that the root of the cell walls is located at the eutectic temperature isotherm for the system. This would certainly be true if $D_S = 0$ for the solute; thus, the length of the cell grooves, L_C, is given for constant $\mathcal{G}_S$ by

$$L_C \approx -m_L(C_{\tilde{E}} - C_t)/\mathcal{G}_S \qquad (6.1)$$

for the case where the liquidus slope, m_L, is constant over the concentration range from the tip of the cell, C_t, to the eutectic concentration, $C_{\tilde{E}}$, at the root of the cell. Because of diffusion in the solid, L_C can be shortened in some cases. For the same C_∞, as $\mathcal{G}_L$ and V are varied so as to cause ρ/λ to decrease, the amount of eutectic forming at the cell roots increases. This occurs because, as the cell tips sharpen, C_t decreases due to lateral diffusion as discussed in Chapter 4 (see Fig. 4.34(b)). Since for a single paraboloidal filament, where the value of k/k_0 decreases towards unity as the Péclet number, $V\rho/2D$, decreases, if we can neglect diffusional overlap from adjacent filaments (only reasonable at $\rho/\lambda \ll 1$), solute conservation allows us to predict the average fractional width,

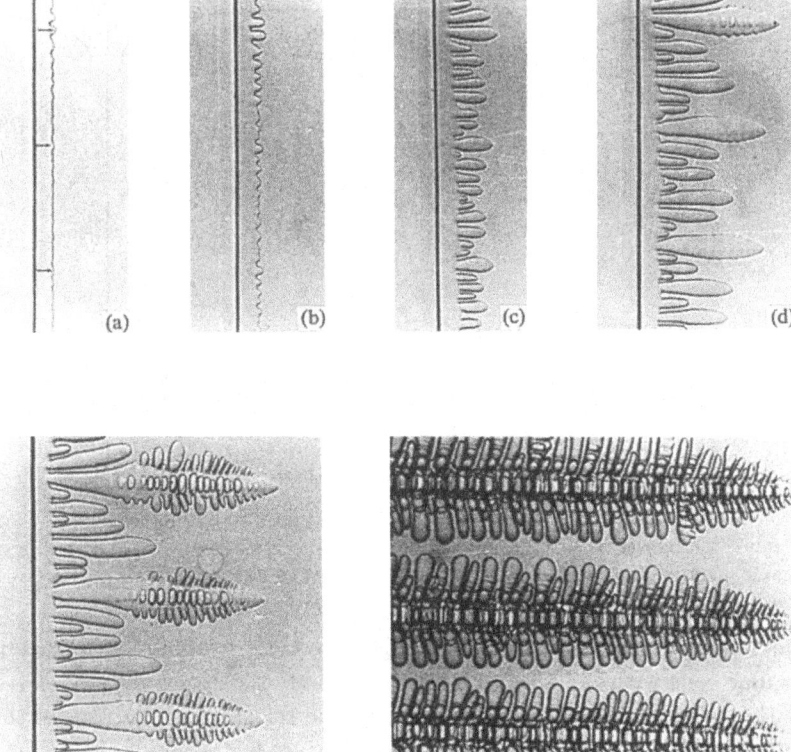

Fig. 6.7. Temporal development of succinonitrile solid/liquid interface shapes with an abrupt change in V from zero to 3.4 μm s^{-1} at $\mathcal{G}_L = 67\,^{\circ}$C cm^{-1}: (a) $t = 50$ s, (b) 55 s, (c) 65 s, (d) 80 s, (e) 135 s and (f) 740 s, all at 41× (succinonitrile–4.0 wt% acetone, courtesy of R. Trivedi).

a/λ, of the eutectic zone at the cell wall; i.e.,

$$k_1 \left(1 - \frac{a}{\lambda}\right) C_\infty + \frac{a}{\lambda} C_{\tilde{E}} = C_\infty \qquad (6.2)$$

where k_1 is the single filament value. Thus, most of the cell has a solute concentration $k_1 C_\infty$ while the wall has a concentration $C_{\tilde{E}}$ to give a segregation ratio, SR $= C_{\tilde{E}}/k_1 C_\infty$. In some systems SR can be as large as 10^4–10^6.

To understand the factors controlling λ, it is useful to consider both a macroscopic view and a microscopic view. From a macroscopic viewpoint, we can define the isotherm that touches the tips of the cells as

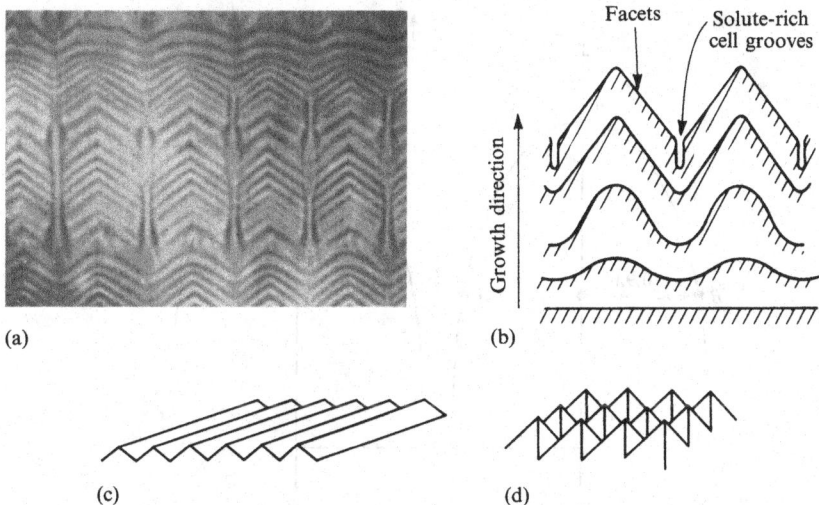

Fig. 6.8. Schematic representation of cell development observed in Ge alloys (faceting system). (a) Successive positions of liquid/solid interface showing development of faceted cells. (b) Schematic diagram of liquid/solid interface at successive times showing the breakdown of the plane front interface to form faceted cells. (c) Elongated cells which form when the growth direction is $\langle 110 \rangle$. Facets on the cell sides are $\{111\}$ planes inclined at $54°74'$ to the growth axis. Facets on cell ends are $\{111\}$ planes parallel to the growth axis. (d) Regular cells which form when the growth direction is $\langle 100 \rangle$. Facets on the cell faces are $\{111\}$ planes $35°46'$ to the growth directions.[4]

being ΔT_E below the liquidus isotherm, $T_L(C_\infty)$, for the bulk liquid. The extremum condition indicates that ρ_t and λ adjust within the limits of their constraints to make ΔT_t a minimum. Looking at the component parts of ΔT_t from the coupling equation (see Eq. (5.1)) and neglecting the ΔT_K contribution, we expect

$$\Delta T_E = \frac{B}{\lambda}; \qquad \Delta T_C = -m_L(C_t - C_\infty) = A\lambda^q \qquad (6.3a)$$

where $B = 2\gamma/\Delta S_F$, A contains $V, \mathcal{G}, k_0, C_\infty$, etc. and q depends on ρ/λ. The ΔT_E term favors large λ while the ΔT_C term favors small λ; thus application of the extremum condition yields

$$\lambda^* = (B/qA)^{\frac{1}{q+1}} \qquad (6.3b)$$

In this approach, the difficult portion of the analysis is an accurate prediction of A and q via the determination of C_t. However, it clearly reveals the important considerations involved.

From a more microscopic but only semiquantitative view, let us con-

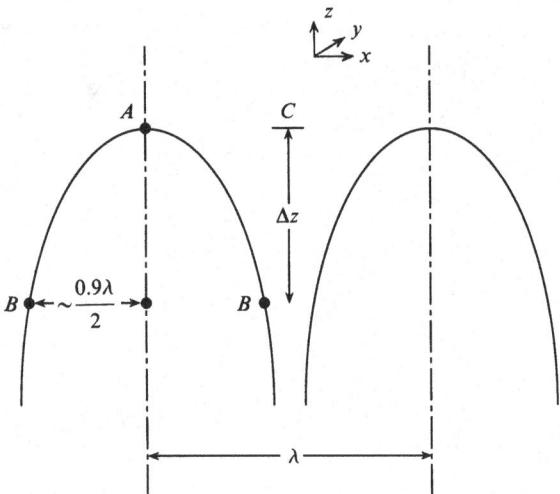

Fig. 6.9. Typical cell cross-section for metal alloys.

sider the cell cross-section in Fig. 6.9. The concentration difference on
the cell cap between the points B and A, $\Delta C_{BA} \sim \epsilon C_t$ where $\epsilon \sim \epsilon' \lambda^p$
(ϵ' and p are constants). From heat flow considerations we expect to
find $-m_L \Delta C_{BA} \sim \bar{\mathcal{G}}_L \Delta z$ (neglecting curvature effects) where $\bar{\mathcal{G}}_L$ is the
average axial temperature gradient in the B–A region. Since lateral dif-
fusion from point A must proceed past point C in the time it takes for
the interface to move from B to C ($t = \Delta z / V$) in order for C_t to be
reduced below C_∞ / k_0, the diffusion scaling law gives

$$\left(\frac{0.9\lambda}{2}\right)^2 = Dt = \frac{D\Delta z}{V} = \frac{-m_L D \Delta C_{BA}}{\bar{\mathcal{G}}_L V}$$

$$= \frac{-m_L D \epsilon' \lambda^p C_\infty (k/k_0)}{\bar{\mathcal{G}}_L V} \qquad (6.4a)$$

since $C_t = (k/k_0)C_\infty$. Rearranging yields

$$\lambda \sim \left[\frac{-5 m_L D \epsilon' (k/k_0) C_\infty}{\bar{\mathcal{G}}_L V}\right]^{\frac{1}{2-p}} \qquad (6.4b)$$

which is consistent with the results of Fig. 6.10 provided $p = 1$ for this
system in this range of cell development. For other systems and other
ranges of ρ/λ, p will not be unity. The general trend to be gathered from
Eq. (6.4b) is that λ decreases for a specific system as V increases, as $\mathcal{G}_L$
increases and as C_∞ decreases. This is consistent with the experimental
findings for metallic systems.

To obtain a more quantitative level of modeling, it is necessary to solve

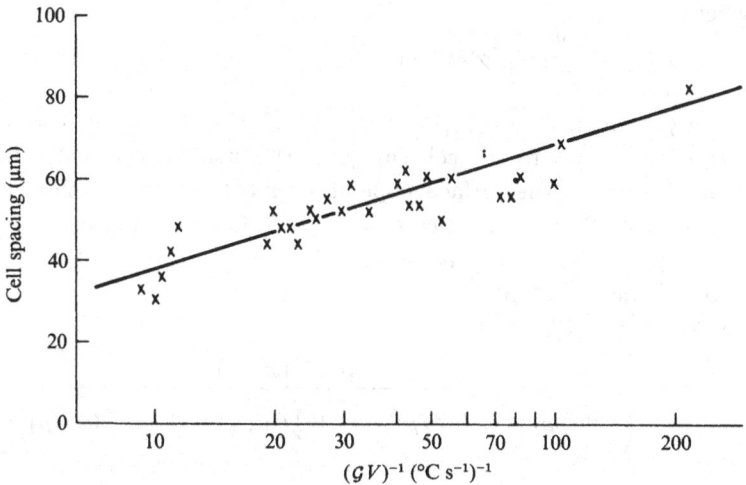

Fig. 6.10. Cell spacings observed in dilute Sn–Pb alloys.[5]

for both the general solute distribution (see Eq. (3.15)) and the general temperature distribution (similar form). To illustrate the procedure for determining the coefficients suppose we truncate each series after the second term. Thus, there are two undetermined coefficients for each of C_L, T_L and T_S plus two undetermined shape parameters ρ and λ. If one assumes a paraboloidal shape for most of the central region of the cell, then the application of solute conservation plus both temperature continuity and heat conservation at two points on the surface would allow all the undetermined coefficients to be fixed in terms of only λ since $\rho \propto \lambda$ for a specified cap shape. Widely separated points would be best and one should be at $x = 0$. The other could be at $x = \lambda/4$.

To understand the relationship between ρ and λ, let us use a continuum picture of the surface and evaluate what the average curvature, $\bar{\mathcal{K}}$, must be by considering first two-dimensional parabolic-shaped cell caps and then three-dimensional paraboloidal-shaped cell caps. One knows that $\Delta G_E = 2\gamma/\lambda$ and about $4\gamma/\lambda$ for these two cases so that $\bar{\mathcal{K}} = 1/(\lambda/2)$ and $2/(\lambda/2)$, respectively. For any two-dimensional surface shape, $Z(x)$, the average curvature between two points a and b on the surface can be written in terms of the arc length, ℓ_{ab}, along the surface; i.e.,

$$\bar{\mathcal{K}} = \frac{1}{\ell_{ab}} \int_a^b \left[\frac{-Z''}{(1 + Z'^2)^{3/2}} \right] d\ell = \frac{1}{\ell_{ab}} \left[\tan^{-1} Z_a' - \tan^{-1} Z_b' \right] \quad (6.5a)$$

where

$$\ell_{ab} = \int_a^b (1 + Z'^2)^{1/2} dx \qquad (6.5b)$$

For a parabola, $Z/\rho = \frac{1}{2}[1 - (x/\rho)^2]$, $Z' = -x/\rho$, we choose $x_a = 0$ and $x_b = x^*$ where the cell cap joins the straight cell walls; thus, $\bar{\mathcal{K}}$ depends only on the surface slope at x^* and the total length of the cell cap between 0 and x^*. Since $\bar{\mathcal{K}} = 1/(\lambda/2)$ for this case, one constraint on the parabolic cell cap parameters is

$$\frac{\rho}{(\lambda/2)} = \frac{\tan^{-1}(x^*/\rho)}{(\ell_{ab}/\rho)}$$

$$= \frac{\tan^{-1}(x^*/\rho)}{\left((x^*/2\rho) \left[1 + (x^*/\rho)^2\right]^{1/2} + \frac{1}{2}\ell n \left\{ x^*/\rho + \left[1 + (x^*/\rho)^{1/2}\right]^2 \right\} \right)}$$

$$(6.5c)$$

Defining the cell boundary width, δ, as $\delta = \lambda/2 - x^*$, one sees that Eq. (6.5c) relates δ/ρ to x^*/ρ so that, for a given x^*/ρ, δ/ρ is determined as is λ/ρ. Because the burden of the needed excess free energy $2\gamma/\lambda$ must come from the average curvature on the cell cap between $x = 0$ and $x^* = (\lambda - \delta)/2$, the larger is δ the smaller must be ρ to provide the average curvature. For small values of δ/λ, Eq. (6.5c) yields no real solution for ρ/λ. The minimum value of δ/λ giving a solution is $\delta/\lambda \approx 0.20$ with $\rho/\lambda \approx 0.25$.

For the three-dimensional paraboloid of revolution cap shape, $Z/\rho = \frac{1}{2}[1 - (r/\rho)^2]$ and

$$\bar{\mathcal{K}} = \frac{1}{A_{ab}} \int_a^b \int_0^{2\pi} \left\{ \frac{2 + (r/\rho)^2}{\rho[1 + (r/\rho)^2]^{3/2}} \right\} d\ell r d\phi \qquad (6.6a)$$

$$= \frac{\pi\rho}{A_{ab}} \left\{ \left(\frac{r^*}{\rho}\right)^2 + \ell n \left[1 + \left(\frac{r^*}{\rho}\right)^2\right] \right\} \qquad (6.6b)$$

with

$$A_{ab} = \int_a^b \int_0^{2\pi} d\ell r d\phi = \frac{2}{3}\pi\rho^2 \left\{ \left[1 + \left(\frac{r^*}{\rho}\right)^2\right]^{3/2} - 1 \right\} \qquad (6.6c)$$

Thus, for this case where $r^* = (\lambda - \delta)/2$, the average curvature constraint equation is

$$\frac{8\rho}{3\lambda} = \frac{\left\{ (r^*/\rho)^2 + \ell n \left[1 + (r^*/\rho)^2\right] \right\}}{\left\{ \left[1 + (r^*/\rho)^2\right]^{3/2} - 1 \right\}} \qquad (6.6d)$$

and the minimum value of δ/λ is about 0.165 occurring at $r^*/\rho \approx 3$ ($\rho/\lambda \approx 0.14$) where $\bar{K}$ is a maximum. One finds that, as r^*/ρ increases from 0.5 to 5.0, $r^*\bar{K}/2$ increases from 0.44 to 0.82 and stays relatively constant at the value of 0.82 until r^*/ρ reaches 3.0 and then falls slowly to 0.80 as r^*/ρ increases to 5.0. We thus see that there is no value of r^*/ρ yielding $\bar{K} = 2/r^*$ as it must if sidewall creation is to be fully accounted for. This appears to be a problem associated with use of the continuum surface approximation.

A more general constraint involves coupling the solute and thermal fields on the interface according to Eq. (5.1b) leading to

$$T_i = T_L(C_\infty) + \bar{m}_L(C_i - C_\infty) - \frac{\gamma}{\Delta S_F}\bar{K} - \Delta T_K \qquad (6.7)$$

Application of Eq. (6.7) at the tip ($x = 0$ or $r = 0$) yields another constraint equation connecting ρ and λ since they both enter in T_i and C_i. Finally, if one chooses not to apply the surface creation constraint, application of the extremum condition $(\partial T_i/\partial \lambda)_{\lambda^*} = 0$ leads to the final condition needed to solve for the three key cell parameters (ρ, δ, λ).

With this approach, the larger is the number of terms taken in the series solutions for C_L, T_L and T_S, the more accurate will be the determination of λ^*. Of course, with this continuum approximation, proper specification of ΔT_K over the surface is needed for a successful point-to-point solution.

A modification of the above approach which is less sensitive to the assumed interface shape is to replace the conditions applied on the interface at $x = \lambda/4$ by more global conditions such as

(1) net solute conservation between $x = 0$ and $x = \lambda/4$ (for a two-dimensional cell)

$$\frac{\lambda}{4}VC_\infty = k_0 V \int_0^{\lambda/4} C_i dx + \int_{Z_i}^\infty \left(-D_L\frac{\partial C_L}{\partial x}\right)_{\lambda/4} dz \qquad (6.8a)$$

(2) net heat conservation between $x = 0$ and $x = \lambda/4$ (for a two-dimensional cell)

$$\frac{\lambda}{4}(K_L\mathcal{G}_L) + K_L \int_{Z_i}^\infty \left(-\frac{\partial T_L}{\partial x}\right)_{\lambda/4} dz + \frac{\lambda}{4}V\Delta H_F$$

$$= \frac{\lambda}{4}(K_S\mathcal{G}_S) + K_S \int_{-\infty}^{Z_i} \left(-\frac{\partial T_S}{\partial x}\right)_{\lambda/4} dz \qquad (6.8b)$$

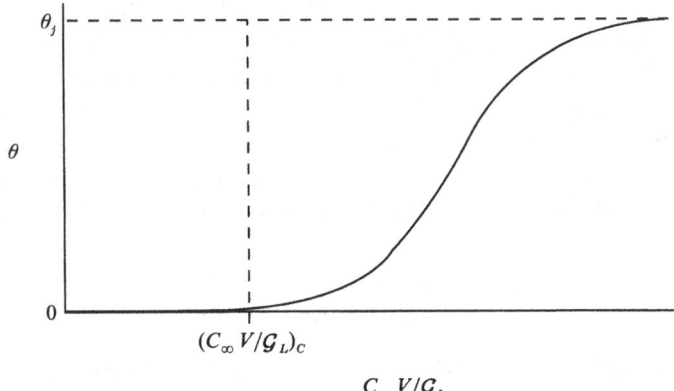

Fig. 6.11. Cell wall angle deviation, θ, from the interface normal direction ($\theta = 0$) as a function of the constitutional supercooling parameter, $C_\infty V/\mathcal{G}_L$ ($\theta_d =$ dendrite direction).

(3) average temperature continuity

$$\frac{4}{\lambda} \int_0^{\lambda/4} T_L(Z_i)\mathrm{d}x = \frac{4}{\lambda} \int_0^{\lambda/4} T_S(Z_i)\mathrm{d}x \tag{6.8c}$$

Such an approach is expected to lead to a much more accurate result when using such strongly truncated series solutions for the solute and thermal fields.

6.2.1   Filament orientation changes with growth conditions

As the crystal growth conditions change so as to increase CSS and cause the filament cross-section to change from the $\rho_t/\lambda \gg 1$ case to the $\rho_t/\lambda \ll 1$ case, the orientation of the filaments relative to the crystal axis also changes. From studies on horizontally grown Sn and Pb crystals, it was observed that the cell boundaries are aligned parallel to the axis of heat flow for small V but that the alignment changed from this direction and asymptotically approached the dendrite direction as V increased. This is illustrated in Fig. 6.11 for the general CSS variable, $C_\infty V/\mathcal{G}_L$. Such a situation develops for constant V by lowering $\mathcal{G}_L$ or by increasing C_∞ and could not occur if the interface were maximally rough and a uniform attachment mechanism operated on the interface. This observation indicates that a layer generation and layer flow mechanism is operating.

There are a number of additional factors that strongly favor the operation of a layer-flow mechanism rather than a uniform attachment mechanism for metals. The first was indicated in Chapter 5 (see page 260) and involves the breakdown to a cellular interface. A continuum interface picture would predict cell formation at smaller C_∞, for fixed $\mathcal{G}_L/V$ for an interface of lower γ. However, experimentally for the Pb–Sn system, this author found that the $Pb_{\langle 111 \rangle}$ interface was stable to higher C_∞^{Sn} than high angle $Pb_{\langle hk\ell \rangle}$ interfaces which would have a higher γ. Likewise, in the cellular domain for the Pb–Sn system, at fixed C_∞^{Sn}, V and $\mathcal{G}_L$, $\lambda_{\langle hk\ell \rangle} \sim 2\text{--}4\ \lambda_{\langle 100 \rangle}$ which is unexpected for a continuum interface, uniform attachment kinetics and the expected γ-plot. However, it does fit with a TLK picture of the interface as we shall see below. Additional experimental evidence is that layer sources are seen on decanted interfaces (see Fig. 6.5) although it has been argued that these result from thermal faceting after decantation. Finally, during unconstrained growth, the dendrite orientation for FCC crystals is sharply in the $\langle 100 \rangle$ direction with negligible variance. All of these pieces of evidence, taken in combination, point to the presence of layer-flow mechanisms operating at the interfaces of metal crystals grown from the melt under normal driving force conditions.

If the growth of an $(hk\ell)$ interface for an FCC crystal is considered from the TLK point of view, as the cell caps sharpen (ρ_t/λ decreases) beyond a certain point, a single layer source can no longer effectively feed the cap and a second layer source must begin to operate on one or more of the other three facet planes where there is sufficient contour to expose the plane and sufficient supercooling for two-dimensional nucleation on the plane. As the contour continues to sharpen, all four of the layer sources must begin to operate.[6] This process is illustrated in Fig. 6.12 for the transition from one to two operating layer sources for two-dimensional cells where we see that, for a particular layer source as the cell cap sharpens, the layer source locations move closer to the center of the cell. In addition, cell cap growth due to the second layer source on cell (1) begins to overhang the cell cap growth due to the primary layer source on cell (2). Thus, the boundary between the two cells must begin to angle towards the right. This process will continue until the caps are growing in a direction that is symmetrically located between the two operating layer sources. For cells in the FCC system, this process continues until the four operating layer sources symmetrically bound the cell cap and it is growing in the $\langle 100 \rangle$ direction. Vestiges of multiple layer sources operating on cell caps are shown in Fig. 6.13 for various decanted interfaces.

When the cell cap contour is sharp enough to require multiple layer

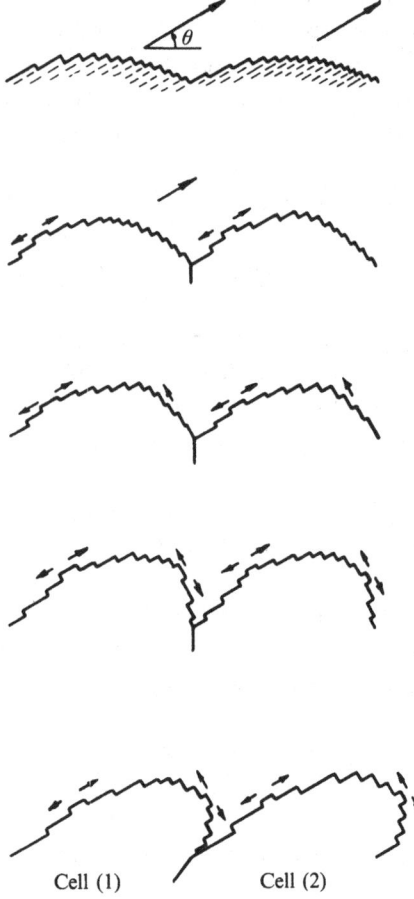

Fig. 6.12. Proposed cell development from layer sources as $C_\infty V/\mathcal{G}_L$ increases.[6]

sources but not sharp enough to allow appreciable overhang, the cell width, λ, for an $\langle hk\ell \rangle$ interface must be larger than the cell width for a $\langle 100 \rangle$ interface under identical growth conditions. This is illustrated in Fig. 6.14. This significant difference in cell size as a function of interface orientation for FCC crystals can be accounted for by the difficulty of the $\{111\}$ layer source system to form a bounded cell cap. The layer sources must each have sufficient supercooling for layer nucleation at the needed rate. For the two orientations considered in Fig. 6.14, if λ were the same in both cases, less solute would be partitioned to the cell boundaries for the $\langle hk\ell \rangle$ orientation because the average cell cap is flatter than for

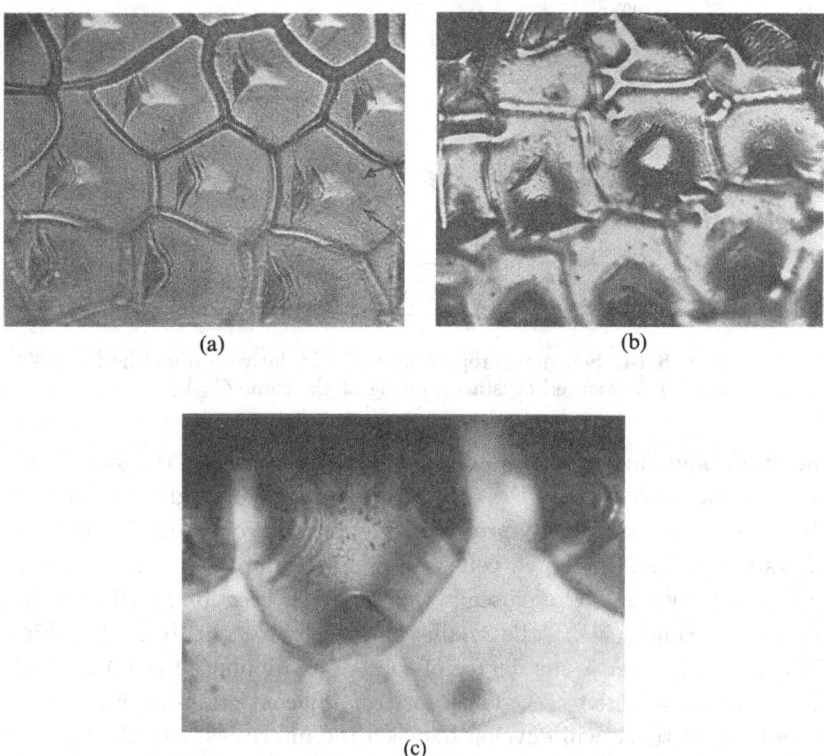

Fig. 6.13. Experimental observation of cell development from layer sources on decanted interfaces of Pb crystals: (a) one layer source operating, (b) two layer sources operating and (c) three layer sources operating (250×).[6]

the $\langle 100 \rangle$ case. Thus, T_i must be lower for the $\langle hk\ell \rangle$ than for the $\langle 100 \rangle$ orientation for fixed λ. Only by increasing λ for the $\langle hk\ell \rangle$ case can ΔT_E be reduced sufficiently to allow $T_i(\langle hk\ell \rangle)$ to approach $T_i(\langle 100 \rangle)$.

The preferred casting texture of metals can be understood from the application of these concepts and such textures are related to either preferred nucleation or preferred growth of grains of certain orientations. As an example, let us consider FCC metals. For very pure metals, the preferred nucleation texture is with the closest packed plane parallel to the mold surface which leads to $\langle 111 \rangle$ as the dominant crystal axis. Competitive growth between this orientation and other orientations depends upon the shape anisotropy of grain boundary grooves and this also favors the $\langle 111 \rangle$ with respect to encroachment onto its neighbors. For slightly impure FCC alloys where cells or dendrites do not form,

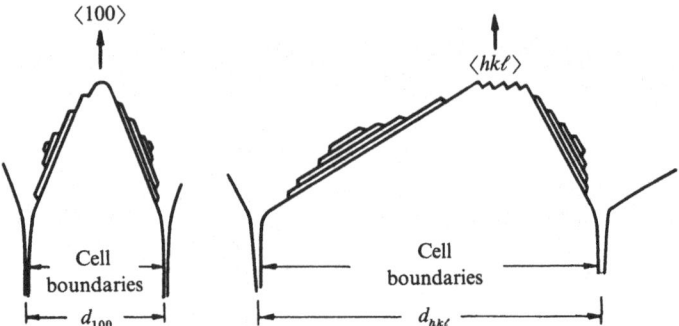

Fig. 6.14. Schematic representation of relative cell widths for $\langle 100 \rangle$ and $\langle hk\ell \rangle$ oriented crystals growing at the same $C_\infty V/\mathcal{G}_L$.

the grain boundary groove becomes a sink for solute and the lateral encroachment mechanism of the $\langle 111 \rangle$ grains onto their neighbors is not so effective. Thus, no strongly preferred growth texture is seen. For impure alloys where well-developed cells and dendrites form, the phenomenon of Fig. 6.11 occurs. The closer is the dendrite direction ($\langle 100 \rangle$) to the interface normal ($\langle hk\ell \rangle$), the smaller is the cell or dendrite size (see Fig. 6.14) so the greater is the degree of solute partitioning to cell walls and the lower is C_t (higher T_i). Thus, for two adjacent grains of different θ, a lead/lag distance will develop between the interfaces with the leading grain laterally encroaching on the lagging grain. This leads to a $\langle 100 \rangle$ growth texture for FCC metals.

6.2.2   *Quantitative TLK picture*

When one moves to considering the more realistic TLK model of an interface, the coupling equation along the cell caps is no longer given by Eq. (6.7) and must be replaced by

$$T_i = T_L(C_\infty) + \bar{m}_L(C_i - C_\infty) - (\delta T_E + \Delta T_K) \tag{6.9}$$

where δT_E is the local surface creation contribution that accounts for the creation, interaction and annihilation of layer edges. The larger is the ledge energy, γ_ℓ, the larger will be δT_E needed for the creation of new layers by two-dimensional nucleation at the facet plane orientations. As described in Chapter 2 of the companion book[7] (see Fig. 2.34), this involves the development of chord-like flats cutting the cell cap. As γ_ℓ increases, for rapid layer edge attachment kinetics, these flats will grow in size until they produce a completely faceted cell cap like that shown in Fig. 6.15.[8] In this case the supercooling needed for two-

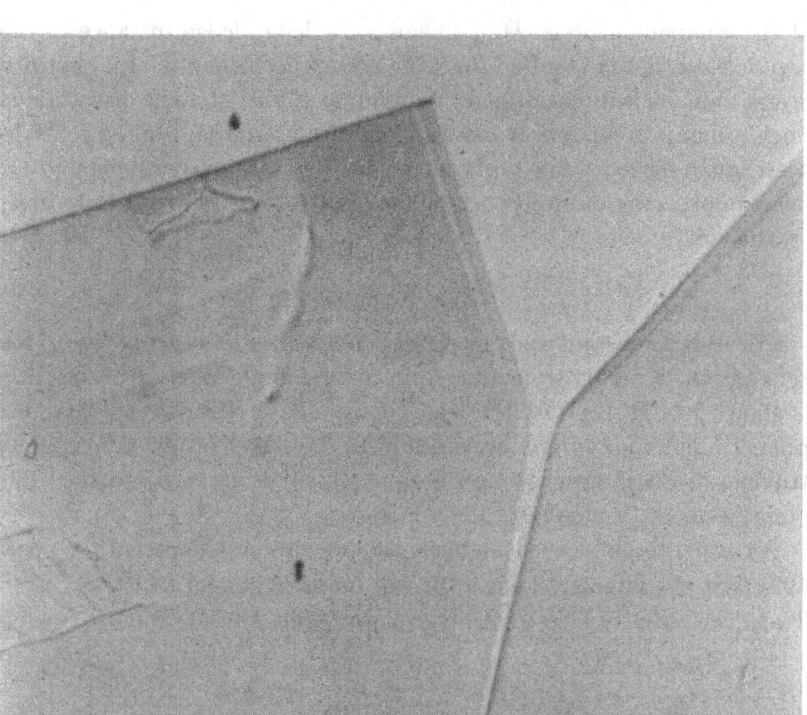

Fig. 6.15. Faceted cells of salol during growth from the melt (60×).[8]

dimensional nucleation is sufficiently large, at the prevailing macroscopic V, that the layer spreading kinetics is extremely rapid and the interface is completely composed of faceted segments. It is likely that no additional ΔT_K is needed to drive the layer edge attachment process because the δT_E magnitude at the nucleation site is sufficiently large that, even at the tip of the cell a distance ΔZ^* away, more than enough supercooling exists to drive the layer edge attachment process ($\delta T_E - \bar{\mathcal{G}}\Delta Z^* > V/\beta_2\Delta S_F$).

6.2.3   Side-branch development and dendrite spacing

On a microscopic scale, it is the δT_E term in Eq. (6.9) that leads to the onset of side-branch development. The location of the initial lateral instability is along the ribs located midway between the operating layer sources that are feeding the tip of the filament (see Fig. 3.2(d) of the companion book[7]). Changing the ledge–ledge distance in this region is all that is required to create a bump sufficiently large that new layers may nucleate locally on the bump at the facet plane orientations. Such a bump is capable of independent lateral growth leading to "side-branching" such as can be observed in Fig. 6.7. Experimentally, the onset of observable side-branching (see Fig. 6.16)[3] is found to be given by the condition

$$\frac{\mathcal{G}_L}{V^{1/2}} \lesssim A' \lambda_\theta \frac{C_\infty}{k_0} \qquad (6.10)$$

where A' is a constant and the cell spacing λ_θ is a function of the crystal orientation, θ. This transition from a cellular to a dendritic interface is shown in Fig. 6.17 for the lead–tin system.[9] The lower breakdown refers to high index $\langle hk\ell \rangle$ orientations in the center of the stereographic triangle (λ_θ is largest) while the upper breakdown refers to the $\langle 100 \rangle$ orientation (λ_θ is smallest).

For a parabolic-shaped filament tip, we saw in Chapter 4 (see page 192) that the lateral velocity, V_x, will be proportional to $V_z^{1/2} = V^{1/2}$. Thus, the onset of CSS in the lateral direction will be given by

$$\frac{\mathcal{G}_{L_x}}{V_x} = \frac{b\mathcal{G}_L}{V^{1/2}} = A\, C_i = bA' \lambda_\theta \frac{C_\infty}{k_0} \qquad (6.11)$$

such as is found experimentally (b, A and A' are constants). All the considerations already discussed for planar front instability now apply to paraboloidal front instability in a lateral direction.

By analogy with the closely similar cellular solidification case, one would expect that the primary dendrite arm spacing would depend on the product $\mathcal{G}_L V$ (which is essentially an effective cooling rate, $\dot{T}$) as does the cell spacing. The experimental data on a wide variety of systems correlates well with the cooling rate parameter over more than an order of magnitude variation in primary dendrite arm spacing. The geometry of the primary and secondary arms is illustrated in Fig. 6.18.[10] Most experimental results are either plotted as a function of average cooling rate during solidification, $\bar{\mathcal{G}}_L V$, or of local solidification time, $t_f = \Delta \tilde{T}_S / \mathcal{G} V$ where $\Delta \tilde{T}_S$ is the non-equilibrium temperature range of solidification. The experimental relationship between the dendrite arm spacing, λ_d, and the thermal variable, t_f, is given by

$$\lambda_d = a t_f^n = b(\mathcal{G} V)^{-n} \qquad (6.12)$$

where the exponent n is in the range of $1/3$–$1/2$ for secondary dendrite arm spacings and generally very close to $1/2$ for primary dendrite arm

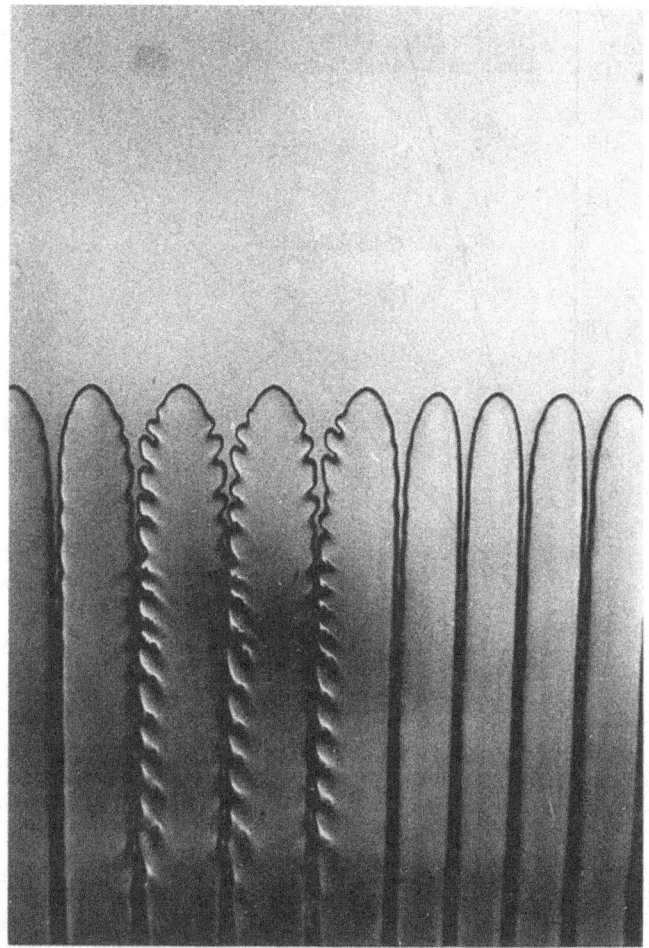

Fig. 6.16. Cells in succinonitrile with secondary side-branches just discernible (cell/dendrite transition, courtesy of R. Trivedi).

spacings ($p \sim 0$ in Eq. (6.4b)). Some experimental data for this situation are presented in Figs. 6.19 and 6.20.[10]

The final dendrite spacing that one sees and measures in a fully solidified casting is usually much coarser than the one that initially forms. The coarsening comes about because some of the arms which form initially became unstable later during the overall solidification process and melt while others continue to grow. This is a type of "Ostwald ripening" where $\Delta G_E = \gamma \mathcal{K}$ is the driving force for the morphological change that

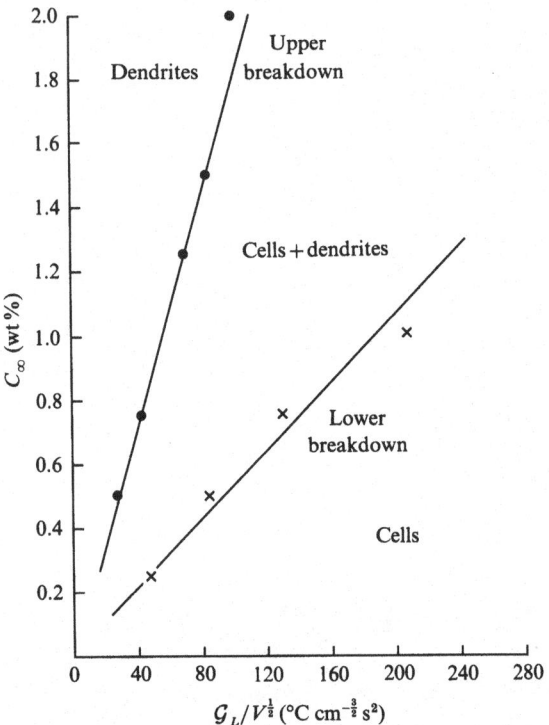

Fig. 6.17. Effect of growth conditions for a Pb–Sn alloy upon the transition from a cellular to a dendritic interface. The upper (100) and lower ($hk\ell$) curves show the limits of the orientation dependence.[9]

leads to $\lambda_S \propto t_c^{1/2}$ (λ_S = secondary arm spacing and t_c = coarsening time).

6.3   Constrained growth, polyphase crystallization

Polyphase crystallization involves the simultaneous formation of two or more new phases from a nutrient phase; i.e., two-phase or three-phase, etc., eutectics, peritectics, monotectics, eutectoids, etc. We shall use the two-phase eutectic to illustrate the principles involved. The competitive processes to be considered are (i) the repeated and somewhat independent nucleation and growth of the two solid phases from the nutrient phase and (ii) the cooperative growth of a duplex phase front from the nutrient. Within the latter category, a variety of patterns for the two phases exist, some of which are stable stationary state patterns

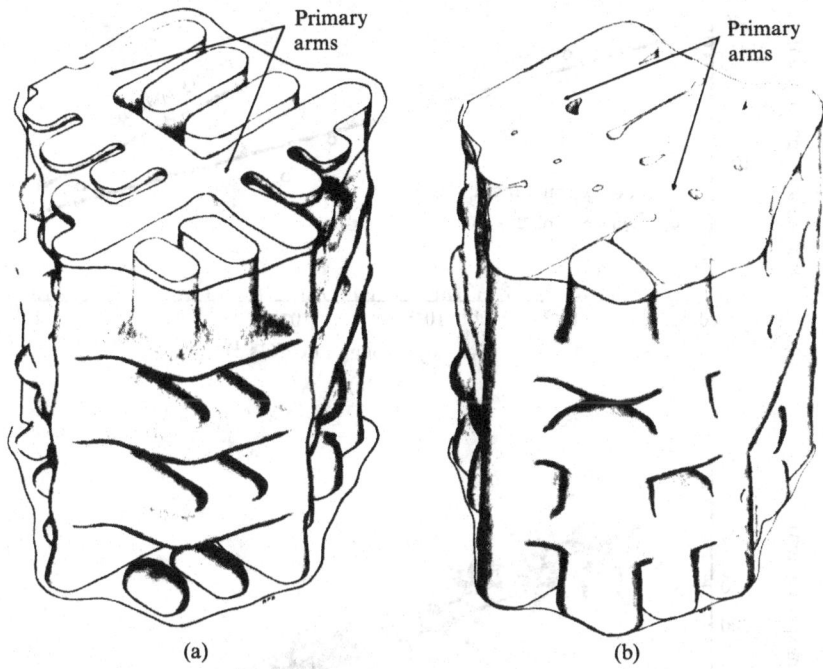

Fig. 6.18. Schematic diagram of the dendrite structure observed in an Al–4.5%Cu alloy at (a) 50% solid and (b) 90% solid.[10]

while others are transitional state patterns. We shall approach the topic by considering the factors involved and the requirements needed for the cooperative growth mode and use the extremum condition to determine the most stable duplex phase pattern.

Experimentally, one finds *continuous microstructures* where the phases can be traced along some unbroken path from the beginning of the crystallization process to its termination so that both phases are continuous in the growth direction as in Fig. 6.21(a). Both lamellar structures and rod structures fulfill this criterion (even some globular structures). The discontinuous lamellae in Fig. 6.21(a) generally have a specific crystallographic relationship with the continuous phase. Mismatch or "faulted" surfaces exist in the volume and it almost looks like the lamellae have undergone a shearing movement normal to their long axis. The second dominant type of microstructure that one finds is the *discontinuous* microstructure where one of the phases of the alloy is dispersed as discrete particles in the matrix phase (see Fig. 6.21(b)). A

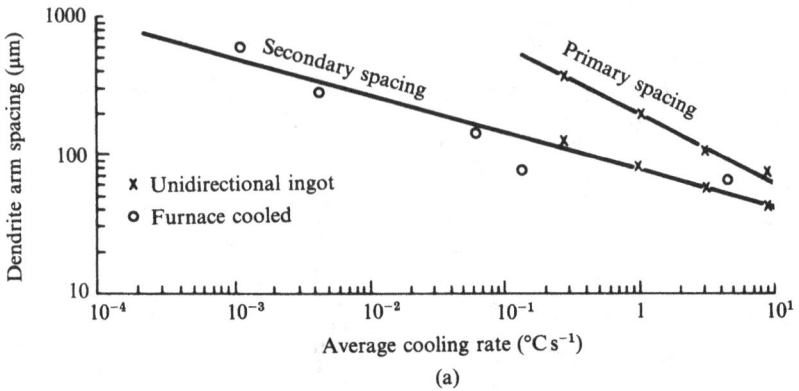

(a)

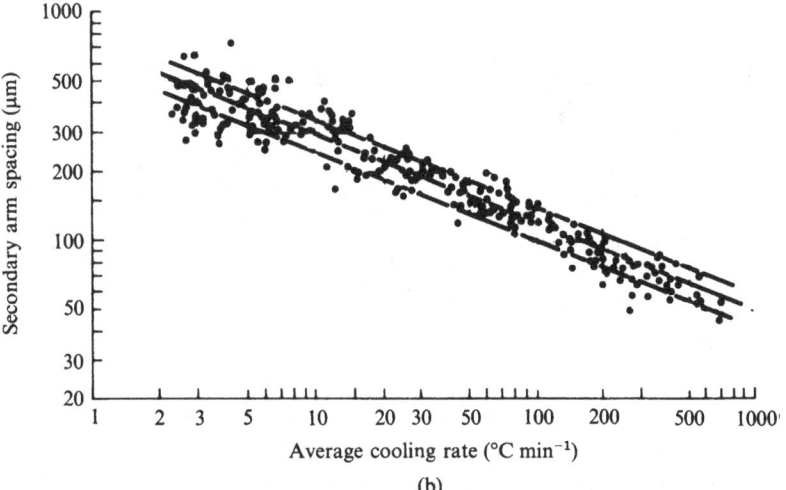

(b)

Fig. 6.19. Some experimental data on dendrite arm spacings in ferrous alloys: (a) Fe–25%Ni and (b) commercial steels containing 0.1–0.9% C.[10]

third type of structure that is observed is the spiral structure illustrated by Fig. 6.21(c).

The preferred microstructure reaction mechanism differs from system to system and, for any given system, it varies with V and $\mathcal{G}_L$ roughly as illustrated in Fig. 6.22.[12] There are exceptions to this general representation, particularly when comparing cases of strongly preferred epitaxial orientation with cases where preferred orientation is absent. For any one

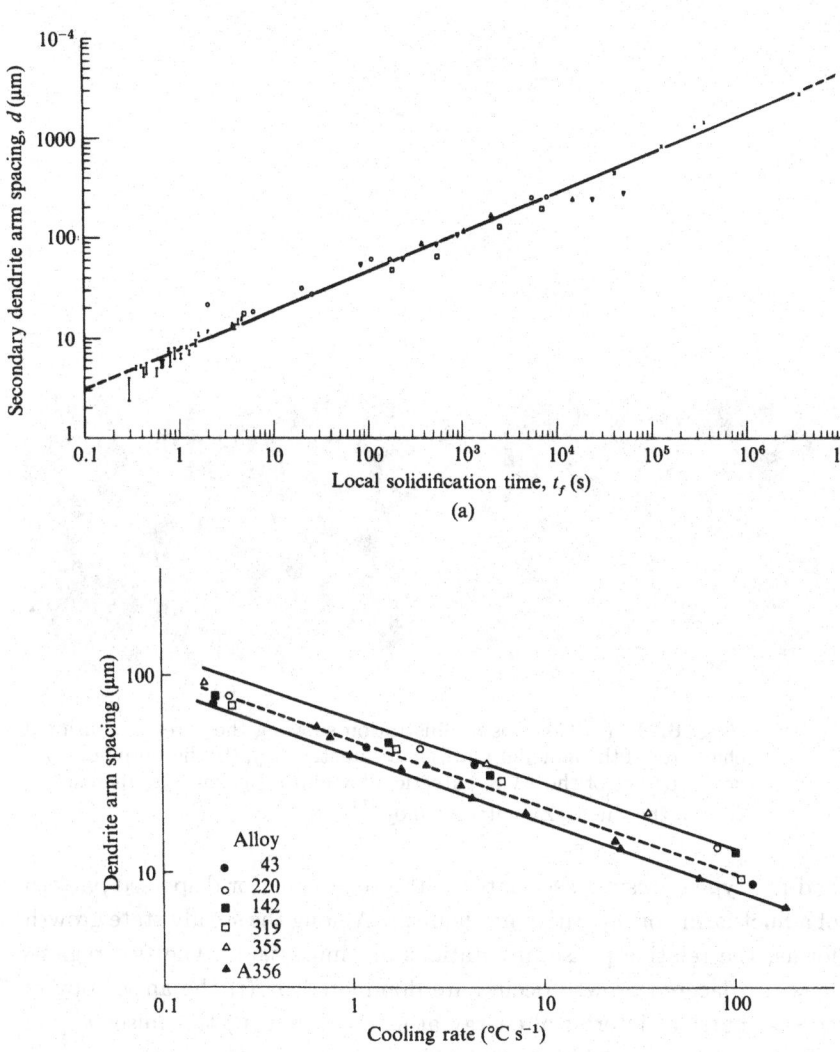

Fig. 6.20. Experimental data on dendrite arm spacings in Aℓ alloys: (a) Aℓ–4.5% Cu and (b) commercial Aℓ alloys.[10]

type of structure, the range of stability varies so widely from system to system that it is rarely possible to illustrate experimentally more than one morphological transition with any one alloy.

With reference to Fig. 6.22, a clear distinction can be made for all alloys between the type of results associated with steady state growth

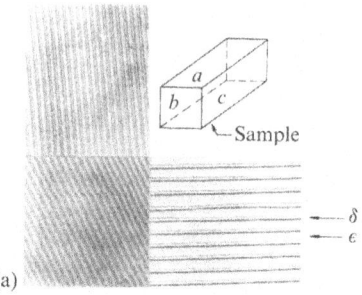

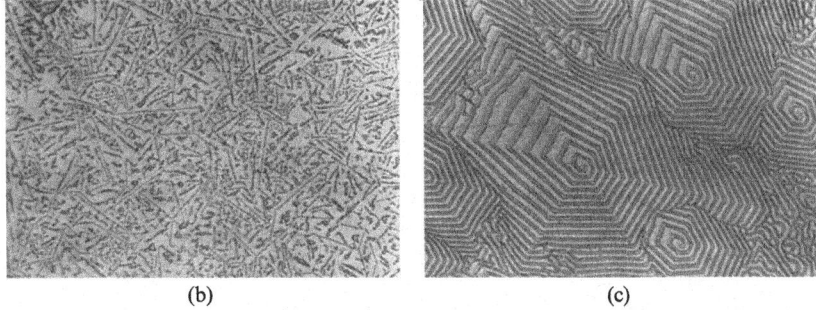

(b)   (c)

Fig. 6.21. (a) Composite illustration showing the three-dimensional character of the lamellar plates in a eutectic alloy, (b) discontinuous micro structure of the Al–Si eutectic alloy and (c) hexagonal microstructure of the Zn–MgZn eutectic alloy.[11]

and the type of results associated with the globular or dispersed pattern of a fluctuating or repetitive mechanism. Among the steady state growth forms, the relative phase orientations are important. The two regions in which faceted growth occurs are dominated by (i) the anisotropy of crystal/nutrient interfacial energy at small V and (ii) the anisotropy of interface attachment kinetics at high V.

On a macroscale, a eutectic sample is made up of individual grains wherein the matrix phase has a single crystallographic orientation. Within a single grain, a cellular or dendritic substructure is often present and a single cell has been called a colony (colony of lamella) as seen in Fig. 6.23. Within a single cell, the lamellae have a certain spacing and continuous length that changes rapidly near the cell walls where one finds a transition to rod and globular morphologies as the discontinuous phase orientation is changing rapidly with distance. Within the lamellar domain, various discontinuities and shear defects are present.

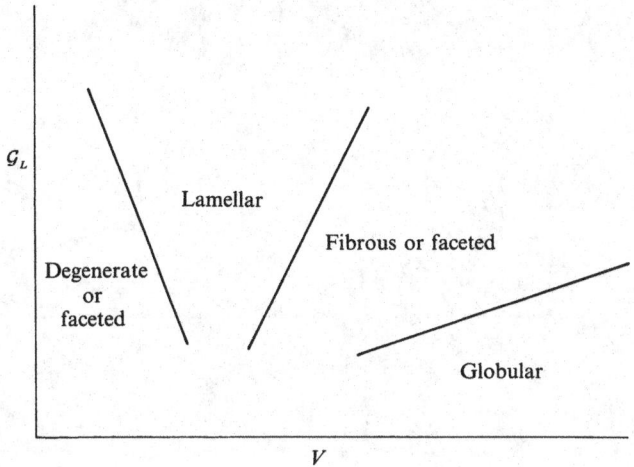

Fig. 6.22. Stability domains for different microstructures in a eutectic alloy.[12]

From the classical eutectic diagram shown in Fig. 6.24(a) one notes that, under equilibrium conditions, two solid phases and one liquid phase coexist at position E. At the eutectic temperature, $T_{\tilde{E}}$, the duplex α/β solid has the proportions given by the lever rule

$$\frac{\text{Proportion of } \alpha\text{-phase}}{\text{Proportion of } \beta\text{-phase}} = \frac{FE}{DE}$$

During crystallization, the proportions of the two phases may be slightly different from this because we may have $T_i < T_{\tilde{E}}$ so we will consider a lamellar eutectic to have the morphology illustrated in Fig. 6.25 with the proportions $\lambda_\alpha/\lambda_\beta$. Using a continuum picture, each phase has a different average curvature $\bar{\mathcal{K}}_\alpha$ and $\bar{\mathcal{K}}_\beta$ so that pseudo liquidus lines for these average curvatures are as illustrated in Fig. 6.24(b) leading to $(C'_{\tilde{E}}, T'_{\tilde{E}})$ as the pseudo eutectic condition. If one adds in an undercooling factor, ΔT_K, to account for molecular attachment kinetic effects, the pseudo eutectic point shifts to $(C''_{\tilde{E}}, T''_{\tilde{E}})$. The actual interface temperature, T_i, which is relatively constant across the interface ($\mathcal{G}_L\lambda \ll 1°\text{C}$), is somewhat below the pseudo eutectic point because of solute partitioning effects at the interface. The lamellar system adjusts its key parameters $(\lambda_\alpha, Z'_\alpha, \ell_\alpha, \lambda_\beta, Z'_\beta, \ell_\beta)$ to maximize T_i, where ℓ_j is the cap length and Z'_j is the surface slope at the triple point for the j-phase.

For steady state growth from a melt of concentration C_∞, overall

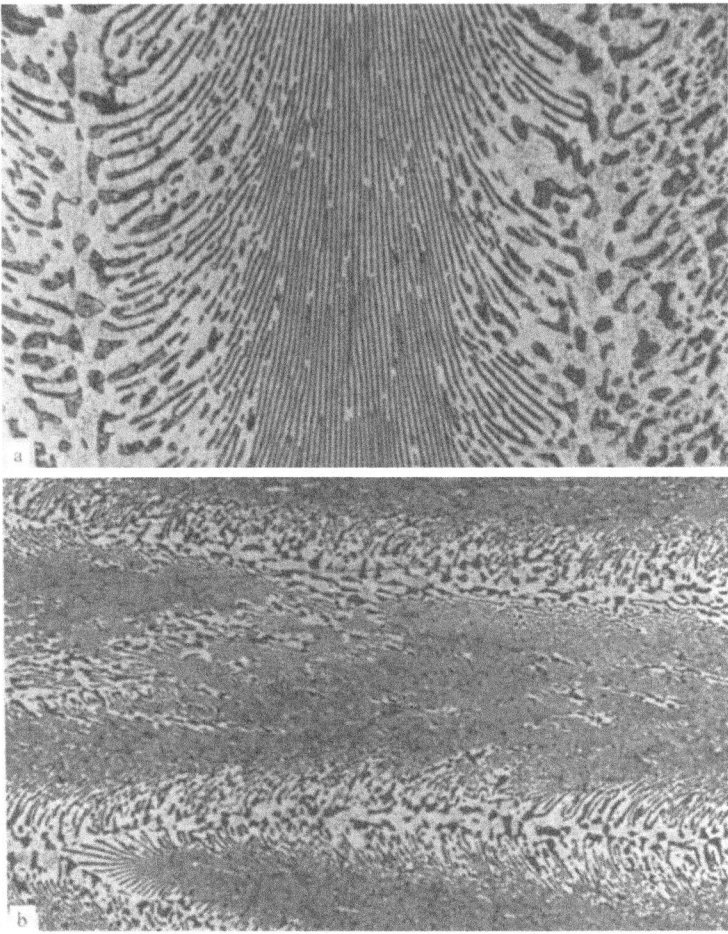

Fig. 6.23. Variation of lamellar morphology within a colony for the Al–Zn phase. Zn is the continuous phase (500×).

solute conservation requires that the ratio $\lambda_\alpha/\lambda_\beta$ be fixed according to

$$\lambda_\alpha k_\alpha \bar{C}_{i\alpha} + \lambda_\beta k_\beta \bar{C}_{i\beta} = (\lambda_\alpha + \lambda_\beta)C_\infty \qquad (6.13a)$$

with $\bar{C}_{i\alpha} \approx \bar{C}_{i\beta} \approx C''_{\tilde{E}}$. As C_∞ departs from $C''_{\tilde{E}}$, the plane wave term in the solute distribution for the liquid increases leading to an increased probability of constitutional supercooling development and subsequent cell (colony) formation. In what is to follow, we shall let $C_\infty \approx C''_{\tilde{E}}$ hold so that the plane wave term is negligible.

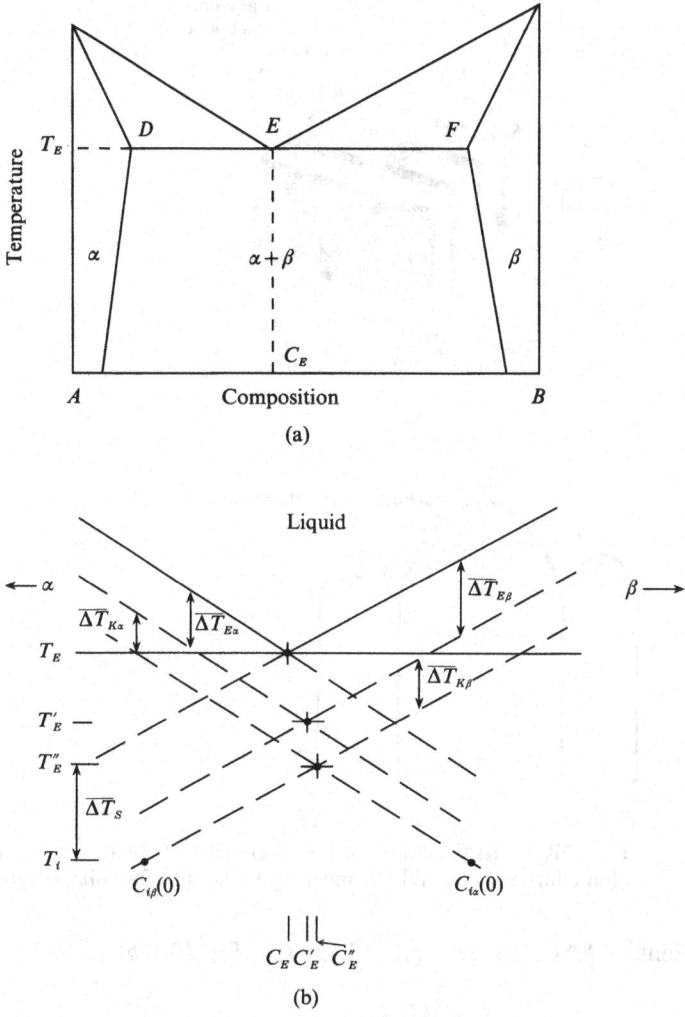

Fig. 6.24. (a) Typical eutectic-type phase diagram and (b) effective eutectic points when γ and β_k-effects are included in the growth analysis plus the key temperatures involved in the growth of a duplex interface.

The net conservation of B atoms at the α-phase interface is given by

$$\frac{\lambda_\alpha}{2}V(1-k_\alpha)\bar{C}_{i\alpha} = -D_B \int_{TP}^{\infty} \left(\frac{\partial C}{\partial x}\right)_{x_{TP}} \mathrm{d}z \approx D_B \frac{(\Delta C_i/2)\epsilon\lambda}{(\lambda/2)} \quad (6.13b)$$

and TP refers to the triple point where $C_i \approx C_E''$, $\epsilon\lambda$ is the total extent of the solute profile in the z-direction with an average lateral concentration

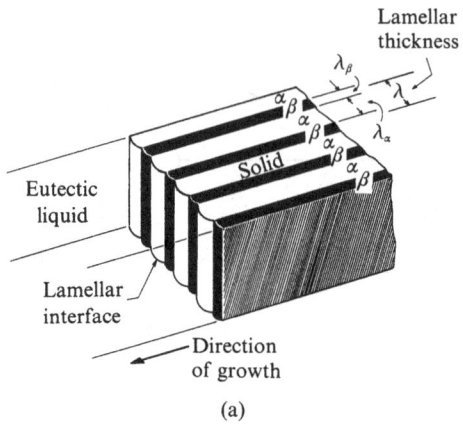

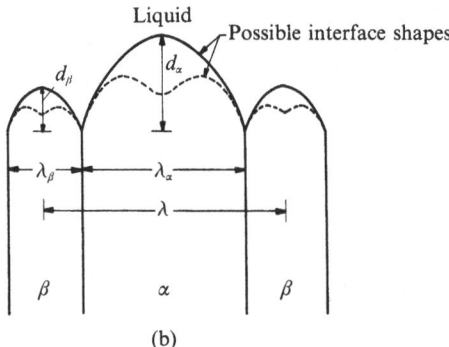

Fig. 6.25. (a) Geometrical representation of steady state growth for a lamellar eutectic and (b) most probable interface shape categories.

gradient of $\frac{1}{2}\Delta C_i/(\lambda/2)$. Since $\bar{C}_{i\alpha} \approx C'_{\tilde{E}}$, Eq. (6.13$b$) yields

$$\Delta C_i = \frac{V(1 - k_\alpha)}{2\epsilon' D_B} C''_{\tilde{E}} \lambda \qquad (6.13c)$$

where $\epsilon' = \epsilon \lambda_\alpha \bar{C}_{i\alpha}/\lambda C''_{\tilde{E}} \sim \lambda_\alpha/\lambda$.

Two additional constraints on the system are (i) the coupling equation constraint

$$(\overline{\Delta T}_E + \overline{\Delta T}_C + \overline{\Delta T}_K)_\alpha = (\overline{\Delta T}_E + \overline{\Delta T}_C + \overline{\Delta T}_K)_\beta \qquad (6.13d)$$

where averages are chosen although point-to-point values hold also and (ii) the excess energy storage constraint

$$\overline{\Delta T}_{E\alpha} + \overline{\Delta T}_{E\beta} = \frac{2\gamma_{\alpha\beta}}{\Delta S_F \lambda} \qquad (6.13e)$$

where $\gamma_{\alpha\beta}$ is the energy of the α/β interphase boundary. A third constraint applies only at the triple point since the concentration is constant at that location; i.e., neglecting ΔT_K effects,

$$\frac{\gamma_{\alpha L}}{\Delta S_{F\alpha}}(\mathcal{K}_\alpha)_{TP} = \frac{\gamma_{\beta L}}{\Delta S_{F\beta}}(\mathcal{K}_\beta)_{TP} \qquad (6.13f)$$

The approach taken in Fig. 6.24(b) allows one to convert ΔC_i from Eq. (6.13c) into $\overline{\Delta T}_C$ via

$$\Delta C_i = \Delta C_{i\alpha} + \Delta C_{i\beta} = \overline{\Delta T}_C \left(\frac{1}{|m_{L\alpha}|} + \frac{1}{|m_{L\beta}|}\right) \qquad (6.13g)$$

If, for the moment, one neglects the ΔT_K terms, the interface undercooling, ΔT_λ, below $T_{\bar{E}}$ is given by

$$\Delta T_\lambda = \frac{A}{\lambda} + B\lambda \qquad (6.14a)$$

where

$$A = 2\gamma_{\alpha\beta}/\overline{\Delta S}_F \qquad (6.14b)$$

and

$$B = \frac{V(1 - k_\alpha)C''_{\bar{E}}}{2\epsilon' D_B\,(1/|m_{L\alpha}| + 1/|m_{L\beta}|)} \qquad (6.14c)$$

Thus, the energy storage term favors a large value of λ while the solute transportation term favors a small value of λ. Applying the extremum condition yields the optimum value, λ^*, where

$$\lambda^* = (A/B)^{1/2} \qquad (6.14d)$$

and

$$\Delta T_\lambda^* = 2(AB)^{1/2} \qquad (6.14e)$$

We note that this result allows the interface undercooling to be equally partitioned between ΔT_E and ΔT_C and that $\lambda^* \propto V^{-1/2}$. The predictions for λ^* and ΔT_λ^* are largely borne out by the experimental results of Fig. 6.26 within the natural uncertainties of the different parameters entering A and B.

Although the extremum condition is the dynamically most stable condition, the eutectic system must have an effective mechanism for achieving the optimum value of λ^*. Halving or doubling λ can be readily achieved using the mechanisms illustrated in Fig. 6.27 wherein a depression can form in the central region of the lagging phase while the adjacent phase develops a lateral branch to intrude into this pocket.[13] Edgewise growth of this side-branch leads to a local halving of λ. If the interface contains a large density of these extrinsic plate faults of the type shown in Fig. 6.27(d), then growth of the fault plate to the left

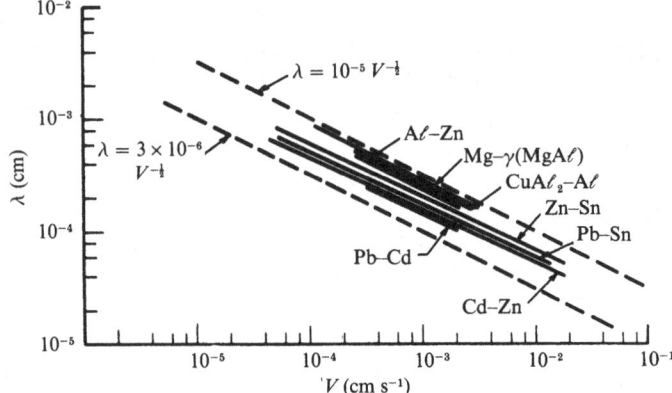

Fig. 6.26. Experimentally determined lamellar spacings, λ, as a function of V for a variety of eutectics grown under constrained growth conditions.

halves the local λ while shrinking of the fault plate to the right doubles the local λ. The presence of such faults is obviously important to each variation in the average value of λ. Given large enough perturbations and long enough time at the steady state growth condition, one might expect the extremum condition of $\lambda = \lambda^*$ to be achieved. However, like the condition of minimum free energy, this condition of marginal stability is an end point that is only approached asymptotically and may not be achieved in a particular experiment. It is interesting to note that, with thin specimens of transparent materials, Hunt and Jackson[8] found that a given value of λ was maintained even when V was significantly increased. Thus, in spite of the driving force for λ reduction, an insufficient fault density was present to cause the expected change and the new fault formation probability was too small to be effective in the time of the experiment. The wall constraints obviously played an important role in this result; however, it suggests that even in bulk material, attainment of the extremum condition is not guaranteed during the lifetime of a particular experiment. Of course, one should also remember that if a TLK model is used, and all of the constraints applied, there may be no parameter left undetermined so the extremum condition would not be applied.

The foregoing applied well to many metallic systems where the interface reaction process is rapid and ΔT_K can be neglected for both phases. When one of the phases has sluggish kinetics, this additional process must be considered as being important in the coupling equation

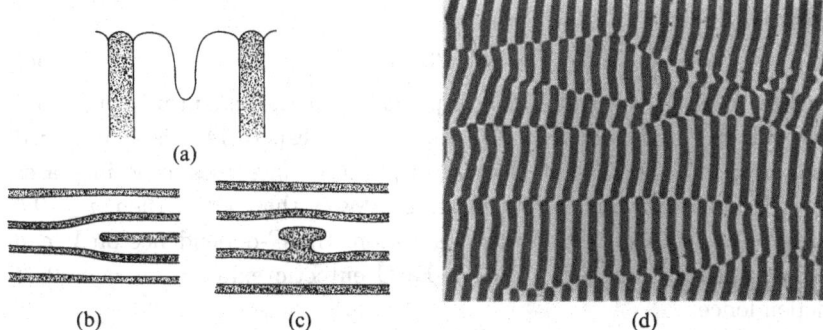

Fig. 6.27. Schematic illustrations of mechanisms whereby plate spacing adjusts to increasing V: (a) new lamellae form in a depressed region of the wide phase, (b) extra lamellae move left across the interface, (c) a plate branches to generate a new lamella moving laterally on the interface in the pocket illustrated in (a) and (d) actual observation of interface shear faults in the $A\ell$–$CuA\ell_2$ eutectic which allow mechanism (b) to function.[13]

(see Eq. (6.13d)). If layer generation is rapid and layer motion is the limiting step, Eqs. (6.14d) and (6.14e) still hold because ΔT_K is independent of λ. However, if the more general case of layer generation-limited kinetics holds then, following Eqs. (5.15) and (5.16) for the α-phase being the sluggish phase, a relationship between $\Delta T_{K\alpha}$ and the nucleation area $A_0 \sim \lambda^2$ is obtained. Including adsorption effects in this nucleation/layer growth process, face adsorption leads to a larger effective value of γ_ℓ and a smaller value of β_2 while edge adsorption leads to a smaller value of γ_ℓ and a smaller value of β_2. We expect the degree of adsorption to increase as V decreases. Using Eq. (5.16b), Eq. (6.13a) must be replaced by

$$\Delta T_\lambda = \frac{A}{\lambda} + B\lambda + \frac{2\bar{\alpha}}{3\ell n(\lambda^2/a)} \tag{6.15a}$$

where $a = (\pi^{1/2}\beta_2\Delta S_F h/V_0)^{2/3}$. Applying the extremum condition leads to

$$\lambda^* = \frac{C}{2B} + \left[\left(\frac{C}{2B}\right)^2 + \frac{A}{B}\right]^{1/2} \tag{6.15b}$$

$$= \left(\frac{A}{B}\right)^{1/2}\left[1 + \frac{C}{2(AB)^{1/2}} + \frac{C^2}{\gamma AB} + \cdots\right] \tag{6.15c}$$

$$\approx \lambda_0^* V^{-n} \tag{6.15d}$$

where

$$C = 4\bar{a}a/3[\ell n(\lambda^2/a)]^2 \qquad (6.15e)$$

In Eq. (6.15d), the value of n depends upon the magnitude of C and the velocity dependence of C. As $C \to 0$, Eq. (6.14d) is reproduced and $n = 0.5$. For no interface adsorption, C increases as V increases and, if it increases at a rate faster or slower than $V^{1/2}$, then $n < 0.5$ or $n > 0.5$, respectively. With adsorption, the C-dependence on V can be significantly altered. The Fe–Fe$_3$C eutectic exhibits an $n \approx 0.34$ dependence.

The foregoing treatment allows one to determine the most important parameter of the eutectic system, its lamellar spacing λ^*. To determine the other interface shape parameters Z'_α, ℓ_α, Z'_β and ℓ_β, one needs to consider Eqs. (6.13e) and (6.13f). In addition, one must recognize that Eq. (6.12f) is only the temperature equilibrium portion (neglecting ΔT_K effects) of the dynamic equilibrium balance occurring at the triple point. The full equilibrium balance also requires what is called the Young's force balance; i.e., considering the triple junction of Fig. 6.28,

$$\gamma_{\alpha L} \sin\theta_\alpha = \gamma_{\beta L} \sin\theta_\beta \qquad (6.16a)$$

and

$$\gamma_{\alpha L}\cos\theta_\alpha + \gamma_{\beta L}\cos\theta_\beta = \gamma_{\alpha\beta} \qquad (6.16b)$$

with

$$\sin\theta_j = (1 + Z_j^{'2})^{-\frac{1}{2}}; \quad \cos\theta_j = Z'_j(1 + Z_j^{'2})^{-1/2}; \quad j = \alpha, \beta \qquad (6.16c)$$

so that Z'_α and Z'_β are determined by these two conditions. Eq. (6.13e) provides one additional equation between ℓ_α and ℓ_β while Eq. (6.13f) provides another in a slightly more complex format. If the cap shapes were parabolic-like (for simplicity), $\mathcal{K}_j = (1/\rho_j)/[1 + (\lambda_j/\rho_j)^2]^{3/2}$ and

$$\frac{\ell_j}{\rho_j} = \frac{\lambda_j}{2\rho_j}\left[1 + \left(\frac{\lambda_j}{\rho_j}\right)^2\right]^{1/2} + \frac{1}{2}\ell n\left\{\frac{\lambda_j}{\rho_j} + \left[1 + \left(\frac{\lambda_j}{\rho_j}\right)^2\right]^{1/2}\right\} \qquad (6.17)$$

so that Eq. (6.13f) provides a relationship between the ρ_j and, using Eq. (6.17), this becomes the additional needed relationship between the ℓ_j. For elliptical-like cap shapes, an analogous relationship exists. Thus, for any simple cap shape, the key parameters may be determined by the procedures presented here.

For a general, non-simple cap shape, instead of using average values in the coupling equation, a point-to-point treatment must be utilized. In addition, one should include the ΔT_T contribution since Fig. 6.22

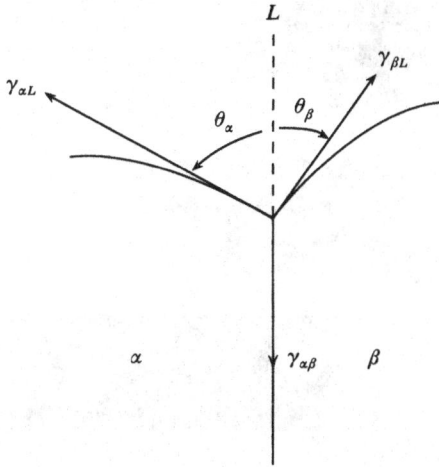

Fig. 6.28. Illustration of the energetic balance of forces operating at the triple junction (α, β, L).

shows that $\mathcal{G}_L$ is an effective experimental variable in the eutectic interface shape determining process. Using the point-to-point approach gives good match between theory and experiment for the case illustrated in Fig. 6.29.[8]

The nice fit between theory and experiment represented by Fig. 6.26 seems to hold for most eutectics involving primarily metallic phases. However, one must expect the picture to be somewhat more complicated for eutectics involving primarily ceramic phases because of the electrostatic factors involved. In a two ceramic phase eutectic, each phase will have a different thermoelectric power so that an important difference should be expected for the eutectic temperature between an isothermal system and a eutectic crystal grown under constrained growth conditions with a temperature gradient $\mathcal{G}_S$. For the isothermal system, only 1–10 mV type variations of electrical potential are expected between the three phases and this is not enough to produce serious changes in the chemical potentials of the two species and thus serious changes in both the eutectic composition and the eutectic temperature.

On the other hand, for a eutectic crystal grown under a specific $\mathcal{G}_S$, each phase is expected to have a different electrostatic potential at the eutectic temperature if electrically isolated from each other (i.e., ϕ_α and ϕ_β). Because of the intimate contact between the phases in the solid, circulating electrical currents are required between the two phases to

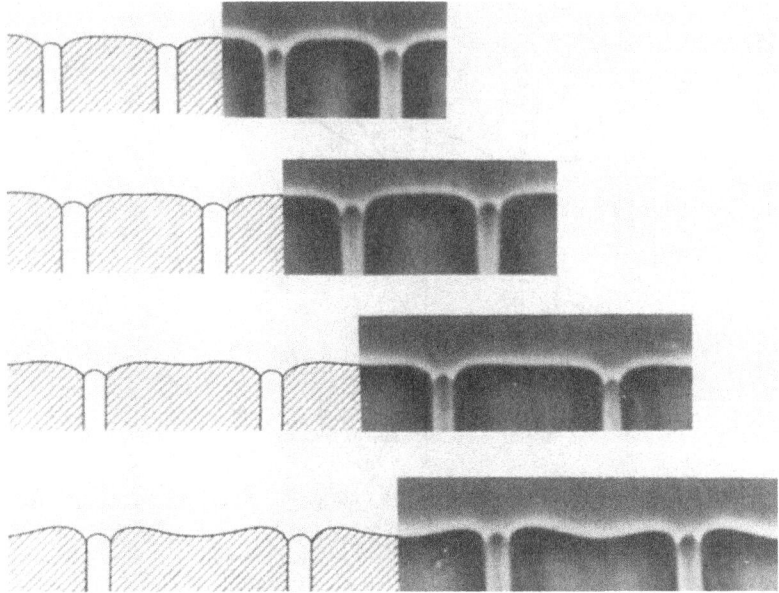

Fig. 6.29. Calculated and observed interface shapes for a transparent organic eutectic.[8]

yield a compromise electric potential, ϕ_i, at the solid/melt interface. Considering just the average value, $\bar{\phi}_i$, the change in electrochemical potential for the j species at the eutectic interface will be of the order $\hat{z}_p^j e(\bar{\phi}_i - \phi_p)$ where p refers to the α, β or L phase and $\hat{z}_p^j$ is the effective valence of the j-species in the p-phase. If $\mathcal{G}_S$ is large, $\bar{\phi}_i - \phi_p$ will be large so that significant shifts in both the eutectic composition and eutectic temperature should be expected. Further, one should expect that the circulating current flow represents a field-drive for lateral transport of chemical species in the interface region so that D in Eq. (6.14c) must be considered here as an effective diffusivity that may be orders of magnitude different from conventional liquid diffusion values.

This electrical potential effect for a phase, which increases as the product of $\mathcal{G}$ and the Seebeck coefficient, π, changes the electrochemical potential for each species so that the liquidus lines in Fig. 6.24(b) will shift leading to a shift in the eutectic point.[14] This can have a number of interesting consequences. For example, suppose that we can set $\pi_L \approx \pi_\beta \approx 0$ and we are growing from a melt with $C_\infty \tilde{<} C_{\tilde{E}}$ for the $\mathcal{G} = 0$ case. Suppose, further, that the sign of π_α is such that the α-

liquidus moves to the left as $\mathcal{G}$ increases. Then, we should find that the α-phase will be the primary phase for small $\mathcal{G}$ but, as $\mathcal{G}$ increases, the new eutectic point will move left on the phase diagram so that, for $\mathcal{G} > \mathcal{G}_c$, the new $C_{\tilde{E}} < C_\infty$ and the β-phase becomes the primary phase. Using similar reasoning, one can readily show that it is possible to stabilize a metastable γ-phase and destabilize the original α-phase if $\pi_\gamma > -\pi_\alpha$ and increasing $\mathcal{G}$ causes the α-liquidus to move to the left. These considerations apply equally well to peritectic reactions.

6.3.1 *TLK viewpoint*

From the more correct TLK picture illustrated in Fig. 6.30(a), layer creation occurs at the facet plane orientations for each phase and sufficient undercooling must be present for layer generation at the rate needed to satisfy V. When γ_ℓ is very small, the nucleation rate is very high and the ledge motion rate is also high because the ledge will be quite diffuse. Thus, only a small facet area, much smaller than λ^2, is needed for the generation of these layers so the cap shape exhibits a rounded appearance. The layer edges, flowing towards the triple point, provide the excess energy for formation of the α/β phase boundary. If no orientation relationships exist such that $\gamma_{\alpha L} + \gamma_{\beta L} > \gamma_{\alpha\beta}$, then a deep liquid groove must develop between the two phases of length ℓ^* and width δ as illustrated in Fig. 6.30(b) such that

$$\gamma_{\alpha L} + \gamma_{\beta L} + \ell^* \delta \overline{\Delta S}_F (\bar{\mathcal{G}}_S \ell^* + \Delta T_\lambda^*) = \gamma_{\alpha\beta} \tag{6.18}$$

Here, the crystallographic orientations of the α and β-phases that lead to the smallest value of $\gamma_{\alpha\beta}$ are the most stable. The crystallographic structure of some phases will not allow symmetrical facets on the lamellar interface and, in some cases, will not be properly oriented to feed the lamellar morphology. Thus, some α–β phase contributions are unable to provide perfectly symmetrical interface shapes of the overall lamellar form because of the TLK requirements.

As γ_ℓ increases, the facet area dominated by the nucleation events increases and the lamellae begin to develop angular morphological characteristics. In this case, the average undercooling associated with this layer creation plus the ΔT_K associated with the layer propagation must always be equal to or greater than $2\gamma_{\alpha\beta}/\overline{\Delta S}_F \lambda$. As this difference increases, the third term on the right in Eq. (6.15a) begins to assume dominance and, from the TLK perspective,

$$\Delta T_\lambda = B\lambda + \frac{2\bar{\alpha}}{3 \, \ell n(\lambda^2/a)} \tag{6.19a}$$

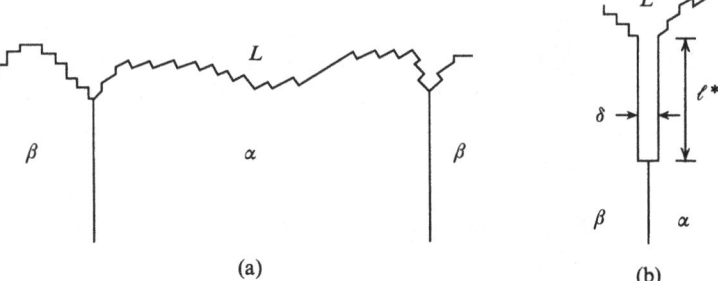

Fig. 6.30. TLK representation of a eutectic interface shape: (a) with easy balance of triple junction forces and (b) with difficult balance of triple junction forces.

where any layer edge interaction effects have been neglected because, at this large value of γ_ℓ, the morphological characteristics are completely angular. In this limit, λ^* is given by

$$\lambda^* = \frac{C}{B} \qquad (6.19b)$$

which approaches a V^{-1} dependence.

6.3.2   Lamellae to rod transition

One expects the transition from lamellae to rods to occur when $\Delta T_\lambda^R < \Delta T_\lambda^L$ via overlap of the lagging phase by the leading phase as illustrated in Fig. 6.31.[11] Since diffusion is more efficient for the two-dimensional lateral transport with rods than the one-dimensional lateral transport with lamellae, $\overline{\Delta T}_C^R/\overline{\Delta T}_C^L = B^R/B^L < 1$ and

$$\Delta T_\lambda^R = \frac{\pi \lambda_\alpha \gamma_{\alpha\beta}}{\Delta \overline{S}_F (\lambda/2)^2} + B^R \lambda \qquad (6.20)$$

which definitely favors the rod form when $\lambda_\alpha/\lambda \overset{\sim}{<} (2\pi)^{-1}$. Transition to the rod form from the lamellar form is favored, for $\lambda_\alpha/\lambda \overset{\sim}{<} (2\pi)^{-1}$, when the lead–lag distance, d, between phases is large because this immediately reduces B and, if the reduction in ΔT_C is greater than the increase in ΔT_E, transition can occur. Reduction of $\mathcal{G}_L$ to increase d favors the transition.

In practice, one expects the phase organization leading to the minimum ΔT_i^* to be such that the phase which has the larger ΔT_i for independent growth at the same V (rather than coupled growth), either α or β, is placed in the most advantageous configuration at the duplex solid–liquid front; i.e., by virtue of being the leading phase or by being the fibrous (rod-like) phase. This occurs because ΔT_C is reduced for

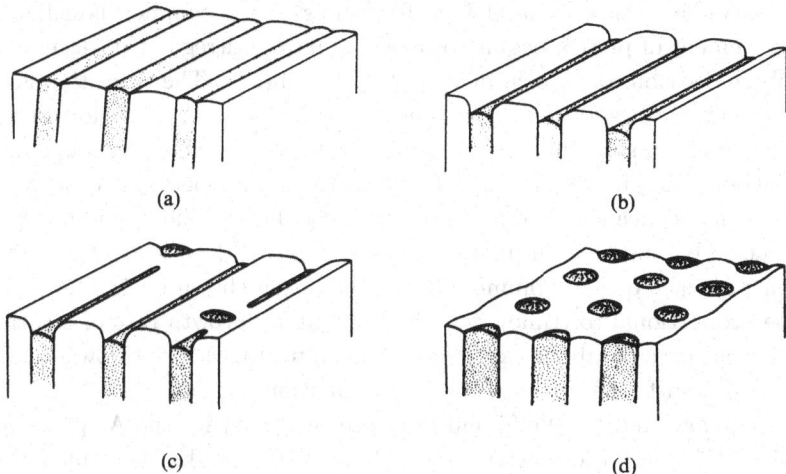

(a)　　　　　　　　　　　　　(b)

(c)　　　　　　　　　　　　　(d)

Fig. 6.31. Schematic illustration of the transition from a lamellar to a rod-like or fibrous structure.[11]

the leading phase, especially if it is fibrous. An important example of such cooperation is found when the attachment kinetic barrier to growth is appreciably greater for one of the phases, such as between $A\ell$–Si or $A\ell$–Ge. Here, the more covalent material (Si or Ge) has a significant γ_ℓ so that the ΔT_K term is dominant and it is observed to be the leading phase at the growth front. This arrangement is preferred because the larger ΔT_K and the smaller ΔT_C of the leading phase balances the smaller ΔT_K and the larger ΔT_C of the lagging phase. A reversal of the phases would be such as to impede the kinetically sluggish phase still further by an increased ΔT_C and a duplex phase front could not be stable.

It is relevant here to notice that, in the foregoing considerations, the volume fraction $\lambda_\alpha/\lambda_\beta$ enters into all alternative phase morphologies. All other factors being equal, that phase with the larger volume fraction will exhibit a larger $\overline{\Delta T_C}$ for growth at V with $d = 0$. Thus, a phase arrangement wherein it is the leading or fibrous phase should lead to the most stable duplex interface. It is also important to notice that, if $\gamma_{\alpha\beta}$ is strongly anisotropic, the lamellar form may be favored even though all other factors favor the rod or fibrous form. Finally, as V increases, lateral diffusion becomes more difficult so the lead–lag distance, d, increases and transition to the rod form is expected to occur. This transition is delayed if one increases $\mathcal{G}_L$ as one increases V.

Below a certain value of $V \sim 10^{-4}$ cm s^{-1}, the common lamellar arrangement of phases begins to break up into an irregular or degenerate phase distribution which, in some cases, is fibrous. The transition is observed to be orientation-dependent, starting in some grains before others. Although, in some instances, this is a totally solid state rearrangement phenomenon, in the Aℓ–Ag$_2$Aℓ system the crystallographic destruction of lamellae (see Fig. 6.32)[15] appears to be due to solid/liquid facet formation by one or both phases. Since the rate of development of a facet on a surface due to volume diffusion has been shown by Mullins[16] to be proportional to (time)$^{-2/3}$ while the lateral growth rate of a surface element for a parabolic cap is proportional to (time)$^{-1/2}$, below a critical $V = V_c$, surface energy driven facet formation should develop.

The low energy solid/liquid faces are the {111} for the Aℓ-phase and the (0001) for the Ag$_2$Aℓ-phase.[15] Above $V = V_c$ the lamellar arrangement is one in which the habit plane is Aℓ\{111\}$\|$Ag$_2$Aℓ(0001) and A$\ell\langle 110\rangle\|$Ag$_2$Aℓ[11$\bar{2}$0]. The volume fractions of the phases are approximately equal and $\lambda \sim 4 \times 10^{-4}$ cm at $V = V_c$. Facet formation on the compound should coincide with the lamellae habit and thus contribute to the stability of the lamellar arrangement. Facet formation on the Aℓ can, however, occur on {111} faces at 54° to the preferred $\langle 110\rangle$ growth axis and traces of such faces on the {110} projection of a transverse section will make an angle of 45° with the axial habit plane of the lamellar arrangement. For this system, it has been observed that degeneration involves encroachment by the Aℓ-phase upon the Ag$_2$Aℓ-lamellae, until the latter are cut into fibers of approximately rectangular section (see Fig. 6.32) with the angular change being very close to the anticipated 45°.

At high V, facet formation again occurs as a dominant morphological characteristic but now it is a kinetically dominated effect. The supercooling ΔT_K needed for layer generation is appreciably larger than that needed for steady layer motion so the interface becomes composed of large flat segments rather than smoothly rounded segments.

6.3.3   *Transition to non-steady state configurations*

Referring to Fig. 6.22, it is seen that failure to grow by a steady state diffusion process occurs at the lower right of the diagram. This involves very high freezing rates or impossibly low temperature gradients for metal–metal systems but, for metal–non-metal systems like Aℓ–Si, Aℓ–Ge, Fe–C or Ni-C, the critical conditions fall within the typical experimental range. At least two possible non-steady state reactions are

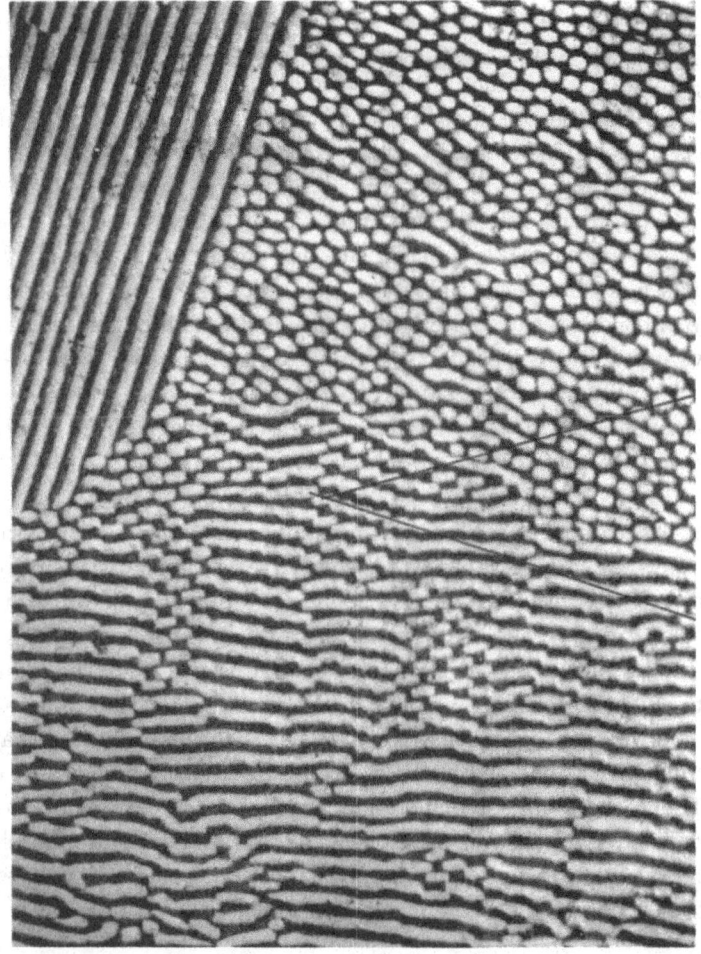

Fig. 6.32. Destruction of lamellae in the $A\ell$–$Ag_2A\ell$ eutectic by solid/liquid facet formation.[15]

available that may be competitive with the steady state reaction. One of these involves repeated nucleation of one phase as independent particles ahead of the macroscopic interface. The other involves a transition of the macroscopic interface from a steady state mode to an oscillatory mode which produces a globular-like appearance but the dispersed phase particles are all interconnected.

The independent nucleation possibility requires a critical growth velocity, V_c, such that the nucleation rate of the dispersed phase, $\dot{N}_\beta$, in the supercooled zone of length $\Delta z^* = \Delta T_i^*/\mathcal{G}_L$ ahead of the interface, is sufficient to disrupt the steady state growth completely; i.e., $\dot{N}_\beta \sim$ $(5\lambda_c^*)^{-3} \sim 10^7$ cm^{-3} s^{-1}. This is a very large heterogeneous nucleation frequency for the value of ΔT_i^* present experimentally ($\Delta T_i^* \sim 1$–$2°$C) and requires the presence of a large number of very effective heterogeneous nuclei. This is not likely in general.

An oscillatory growth mode will occur naturally when the lead–lag distance, d, becomes large (occurs at high $V/\mathcal{G}_L$). Continued side-branching of the leading phase (kinetically sluggish phase) can lead to a $\overline{\Delta T}_C(\lambda, t)$ that is less than the steady state value because of an initial transient type of effect for the solute redistribution. In addition, since the surface to volume ratio is least for sphere-like or globular particles, one might expect $\overline{\Delta T}_E(\lambda, t)$ to be smaller in the small λ range also; i.e., just as the transition from lamellae to rods caused a change from λ to $\lambda' > \lambda$, a transition from rods to globules causes a change from λ' to $\lambda'' > \lambda'$ so that $\Delta T_E(\lambda'', t) \stackrel{\sim}{<} \overline{\Delta T}_E(\lambda') \stackrel{\sim}{<} \overline{\Delta T}_E(\lambda)$.

In the case of Aℓ–Si, the Si crystals grow as small dendrites with a preferred $\langle 001 \rangle$ axis and $\{100\}$ side plates and many appear to be connected because they have the same crystallographic orientation. In Aℓ–Ge, Fig. 6.33 illustrates what appears to be an oscillatory growth mode and, although the outline of the Ge is irregular, the twin traces in the *Ge* particles are aligned in the growth direction and can often be followed through successive particles.[12] This indicates that these seemingly disconnected particles are, in fact, parts of the same crystal. In Fe–C, the behavior of nodular graphite in eutectic alloys seems to provide an unmistakable example of the repeated nucleation mechanism, while the wavy form of flake graphite must be considered as an example of the oscillatory duplex interface mechanism.

6.3.4   *Ternary solute addition effect*

It is possible to modify the morphological transitions discussed above significantly by the addition of a small concentration, C_∞^j, of a special solute element j, to the melt. This element j must be such that m_α^j/k_α^j is very large in magnitude and very different from m_β^j/k_β^j. Thus, the presence of j, at concentration C_∞^j, can not only change which phase is the leading phase but it can also greatly increase d so that the transition from lamellae to rods to globules can be made to occur at a relatively low value of $V/\mathcal{G}_L$.[17] Conversely, the ternary addition may be such as to stabilize the lagging phase, reduce the lead–lag distance and maintain the lamellar or rod form to larger values of $V/\mathcal{G}_L$. Thus, the differential solute build-up of this third constituent ahead of the α and

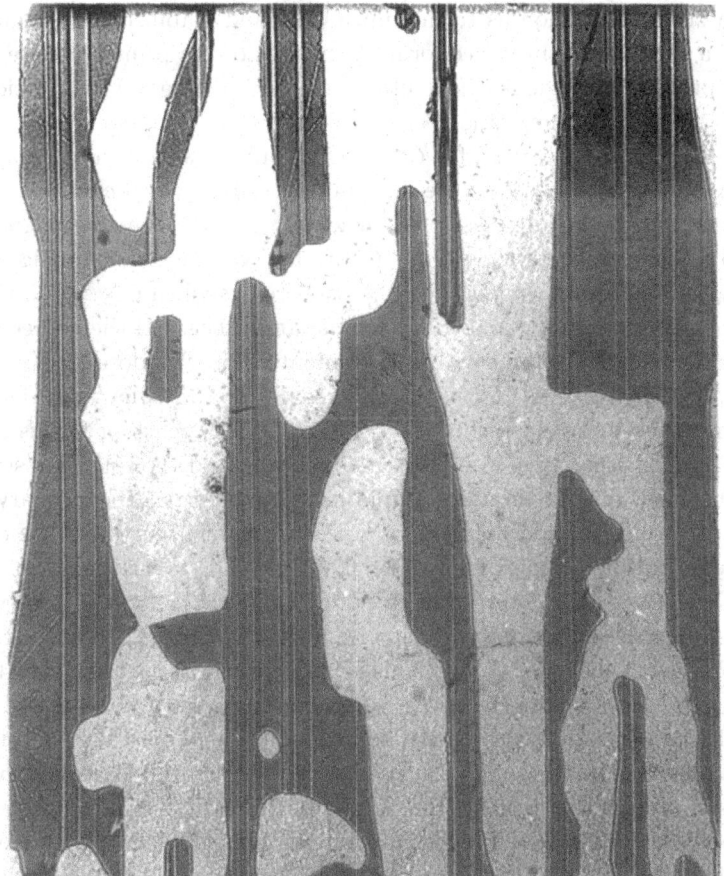

Fig. 6.33. Phase particle morphology in the Aℓ–Ge eutectic illustrating that the dark phase (Ge), although seeming discontinuous, is a continuous phase because of the continuity of the twin lamellae between particles.[12]

β-phases is an additional important factor in morphology dominance considerations.

6.3.5   *Off-eutectic compositions*

Structures, fully eutectic in morphology, may be obtained under high $\mathcal{G}/V$ growth conditions even at substantial compositional deviations from the binary eutectic composition. So long as a cellular interface does not form and the interface remains macroscopically planar, a lamellar or rod-like duplex interface can adjust its relative volume fraction, $\lambda_\alpha/\lambda_\beta$, in such a direction as to minimize CSC and stabilize the planar front. If a filamentary interface forms, dendrites of the primary phase appear separated by regions of eutectic phase which have a $\lambda_\alpha / \lambda_\beta$ ratio closer to the equilibrium value.

As discussed earlier, for $C_\infty \neq C_{\tilde{E}}$, a plane wave term of magnitude ΔC_p develops in the solute distribution profile which can lead to CSC. If $\lambda_\alpha / \lambda_\beta$ is the same as it was when $\bar{C}_S = C_{\tilde{E}}$, then the magnitude of the plane wave term is $C_{\tilde{E}} - C_\infty$. However, referring to Fig. 6.24, if $\lambda_\alpha / \lambda_\beta$ increases when $C_\infty < C_{\tilde{E}}$ or decreases when $C_\infty > C_{\tilde{E}}$, then the magnitude of the plane wave term is diminished and the degree of CSC is decreased. It can be almost eliminated for C_∞ close to $C_{\tilde{E}}$ because $C''_{\tilde{E}} \neq C_{\tilde{E}}$ but, as $|C_\infty - C_{\tilde{E}}|$ increases beyond a few per cent, ΔC_p increases almost linearly with $|C_\infty - C_{\tilde{E}}|$. This behavior is illustrated in Fig. 6.34 where $\mathcal{G}_L / V$ is plotted versus C_∞ and the solid line separates the fully duplex interface formation domain from the primary phase dendrite formation domain.[18] The dashed line represents the onset of constitutional supercooling (for fixed m_L / D_L) assuming that $\Delta C_p = C_{\tilde{E}} - C_\infty$. It is interesting to note that, for C_∞ near $C_{\tilde{E}}$, a typically lamellar structure obtains whereas, for $C_\infty \sim C_{\tilde{E}}/2$, a rod-like structure obtains at the same $\mathcal{G}_L / V$. This is fully consistent with the view that, as $\lambda_\beta / \lambda_\alpha$ decreases below $(\lambda_\beta / \lambda_\alpha)_c$, the rod morphology is more stable than the lamellar morphology. An additional piece of data that is consistent with the view that $\lambda_\alpha / \lambda_\beta$ may adjust to eliminate ΔC_p for C_∞ near $C_{\tilde{E}}$ is shown in Fig. 6.35.[19] For $\mathcal{G}_L \sim 100\,°\text{C cm}^{-1}$ and C_∞ within a few per cent of $C_{\tilde{E}}$, the Pb–Sn eutectic can be solidified at extremely high rates and still maintain a fully duplex interface of the lamellar composite form. In this range of V, $\lambda_{exp} \sim 0.1\ \mu\text{m}$.

6.4   Unconstrained growth, dendrite and eutectic morphologies

6.4.1   *Pure melt dendrites*

The freely growing dendrite in a pure melt exhibits the velocity characteristics of a steady state system; i.e., $V = $ constant. However, although the tip is experimentally found to be a paraboloid of constant radius of curvature (see Fig. 6.36(a)) during growth, at a distance $r = r_p$ from the dendrite axis, periodic development of side-branches originates which should periodically perturb the overall thermal field of the growing filament. Within the accuracy of presently available experimental techniques, no periodic oscillation in V has been discriminated.

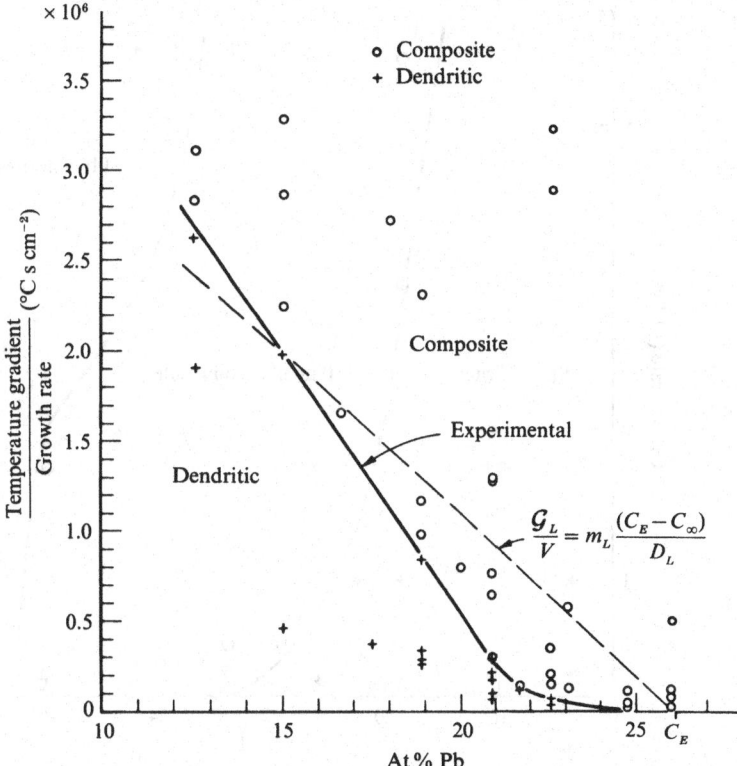

Fig. 6.34. Planar front stability conditions for directionally solidified Sn–Pb alloys near the eutectic concentration. (Courtesy of M. Fleming.)

The first step in modeling a growing dendrite was to treat it as an isothermal parabolic needle growing axially at constant velocity, V, and Ivantsov[20] found, for $\gamma = 0$ and $\beta_2 = \infty$, a shape-preserving solution for the paraboloid of revolution of the form

$$\Delta\theta = \Delta T_\infty(c/\Delta H_F) = \hat{P}_T \exp(\hat{P}_T)E_1(\hat{P}_T) \qquad (6.21)$$

where $\Delta\theta$ is the non-dimensional supercooling (the Stefan number), $\hat{P}_T$ is the thermal Péclet number, $\hat{P}_T = V\rho_t/2D_T$ where ρ_t is the radius of curvature of the filament tip, and E_1 is the exponential integral. The inverse of Eq. (6.21), $V\rho_t = f(\Delta\theta)$, provides an infinite range of solutions for V and ρ_t for a given value of $\Delta\theta$.

A few years later, Temkin[21] and Bolling and Tiller[22] considered the growth of a non-isothermal parabolic needle from a continuum surface

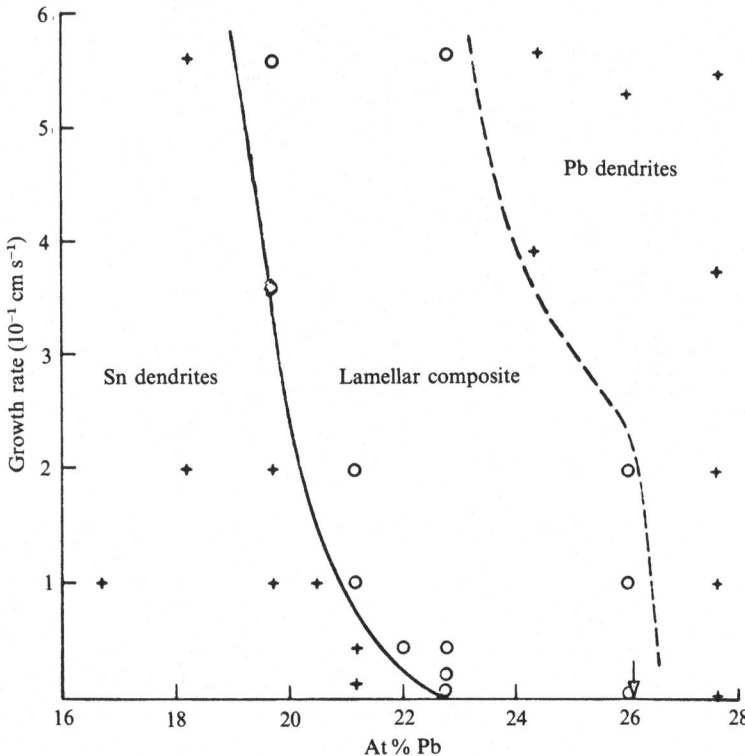

Fig. 6.35. V versus C_∞ plot for Sn–Pb alloys showing the concentration domains over which various microstructures are observed at $\mathcal{G}_L = 100\,^\circ\text{C cm}^{-1}$. (Courtesy of M. Fleming.)

point of view with $\gamma \neq 0$ and $\beta_2 \neq 0$. They found it to grow more slowly than the isothermal needle but that the multiple operating states of V and ρ_t lay along a curve with a maximum in V as illustrated in Fig. 6.36(b). Based upon the approximations of the modeling and the extremum condition, this (V_{max}, ρ_t^*) point was taken as the single operating point for a given $\Delta\theta$. From our present perspective, 30 years later, we recognize that this evaluation does not incorporate the surface creation constraint nor does it incorporate the constraint associated with layer flow kinetics. It is, therefore, not surprising that V_{max} in Fig. 6.36(b) did not agree exactly with later experiments. Unfortunately, many people took this discrepancy to indicate that the extremum condition was not a valid operating principle to use for such growing crystal forms.

Carefully gathered experimental data on the ice/water system and on

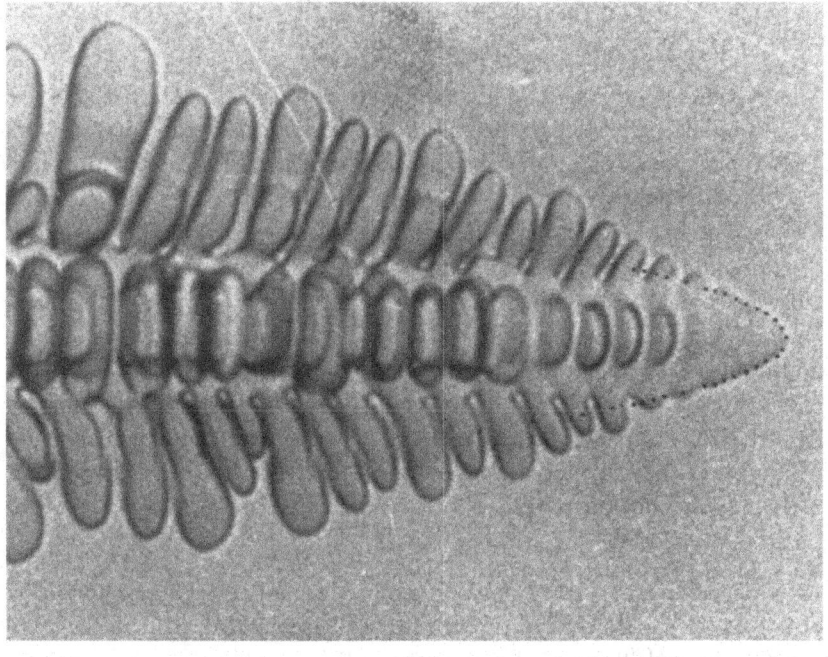

(a)

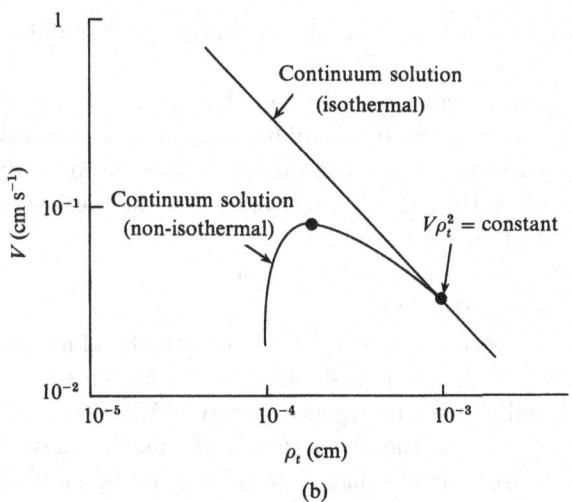

(b)

Fig. 6.36. (a) Enlarged view of the steady state dendrite tip shape in the succinonitrile–acetone system. The dotted line is a standard parabola with the same radius of curvature as that of the dendrite tip (580×) and (b) V versus tip radius, ρ_t, calculation plots at $\Delta T_\infty = 1.20\,^\circ\mathrm{C}$.

succinonitrile by Glicksman and associates[23] yielded data of the form

$$V = \beta \mathcal{L}(\Delta\theta)^b \tag{6.22}$$

where β and b are numerical coefficients and $\mathcal{L} = 2D_T\Omega\Delta S_F\Delta H_F/\gamma c$ is a lumped material parameter. The value $b \sim 2.5$ agreed well with theory but β was in serious disagreement with almost all theories; i.e., V_{max} theories gave values of β too large by 3 and 7 times, respectively, for ice/water and succinonitrile. However, a numerical analysis method of Oldfield[24], which yielded $V\rho_t^2 \approx 100 D_T\gamma\Omega c/\Delta S_F\Delta H_F$, gave fairly close agreement. Subsequent theoretical modeling by Langer and Müller-Krumbhaar[25] proposed that the onset of side-branching provided an additional constraint to the system and postulated that the wavelength of the marginally stable perturbation, $\lambda^* = 2\pi\omega^{*-1}$, sets the scale for the dendrite tip radius ρ_t^*; i.e., $\rho_t^* = \lambda^*$. For a spherical front, the condition of marginal stability for a pure material growing from its supercooled melt may be shown to be

$$\omega^{*2} = -\bar{\mathcal{G}}/\tilde{\Gamma} \tag{6.23}$$

where $\tilde{\Gamma} = \Omega\gamma/\Delta S_F$ and $\bar{\mathcal{G}}$ is the average thermal gradient (weighted by the thermal conductivities of each phase). Evaluating $\bar{\mathcal{G}}$ for a dendrite tip, Glicksman[26] found the growth condition to be

$$V\rho_t^2 = 8\pi^2 D_T c\tilde{\Gamma}/\Delta H_F \tag{6.24}$$

which agrees with Oldfield's[24] result and the experimental data within 20%.

From our present perspective, we know that the constraint of surface creation, applied in Eqs. (6.5) and (6.6), must also be applied in this case especially when we use the continuum surface approximation. For a paraboloid of revolution, Eqs. (6.6a)–(6.6c) give the average curvature over the filament of radius r^* to be

$$\bar{\mathcal{K}} = \frac{3}{2\rho_t}\left\{ \frac{(r^*/\rho_t)^2 + \ell n[1 + (r^*/\rho_t)^2]}{[1 + (r^*/\rho_t)^2]^{3/2} - 1} \right\} \tag{6.25}$$

while the surface creation requirement for the cylindrical filament growing radially is $r^{*-1} = \bar{\mathcal{K}}$ which leads to $r^*/\rho_t = 0.585$. If $r^*/\rho_t > 0.585$, then $\bar{\mathcal{K}} > r^{*-1}$ and excess driving force is available for surface creation which can be utilized for the development of side-branches. Thus, it seems reasonable to conclude that $r^* = 0.585\rho_t$ is the location on the needle tip where the tip is marginally stable relative to perturbation development. To gain an independent evaluation for this unique value of r^*, we recall that the axial growth of a paraboloidal filament gives an identical thermal field to a growing cylinder and thus select r^* to be the marginally stable radius for a growing cylinder; i.e., $r^* = R_C$, the critical radius at which a cylinder begins to develop surface contour oscillations of the type illustrated in Fig. 5.19. To gain a measure of R_C, let us first consider the perturbation analysis findings for a growing cylinder from a pure melt with $\beta_2 \to \infty$. From such an analysis,[27] one finds for small $\hat{P}_T = R\dot{R}/2D_T$ that

$$\left(\frac{\dot{\delta}/\delta}{\dot{R}/R}\right) = A - \frac{\gamma}{\Delta S_F}(\ell^2 + m^2 - 1)\frac{B}{R}\hat{P}_T \tag{6.26a}$$

and

$$\Delta T_\infty = \frac{\gamma}{\Delta S_F R} - \frac{\Delta H_F}{c}\ell n(1.78\,\hat{P}_T)\hat{P}_T \tag{6.26b}$$

where

$$B = [F_1 + (K_S/K_L)F_2]/(2\Delta H_F/c) \tag{6.26c}$$

$$A = F_1 - 1 \tag{6.26d}$$

$$F_1 = [mK^*_{\ell-1}(m) + \ell K^*_\ell(m)]/K^*_\ell(m) \tag{6.26e}$$

and

$$F_2 = [mI^*_{\ell+1}(m) + \ell I^*_\ell(m)]/I^*_\ell(m) \tag{6.26f}$$

In Eqs. (6.26), I^* and K^* are modified Bessel functions of the first and second kinds, respectively, while c is the specific heat per unit volume. At the marginal stability condition for the (ℓ, m) perturbation, $(\dot{\delta}/\delta)/(\dot{R}/R) = 1$ and $R = R_C^{\ell m}$ while $\hat{P}_T = \hat{P}_{Tc}^{\ell m}$. Thus, from Eqs. (6.26a) and (6.26b)

$$\hat{P}_{Tc}^{\ell m} = \left[\frac{\gamma}{\Delta S_F}(\ell^2 + m^2 - 1)\frac{B}{(A-1)}\right]\frac{1}{R_C^{\ell m}} \tag{6.27a}$$

and

$$R_C^{\ell m} = \frac{\gamma}{\Delta S_F \Delta T_\infty}\left[1 - (\ell^2 + m^2 - 1)\frac{B(\Delta H_F/c)\,\ell n(1.78\,\hat{P}_{Tc}^{\ell m})}{(A-1)}\right] \tag{6.27b}$$

Plots of calculated R_C versus ΔT_∞ for Ni are shown in Fig. 5.20. Recognizing that $\hat{P}_{Tc} = \dot{R}R_C/2D_T = V\rho_t/2D_T$ and that $R_C/\rho_t = 0.585$, Eq. (6.27a) can be rearranged to yield

$$V\rho_t^2 = \beta_{\ell m}\frac{D_{Tc}}{\Delta H_F}\left(\frac{\gamma}{\Delta S_F}\right) \tag{6.27c}$$

where

$$\beta_{\ell m} = 1.71\left[\frac{F_1 + (K_S/K_L)F_2}{F_1 - 1}\right](\ell^2 + m^2 - 1) \tag{6.27d}$$

Eq. (6.27c) bears an interesting similarity to Eq. (6.24). The data plotted

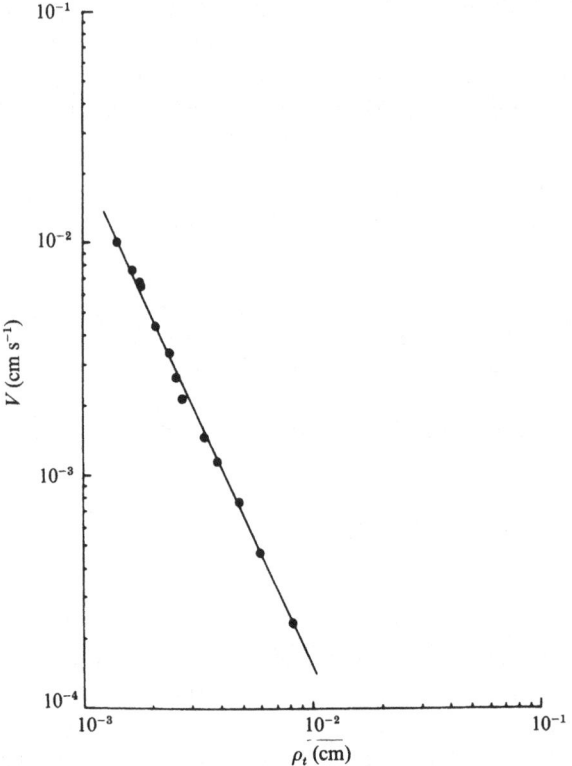

Fig. 6.37. V versus ρ_t at various ΔT_∞ for pivalic acid.[26]

in Fig. 6.37 for pivalic acid clearly show that $V\rho_t^2 = $ constant for this system.[26] For the case $m = 0$ and $\ell = 4$, $F_1 = F_2 = 4$ and it is the magnitude of K_S/K_L that determines how closely β_{40} approximates $8\pi^2$ in Eq. (6.24). If $K_S/K_L = 0.5$ as we expect to find for many semiconductors, $\beta_{40} = 51.5$. However, if $K_S/K_L = 2$, as we expect to find for metals, $\beta_{40} = 103.3$. For succinonitrile, $K_S/K_L \approx 1$, so $\beta_{40} \approx 70$ and we see that the surface creation constraint is just that needed to yield the experimental operating point in Fig. 6.36(b). The value of V for this dendrite filament in the $\hat{P}_T \ll 1$ regime is thus given by

$$V = \frac{2.94\,\beta_{\ell m}D_T(c/\Delta H_F)\Delta T_\infty^2}{(\gamma/\Delta S_F)\left\{1 - (\ell^2 + m^2 - 1)\left[B(\Delta H_F/c)/(A-1)\right]\ell n(1.78\,\hat{P}_{Tc}^{\ell m})\right\}}.$$
(6.27e)

Since R_C decreases and $\hat{P}_{Tc}$ increases as ΔT_∞ increases, the exponent of ΔT_∞ in Eq. (6.27e) is greater than 2 and generally reaches ~ 2.5.

There is room for argument concerning the exactitude of the above expression for V relative to the experimental value, since it is based upon the thermal field for single filament growth while the experimental value involves the interaction with neighboring filaments. However, the important point to be made here is that the necessary addition of the surface creation constraint completely specifies the free dendrite growth problem and no extremum is possible for V with respect to variations of ρ_t. An extremum condition enters the freely growing continuum filament problem only with respect to variations of V with axial orientation.

It is from the TLK picture of nucleating and flowing layers, as discussed in Chapter 3 of the companion book,[7] that the optimum orientations may be predicted. Because new layers generally nucleate at crystallographic poles located some distance back from the filament tip and flow towards the major annihilation location at the tip, the temperature variation along the surface of the filament is the reverse of that predicted from the continuum approximation. This causes the change in the local heat field at the filament tip based on the TLK picture, relative to the isothermal filament, to be opposite to that for the continuum approximation. Thus, instead of the non-isothermal tip filament slowing down relative to the isothermal tip filament, as expected from the continuum approximation, it speeds up because of the surface creation and annihilation requirements imposed by the TLK picture of the tip microstructure (see Fig. 3.2 of the companion book[7]). The possible orientations for new layer generation depend upon the shape of the γ-plot at the growth temperature and the alloy condition of the nutrient phase. Only significant cusp orientations can develop facet planes and, thus, only these crystallographic poles on the tip surface can be sites for layer creation.

Straight filaments require a symmetrical array of layer sources (primary, secondary, etc.) and, for a particular material, only a few possibilities exist wherein the layers form a closed roof-type structure over the tip as illustrated in Fig. 6.38. The convergence of layers at the filament tip raises the temperature of the tip compared to a tip of the same ρ_t but with no layer convergence and thus no layer edge annihilation. This increases the non-isothermal tip temperature gradient effect and thus increases V. Further, of the set of possible closed roof structures that might form in a particular material, that member having the maximum number of different layer sources forming the roof structure leads to the extremum in V and that is the dominant dendrite orientation that will be observed experimentally. Of course, if the solid exhibits anisotropy

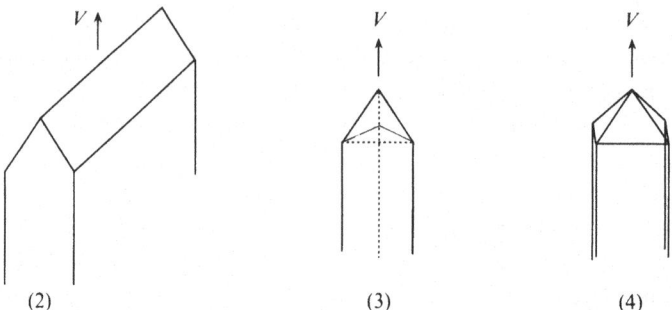

(2)   (3)   (4)

Fig. 6.38. Schematic illustration of the number of layer planes operating to form the roof structure for different dendrite tips.

of thermal conductivity, this factor also will be a consideration in determining the extremum in V. For crystal growth under very different conditions of (T_i, C_i, P_i, ϕ_i), the orientation, θ, producing the extremum in V may change abruptly from one orientation to another of the closed roof-forming set.

6.4.2   *Alloy melt dendrites*

The experimental V for freely growing dendrites exhibits an increase in V at fixed ΔT_∞ with the addition of a small amount of solute to the melt followed by a decrease in V at larger values of C_∞.[28] Some experimental results are shown for ice dendrites in Fig. 6.39. From a cursory consideration of the coupling equation, one would expect V to decrease as C_∞, and thus ΔT_C, increases because less driving force is available for ΔT_T. However, from the previous section, we have also seen that $r^* = R_C$ is an important consideration in the determination of V so that, even though ΔT_T decreases as C_∞ increases, perhaps R_C initially decreases sufficiently with C_∞ to more than offset the decrease in ΔT_T. To assess this idea, one could consider the perturbation analysis findings for a cylinder growing from a supercooled alloy melt. However, it is simpler to utilize the findings of the last section that the surface creation constraint leads to the Langer and Müller-Krumbhaar tip stability criterion[25] wherein the observed tip radius, ρ_t, is equal to the shortest wavelength, λ^*, which can form under the local conditions at the tip; i.e.,

$$\rho_t = \lambda^* \tag{6.28a}$$

Following Trivedi,[29] this stability condition can be expressed as

$$\rho_t = [\tilde{\Gamma}/\sigma^*(m\bar{\mathcal{G}}_C - \bar{\mathcal{G}}_T)]^{1/2} \tag{6.28b}$$

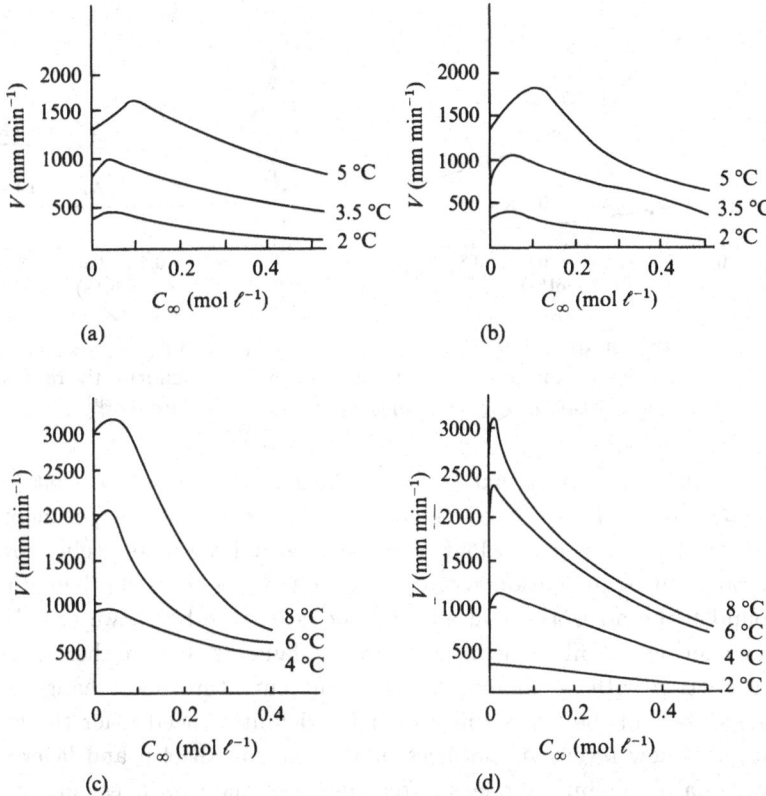

Fig. 6.39. V for ice dendrites grown from various aqueous solutions as a function of C_∞: (a) NaOH, (b) HCℓ, (c) sugar and (d) NaCℓ.[28]

where $\bar{\mathcal{G}}_C$ and $\bar{\mathcal{G}}_T$ are the average concentration and temperature gradient seen by a perturbation, respectively, evaluated over the parabolic filament tip and σ^* is a stability constant ($\sigma^* \approx 0.025$).

Using Eq. (4.21e) for $\phi(\hat{P}_C)$ and an analogous expression for $\phi(\hat{P}_T)$, the coupling equation at the filament tip can be immediately written as

$$\Delta T_\infty = \frac{2(\gamma/\Delta S_F)}{\rho_t} + \frac{\Delta H_F}{c}\phi(\hat{P}_T) + m_L C_\infty \left[1 - \frac{1}{1-(1-k_0)\phi(\hat{P}_C)}\right]$$
$$(6.29a)$$

with $\hat{P}_C = (D_T/D_C)\hat{P}_T$. From solute and heat conservation at the filament tip, both $\mathcal{G}_C$ and $\mathcal{G}_T$ can be readily evaluated. To obtain the average values of these quantities a number of approximations are usually made: (i) the tip is assumed to be isothermal and isoconcen-

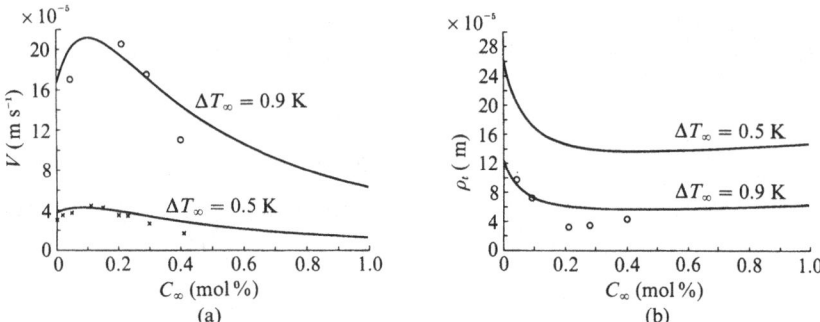

Fig. 6.40. (a) V versus C_∞ at two values of ΔT_∞ for succinonitrile dendrites from acetone solutions and (b) their dendrite tip radius, ρ_t, as a function of C_∞ (o,x, measured and – calculated).[29]

trate, (ii) $D_S = 0$ so that $\bar{\mathcal{G}}_C = \mathcal{G}_C/2$ and (iii) $\mathcal{G}_S = 0$ so that $\bar{\mathcal{G}}_T = K_L\mathcal{G}_T/(K_S + K_L)$. Even though the filament tip is experimentally observed to be a paraboloid of revolution and a single freely growing paraboloid of revolution with $\gamma = 0$ and $\beta_2 \to \infty$ is mathematically found to be an isothermal and an isoconcentrate body, we are always considering one filament in an array of filaments so that there is some interaction with neighbors. Further, the curvature does change along the surface of the tip, a supercooling is definitely needed for the nucleation of new layers at the facet plane poles on the tip and layer edge annihilation definitely releases free energy at the tip. These factors are additional contributions to Eq. (6.29a). However, in the approximation that these additional factors may be neglected, Eq. (6.28b) becomes

$$\rho_t = \frac{\left(\frac{\gamma}{\Delta S_F \sigma^*}\right)}{\frac{2\hat{P}_T \Delta H_F}{(1 + K_S/K_L)c} - \frac{m_L(1 - k_0)\hat{P}_C, C_\infty}{1 - (1 - k_0)\phi(\hat{P}_C)}} \tag{6.29b}$$

For succinonitrile–acetone solutions, $K_S/K_L \approx 1$ and other properties of these mixtures are given in Table 6.1. Plots of V versus C_∞ and ρ_t versus C_∞ are given in Fig. 6.40 for $\Delta T_L = 0.9$ K and 0.5 K where we see that reasonable conformance with the experimental results is found. The partitioning of ΔT_∞ into the three subdriving forces is shown as a function of C_∞ (acetone) in Fig. 6.41. Finally, $V\rho_t^2$ is plotted as a function of C_∞ (acetone) in Fig. 6.42 to show that the predictions neatly span between the purely thermal case at $C_\infty = 0$ and the purely solutal case as C_∞ becomes large.[29]

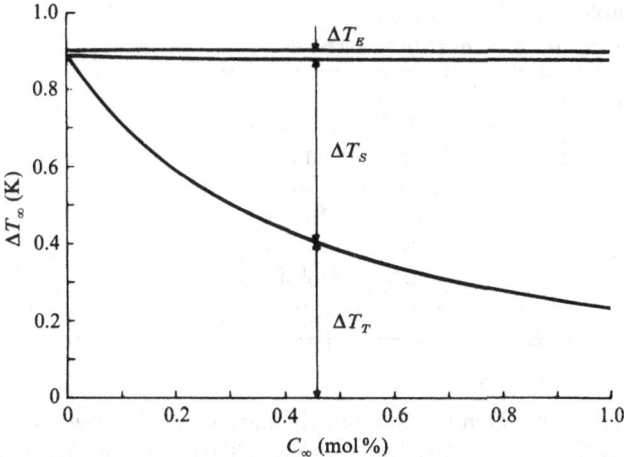

Fig. 6.41. Calculated magnitudes of ΔT_E, ΔT_C and ΔT_T for dendrite filament growth at a given ΔT_∞ and C_∞ in the succinonitrile–acetone system.[29]

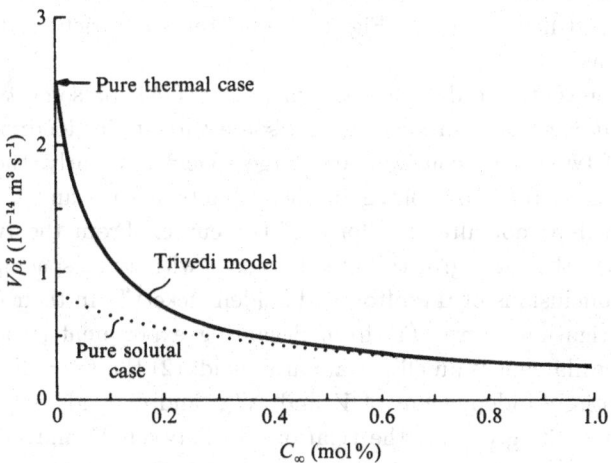

Fig. 6.42. Calculated $V\rho_t^2$ product as a function of C_∞ at $\Delta T_\infty = 0.9\,\mathrm{K}$ for dendrite filament growth in the succinonitrile–acetone system.[29]

6.4.3   Pulled dendrites

Technological harnessing of the freely growing dendrite can be accomplished by pulling the dendrite from a supercooled bath at exactly the rate of its natural V for that value of ΔT_∞. This leaves the growing dendrite tip in a stationary state condition a short distance below the

Table 6.1. *Properties of*
succinonitrile–acetone mixtures.

T_M	331.24 K
c	1937.5 J kg^{-1} K^{-1}
ΔH_F	46.26×10^3 J kg^{-1}
a	1.14×10^{-7} m^2 s^{-1}
$\tilde{\Gamma}$	6.62×10^{-8} K m
D_L	1.27×10^{-9} m^2 s^{-1}
m_L	-2.16 K (mol.%)$^{-1}$
k_0	0.103 mol.% (mol.%)$^{-1}$

surface of the melt. Dendrites of Sn, Ge and Si and other semiconductors have been produced in this fashion and it is the latter two that are of interest to us here because they are twinned ribbon-like crystals with {111} flat surfaces growing by a reentrant corner mechanism.[7] For these semiconductor systems, facet formation is substantial and one observes marked periodicities and edge serrations because of the dominant layer formation and flow kinetics. Fig. 6.43 illustrates a variety of twinned growth forms.

Some representative data are shown in Fig. 6.44 for some two-twin dendrites at various twin spacing and some three-twin dendrites with one pair of twins at a constant and large spacing. Clearly, the larger twin spacing is rate controlling in the three-twin case and the additional twin does not alter the form of the curve. From the two-twin data, we see that the growth rate decreases with twin spacing. The principal conclusions of Hamilton and Seidensticker[30] in their Ge dendrite investigations were: (1) from decanting experiments, the tip of the growing filament is an elliptical paraboloid; (2) the tip radius seems to be essentially independent of V and ΔT_∞ and depends only on the twin spacing, (3) given ρ_t, the relationship between V and ΔT_∞ can be computed with considerable accuracy and is found to agree closely with experiment, (4) the dominant growth mechanism appears to be layer flow with a kinetic rate constant of $\beta_2 \Delta S_F \tilde{>} 3$ cm s^{-1} °C^{-1} and (5) best agreement between theory and experiment occurs with the use of γ between 45 and 90 erg cm^{-2}.

The observed external shapes of the Ge dendrites can be rationalized by imagining the following sequence of events for a two-twin dendrite that is being pulled from a supercooled melt with the temperature having a cylindrical symmetry and increasing slowly in the radial direction. At

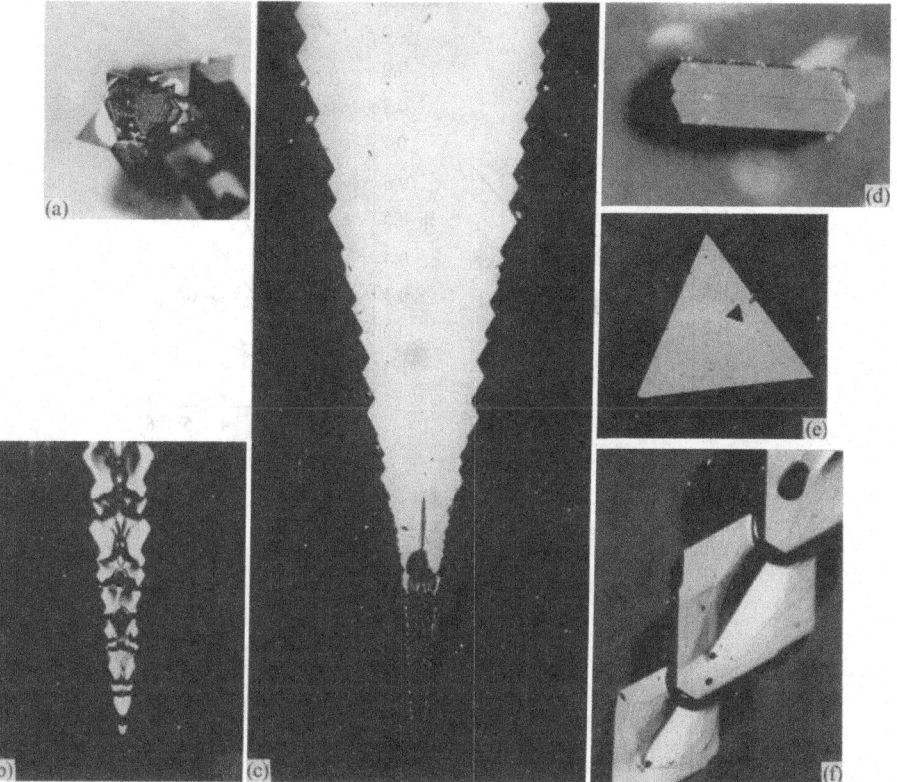

Fig. 6.43. Some twinned growth forms: (a) solution-grown, five-twinned star-shaped Si crystal, (b) melt-grown Ge dendrite tip, (c) melt-grown Ge dendrite, (d) solution-grown InAs crystal, (e) vapor-grown Ge plate crystal and (f) melt-grown Ge dendrite body.

time t, the dendrite with geometry illustrated in Fig. 6.45 starts to grow into the supercooled melt.[31] Although it is growing continuously, we will imagine it to be growing in steps for simplicity. The fastest direction of growth will be in the axial direction utilizing the 141° reentrant angle so that segment 1 becomes filled in and the side facets taper inwards. Next, the side facets and the axial facet see a slightly reduced supercooling and grow to fill in segments 2. Finally, at still further reduced supercooling, segments 3 fill in. After completing these three growth cycles, at time t_1 the dendrite is pulled upwards until the bottom of the dendrite is at the upper surface of the melt. Repetitions of this process give a saw-toothed appearance to the sides of the dendrite. Triangles 3

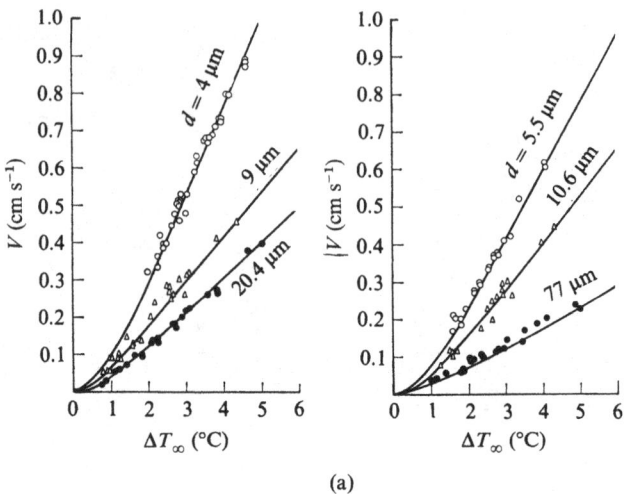

(a)

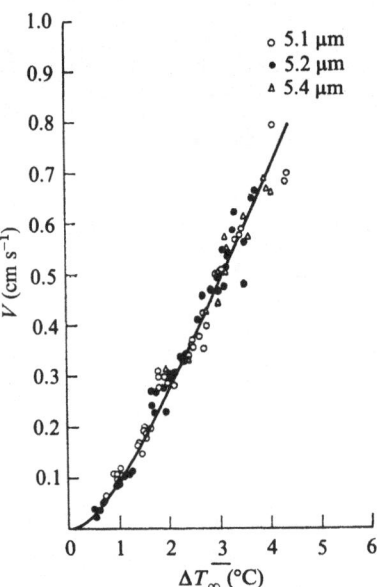

Fig. 6.44. V versus ΔT_∞ for (a) doubly twinned Ge dendrites at several spacings, d, and (b) results for triply twinned Ge dendrites at three different sets of spacings. The larger twin spacing is controlling in this latter case.[30]

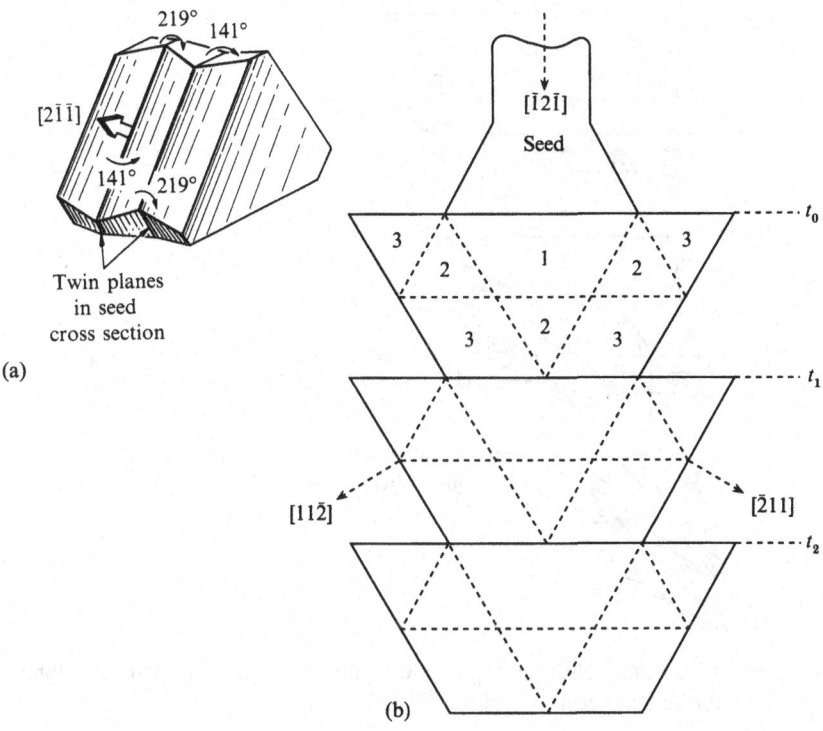

Fig. 6.45. Sequenced growth of a doubly twinned Ge or Si dendrite in the $[\bar{1}2\bar{1}]$ direction via steps 1, 2, 3, etc.[31]

may not develop fully during a proper growth procedure since they require backward growth into a somewhat heated fluid; however, this is certainly a possibility if sufficient supercooling exists in that direction.

Such a process can clearly account for the edge morphology of the dendrite shown in Fig. 6.43(c), allowing growth in both the length and width directions. These dendrites also grow in thickness, presumably by surface nucleation, and such a process would be required to produce the overlapping platelet effect of Fig. 6.43(f). The thicker portion of the dendrite grows towards the tip in a layer fashion and overgrows the underlying surface; however, as it does so, it is constrained by the lateral {111} facets to develop a shorter and shorter front.

On a more microscopic scale, a short distance behind the advancing

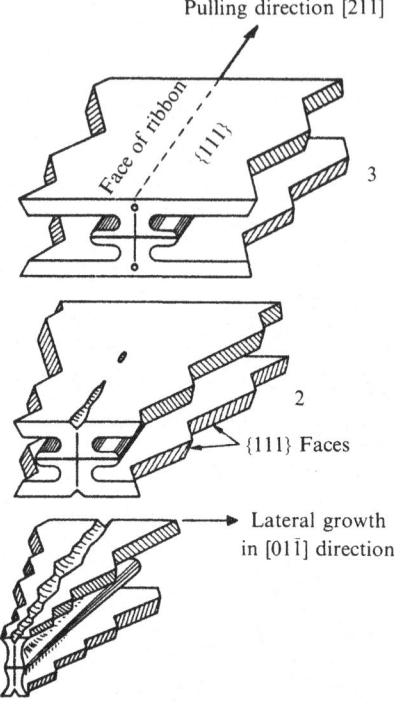

Fig. 6.46. Three stages of development of a Si or Ge dendrite ribbon during growth from a melt.[32]

tip, a facet system of {111} planes provides a series of reentrant edge systems allowing growth to proceed outward to give an H-shaped cross-section as illustrated in Fig. 6.46.[32] The heat rejected from the outward growth of the arms retards the growth of the region between the twin core and arms. The freezing-in of the volume is the last step in the formation of the ribbon. Fig. 6.46 illustrates how dislocations and microsegregation patterns can be formed by the freezing of these trapped liquid pockets.

6.4.4   *Dendritic versus spherulitic crystallization*

The classical picture of a spherulite is shown in Fig. 1.1 of the companion book[7] for a blend of isotactic and atactic polypropylene. Another spherulite is shown in Fig. 6.47 for the malonamide-*d*-tartaric acid system. These spherulite patterns appear to have a filamentary array character in common with dendrite patterns; however, they do not exhibit the crystallographic side-branching characteristic of the dendrite

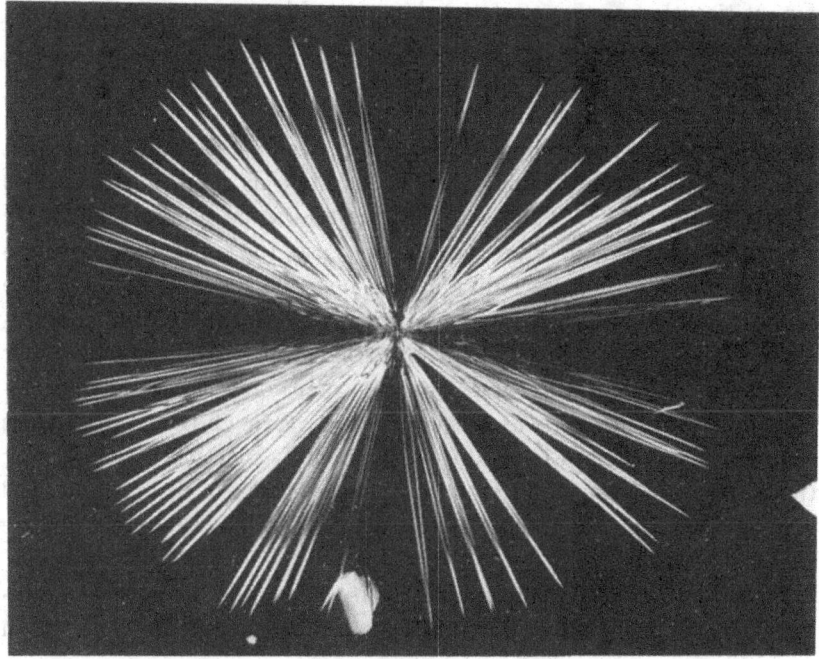

Fig. 6.47. The filamentary pattern of spherulitic growth in malon-amide containing 10% *d*-tartaric acid.

array. Because of this, many people have thought that the spherulite morphology is primarily a consequence of the complex molecular configurations assumed by the polymers. However, data exist to suggest that the envelope morphology of a growing array of dendrites can adjust its shape, in response to the detailed environmental conditions prevailing in the fluid surrounding the growing array, to produce similar morphologies.

Several decades ago, Luyet,[33] studying the freezing of ice in thin films of concentrated aqueous solutions, observed ice crystal morphologies of distinctly hexagonal and distinctly spherulitic form. The filamentary nature of both these forms was conclusively revealed using an evaporation technique to show individual filaments measuring only a few hundred angstroms in diameter. However, because the spherulite diameters were an order of magnitude larger than the film thickness, perhaps it was only an artifact associated with the two-dimensional constraint and was not truly representative of three-dimensional crystallization. A large body of subsequent work on polymers by Keith and Padden[34] strengthened the

supposition that a strong relationship existed between dendritic growth
and spherulitic growth.

Studies by Geering[35] on the water/glycerol system in bulk quantities
have confirmed the three-dimensional character of the original Luyet [33]
observations. By controlling ΔT_∞ and C_∞, the filament array expansion
rate, V, the filament spacing, λ, and envelope morphology transitions
were investigated.[35] These experiments and some comparison theory
strongly supported the view that, in this case, the spherulitic morphology resulted largely from individual filament interfacial response to the
thermal, solute and hydrodynamic potential differences in its environment. The latter is strongly involved in various systems that either
expand or contract on freezing. For the expansion case, as in ice formation, fluid must flow outwards from the array through the interdendritic
channels in an attempt to produce an ultimately spherical flow field. For
the contraction case, fluid must flow inwards from a distant spherical
flow field through the interdendritic channels to feed the interdendritic
channel growth.

From the Hagen–Poiseuille equation (see Eq. (2.10)) the pressure differential needed to convey liquid through a tube of diameter λ and length
L at velocity u_f is given by

$$\Delta P = 8\rho_L \nu_L L u_f / \lambda^2 \qquad (6.30a)$$

where ρ_L is the density of the liquid. Considering a cubic dendrite array
with $90°$ side-branching of total volume L^3, the individual cell volume
is λ^3 and each cell is bordered by 12 filaments of radius $R(t)$ that grow
radially at $\dot{R}(t)$. Thus, from fluid conservation, the change in u_f across
a single cell in the radial direction is

$$\Delta u_f = 6\pi R \dot{R} \Delta \rho / \rho_L \lambda \qquad (6.30b)$$

where $\Delta \rho$ is the density change on freezing. Since $u_f \approx 0$ and $u_f = u_f^{max}$
at the center and periphery of the array, respectively, $u_f^{max} = \Delta u_f (L/\lambda)$.
Applying Eq. (6.30a) with $u_f \approx u_f^{max}/2$, we have

$$\Delta P \approx 24\pi(1 - 2\pi R/\lambda)\nu_L \Delta \rho \, V \rho_t L^2 / \lambda^4 \qquad (6.30c)$$

where $V\rho_t = R\dot{R}$ for the cells at the envelope surface and the factor
$1 - 2\pi R/\lambda$ corrects for the fact that the cell wall is not a solid tube
through which fluid flows but is only fractionally solid and thus offers
only a fractional drag to the fluid flowing through the cell. Since this
equation neglects any lateral mixing of fluid between the cells, it is more
reasonable to use an undetermined parameter β in place of $24\pi(1 - 2\pi R/\lambda)$ in Eq. (6.30c).

The important point to note about ΔP in Eq. (6.30c) is that (1) it increases strongly as ΔT_∞ increases ($V\rho_t$ increases while λ decreases), (2) it increases strongly as C_∞ increases ($V\rho_t$ decreases but λ strongly decreases and ν_L increases generally) and (3) it increases strongly with time ($L = Vt$). Thus, ΔP acts as a bending pressure between the tip and the root of each filament in the array to turn it towards a radial direction. If the bulk modulus of the filament is small and the aspect ratio, $L/\bar{R}$, of the filament is large, such bending will readily occur. Slip in the filaments gives a permanent set to this radial configuration.

Geering's results can be best represented by considering Figs. 6.48, 6.49 and 6.50. The spherulite morphogenesis was observed to be the following (see Fig. 6.48): (i) after initial nucleation of an ice crystal, a single crystal begins to grow in the shape of a solid polyhedron; (ii) this polyhedron becomes unstable to perturbations and ripples develop on its surface; (iii) the ripples amplify in magnitude and become dendritic filaments growing into the liquid in low index crystallographic directions; (iv) the filamentary array develops a polyhedral envelope and continues to grow in that shape; (v) at a particular envelope diameter, $\tilde{D}_T$, the filaments begin to bend into the radial direction and the polyhedral envelope begins to round off, becoming more spherical. This model was tested in early experiments by an observation made on heating an ice spherulite, growing from a glycerol/water solution at $T = -45\,°C$, to a temperature of $T = -35\,°C$. At the higher temperature, after an abrupt reduction in growth rate, filament arrays with much larger spacing began to appear at random points on the surface of the spherulite. As these arrays grew and branched, they developed into filamentary bodies with polyhedral outlines. Eventually, the spherulite developed a set of hexagonal prisms clustered about its periphery; i.e., the transformation *back* from a spherulitic to a crystallographic array had been achieved.

A summary of the array envelope shapes and ice crystallographic structures observed in Geering's[35] isothermal solidification experiments is presented in Fig. 6.49. These three regions at increasing ΔT_∞ and constant C_∞ are: I, classical dendrite prisms of hexagonal ice; II, envelope morphology transition regime of hexagonal ice; IIIA, spherulite regime of hexagonal ice; IIIB, spherulite regime of cubic ice; and IIIC, spherulite regime of non-crystalline ice. On theoretical grounds, at values of ΔT_∞ approaching $80\,°C$, one might expect that both a diffusionless transformation ($k = 1$) and an isenthalpic transformation ($\Delta H_F \Delta T_\infty/c = 1$) should occur. Experimentally, structureless, evanescent spheres of amorphous ice are observed to grow in this ΔT_∞ range.

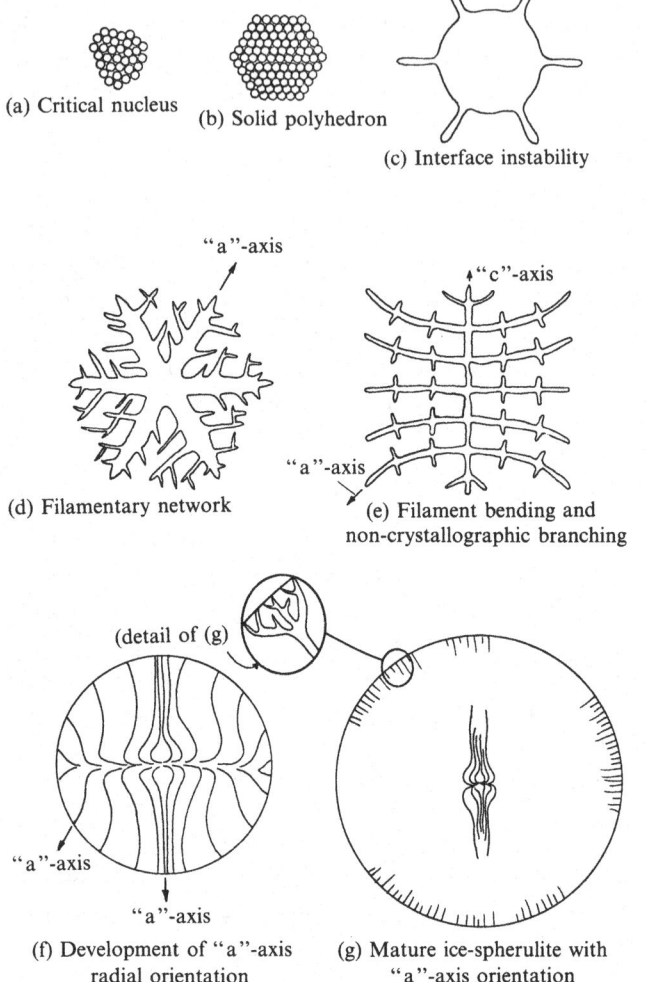

Fig. 6.48. The various stages in the genesis of an ice spherulite during crystallization from a water/glycerol solution.[35]

From the theoretical arguments leading to Eq. (6.30c), one expects the transition from a polyhedral to a spherical array envelope to occur at a critical pressure $\Delta P = \Delta P_c$ and that the array diameter, $\tilde{D}_T$, at which transition occurs should be given from Eq. (6.30c) by

$$\tilde{D}_T \propto \nu_L^{-1/2}\lambda^2 \propto \nu^{-1/2}V^{-2} \qquad (6.31)$$

since $\hat{P}_T^{1/2}$ is only a slowly varying function of ΔT_∞ and the data of

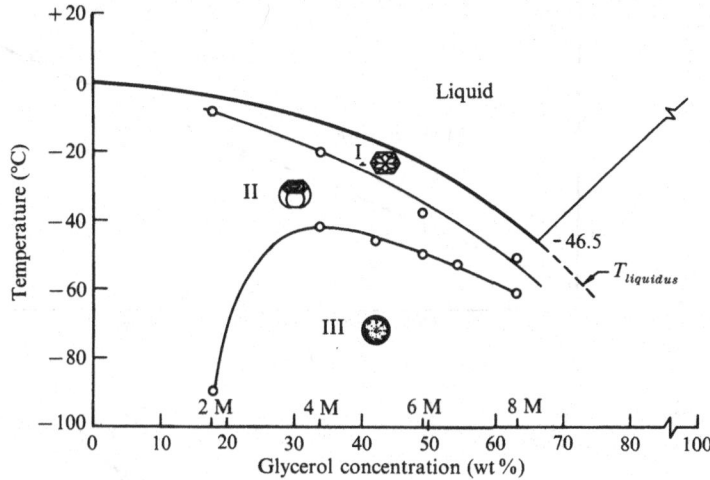

Fig. 6.49. Major solid morphology regions for ice growing in the water/glycerol system.[35]

Fig. 6.50(a) show $\lambda = a/V$. Fig. 6.50(b) shows excellent agreement with Eq. (6.31).

To conclude this section, we see that, when the hydrodynamic pressure forces are considered during the growth of a dendrite array, a natural transition to a spherulitic filament array can occur when the fluid viscosity is sufficiently high and the interfilament spacing is sufficiently low. It is interesting to note that pre-spherulitic crystallographic arrays of filaments have also been observed in polypropylene and in several carbonates.

6.4.5 Dendritic crystallization of highly oriented polymer melts

The products of many commercial processes are highly oriented semicrystalline homopolymeric solids. In most of these cases, the development of chain orientation is believed to precede that of crystallization. Examples of such processes are melt-drawn fibers and films, blown film and, to some extent, injection-molded pieces. Under such high strain conditions, the crystallization process must be controlled by energy flows, particularly heat diffusion, as opposed to interface control normally assumed in theories of polymer crystallization.

For crystallization from strained melts, the basic growing object is a spherulite where, in most cases, the polymer chains in a particular filament are folded and lie approximately tangent to the spherulite growth

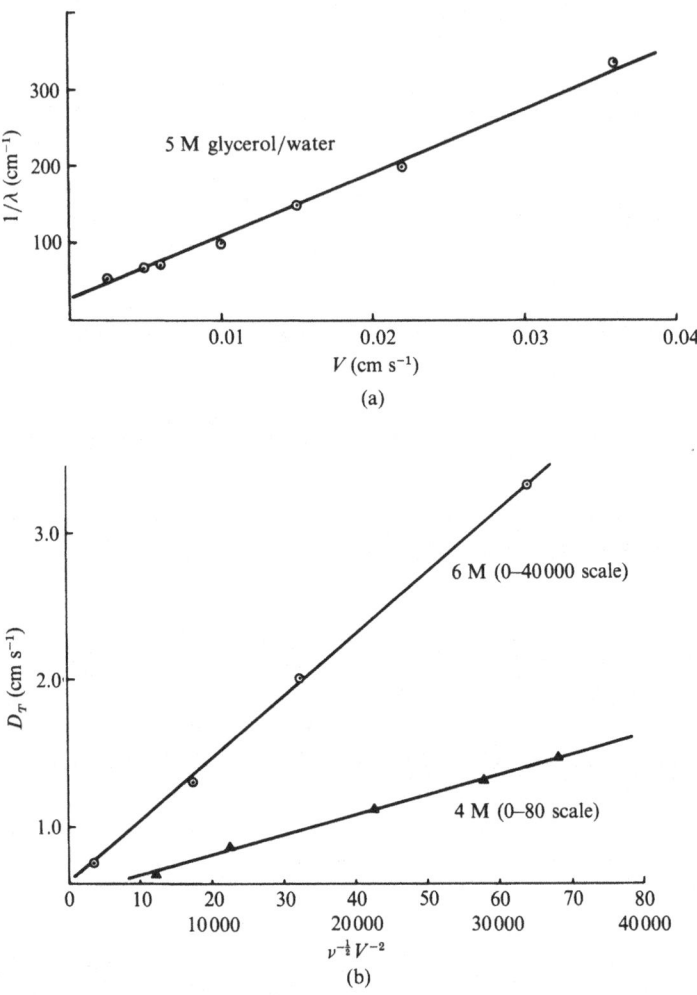

Fig. 6.50. Ice crystal growth from water/glycerol solutions: (a) inter-filament spacing versus crystal growth rate and (b) transition diameter, $\tilde{D}_T$, to spherulitic form as a function of the parameter $\nu^{-1/2}V^{-2}$.[35]

front. For crystallization under high strain, a very thin (~ 100 Å) core filament with chain and filament axis coincident is observed. Transmission electron micrographs reveal such fibrillar crystals in films of polystyrene, polybutene-1, polypropylene, etc. Melt stretched thin films are observed to crystallize at rates ~ 1–10 cm s^{-1} while kinetic studies of the melt-spinning of fibers puts the crystallization rates in the range

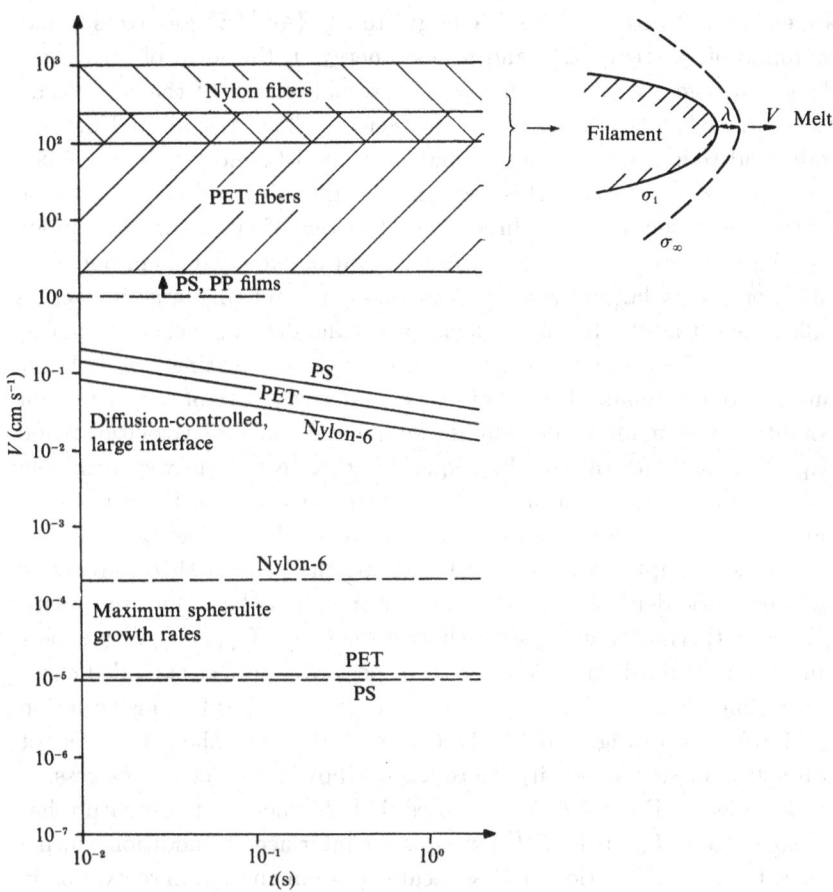

Fig. 6.51. V versus t plots for polystyrene (PS), polyethylene tereph-
thalate (PET) and nylon-6 crystals. (1) Lowest curves give largest
reported spherulitic growth rates, (2) central curves give heat diffusion-
limited growth of a planar front and (3) upper curves give growth rates
in known oriented states.[36]

$V \sim 10^2$–10^3 cm s^{-1} (see Fig. 6.51), many orders of magnitude faster
than the maximum observed spherulite growth rates.[36]

These very large rates of transformation in strained systems are due
in large part to the lowered entropy of the melt. As the strain on a
molecule in the melt is increased, its possible number of configurations
decreases; i.e., the melt entropy is lowered. Consequently, the free energy increases, for a given ΔT_∞, and the driving force for crystallization
decreases because ΔS_F decreases. In addition, the growing filaments

are expected to contain elastic energy storage ($\frac{1}{2}\sigma^2/E^*$ for stress σ and modulus of elasticity E^*) and excess energy in the form of structural defect storage, ΔE_d. This increased internal energy of the solid leads to a reduced latent heat for the filaments. Thus, the altered driving force and reduced latent heat allow very rapid filament growth. Tiller and Schultz[36] analyzed this free dendrite growth problem for polymer systems and showed that three characteristics of crystallization under very high stresses are (i) a filamentary habit of extended chain polymer, (ii) enormously large crystal growth velocities and (iii) large intracrystalline defect levels. In this problem, V, ρ_t and defect concentration, C_d, were the variables and mathematical expressions relating ΔT_∞, V, ρ_t and C_d to a normalized tensile force were derived. Use of the extremum condition was made to determine the general range of velocities to be expected and, with the coupling equation, three equations were available to solve for the three variables. The predicted values of V for isotactic polystyrene fell within the experimental range of observation.

In this example, it is useful to describe the general thermodynamic picture of the dendrite growth when strain is included. One starts with a bulk isothermal liquid system characterized by $(T_\infty, \sigma_\infty)_L$ at a departure from equilibrium of ΔG_∞ with respect to a perfect extended chain crystalline phase at $(T_\infty, 0)_{\hat{c}}$. The free energy profiles for this situation are illustrated in Fig. 6.52 for the two cases (i) where the entropy is not a function of stress and (ii) where the entropy is a function of stress.

Considering Fig. 6.52(a), one notes that the melting temperature has changed from $T_M(0)$ to T^{*P} for a planar interface. In addition, during the actual crystallization process, heat evolution and strain relaxation in the interface region cause the state variable change from $(T_\infty, \sigma_\infty)_L$ to $(T_i, \sigma_i)_L$ in the liquid at the interface and to $(T_i, \sigma_i', \Delta E_d)_{\hat{c}}$ in the crystal at the interface. Here, ΔE_d is the total stored defect energy for various types of non-equilibrium defects grown into the crystal having a state of elastic strain σ'. Thus, for the case of zero interface curvature, the free energy driving force available at the interface, relative to defect-free crystal at T_i, is

$$\begin{aligned}
\Delta G_i &= G_L(T_i, \sigma_i) - G_{\hat{c}}(T_i, 0) \\
&= \Delta S_F(0)[T_M(0) - T_i] + \sigma_i^2/2E_L^*
\end{aligned} \tag{6.32a}$$

The overall coupling equation for the case of filamentary crystal growth with a tip radius of curvature ρ_t is

$$\Delta G_\infty = \Delta G_{sv} + \Delta G_E + \Delta G_K$$

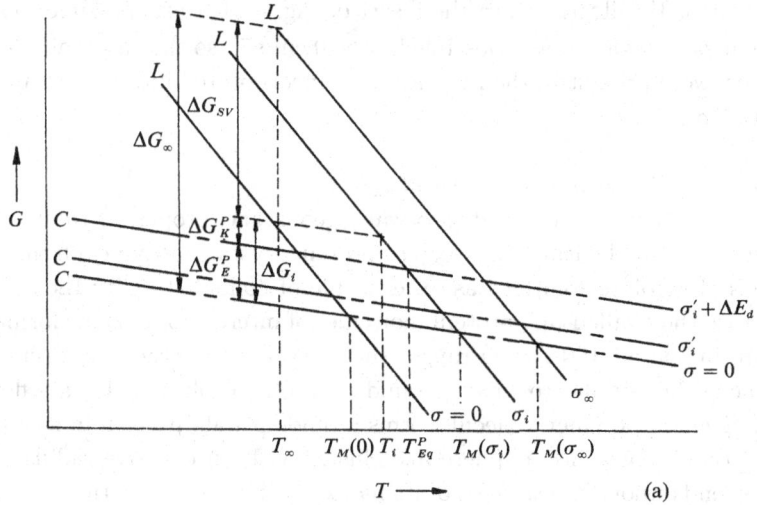

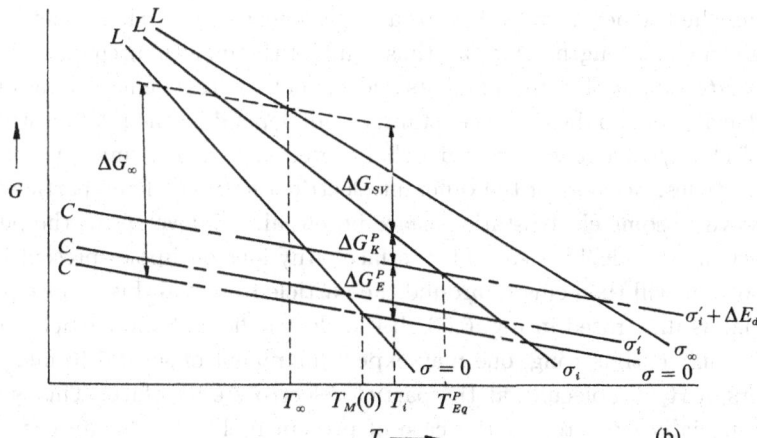

Fig. 6.52. Free energy versus temperature plots pertinent to polymer crystallizations (a) ignoring conformational entropy changes in the melt and (b) including conformational entropy changes in the melt.[36]

where

$$\Delta G_{sv} = [G_L(T_\infty, \sigma_\infty) - G_L(T_i, \sigma_i)] - [G_{\hat{c}}(T_\infty, 0) - G_{\hat{c}}(T_i, 0)] \quad (6.32b)$$

and

$$\Delta G_E = \frac{\sigma_i^2}{2E_{\hat{c}}^*} + \Delta E_d + \frac{2\gamma v_m}{\rho_t} \quad (6.32c)$$

By considering viscoelastic relaxation of the zone of width λ ahead of the filament tip illustrated in the insert of Fig. 6.51 it was possible to show that no relaxation was possible if $V \overset{\sim}{>} 10$ cm s^{-1} so that all chain defects that were present in the melt must be incorporated into the crystal and $\sigma_i = \sigma_\infty$.

6.4.6   *Extensions to biological systems*

In many ways, it is advantageous and proper to think of the growth of biological solids such as protein crystals (enzymes) from aqueous electrolyte solutions as an example of colloid stability theory. Because the "balled-up" protein molecules of interest for enzyme formation are in the 50–100 Å size range, they also have interaction effects with each other similar to what is found in a colloid solution. Even a flexible long chain polymer molecule forms a random walk pattern in its solvent so that it becomes a sphere-like cluster with an effective radius which depends upon the strength of its molecular interaction with the solvent. The better the chain is wetted by the solvent, the more open and less bunched it becomes leading to a larger effective particle diameter for a given chain length. One can thus think of the molecular chain as a *composite colloid* of chain segments and solvent where its effective radius, R, density, $\bar{\rho}$, and dielectric constant, ε_*, all depend on the relative strength of chain/solvent versus chain/chain interactions. For any type of such particles, because of the different electrical natures of the particles and solvent, some electrostatic surface potential, ϕ_s, develops at the periphery or particle/bulk solvent interface. Any ions or dipoles present in the solvent will then cluster around the particle to screen this surface potential as illustrated in Fig. 6.53. If the electric field at the particle surface is sufficiently strong, one may expect alignment of several to many layers of H_2O molecules at the particle/electrolyte interface. This is to be especially expected in the case of protein molecules. In all cases, one can consider the system as a solution consisting of colloid-like entities surrounded by space charge sheaths of ions and dipoles that move as a unit via Brownian motion in a vast sea of solvent electrolyte. Let us now consider the interplay of forces that lead to either a stable dispersion of such particles or an agglomeration process. Since crystallization is a type of agglomeration reaction, one is interested in the type of conditions that would allow controlled agglomeration to a large surface. Let us first look at the qualitative picture.

Two major forces determine the agglomeration tendency of colloidal suspensions: (1) electrodynamic forces of the van der Waals dispersion type which favor agglomeration and (2) electrostatic forces which favor

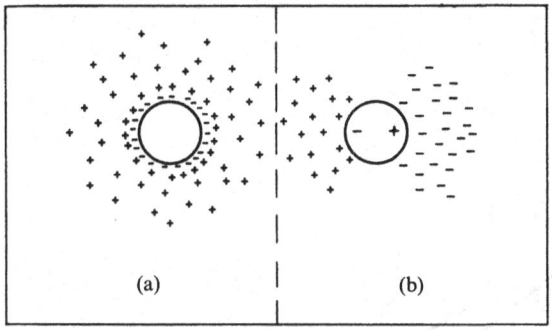

(a)   (b)

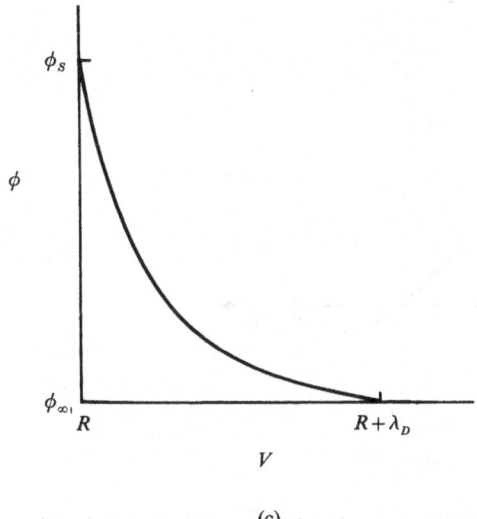

(c)

Fig. 6.53. Schematic illustration of (a) the ionic double layer around a colloidal particle, (b) the neutralized electric dipole of a large molecule and (c) the electrostatic potential distribution, ϕ, varying from ϕ_s at the particle surface to ϕ_∞ at the edge of the space charge layer ($r = R + \lambda_D$) in the solution.

dispersion. Fig. 6.54 illustrates the total interaction energy, $U(H)$, and the resultant force, $F(H)$, between two colloidal particles at a separation distance H between their surfaces. $U_m(H)$ is the electrodynamic part while $U_i(H)$ is the electrostatic part. The principal features of the $U(H)$ curve are: (i) the presence of a primary potential minimum at the particle–particle contact condition of $H = 0$ (requires stripping away the space charge layer), (ii) a secondary potential minimum of small magni-

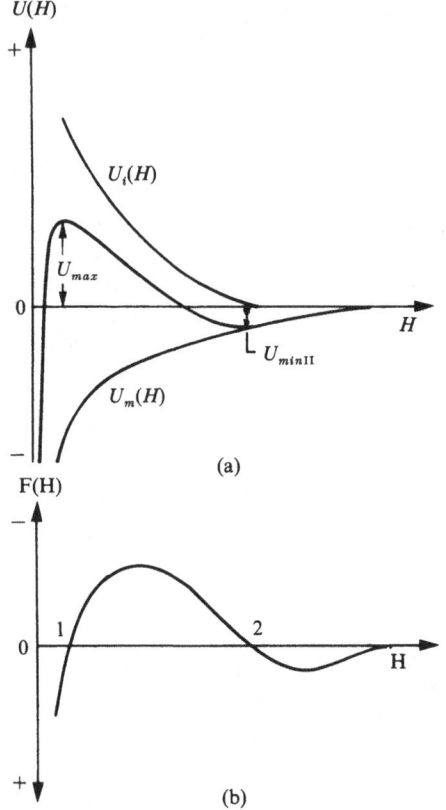

Fig. 6.54. (a) Interaction energy, $U(H)$, and (b) resultant force, $F(H)$, between two electrically charged particles as a function of separation distance H between the surfaces of the particles.

tude at relatively large distance ($H \sim 10^2$–10^3 Å), and (iii) a potential barrier separating the two minima.

For a given situation, the height of the barrier and the depth of the secondary minimum increase with particle radius, a. Thus, one has the picture of colloids in a dilute solution moving randomly via Brownian motion, drifting diffusionally in a long range electrochemical potential gradient (including $U(H)$) and temporarily coalescing into weakly bound clusters where the interparticle separation distances place them in their secondary potential minima. Thermal fluctuations allow relatively easy separation to larger separation distances, because the colloid–colloid binding energy is only a few κT, but difficult passage to smaller separations and the primary potential well minimum unless U_{max} in Fig. 6.54 is small.

For monodisperse colloid sizes and the proper solution concentration of colloids, very large ($\sim$ centimeters), stable and ordered three-dimensional arrays of these particles ($\sim 10^{12}$ particles) bound in the secondary minimum can develop. The small crystallites that are weakly bound to each other in these secondary minima have the ability to freely rotate relative to each other until close crystallographic alignment, and thus a minimum free energy, has been achieved. Under small barrier conditions, a local thermal fluctuation can occur to allow coalescence of a number of these crystallites to form a larger crystal that is a single crystal. We shall see below that a larger crystal (larger a) leads to a larger electrostatic barrier, for constant electrolyte conditions, so that further growth occurs more slowly. At some critical crystal size $a = a^*$, for constant electrolyte conditions, the barrier is so high that no further particle passage across the barrier can occur so the crystal growth stops. In conventional practice, as a increases, the electrolyte double layer from attaching particles is partially or completely returned to the bulk solution so that the bulk electrolyte concentration increases. This decreases the average Debye length around the remaining particles and decreases U_{max}. Because of this factor, a^* would be increased if independent colloid–colloid agglomeration could be stopped; however, it cannot so that the ultimate a^* is reduced.

Experimentally, one can alter $U(H)$ in Fig. 6.54 via two major electrostatic effects which can be illustrated by considering Fig. 6.53(c). The electrostatic repulsive force can be altered by either (i) the addition of a surfactant to raise or lower the surface potential, ϕ_s, or (ii) by altering the electrolyte concentration to increase or decrease the Debye length, λ_D. Increased electrostatic repulsion force occurs with increased $|\phi_s|$ and increased λ_D. Changing the temperature also changes ϕ_s, λ_D, the magnitude of the electrodynamic force and the ability to obtain the energy fluctuation for surmounting U_{max}.

From a simple quantitative viewpoint, the non-retarded electrodynamic force interaction is given by

$$\text{sphere–sphere:} \quad F_1 = \tilde{A}a/12H^2 \qquad (6.33a)$$

$$\text{flat–sphere:} \quad F_2 = -\tilde{A}a/6H^2 \qquad (6.33b)$$

where flat refers to a material half-space with a flat surface (like a crystal) interacting with a sphere (colloid particle) and $\tilde{A}$ is the Hamaker constant given by

$$\tilde{A} = \frac{3h\bar{w}}{4\pi} \tag{6.33c}$$

Here $\bar{w} = \bar{w}_{132}$ is a frequency that combines the dielectric properties of the three materials and is given by Eq. (7.18b) of the companion book;[7] i.e. $1 \equiv$ particle, $2 \equiv$ crystal and $3 \equiv$ separating medium. The electrostatic forces are given, for constant ϕ_s, by

$$\text{sphere–sphere:} \quad F_2 = \frac{1}{2} \frac{\varepsilon_* a \phi_{s1} \phi_{s2} \kappa_D \exp(-\kappa_D H)}{1 + \exp(-\kappa_D H)} \tag{6.34a}$$

$$\text{flat–sphere:} \quad F_2 = \frac{\varepsilon_* a \phi_{s1} \phi_{s2} \kappa_D \exp(-\kappa_D H)}{1 + \exp(-\kappa_D H)} \tag{6.34b}$$

where $\kappa_D = \lambda_D^{-1}$, and $\varepsilon_* = \varepsilon_{*30}$ is the static dielectric permeability of the solution.

The total force for these two cases is

$$F_T = F_1 + F_2$$
$$= \beta a \left[-\frac{\tilde{A}_{132}}{6H^2} + \frac{\varepsilon_{*30} \phi_{s1} \phi_{s2} \kappa_D \exp(-\kappa_D H)}{1 + \exp(-\kappa_D H)} \right] \tag{6.35}$$

where $\beta = 1/2$ for the sphere–sphere interaction and $\beta = 1$ for the flat–sphere interaction. Considering Eq. (6.35), if materials 1 and 2 are identical and U_{max} is large enough that the sphere–sphere interaction leads to a stable dispersion, then F_T for the flat–sphere interaction is twice as large and no crystal growth is possible. If U_{max} is reduced sufficiently for the flat–sphere interaction to allow crystal growth, then a competing reaction is the sphere–sphere agglomeration. On the surface it looks like an impossible situation for controlled crystal growth.

What one wishes to achieve is the ideal potential state illustrated in Fig. 6.55 so that crystal growth can occur without agglomeration also occurring randomly within the solution. There appear to be several chief ways wherein this condition might be achieved for some systems: (1) the flat contains channels of solvent of sufficient volume fraction that $\tilde{A}_{132} \gg \tilde{A}_{131}$ where 1 refers to the sphere and 2 refers to the flat, (2) the flat is actually partially crystalline so that now there is at least some low frequency correlation and perhaps some high frequency correlation between the electromagnetic waves emanating from the crystal (this will definitely increase $\tilde{A}_{132}$), (3) the electrostatic potential, ϕ_s, at the surface of the flat becomes appreciably reduced during growth compared to that for the sphere and (4) the dynamic value of κ_D at the crystal/solution interface, at crystal growth rate V, is reduced compared to that for the sphere/solution interface. These possibilities are not mutu-

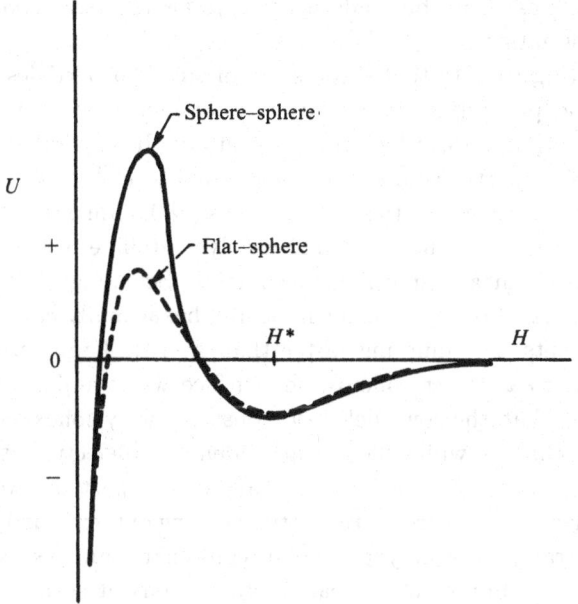

Fig. 6.55. Relative interaction energies, U, with surface separation distance, H, for the cases of sphere–sphere and sphere–flat (colloid–crystal) interactions.

ally exclusive. This coherent component of the electrodynamic force is a new and important concept. The important point for us to recognize is that the radiation intensity for the coherent pattern can be significantly increased by constructive interference so that $\tilde{A}_{132}$ can be appreciably increased. This increase of $\tilde{A}_{132}$ will also increase with the size of the crystal up to some limiting crystal size. One would also expect $\tilde{A}_{132}$ to decrease if angular orientation of the protein molecules were not exact. Thus, lattice point alignments, the molecular orientation alignments and crystal size are all important to the magnitude of $\tilde{A}_{132}$.

In closing this section on biological crystallization of long chain molecules, we find that a delicate balance exists between two competitive processes, sphere–sphere agglomeration and sphere–crystal agglomeration. Unless one continuously adjusts the bulk solution electrolyte content, pH and polymer particle content, rejection of electrolyte and H^+ ions at the growing crystal interface will upset the delicate balance towards favoring the sphere–sphere agglomeration reaction which will effectively stop the crystallization reaction. Interface morphology studies on such crystals can only be made if the crystals are large enough for practical investigation.

It is interesting to note that dispersions of colloidal particles exhibit thermodynamic properties similar to those of molecular systems, including a hard-sphere disorder–order transition. In experiments with organophilic SiO_2 particles in cyclohexane, gravity settling was used to concentrate the particles. With small particles ($\sim 0.3\ \mu m$), the slow sedimentation permits rearrangement into a polycrystalline phase whereas larger particles form an amorphous sediment. Scanning electron microscopy of the crystalline sediment indicated hexagonally close-packed layers. Sediments accumulating faster than a critical rate produce a glass. For these particular colloids, the surface was stabilized with 1-octadecanol so that the low dielectric constant of cyclohexane eliminated electric charges while their nearly identical dielectric constants suppressed the van der Waals forces. Therefore, the particles were able to behave as hard spheres. Since the lattice spacing of these particles was in the optical range, the polycrystalline (columnar) samples exhibited irridescence very reminiscent of opal crystals. Thus, it seems that nature produces a variety of different crystals from solution using colloids as the soluble constituents.

6.4.7   Eutectics

With the unconstrained growth case, one must consider the macroscopic crystal shape as well as the microscopic duplex phase morphology. The macroscopic crystal shape determines the heat transfer part of the problem; i.e., ΔT_T, for a melt with bulk composition of $C_\infty = C_{\tilde{E}}$. When the bulk melt composition is off the eutectic composition, then the macroscopic shape also determines ΔT_C (plane wave type of term). Just as in the constrained growth case, it is the microscopic morphology and spacing that determine ΔT_i and one can utilize the extremum condition to produce λ^* in exactly the same fashion as Eqs. (6.13)–(6.15). For this case we find that

$$\Delta T_i = \Delta T_\infty - \Delta T_T - \Delta T_C \qquad (6.36)$$

In Eq. (6.36), only ΔT_i is a function of λ so that application of the extremum condition to minimize ΔT_i leads to a maximum value of ΔT_T (ΔT_C is proportional to ΔT_T) and thus a maximum V for a given particle radius. All of the microstructure considerations given earlier apply here to produce $\lambda^* = f(V)$ but now V is determined by the solution of the macroscopic heat flow and matter transport leading to $\hat{P}_T = g(\Delta T_\infty)$ where f and g are two functions and $\hat{P}_T$ is the thermal Péclet number.

It can be readily shown that macroscopically isothermal and isoconcentrate bodies yield $\Delta T_T \propto \Delta T_C$ by considering the dimensionless heat and mass transport equations with their associated boundary conditions. Defining $\psi = C/C_\infty$ or T/T_∞, the appropriate Péclet number as $\hat{P}_u = u\rho/2D_C$ or $u\rho/2D_T$ for the relative velocity, u, and the coordinate transformations $t' = D_C t/\rho^2$ or $D_T t/\rho^2$, $x' = x/\rho$, $y' = y/\rho$, $z' = z/\rho$ and $n' = n/\rho$, the appropriate equations are of the same form and are given by

$$\nabla'^2\psi + 2\hat{P}_u\nabla'\psi = \frac{\partial\psi}{\partial t'} \tag{6.37a}$$

with the BCs

$$\psi(x',y',z',0) = 1 \tag{6.37b}$$

$$\psi(\infty,\infty,\infty,t) = 1 \tag{6.37c}$$

$$g\hat{P}_u\psi_i = -\left(\frac{\partial}{\partial n'}\psi\right)_i \tag{6.37d}$$

where $g = 1 - k_0$ or $\Delta S_F/c$ and $\hat{P}_u = V_n\rho/2D_C = \hat{P}_C$ or $V_n\rho/2D_T = \hat{P}_T$ for the solute and heat problem, respectively. Eq. (6.37d) is not exactly correct for the heat flow problem; however, the per cent error for a pure material is only $(T_M - T_\infty)/T_\infty$ which is small in most cases of interest.

For the simple shapes of interest, the solution to Eqs. (6.37) has the form

$$\frac{\psi - 1}{\psi_i - 1} = \frac{\phi(\omega)}{\phi(\hat{P}_u)} \tag{6.38a}$$

where $\omega = \hat{P}_u(r/\rho)$, $\hat{P}_u = \hat{P}_T$ or $\hat{P}_C$ for the heat and solute profile, respectively, $\phi(\hat{P}_u)$ is of the form given by Eqs. (4.21) for the different simple shapes and ψ_i is given by

$$\psi_i - 1 = g\phi(\hat{P}_u) \tag{6.38b}$$

Using Eq. (6.38b), we have

$$\frac{\Delta T_C}{\Delta T_T} = \frac{-m_L C_\infty(\psi_i - 1)_C}{T_\infty(\psi_i - 1)_T} \tag{6.39a}$$

$$= \frac{-m_L C_\infty[(1 - k_0)\phi(\hat{P}_C)]}{T_\infty[(\Delta S_F/c)\phi(\hat{P}_T)]} \tag{6.39b}$$

In Fig. 6.56, the $\hat{P}_u$ versus $\varepsilon\phi_i$ plot for various ellipsoidal shapes refers to the matter transfer case when $\varepsilon = (1 - k_0)^{-1}$ and to the heat transfer case when $\varepsilon = c/\Delta S_F$. Thus, $\varepsilon\phi_i$ becomes $\Delta C_i/C_\infty(1 - k_0)$ or

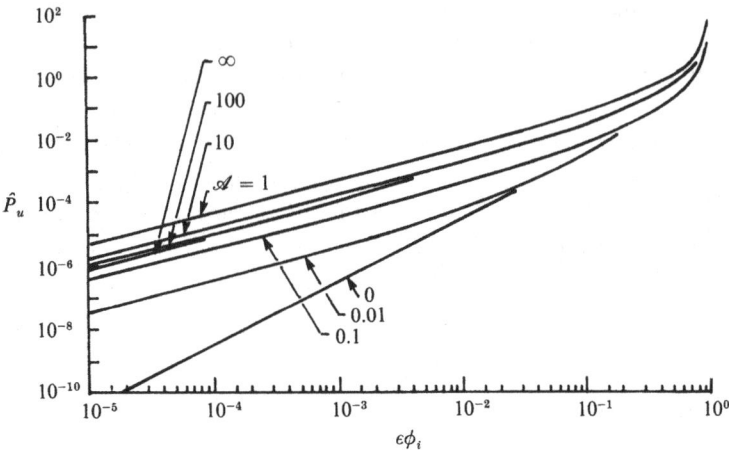

Fig. 6.56. Relationship between Péclet number, $\hat{P}_u$, and the dimensionless interface potential ϕ_i of a spheroid for various aspect ratio parameter $\mathcal{A} = [(\bar{a} + \hat{P}_u)/(a + \hat{P}_u)]^{1/2}$. $\mathcal{A} = 1$ represents the sphere $(a = \bar{a} = 0)$, $\mathcal{A} = \infty$ represents the circular cylinder $(a = 0, \bar{a} = \infty)$ and $\mathcal{A} = 0$ represents the plane $(a = \infty, \bar{a} = 0)$.

$(c/\Delta H_F)\Delta T_i$ for these two cases so that values of C_i, T_i, ΔC_i and ΔT_i can be readily determined for the sphere, cylinder, plate, etc., shape of the growing crystal as a function of the appropriate Péclet number. The asymptotic limits of these plots are often very useful and are given by

sphere: $\qquad \hat{P}_u \to \infty, \quad \phi(\hat{P}_u) \to 1 - 3/2\hat{P}_u + \cdots$ $\qquad\qquad$ (6.40a)

$\qquad\qquad\;\; \hat{P}_u \to 0, \quad \phi(\hat{P}_u) \to 2\hat{P}_u - 2\pi^{1/2}\hat{P}_u^{3/2} + \cdots$ $\qquad\quad$ (6.40b)

cylinder: $\quad \hat{P}_u \to \infty, \quad \phi(\hat{P}_u) \to 1 - 1/\hat{P}_u + \cdots$ $\qquad\qquad\;$ (6.40c)

$\qquad\qquad\;\; \hat{P}_u \to 0, \quad \phi(\hat{P}_u) \to \hat{P}_u \ell n(1.78\hat{P}_u)$ $\qquad\qquad\;\;\,$ (6.40d)

slab: $\qquad\;\; \hat{P}_u \to \infty, \quad \phi(\hat{P}_u) \to -1/2\hat{P}_u + \cdots$ $\qquad\qquad\;$ (6.40e)

$\qquad\qquad\;\; \hat{P}_u \to 0, \quad \phi(\hat{P}_u) \to (\pi\hat{P}_u)^{1/2} - 2\hat{P}_u + \pi^{1/2}\hat{P}_u^{3/2} + \cdots$

$\qquad\qquad\qquad\qquad\qquad\qquad\qquad\qquad\qquad\qquad\qquad\qquad$ (6.40f)

To make these results more tangible, let us consider the independent particle growth mode in the eutectic region of the Fe–C system. In this case, one is interested in comparing the graphite sphere versus plate growth (nodular versus flake graphite), the graphite vs. cementite growth (since $T_{\hat{E}}$ for Fe–Fe$_3$C is only a few degrees below $T_{\hat{E}}$ for Fe–C) and the effect of ternary additions on the dominant particle morphology. We shall use the parameter information listed in Table 6.2.

Table 6.2. *Material parameters for the Fe–C system.*

Parameter	Graphite (C)	Cementite (Fe₃C)	Austenite (Fe)	Fe–C	Fe–Fe₃C	Liquid
k_0^C	5.9	1.45		1.0	1.0	
k_0^{Si}				1.6		
$m(°C\ (\text{at}\%)^{-1})$	139	12	−27			
$T_E(°C)$				1135	1130	
C_∞^C	100	25				17.1
$D_C^C(\text{cm}^2\ \text{s}^{-1})$						10^{-4}
$D_C^{Si}(\text{cm}^2\ \text{s}^{-1})$						10^{-4}
$D_T(\text{cm}^2\ \text{s}^{-1})$						0.5
$\Delta H_F(\text{cal cm}^{-3})$	1500	550	500	600	550	
$\Delta H_F/c(°C)$	7.5×10^3	3×10^3				
$\gamma_{\alpha\beta}(\text{erg cm}^{-2})$				500	500	
mC_∞	2.4×10^3	0.2×10^3				

In Fig. 6.57(a), the basic sphere growth information is presented for the specific diffusivity ratio $D_T/D_C = \hat{P}_C/\hat{P}_T = 5 \times 10^3$. The effect of a different value of D_T/D_C can be seen with $k_0 = 1.6$ and $D_T/D_C = 10^3$. In Fig. 6.57(b) the basic information is given for a laterally growing platelet shaped like a parabolic cylinder. Since $|mC_\infty|$ is so large for this system, we expect $\Delta T_C \gg \Delta T_T$ and $\hat{P}_T \ll 1$. Thus, neglecting ΔT_K effects and using Eq. (6.39b) with $|m_L C_\infty| \approx 2.4 \times 10^3$ and Eq. (6.40b), we expect $\Delta T_\infty \approx \Delta T_C \approx 12 \times 10^7 \hat{P}_T$ leading to $\hat{P}_T \sim 10^{-7}$ for $\Delta T_\infty \sim 10\,°C$. Since $D_T \approx 0.5$ for this melt, this leads to $R\dot{R} \sim 10^{-7}$ for a growing graphite sphere. Taking the approximation of applying the extremum condition to the growing graphite plate, V^* and ρ^* are given in Fig. 6.58. We note that $V^* \sim 10^{-5}$ cm s^{-1} and $\rho^* \sim 5 \times 10^{-6}$ cm for $\Delta T_\infty \sim 10\,°C$ (applying the surface creation constraint instead of the extremum condition would have increased ρ^* and decreased V^*). For the growing sphere, it attains this velocity of 10^{-5} cm s^{-1} at $R \sim 10^{-2}$ cm and thereafter grows more slowly. Since $2R\dot{R} = d(R^2)/dt \sim 2 \times 10^{-7}$, integration shows us that this value of R is attained in ~ 500 s if no appreciable convection exists in the melt. After this time, the lateral growth of the plate would be expected to dominate.

From experimental observations, one generally observes flakes and not spheres from the Fe–C system and this indicates that attachment kinetics

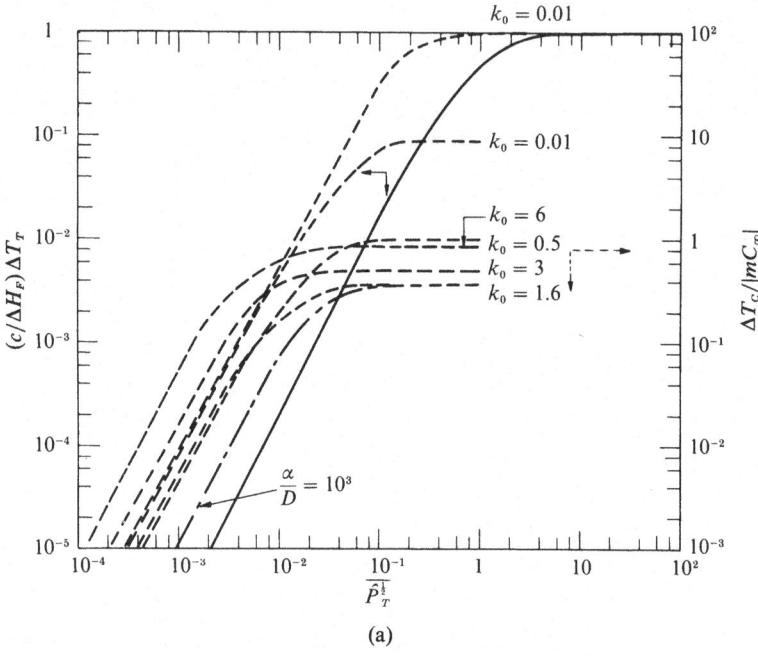

Fig. 6.57. Plots of dimensionless supercoolings $(c/\Delta H_F)\Delta T_T$ and $\Delta T_C/|mC_\infty|$ as a function of thermal Péclet number, $\hat{P}_T^{1/2}$ for several values of k_0 and with $D_T/D_C = 5 \times 10^3$: (a) spherical particle shape.

are a very important factor in making the C-axis growth become very sluggish so that transition from the sphere to the plate form occurs early in the particle's growth. This is thought to be associated with the adsorption of S on the C-face. In addition, one expects that fluid convection is playing a role in mass transport because $\hat{P}_T$ is so small and because the calculated value of V^* leads to too small a flake diameter in reasonable casting times.

Comparing cementite and graphite plates, Fig. 6.58 shows us that, for the same ΔT_∞, $V_{cem} \gg V_{graph}$. Thus, so long as $\Delta T_C(cem) \overset{\sim}{>} 0.04\Delta T_\infty$ (graph), $V_{cem} > V_{graph}$ and independent cementite plate growth will dominate over independent graphite plate growth. In general, when comparing these two possibilities, the graphite plate growth will only dominate until enough supercooling develops for cementite to be a thermodynamically stable phase. Once cementite nucleates, it will grow very rapidly and dominate the phase transformation.

When one includes the cooperative eutectic morphology, one finds

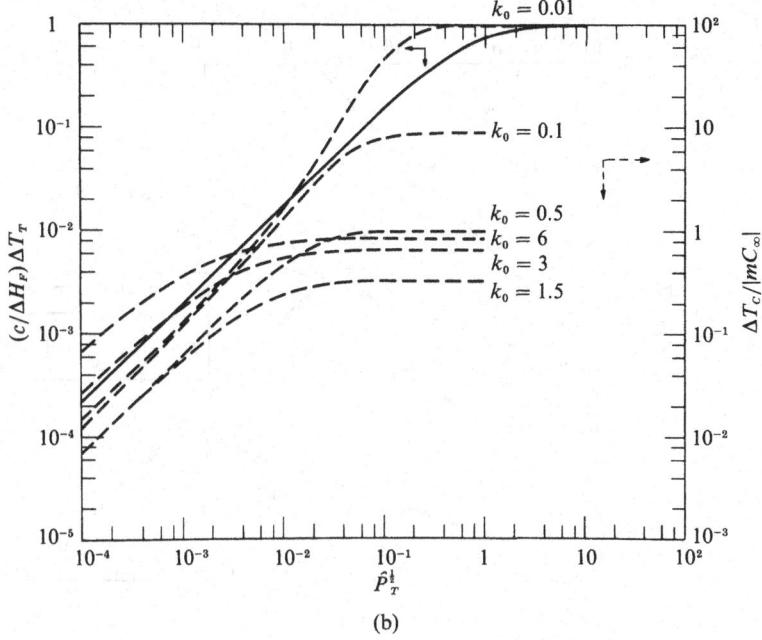

Fig. 6.57. (b) Parabolic cylinder (flake) particle shape.

that, provided $\Delta T_K \ll \Delta T_\infty$, the spherical form of graphite is the dominant form at small ΔT_∞ ($\Delta T_\infty \sim 1\,°C$) and the domain of its dominance can be extended to larger ΔT_∞ by going to hypereutectic concentrations. At larger values of supercooling, the cooperative growth of the Fe–C lamellar eutectic is the dominant form ($\sim 1\,°C < \Delta T_\infty \stackrel{<}{\sim} 7\,°C$). Beyond $\Delta T_\infty \sim 7\,°C$, the cooperative growth of the Fe–Fe$_3$C lamellar eutectic is the dominant form.[37] When adsorption occurs to make ΔT_K a significant factor, the flake graphite form replaces the spherical form and the Fe–Fe$_3$C eutectic begins to be dominant at temperatures just below its eutectic temperature.

From studies on Ni–C eutectics, it has been found that the addition of radioactive S stabilizes the plate-like form of graphite and a high concentration of S was found in the flakes. With the addition of Mg to the melt, all of the S was taken up in the form of particles of an Mg–S compound, the spherical form of graphite developed and no S was found in the spheres. If insufficient Mg was added to scavenge all the S from the melt, the plate-like form of graphite again became the stable form and the plates contained a high S-content.

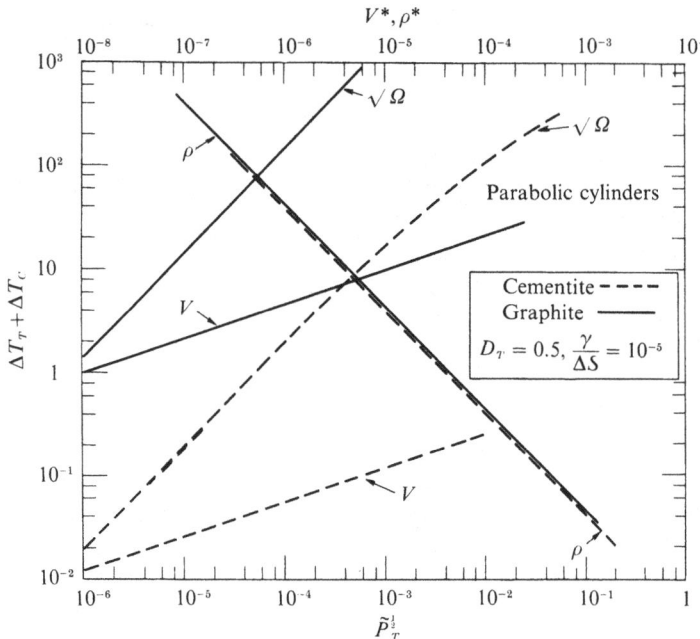

Fig. 6.58. Comparison of V, ρ and $\hat{P}_T^{1/2}$ as a function of partial supercooling, $\Delta T_T + \Delta T_C$, for plate-like particles of cementite and graphite.

The addition of a ternary element to an Fe–C melt can affect the growth morphology and phase constitution in one of several ways. It may change the number of potential nucleus-forming centers; it adds its own ΔT_C^m contribution to ΔT_∞; it may change ΔT_i via a change in $C_{\tilde{E}}$ or the ks; it may change the state of surface adsorption and thus ΔT_K and it may change the difference $\Delta T_{\tilde{E}}$ between the Fe–C and the Fe–Fe$_3$C eutectic temperatures. One must determine exactly which of these five possible effects are important in a particular instance in order to control the kinetics and morphology of particle growth meaningfully and thus the resulting structure of the solid.

7

Physical defect generation during bulk crystal growth

7.1   Dislocation formation mechanisms

A variety of mechanisms exist to account for the introduction of dislocations into crystals grown from the melt: (1) thermal stresses, (2) mechanical stresses, (3) vacancy and interstitial supersaturation and (4) constitutional stresses. The density of dislocations in the growing crystal can also be augmented by certain multiplication and annihilation mechanisms and their interaction can lead to the formation of dislocation networks and boundaries in the crystal.

No matter what the mechanism of stress, σ, introduction into the crystal, if dislocation sources are readily available, the stress converts some of the total strain, $\tilde{\varepsilon}$, to plastic strain depending upon the yield stress, σ_y, of the material. The amount of plastic strain which leads to the introduction of dislocations is given by $\tilde{\varepsilon}_p = (\sigma - \sigma_y)/E^*$, where E^* is Young's modulus, so that the number of dislocations per square centimeter, n_D, introduced by the plastic strain is

$$n_D \sim \frac{\tilde{\varepsilon}_p}{b_*} = 2\frac{(\sigma - \sigma_y)}{E^* b_*} \qquad (7.1)$$

where b_* is the Burger's vector of the dislocations and the factor of 2 enters because a crossed grid is needed to satisfy the strain over a certain surface area. Since the yield stress near the melting point, T_M, is approximately zero for metals but may be reasonably large for covalent or ionic solids, as illustrated in Fig. 7.1, many more dislocations will be introduced into metal crystals by a particular stress, σ, than into non-metal crystals. Of course, if no dislocation sources are present for the

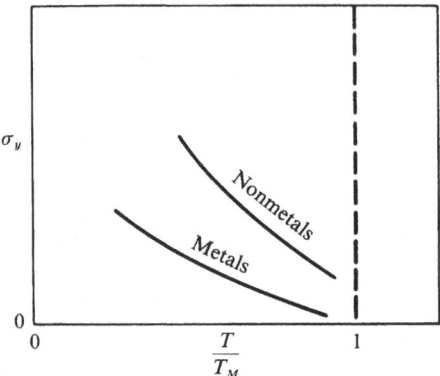

Fig. 7.1. Schematic variation of the yield stress for metals and non-metals with respect to the reduced temperature T/T_M.

generation of dislocations, σ_y may assume very large values and the elastic stress, σ, may not be relieved by plastic flow.

7.1.1   *Thermal stresses*

If the crystal has a temperature distribution such as illustrated in Fig. 7.2, the outside of the crystal wants to contract but cannot do so fully because of the higher temperature interior; thus, the outside will be in tension while the inside will be in compression. If the temperature difference between the inside and outside is ΔT^*, the thermal stress is $\bar{\alpha}^* \Delta T^* E^*$ where $\bar{\alpha}^*$ is the thermal expansion coefficient.

For Si crystal growth in the $\langle 111 \rangle$ direction, three illustrative temperature distributions were considered by Penning[1]: (i) the zero stress case ($\sigma = 0$) occurs when $\{\partial T_S/\partial r = 0$ and $\partial T_S/\partial z = \text{constant}\}$, (ii) the crystal axis-symmetric stress case occurs when $\{\partial T_S/\partial r \neq 0$ and $\partial T_S/\partial z = 0\}$ which would lead to hexagonal slip traces on $\langle 111 \rangle$ slices and zero perpendicular slip traces on longitudinal crystal slices containing the crystal axis (see Fig. 7.3) and, (iii) the non-symmetric stress case occurs when $\{\partial T_S/\partial r \neq 0$ and $\partial T_S/\partial z \neq \text{constant}\}$ which would lead to perpendicular slip traces on longitudinal sections and a trend towards triangular slip traces on the $\langle 111 \rangle$ slices (see Fig. 7.3). The thermal stress increases as the thermal conductivity of the solid decreases. This dislocation generation mechanism is especially important for metal crystals and can lead to values of $n_D \sim 10^4$–10^5 cm^{-2} in Si if dislocation sources are present.

Penning[1] analyzed the stress distribution in a cylindrical crystal of

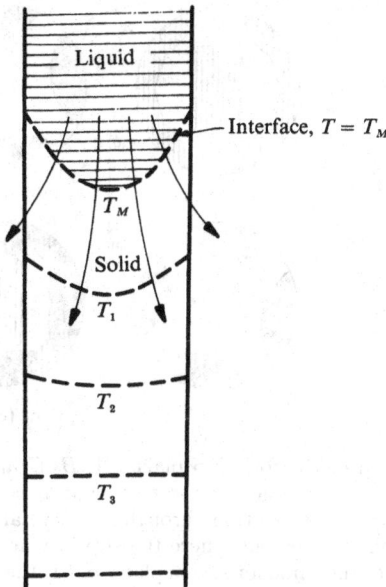

Fig. 7.2. Possible isotherm shape in a crystal during freezing and cooling.

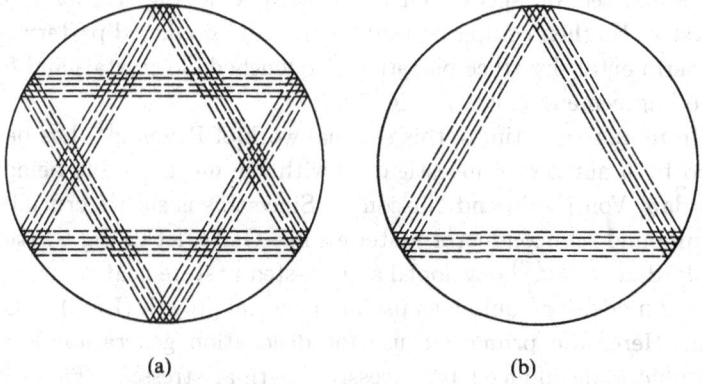

Fig. 7.3. Predominant slip traces on a section perpendicular to a ⟨111⟩ specimen axis for either a DC or FCC crystal: (a) non-zero radial temperature gradient and constant axial gradient and (b) non-zero radial gradient and non-linear axial gradient.

diamond cubic material that was quenched by cooling the surface. Then, for different crystal axes, he evaluated the relative amounts of slip that would occur on the various slip planes so that he could predict the

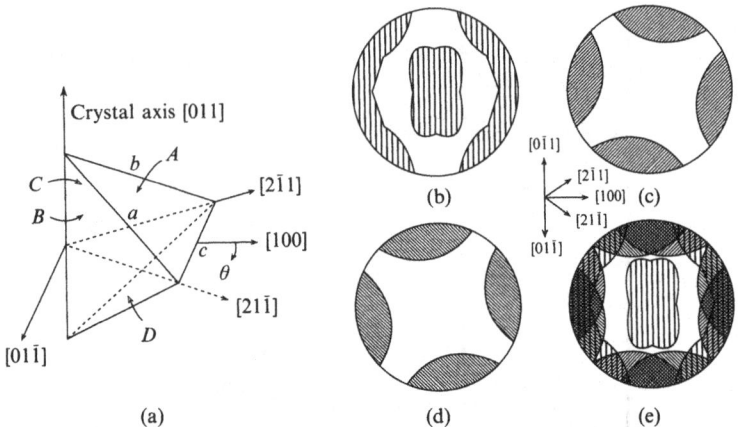

Fig. 7.4. (a) Orientation of the four slip planes A, B, C and D for the $\langle 110 \rangle$-axis DC crystal: the edges of the tetrahedron correspond to the slip directions, (b) a cross-section through the crystal wherein the shaded areas represent the regions where the slip in plane A or D exceeds a certain value, (c) the same for slip in plane B, (d) the same for slip in plane C and (e) superposition of the (a), (b) and (c) results.[1]

pattern and density of slip traces on both cross-sectional slices and longitudinal section slices. This is illustrated in Fig. 7.4 for a [011]-axis crystal. He then compared the theoretically predicted patterns with the experimental slip trace patterns for quenched Ge crystals and found the good agreement shown in Fig. 7.5.

In more recent times, this seminal work of Penning[1] has been revisited by a number of investigators with the most notable being that by Jordan, Von Neida and Nielson[2]. Since this is such a crucial issue for compound semiconductor materials like GaAs, let us dig a little deeper.

Jordan *et al.*[2,3] developed a quasi-steady state heat transfer/thermal stress model for liquid encapsulated crystal growth (LEC) of GaAs and InP. Here, the primary cause for dislocation generation is crystallographic glide induced by excessive thermal stresses. They developed a closed-form temperature profile in a growing CZ crystal boule under the assumptions: (1) a cylindrical crystal of length L and radius R_c growing in time t at a velocity V with a planar interface at temperature T_i and (2) at the top and lateral surfaces, cooling is via natural convection and radiation so the heat flux is proportional to $T_s - T_a$ where T_a is the ambient temperature. A separation of variables solution for $T(r, z)$ was developed where z is the distance into the crystal from the

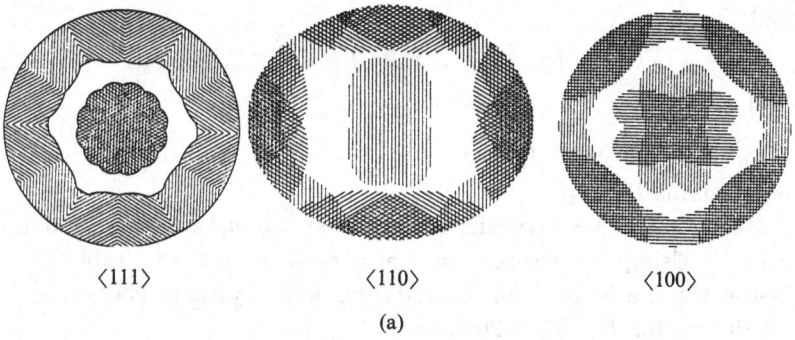

⟨111⟩   ⟨110⟩   ⟨100⟩

(a)

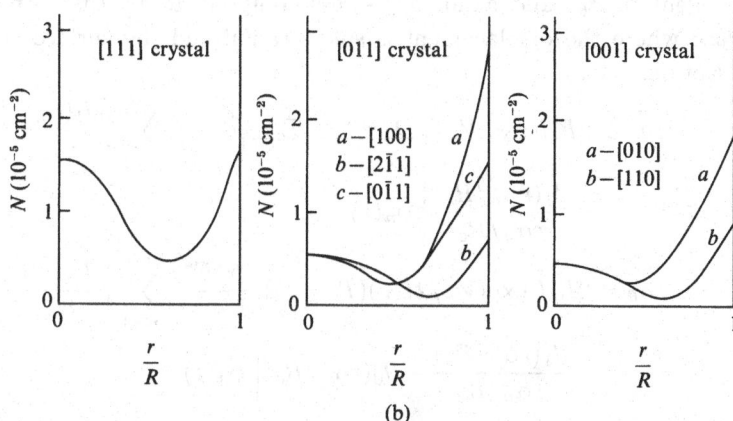

(b)

Fig. 7.5. (a) Calculated patterns of slip for the ⟨110⟩, ⟨111⟩ and ⟨100⟩, assuming that the major part of the total thermal strain during radial quenching is elastic and (b) density of etch pits in Ge crystals after (a).[1]

interface. It is given by

$$T - T_a = 2R_c \tilde{h}(T_i - T_a) \exp(Vz/2D_T) \sum_{n=1}^{\infty} J_0 \left(\frac{r}{R_c} \alpha_n \right) \phi_n(z) \quad (7.2a)$$

where

$$\phi_n(z) = \frac{1}{J_0(\alpha_n)[(R_c\tilde{h})^2 + \alpha_n^2]}$$

$$\times \left\{ \frac{\beta_n \cosh\left[(\beta_n/R_c)(Vt - z)\right] + \hat{\Gamma} \sinh\left[(\beta_n/R_c)(Vt - z)\right]}{\beta_n \cosh\left[(\beta_n/R_c)Vt\right] + \hat{\Gamma} \sinh\left[(\beta_n/R_c)Vt\right]} \right\}$$

$$(7.2b)$$

and

$$\beta_n^2 = \alpha_n^2 + \hat{P}_T \quad ; \quad \hat{\Gamma} = R_c \tilde{h} + \hat{P}_T \tag{7.2c}$$

with

$$R_c \tilde{h} J_0(\alpha_n) = \alpha_n J_1(\alpha_n) \tag{7.2d}$$

determining the α_n.

Here, $\tilde{h}$ is the heat transfer coefficient at the surfaces and $z < 0$ in the crystal. Using the selected physical parameters given in Table 7.1, the isotherms in a 5 cm diameter and 5 cm long crystal of GaAs are given in Fig. 7.6 for $T_i - T_a = 200$ and 20 K.

Classical thermoelastic theory was then used to provide radial, σ_r, tangential, σ_{θ_1} and axial, σ_{z_1} stress components for the plane strain case where the displacement is solely radial and the surface is free of tractions; i.e.,

$$\sigma_r = 2R_c \tilde{h} \exp(Vz/2D_T)(T_i - T_a) \frac{\bar{\alpha}^* E^*}{(1 - \sigma^*)} \sum_{n=1}^{\infty} \left[\frac{J_1(\alpha_n)}{\alpha_n} \right.$$

$$\left. - \frac{J_1(r\alpha_n/R_c)}{r\alpha_n/R_c} \right] \phi_n(z) \tag{7.3a}$$

$$\sigma_\theta = 2R_c \tilde{h} \exp(Vz/2D_T)(T_i - T_a) \frac{\bar{\alpha}^* E^*}{(1 - \sigma^*)} \sum_{n=1}^{\infty} \left[\frac{J_1(\alpha_n)}{\alpha_n} \right.$$

$$\left. + \frac{J_1(r\alpha_n/R_c)}{r\alpha_n/R_c} - J_0(r\alpha_n/R_c) \right] \phi_n(z) \tag{7.3b}$$

and

$$\sigma_z = 2R_c \tilde{h} \exp(Vz/2D_T)(T_i - T_a) \frac{\bar{\alpha}^* E^*}{(1 - \sigma^*)} \sum_{n=1}^{\infty} \left[\frac{2J_1(\alpha_n)}{\alpha_n} \right.$$

$$\left. - J_0(r\alpha_n/R_c) \right] \phi_n(z) \tag{7.3c}$$

Slip activated in the $\{111\}$, $\langle 1\bar{1}0 \rangle$ system leads to 12 permissible glide operations when the resolved shear stress, σ_{RS}, $> \sigma_{CRS}$, the critical resolved shear stress which is an intrinsic material property. Of course, this assumes the existence of readily available dislocation sources.

To calculate σ_{RS} requires the conversion of σ_r, σ_θ and σ_z to σ_{xy}, σ_{xz} and σ_{yz} where the pulling orientation is identified with the z-axis. A standard tensor transformation is used to derive σ_{RS} on a particular slip plane in a given direction. Table 7.2 gives the orientation dependence of σ_{RS} for the $\langle 100 \rangle$ and $\langle 111 \rangle$ pulling directions. Here, θ denotes the angle between an arbitrary vector in the plane of the wafer and the $\langle 100 \rangle$ or the $\langle 1\bar{1}0 \rangle$ for the $\langle 001 \rangle$ and $\langle 111 \rangle$ growth, respectively. It was assumed,

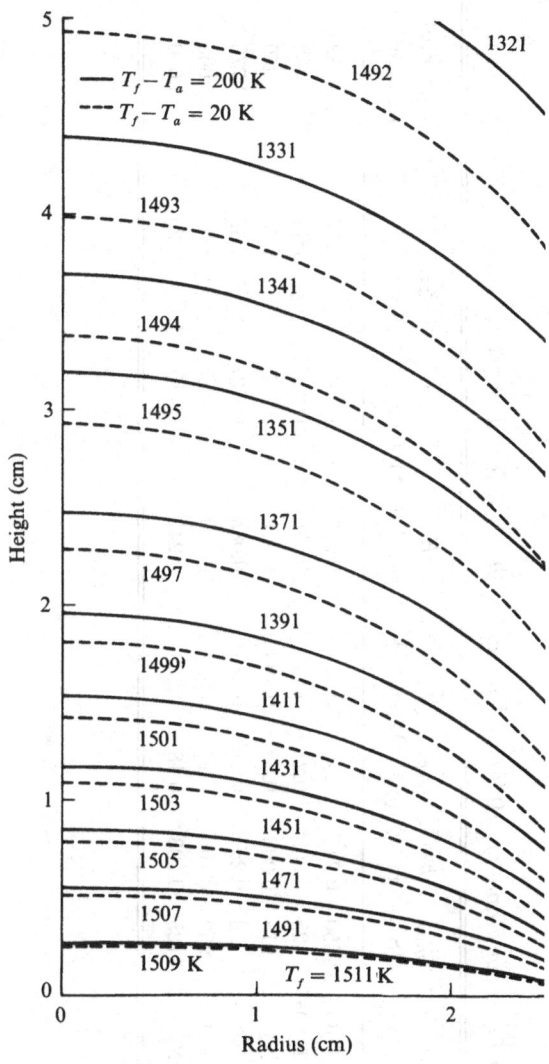

Fig. 7.6. Isotherms in LEC GaAs boule pulled at $V = 4 \times 10^{-4}$ cm s^{-1} ($L = 2R_c = 5$ cm).[2]

as did Penning[1], that most of the stress was elastic with only a small constant fraction being released by plastic flow. Thus, for any one of the 12 slip systems, the generated dislocation density is proportional to the amount of slip or glide strain and the total generated dislocation density is proportional to the total glide strain which is proportional to

Table 7.1. *Parameters for 4 cm diameter, 10^{16} cm^{-3} n-type boules at selected temperatures.*[2]

Crystal	T_a (K)	$T_i - T_a$ (K)	Ambient	t_{max} (S)	$\tilde{h}_c$ (cm^{-1})	$\tilde{h}_r$ (cm^{-1})	$\tilde{h}_{tot}$ (cm^{-1})	K (W cm^{-1} K^{-1})	D_T (cm^2 s^{-1})	$\frac{\bar{\alpha}E^*}{1-\sigma^*}\{111\} \times 10^7$ (dyn cm^{-2} K^{-1})	$\sigma_{\text{CRS}} \times 10^7$ (dyn cm^{-2})
GaAs	1511	40	B$_2$O$_3$	650	0.24	0.57	0.81	0.073	0.040	1.53	0.583
InP	1343	40	B$_2$O$_3$	825	0.16	0.28	0.44	0.096	0.055	1.11	0.362
Si	1683	40	N$_2$	975	0.003	0.29	0.293	0.213	0.110	1.30	1.850
GaAs	1511	200	B$_2$O$_3$	750	0.24	0.34	0.58	0.083	0.040	1.48	0.775
InP	1343	200	B$_2$O$_3$	950	0.16	0.16	0.32	0.110	0.055	1.06	0.665
Si	1683	200	N$_2$	1125	0.003	0.18	0.183	0.242	0.110	1.20	3.610

the total excess shear stress, σ_{ex}, given by

$$\langle 100 \rangle\text{-axis}: \quad \sigma_{ex} = 4\sigma_1^e + 2(\sigma_2^e + \sigma_3^e + \sigma_4^e + \sigma_5^e) \quad (7.4a)$$

$$\langle 111 \rangle\text{-axis}: \quad \sigma_{ex} = \sigma_1^e + \sigma_2^e + \sigma_6^e + 2(\sigma_3^e + \sigma_4^e + \sigma_5^e) \quad (7.4b)$$

where

$$\sigma_i^e = |\sigma_i| - \sigma_{CRS} \quad \text{for} \quad |\sigma_i| > \sigma_{CRS} \quad (7.4c)$$

$$= 0 \quad \text{for} \quad |\sigma_i| \stackrel{\sim}{<} \sigma_{CRS} \quad (7.4d)$$

This procedure leads to dislocation density contours of the form $r = f(z, \theta, t, \sigma_{ex})$, some of which are illustrated in Fig. 7.7 for the top wafer of a $\langle 100 \rangle$GaAs crystal at $T_i - T_a = 200$ K. For such $\langle 001 \rangle$ crystals, we note in agreement with experiment, (a) a four-fold symmetry in n_D, (b) a maximum in n_D at the $\langle 100 \rangle$ edge, (c) a minimum in n_D on the annulus at the $\langle 110 \rangle$ sites and (d) intermediate n_D at the crystal axis. Some of this information is present in Fig 7.5.

In this modeling, $\tilde{h}$ has a profound effect on n_D, especially at the crystal periphery where the dislocations first appear to arise. From Table 7.1, we note that $\tilde{h}_{conv}(\tilde{h}_c)$, due to the B_2O_3 encapsulant, is a significant portion of the total $\tilde{h}$ (suggesting the need for an afterheater to increase the effective T_a). One very important finding was that σ_{ex} at the top wafer went to a maximum after only ~ 1 cm growth had occurred. Further growth led to a reduction of σ_{ex} at the top but the already frozen-in dislocation density at the maximum stress level will not decline with time even though σ_{ex} in that region declines. For different orientations and growth conditions, one can define a critical time, t_c, at which σ_{ex} becomes a maximum and, as $\tilde{h}$ decreases, σ_{ex} will also decrease because $\mathcal{G}_S$ decreases and t_c will increase. It is interesting to note that, for InP$\langle 111 \rangle$, raising the B_2O_3 column height from 7 to 30 mm lowered the radial temperature gradient from 80 to 20 K cm^{-1} and n_D along the $\langle 211 \rangle$-direction was lowered from $\sim 10^5$ to $\sim 10^4$ cm^{-2}. To reduce $\tilde{h}_{conv}$ via the B_2O_3, one should consider (1) either coloring the B_2O_3 with an additive, which should reduce radial temperature gradients and thus natural convection, or using a stiffer solution such as B_2O_3–SiO_2, (2) adding some low conductivity colloidal material to increase the effective viscosity of the B_2O_3 without damaging the GaAs surface and (3) using a magnetic field to constrain movement in the B_2O_3 because it is probably a good ionic conductor at these temperatures.

It is found that, for small diameter crystals, $\sigma_{ex} \propto R_c$ but in the larger crystal diameter ranges ($R_c > 2.5$ cm), σ_{ex} becomes sublinear. Surprisingly, the pull rate, V, has only a small effect on σ_{ex} and n_D

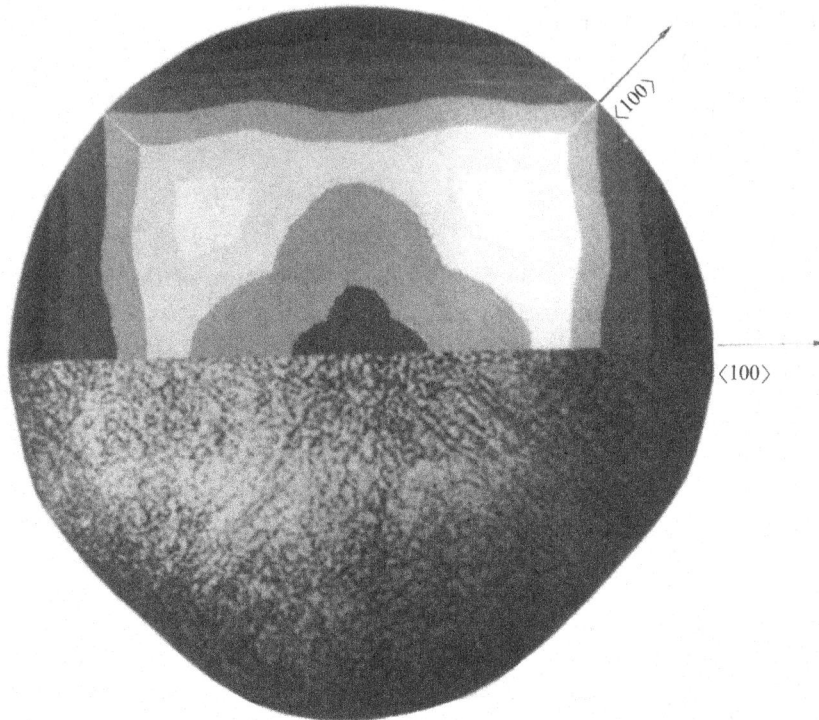

Fig. 7.7. Grey scale representation of dislocation density for the top wafer of a $\langle 100 \rangle$ GaAs boule (top) and macrophotograph (bottom) of a 3.75 cm diameter KPH-etched $\langle 100 \rangle$ GaAs wafer grown by the LEC technique ($\sim 3.5\times$). (Courtesy of A. Jordan.)

in that GaAs boules grown at a slower rate are under only $\sim 15\%$ less stress than one pulled at a faster rate. As $T_i - T_a \rightarrow 20$ K, substantial defect-free areas appear centrally in both 5 cm and 7.5 cm diameter GaAs boules even though n_D at the periphery rises with increase of R_c. For these low gradients, the larger is R_c, the larger is the size of the dislocation-free core. In LEC InP, small diameter crystals can be maintained dislocation-free but above a critical diameter, d_c, which is characteristic for a given ambient temperature gradient, $\mathcal{G}_a$, dislocations are always generated at the outer surface ($d_c = 0.9$ and 1.5 cm for $\mathcal{G}_a = 175$ and 55 K cm^{-1}, respectively). At any diameter, lowering $\mathcal{G}_a$ lowers n_D nearly proportionately; i.e., for $\mathcal{G}_a \tilde{>} 150$ °C cm^{-1}, $n_D \sim (2\text{--}8) \times 10^4$ cm^{-2} and for $\mathcal{G}_a \sim 20$ °C cm^{-1}, $n_D \sim (5\text{--}8) \times 10^3$ cm^{-2} along the $\langle 110 \rangle$ at $R_c = 2.5$ cm. For appreciably lower values of $\mathcal{G}_a$, diameter

Table 7.2. *Orientation dependence of resolved shear stress.*[2,3]

Pulling direction		⟨111⟩	⟨100⟩
Slip plane	Slip direction	Resolved shear stress	Resolved shear stress
(111)	[$\bar{1}$10]	0	$\sigma_1 = -\dfrac{\sqrt{6}}{6}\bar\sigma_r \cos 2\theta$
(111)	[10$\bar{1}$]	0	$\sigma_4 = -\dfrac{\sqrt{6}}{6}\left[\bar\sigma_z - \bar\sigma_r \dfrac{2}{\sqrt{2}}\cos\theta\sin\left(\theta+\dfrac{\pi}{4}\right)\right]$
(111)	[0$\bar{1}$1]	0	$\sigma_2 = \dfrac{\sqrt{6}}{6}\left[\bar\sigma_z - \bar\sigma_r \dfrac{1}{\sqrt{2}}\sin\theta\sin\left(\theta+\dfrac{\pi}{4}\right)\right]$
($\bar{1}$11)	[0$\bar{1}$1]	$\sigma_1 = \dfrac{\sqrt{6}}{9}\bar\sigma_r\left[-\dfrac{3}{2}+2\sqrt{3}\sin\theta\sin\left(\theta+\dfrac{\pi}{6}\right)\right]$	$\sigma_3 = \dfrac{\sqrt{6}}{6}\left[\bar\sigma_z - \bar\sigma_r \dfrac{2}{\sqrt{2}}\sin\theta\sin\left(\theta-\dfrac{\pi}{4}\right)\right]$
($\bar{1}$11)	[110]	$\sigma_3 = \dfrac{\sqrt{6}}{9}\left[\bar\sigma_z - 2\bar\sigma_r \sin\theta\left[\sqrt{3}\sin\left(\theta+\dfrac{\pi}{6}\right)-\sin\theta\right]\right]$	σ_1
($\bar{1}$11)	[$\bar{1}$0$\bar{1}$]	$\sigma_5 = -\dfrac{\sqrt{6}}{9}\left[\bar\sigma_z - 2\bar\sigma_r \left[\dfrac{3}{4}-\sin^2\theta\right]\right]$	$\sigma_5 = -\dfrac{\sqrt{6}}{6}\left[\bar\sigma_z + \bar\sigma_r \dfrac{2}{\sqrt{2}}\cos\theta\sin\left(\theta-\dfrac{\pi}{4}\right)\right]$
(1$\bar{1}$1)	[011]	$-\sigma_5$	σ_3
(1$\bar{1}$1)	[10$\bar{1}$]	$\sigma_2 = \dfrac{\sqrt{6}}{9}\bar\sigma_r\left[\dfrac{3}{2}-2\sqrt{3}\sin\theta\sin\left(\theta-\dfrac{\pi}{6}\right)\right]$	σ_3
(1$\bar{1}$1)	[1$\bar{1}$0]	$\sigma_4 = -\dfrac{\sqrt{6}}{9}\left[\bar\sigma_z - 2\bar\sigma_r \sin\theta\left[\sqrt{3}\sin\left(\theta-\dfrac{\pi}{6}\right)-\sin\theta\right]\right]$	σ_3
(111$\bar{}$)	[011]	$-\sigma_3$	σ_2
(11$\bar{1}$)	[10$\bar{1}$]	$-\sigma_4$	σ_4
(11$\bar{1}$)	[1$\bar{1}$0]	$-(\sigma_1+\sigma_2)=\sigma_6$	σ_1

$$\bar\sigma_r = \sigma_r - \sigma_\theta$$
$$\bar\sigma_z = \sigma_z - \sigma_\theta$$

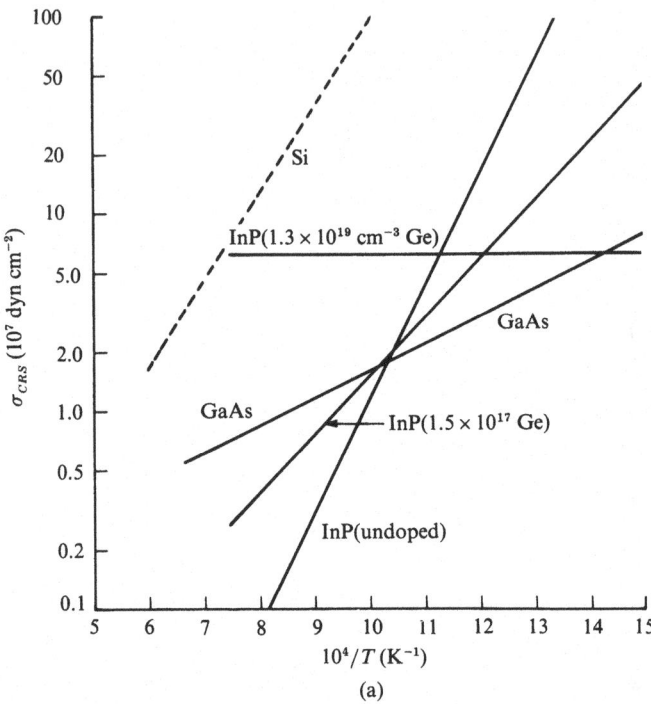

Fig. 7.8. (a) Extrapolated critical resolved shear stress, σ_{CRS}, as a function of temperature, T. (Courtesy of A. Jordan.)

control stability is lost so that, for practical growth conditions, it is difficult to reduce n_D below $\sim 10^3$ cm^{-2}.

In contrast to Si, the suppression of dislocation generation is inherently much more difficult in GaAs and InP. This is partially due to the lower values of K_S (higher $\mathcal{G}_S$), so that growth becomes more radial heat-transport limited, and partially due to the smaller σ_{CRS} near T_i for these systems. Neither of these factors would be a problem if it were intrinsically difficult to nucleate dislocation sources at the surface of dislocation-free crystals. However, for these systems, B$_2$O$_3$ is in contact with the surface and this lowers the interfacial energy which makes it easier to nucleate dislocation half loops at the surface. Perhaps one could make an additive to the B$_2$O$_3$ that would adsorb onto the GaAs crystal surface and provide a very thin layer of compressive material that would impede such dislocation nucleation at the surface. One other factor that seems to have been neglected in the GaAs and InP cases is that such materials have strong piezoelectric coefficients so that

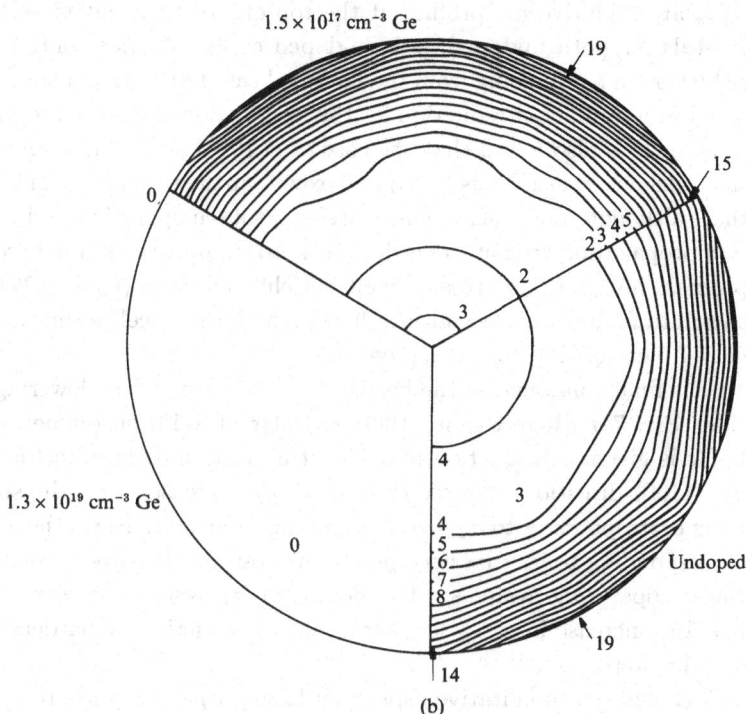

Fig. 7.8. (b) Dislocation density distribution at the top end of an undoped InP⟨111⟩ crystal LEC-pulled at $T_a = 1135$ K $(T_i - T_a = 200$ K$)$. Similar patterns are shown for boules containing 1.5×10^{17} cm^{-3} Ge and 1.3×10^{19} cm^{-3} Ge. Multiplication of the contour labels by 10^7 dyn cm^{-2} gives σ_{ex} for the $\{111\}, \langle 1\bar{1}0 \rangle$ slip system in absolute units. $(2R_c = 2.5$ cm, $V = 4 \times 10^{-4}$ cm s^{-1} and $t_c = 1600$ s.) (Courtesy of A. Jordan.)

radial stress gradients will produce radial electric field gradients causing charged species in both the B_2O_3 and the GaAs to migrate in such a direction as to try and neutralize this field. It may be the corrosive action of such constituents at the GaAs surface that allows such easy dislocation nucleation.

Most studies to date have adopted the viewpoint that solute hardening can be effective in reducing n_D. Incorporation of $> (2$–$4) \times 10^{18}$ cm^{-3} of Si or Te in GaAs and Ge or S in InP were found to curtail drastically the formation of dislocations via a hardening mechanism that increases σ_{CRS}. Fig. 7.8(a) shows extrapolated values of σ_{CRS} as a function of temperature for GaAs and InP compared to Si while Fig. 7.8(b) shows calculated relative n_D profiles at the top end of LEC-pulled InP $\langle 111 \rangle$ crystals for both undoped and Ge-doped cases. We note only a slight reduction in n_D in going from the undoped case to the moderately doped case but a significant reduction for the heavily doped case where n_D goes to zero. It is also found that the case of $\sim 10^{19}$ cm^{-3} In in GaAs for an $R_c = 3.5$ cm crystal leads to $n_D = 0$ except at the periphery. This shows that, although loop nucleation is present, source operation and dislocation propagation are not. This does seem to support the solute hardening process; however, progressive melt enrichment in In ($k_0 \approx 0.12$) leads to solute redistribution and interface breakdown to cell formation after only 7.5 cm of single crystal growth.

As a final comment on the hardening mechanism of n_D lowering, consider Fig. 7.9 where σ_{ex} for the top wafer of a 15 cm diameter $\langle 100 \rangle$ boule is given as a function of σ_{CRS} at a location 3 cm from the center in a $\langle 100 \rangle$ direction. We see that, as σ_{CRS} increases, first the slip systems corresponding to $\sigma_2 + \sigma_3$ become unimportant. Next, the systems corresponding to $\sigma_4 + \sigma_5$ become unimportant and we are left with only the σ_1 operating systems. The dominant σ_1 systems survive up to a five-fold increase in σ_{CRS} leading to an n_D reduction by hardening as a step-by-step process.

Although the qualitative aspects of basing n_D on σ_{ex} are reasonable, several features of this approach show its weakness for quantitative coupling of the crystal growth conditions to n_D. The primary problem is that the critical resolved shear stress method neglects the important rate dependences of the micromechanical behavior of crystals even at low dislocation densities. But dislocation motion is a thermally activated process so that a dynamic model is needed. Estimating n_D to be proportional to σ_{ex} assumes that dislocations are produced by shear-induced, crystallographic glide and that plastic deformation of the crystal is not an important mechanism for stress relief. Recently,[4,5] people have applied the Haasen[6] model, which relates plastic deformation in a crystal with dislocation dynamics and dislocation multiplication, to dislocation generation during crystal growth of InP, GaAs and Si. This work assumes the ready availability of dislocation sources and thus is not really applicable to dislocation-free Si.

Maroudas and Brown[5] developed an asymptotic analysis for describing the evolution of n_D as a function of the magnitude of the plastic

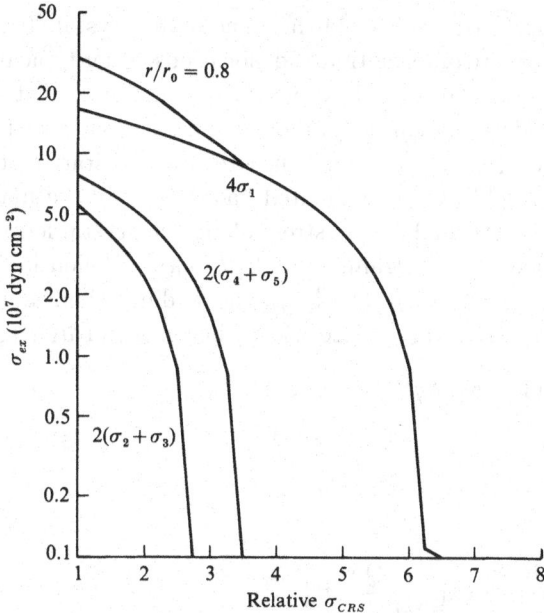

Fig 7.9. Contributions to σ_{ex} at t_{max} as a function of relative σ_{CRS} for the top wafer of a $\langle 100\rangle$ boule ($2R_c = 7.5$ cm) at a location 3 cm from the center in a $\langle 100\rangle$-direction.

strain, $\tilde{\varepsilon}_p$, leading to

$$\frac{dn_D}{d\tilde{\varepsilon}_p} = \frac{K'}{\phi b_*}\left[\sigma_e - \frac{\tilde{G}\tilde{\varepsilon}_p}{\phi} - A_{sh}(n_D)^{1/2}\right] \tag{7.5a}$$

Here, the first term in parentheses is the elastic stress, the second is the stress relief caused by plastic strain while the third is the back stress due to dislocation interactions. In Eq. (7.5a), K' is a constant determined from deformation experiments under constant load, ϕ is a geometrical factor that relates shear to compression and tensile strain and A_{sh} is the strain hardening factor.

The square root of the second invariant of the stress tensor is usually called the von Mises stress, σ_{VM}, after the von Mises condition that is used in continuum plasticity theory[7] ($\sigma_{VM} \equiv \sigma_e$). Defining σ_{0VM} as the characteristic size of the von Mises stress, a dimensionless von Mises stress, $\bar{\sigma}_{VM} = \sigma_{VM}/\sigma_{0VM}$, is used to characterize the dislocation generation which is found to depend upon whether or not $\bar{\sigma}_{VM}$ is a constant or varies spatially with the dimensionless distance, $\zeta = z/R$. One key advantage of the asymptotic analysis[5] is that the behavior of n_D for CZ and LEC crystal growth can be predicted *without* detailed calculation of the elastic stress field throughout the crystal. It is found that the very rapid rate of formation of dislocations under typical crystal growth conditions leads to equilibration of n_D with the local stress field after only a short development length, or boundary layer, near the crystal/melt interface. Because of this structure, the qualitative structure of the dislocation field can be estimated based on only the magnitude and gradient of the thermal elastic stress along the crystal length.

Defining a scaled plastic strain, $x = \tilde{G}\tilde{\varepsilon}_p/\phi\sigma_{0VM}$, dimensionless dislocation density, $y = \tilde{G}b_* n_D/K'\sigma_{0VM}^2$, and a dimensionless material parameter, $\delta = (A_{sh}^2 K'/\tilde{G}b_*)^{1/2}$, Eq. (7.5a) becomes, for $\bar{\sigma}_{VM}(\zeta) = 1$,

$$\frac{dx}{d\zeta} = \hat{D}(1 - x - \delta y^{1/2})^{\tilde{m}} y \tag{7.5b}$$

and

$$\frac{dy}{d\zeta} = \hat{D}(1 - x - \delta y^{1/2})^{\tilde{m}+1} y \tag{7.5c}$$

where

$$\hat{D} = \frac{K'B_0 R_c}{V_p} \exp\left(\frac{-Q}{kT}\right) \sigma_{0VM}^{\tilde{m}+1} \tag{7.5d}$$

is a dimensionless group called the dislocation Damköher number. Here, Q is related to the Peierls barrier that must be overcome for the dislocation to glide, B_0 is an empirically measured constant related to the dislocation mobility,[6] V_p is the crystal pull velocity, $\tilde{m}$ is the effective stress exponent in the dislocation mobility[6] while $\delta = 0.11$ and 0.019 for InP and Si, respectively. The constant $\hat{D}$ scales the time interval for generating dislocations by the time interval for motion of the crystal through the region with applied stress, σ_{0VM}, and temperature T. For $\hat{D} \gg 1$, dislocation density caused rapid plastic stress relief, which eventually halts the formation of more dislocations. For $\hat{D} \ll 1$, dislocations are formed slowly relative to the motion of the crystal. A thin boundary, $\Delta\zeta$, is expected wherein x and y vary from the interface values to the equilibrium values. The maximum value of n_D is given by

$$n_D(\infty) = K'\sigma_{0VM}^2/2\tilde{G}b_* \tag{7.5e}$$

Using the parameters given in Table 7.3, for InP with $T = T_M$, $R_c = 1.5$ cm and $V_p = 1$ cm h^{-1} given $\hat{D} \sim 150$ and ~ 37050 if $\sigma_{0VM} = 0.1$ MPa and 1.0 MPa, respectively. Figure 7.10(a) shows the variation of n_D with distance from the interface for InP at $\sigma_{VM} = 0.1$ MPa.

For the more general case of a spatially varying von Mises stress and, in particular, for the case where the von Mises stress increases with

Table 7.3. *Parameters of the Haasen*[6] *model.*[5]

	Material		
	Si	GaAs	InP
$T_M(\text{K})$	1683	1511	1335
$Q(\text{eV})$	2.2	1.5	1.0
$\tilde{m}$	1.1	1.7	1.4
$b_*(\text{m})$	3.8×10^{-10}	4.0×10^{-10}	4.2×10^{-10}
$A_{sh}(\text{N m}^{-1})$	4.0	–	3.0
$K'(\text{m N}^{-1})$	3.1×10^{-4}	–	1.2×10^{-2}
$B_0(\text{m}^{2+\tilde{m}}\ \text{N}^{-\tilde{m}}\text{s}^{-1})$	8.6×10^{-4}	–	1.36×10^{-8}
$\tilde{G}(\text{N m}^{-2})$	$3.51 \times 10^{10}(T_M)$	$4.0 \times 10^{10}(T_M)$	$2.36 \times 10^{10}(1200\ \text{K})$

distance from the crystal/melt interface, one finds that

$$n_D(z) = \frac{\delta K'[\sigma(z)]^2}{\tilde{G}b_*} \tag{7.5f}$$

Fig. 7.10(b) shows the maximum value of n_D as a function of the maximum von Mises stress in the crystal. These results are in excellent agreement with the calculations of Völkl and Müller[4] and the good quantitative agreement with etch pit densities gives confidence that the Haasen model[6] and the asymptotic analysis lead to a direct interpretation of the influence of the stress gradient in the crystal on n_D.

7.1.2 Mechanical stresses

The presence of insoluble particles in the melt may lead to complex dislocations with a large screw component as shown in Fig. 7.11 when such particles are grown into the crystal. When the edge of a growing crystal layer captures a particle in its path, a shearing of the layer edge may occur as a result of the epitaxial forces between the crystal and particle trying to produce a semicoherent interface. It has been observed experimentally for C particles in naphthalene, that the probability of forming a spiral dislocation, P_{sp}, by particle capture increased to a constant value approaching unity as the crystal growth rate, V, increased. The value of n_D produced by this process should depend upon the type (epitaxial forces), size and concentration of particles as well as upon V. Clearly, at small V, effective repulsion forces exist to inhibit trapping.

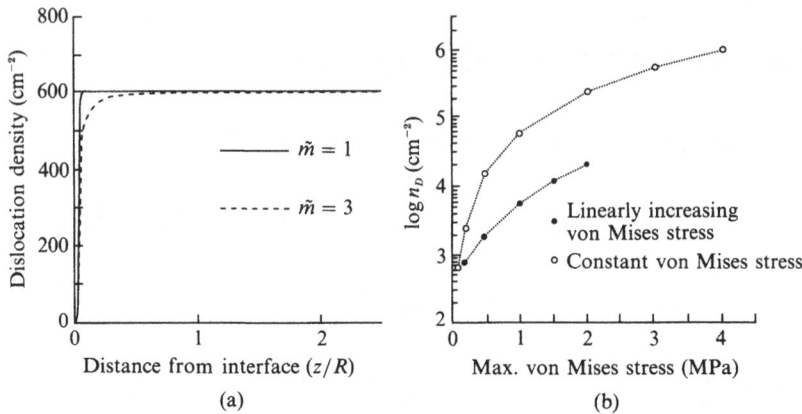

Fig. 7.10. (a) Variation of dislocation density, n_D, with distance into the crystal from the interface predicted by asymptotic theory. The results are for InP at a constant von Mises stress of 0.1 MPa. The curves provide upper ($\tilde{m} = 1$) and lower ($\tilde{m} = 3$) bounds for the axial variation of n_D for Si ($\tilde{m} = 1.1$), GaAs ($\tilde{m} = 1.7$) and InP ($\tilde{m} = 1.4$) and (b) maximum value of n_D as a function of the maximum von Mises stress in the crystal. A minimum von Mises stress, $\sigma_{0VM} = 0.1$ MPa, at the crystal/melt interface has been used for the analysis with axially increasing stress.[5]

A second type of mechanical stress that may put dislocations into a crystal arises only in materials that expand upon freezing and contain deep cell-boundary grooves. At the root of the grooves, the expanding solid forces the liquid to flow outwards to the bulk melt. This is essentially Bernouilli flow (see Eq. (6.30)) so that the pressure differential, ΔP, between the roots and the tips of the cells is given by $\Delta P = 8\nu_L \Delta\rho V L/\lambda^2$, where L is the length of the cells, $\Delta\rho$ is the density difference and λ is the fluid channel width. Thus, ΔP can grow to be ~ 1000 atm for $L \sim 1$ cm, $\lambda \sim 1$ μm, and $V \sim 1$ cm s^{-1}. This ΔP causes the cell channel roots to act like notches and become slip sources for dislocation generation in the crystal near the roots.

A somewhat similar mechanical stress condition can occur near a crucible wall for crystals that expand upon freezing during Bridgeman crystal growth. If the crystal does not wet the crucible wall so that a re-entrant contact angle exists, mechanical stresses may develop for the freezing of the fluid contacting the wall so that twin formation may develop at this location.

A final and novel mechanical stress effect can be seen when one con-

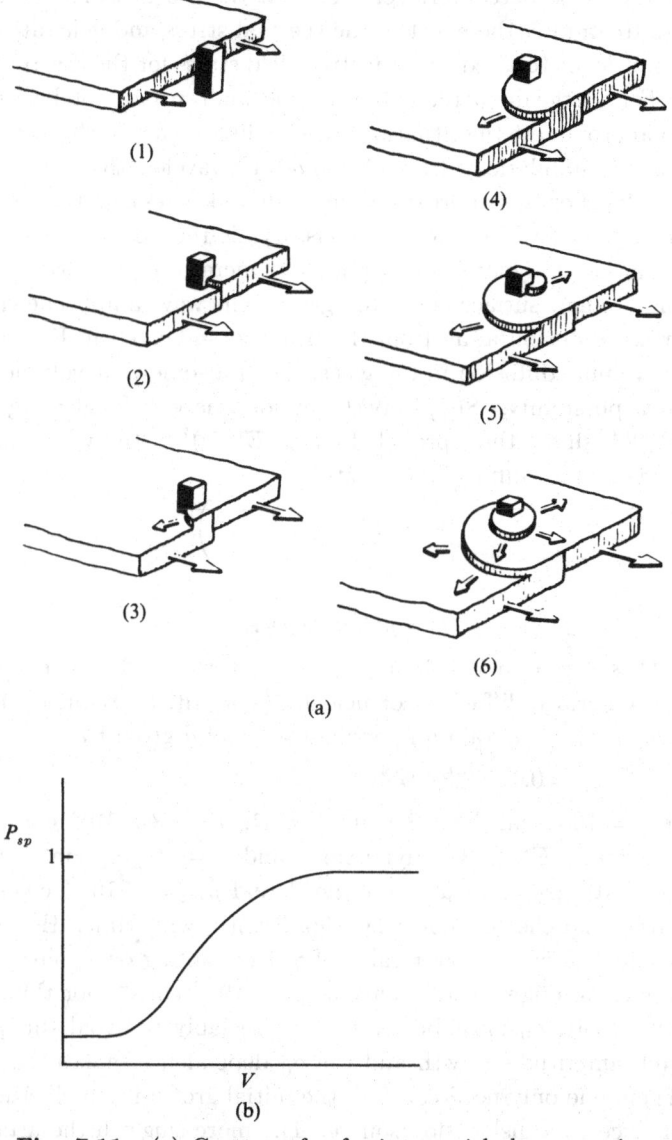

(1)

(4)

(2)

(5)

(3)

(6)

(a)

(b)

Fig. 7.11. (a) Capture of a foreign particle by a growing crystal resulting in the formation of a complex screw dislocation and (b) the probability, P_{sp}, of forming a spiral dislocation as a function of crystal growth rate, V.

siders the operating layer sources in the meniscus rim of a CZ crystal.[8] If the layer height is relatively large (0.1–1.0 μm) the cylindrical surface

of the crystal is detectably serrated and these serrations can act like
notches to amplify the effect of the thermal stress and generate slip. In
essence, it leads to a reduced effective yield stress for the crystalline material. For a large diameter crystal with a small surface notch amplitude,
we can approximate the situation as being like a plane surface containing
a harmonic undulation of amplitude a and wavelength λ as illustrated
in Fig. 5.23. For this approximation, with bulk stress σ_0, the local stress
tensor is given by Eqs. (5.35). Thus, σ_{xx} is reduced as a increases so
there is a thermodynamic driving force to increase a provided the cost of
forming the new surface is not too great. The wave amplitude can grow
by surface diffusion away from the peaks as indicated in Fig. 5.23. Of
course, volume diffusion in the gas space of a somewhat volatile species
is also a possibility. Simple perturbation analysis (neglecting surface
creation)[9] shows that, provided $\sigma_0 > (E^*\gamma\omega)^{1/2}$, the wave amplitude
grows via surface diffusion at a rate

$$\dot{a} = B\left(\frac{\omega^3\sigma_0^2}{E^*} - \gamma\omega^4\right) \tag{7.6a}$$

where

$$B = a\Omega^2 n_s D_s/\kappa T \tag{7.6b}$$

Here, Ω is the atomic volume, n_s is the surface adatom density and
D_s is the surface diffusion coefficient. At a critical frequency of $\omega^* =
(3/4)\,\sigma_0^2/E^*\gamma$, we obtain a maximum value of $\dot{a}$ given by

$$\dot{a}_{max} = 0.1B\sigma_0^8/\gamma^3 E^{*4} \tag{7.6c}$$

Using $a = 10^{-4}$ cm, $\Omega = 2 \times 10^{-23}$ cm^3, $n_s = 3 \times 10^{15}$ cm^{-2}, $D_s =
10^{-5}$ cm^2 s^{-1}, $E^* = 10^{12}$ dyne cm^{-2} and $\gamma = 10^3$ erg cm^{-2}, we find,
for $\sigma_0 = 10^4$ psi, that $\omega^* \sim 1$ μm^{-1} and $\dot{a}_{max} \sim 10^{-9}$ cm s^{-1} and
the surface notches do not grow significantly with time. However, for
softer material with a lower value of γ, like GaAs grown using the LEC
technique, we might readily have $\dot{a}_{max} \sim 10^{-4}$ cm s^{-1} for this value of
σ_0. Obviously, σ_0 must be reduced appreciably to avoid this problem
of notch amplitude growth and the σ_0^8 dependence makes this easy to
do. Then, one only need consider the initial grown-in notch size and its
importance as a dislocation source. The more sluggish the attachment
kinetics, the larger will be a.

7.1.3   Vacancy and interstitial supersaturation

The main difference between these two species is that the former
will form intrinsic dislocation loops while the latter will form extrinsic
loops. At any temperature, the equilibrium vacancy and interstitial con-

centration in Si, $n^*_{V_{Si}}$ and $n^*_{Si_I}$, respectively, are given for independent creation by

$$n^*_{V_{Si}} = \beta_{V_{Si}} \exp\left(-U^f_{V_{Si}}/\kappa T\right) \; ; \; n^*_{Si_I} = \beta_{Si_I} \exp\left(-U^f_{Si_I}/\kappa T\right) \quad (7.7)$$

where $\beta_{Si_I} > \beta_{V_{Si}} \sim 10\text{--}10^2$ and U^f is the formation energy. Since $U_{Si_I} > U_{V_{Si}}$ is the usual approximation made, $n^*_{V_{Si}} > n^*_{Si_I}$ at any temperature. In addition, since the activation energy for the migration of Si_I is thought to be much smaller than the activation energy for the migration of V_{Si}, any supersaturation of Si_I and V_{Si} will lead to recombination events via the high mobility of Si_I. Thus, although one may generate a supersaturation of vacancies by cooling, it is unlikely that one can do so for interstitials because recombination may readily occur. In fact, as the supersaturation of vacancies increases, volume recombination with interstitials proceeds to make Si_I undersaturated until the local two-species metastable equilibrium condition

$$\partial\mu_{Si_I}/\partial n_{Si_I} = -\partial\mu_{V_{Si}}/\partial n_{V_{Si}} \quad (7.8a)$$

is obtained ($\partial n_{Si_I} = -\partial n_{V_{Si}}$ during recombination). This is perhaps one of the reasons why it has been so difficult to gain clear experimental evidence for the presence of Si_I species in Si.

The foregoing is an overly simplistic picture of the point defect situation in Si which is useful for pedagogical purposes but it is in no way complete. First, the Si_I species can be in two sites: (i) the hexagonal site which is at the center of a ring of six Si lattice atoms and (ii) the tetrahedral site which is centrally located between four of the six rings. Second, there are also bond chain types of interstitial defects, called interstitialcies by some. We will denote these $Si_{\hat{i}}$ and the two types are (1) the site pair where two Si atoms share a site with a double bond between them and (2) the bond-centered $Si_{\hat{i}}$ species which is likely to be in the positive state, having donated an electron to one of its bonding neighbors so that it can make two bonds with one (giving it a negative charge) and one bond with the other. In addition, Frenkel defect formation reactions will exist between V_{Si} and Si_I or $Si_{\hat{i}}$ species and these are all expected to be thermally activated with different activation energies for formation. For the reverse reaction of annihilation, the level of activation depends also on the degree of supersaturation of the point defects in the local environment. Because of the multitude of species involved in the complete reaction picture, we shall return to the overly simplistic picture which neglects the different Si_I sites and the two Si_I species in what is to follow. These other species and states are important in the overall description but they are beyond the level of our current discussion.

Although the Frenkel defect (FD) equilibrium is expected to hold in the volume of Si, the specific equilibrium for Si_I and V_{Si} will also depend upon the presence of dopants, dislocations and interfaces as follows.

Case (a): dislocation-free, pure and infinite volume Si so only the FD reaction can hold

$$Si(s) \rightleftharpoons Si_I + V_{Si} \; ; \quad K_0 = [Si_I][V_{Si}] = [Si_I]^2 \qquad (7.8b)$$

giving

$$C^*_{Si_I} = C^*_{V_{Si}} = K_0^{\frac{1}{2}} \qquad (7.8c)$$

Case (b): the presence of a dopant, D, that interacts with the Si_I to produce an association complex, D–Si_I, but which does not change the Fermi level

$$D + Si_I \rightleftharpoons D\text{–}Si_I \; ; \qquad K_1 = [D\text{–}Si_I]/[D][Si_I] \qquad (7.8d)$$

so that

$$C^*_{V_{Si}} = K_0 K_1 \frac{C_D}{C_{D\text{–}Si_I}} \; ; \qquad C^*_{Si_I} \approx K_1^{-1} \frac{C_{D\text{–}Si_I}}{C_D} \qquad (7.8e)$$

If the dopant concentration is large, the Si may be extrinsically conducting, the Fermi level will shift and the population of charged point defects may greatly increase. Fig. 7.12 illustrates the relative equilibrium populations of neutral and charged vacancies at 750 °C as a function of Fermi energy, E_F, when one assumes that only neutral Si_I are formed.[10] The net consequence of the extrinsic conduction condition is that the effective value of $U^f_{V_{Si}}$ will change (strongly decrease for heavily-doped n-type). Analogous equations to Eq. (7.8e) can be generated for the case where many dopants are present and partially associated with both Si_I and V_{Si}.

Case (c): the presence of dislocations provides a stress field with binding states for Si_I, V_{Si} and dopants so that complexes of the following sort can form

$$Si_d + Si_I \rightleftharpoons Si_d\text{–}Si_I \; ; \qquad K_d = \frac{[Si_d\text{–}Si_I]}{[Si_d][Si_I]} \qquad (7.8f)$$

where $[Si_d] \approx \beta_d n_D$ and β_d may vary from $\sim 1 - 10^2$ per atomic site along the dislocation line and we thus expect that

$$C^*_{Si_I} \approx K_d^{-1} \frac{C_{Si_d\text{–}Si_I}}{\beta_d n_D} \; ; \qquad C^*_{V_{Si}} \approx \frac{K_0 K_d \beta_d n_D}{C_{Si_d\text{–}Si_I}} \qquad (7.8g)$$

When internal or external surfaces are present in the Si, each atomic

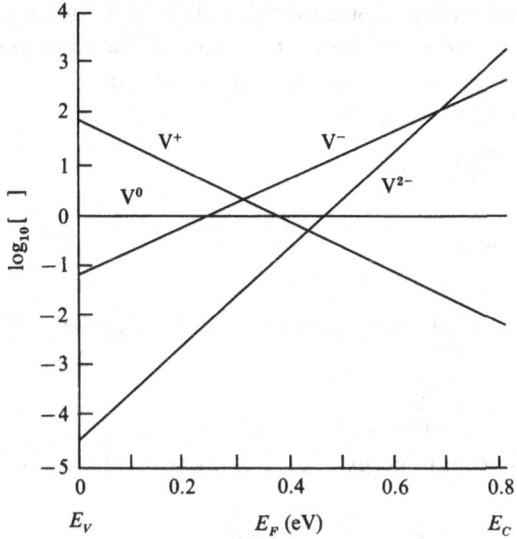

Fig. 7.12. Relative population of various monovacancy species in Si as a function of Fermi energy at 750 °C.[10]

site of interfacial area has β_s potential binding sites so that, once again, association-type reactions can occur to alter the local equilibrium values of $C^*_{Si_I}$ and $C^*_{V_{Si}}$ in the vicinity of such interfaces.

The point of this excursion into the many varieties of point defect equilibria that can prevail in Si and more generally in our crystals and films is that we must use empirical values for β and U^f in Eq. (7.7) because we don't know which type of equilibrium reaction is dominating. We only know that, if all types of reaction are present, each binds a certain percentage of the species in its local potential wells and is in equilibrium with the far-field equilibrium concentration that may be significantly perturbed by the types of structural defects present. Thus, although certain specific and idealized cases will be discussed, one should keep in mind the types of complexities that may arise in a real situation.

Focusing our attention on just the vacancies grown into a dislocation-free Si crystal at T_M, the supersaturation at any temperature T will be given by

$$\frac{n^*_{V_{Si}}(T_M)}{n^*_{V_{Si}}(T)} \approx \exp\left[-\frac{U^f_{V_{Si}}}{\kappa}\left(\frac{1}{T_M} - \frac{1}{T}\right)\right] \qquad (7.9)$$

where, for Si, $U^f_{V_{Si}} \approx 2.4$ eV. These vacancies tend to cluster into discs whose kinetics of formation is limited by a two-dimensional nucleation event which may be heterogeneously catalyzed. Under the proper conditions, this vacancy disc can collapse to form a dislocation loop. The constrained equilibrium between an unfaulted dislocation loop and the V_{Si} species is given, for $\Omega = b_*^3$, by

$$\frac{\partial E_R}{\partial R} + \mu_{V_{Si}} \frac{\partial C_{V_{Si}}}{\partial R} = 0 \qquad (7.9b)$$

where

$$\frac{\partial C_{V_{Si}}}{\partial R} = -\frac{2\pi R}{b_*^2} \qquad (7.9c)$$

Incorporating the standard expression for $\mu_{V_{Si}} = kT \, \ell n(C_{V_{Si}}/C_{V_{Si}}^*)$, we have

$$\ell n \left(\frac{C_{V_{Si}}}{C_{V_{Si}}^*} \right) = \frac{b_*^2}{2\pi R \kappa T} \frac{\partial E_R}{\partial R} \qquad (7.9d)$$

for a loop of radius R and energy E_R. The theoretical expression for a faulted loop is

$$E_R = \frac{\tilde{G} b_*^2 R}{2(1-\sigma^*)} \left[\ell n \left(\frac{8R}{R_c} \right) - 1 \right] + \frac{2\pi R}{b_*^2}(\pi R_c L) + \pi R^2 \gamma_{st} \qquad (7.9e)$$

where R_c is the core radius of the dislocation, L is the average strain energy per atom of the dislocation core, σ^* is Poisson's ratio, $\tilde{G}$ is the shear modulus and γ_{st} is the stacking fault energy per unit area.

Combining the first and second terms leads to

$$\frac{\partial E_R}{\partial R} = \frac{\tilde{G} b_*^2}{2(1-\sigma^*)} \left[\ell n \left(\frac{R}{b_*} \right) + Z + 1 \right] + 2\pi R \gamma_{st} \qquad (7.9f)$$

where $Z = 1.98$, 1.48 and 1.96 for $R_c/b_* = 2$, 1 and $\frac{1}{2}$, respectively. We shall choose $Z \approx 1.8$ as being representative. From Eq. (7.6d) we have

$$\ell n \left(\frac{C_{V_{Si}}}{C_{V_{Si}}^*} \right)_{Eq} = \frac{\tilde{G} b_*^3}{4\pi (1-\sigma^*)\kappa T(R/b_*)} \left[\ell n \left(\frac{R}{b_*} \right) + 2.8 \right] \qquad (7.9g)$$

where

$$C_{V_{Si}}^{'*} = C_{V_{Si}}^* \exp(b_*^2 \gamma_{st}/\kappa T) \qquad (7.9h)$$

For an unfaulted loop, we choose $\gamma_{st} = 0$ and Eq. (7.9g) gives us the vacancy supersaturation needed to keep a loop of size R from growing or shrinking.

It has been estimated that dislocation loops will nucleate at a loop radius of $\sim 3b_*$–$5b_*$ for metals and several times this for semiconductors or non-metals. Rather than focus on the complex nucleation problem, one can ask what is the relative undercooling, $\Delta T/T_M$, of a vacancy supersaturated crystal required to keep a dislocation loop of radius R from shrinking. Results for this computation are given in Fig. 7.13

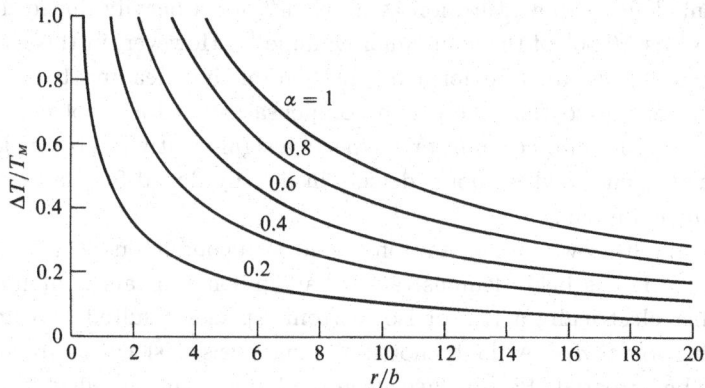

Fig. 7.13. Relative undercooling, $\Delta T/T_M$, of a vacancy supersaturated crystal required to keep a dislocation ring of radius R/b_* from shrinking for several values of the parameter α.[11]

for several values of the important material parameter $\alpha = \tilde{G}b_*^3/6(1 - \sigma^*)U_{V_{Si}}^f$.[11] A typical value of α for metals is 0.6 so that dislocation rings of $r = 15b_*$ are growing only at temperatures below $\Delta T/T_M \approx 0.2$. For Ge and Si one finds that, because $\tilde{G}$ is larger for these materials than for metals, $\Delta T/T_M$ is increased from 0.2 to about 0.5 for $r = 15b_*$. This makes it difficult for dislocation loops to grow to a large size in these materials via vacancy supersaturation forces. However, if $\sim 10^{16}$ vacancies per cubic centimeter are frozen into the lattice at T_M and if most of them form vacancy discs one atom thick and ~ 0.1 μm in diameter, the resulting total length of dislocation line per cubic centimeter in the crystal would be about 10^7 cm. Thus, an initially dislocation-free crystal may lead to the generation of rather large total dislocation lengths in the crystal via the vacancy supersaturation mechanism. The initial presence of $\sim 10^2$–10^3 dislocation lines per square centimeter would have served as ready sinks for this vacancy supersaturation and none of the additional dislocation loops would have formed.

If instead of vacancy supersaturation, we have Si_I supersaturation, then extrinsic stacking fault loops are expected to form and Eqs. (7.9g) and (7.9h) would give the supersaturation, $C_{Si_I}/C'^*_{Si_I}$, needed to keep the extrinsic fault of size R/b_* from either growing or shrinking at temperature T. Using an equation for $C^*_{Si_I}$ analogous to Eq. (7.9a), one can tell at any particular temperature, T, whether such a fault of size R/b_* is stable. Because the effective values of $C^*_{V_{Si}}$ and $C^*_{Si_I}$ will depend upon the doping levels, one must generally expect to find a reduced point defect supersaturation at a given T for a heavily doped material (see Eqs. (6.50) of the companion volume)[8] However, in this situation, charged jogs can also form on dislocation lines leading to a reduced value of E_R so that there is a compensating factor involved. Taking both factors into account, one expects a significantly altered dislocation density generated by point defects in heavily doped Si, GaAs or other semiconductors.

In melt-grown, undoped compound semiconductors, $n_D \sim 10^3$–10^6 cm^{-2}. It has been demonstrated that, in some instances, doping with either electrically active or isoelectronic species resulted in a dramatic reduction of n_D. At high enough doping levels, dislocation-free crystals can be prepared. People have attempted to explain this effect in terms of solute hardening but this doesn't seem to be correct in terms of the high temperature hardness measurements for Si and In in GaAs. Experiments have clearly shown that, although both of these solutes are very effective in reducing n_D, they have no significant effect on the high temperature critically resolved shear stress. A much more probable mechanism of n_D reduction for the GaAs experiments is via the concept of amphoteric native defects; i.e., defects whose identity and formation energy depend on the position of the Fermi energy as already discussed for heavily doped Si crystals.

7.1.4   Constitutional stresses

Because substitutional solutes generally have a different effective size than the solvent, a one-dimensional concentration gradient gives rise to a corresponding strain gradient because the high concentration region prefers to have a different lattice parameter than the low concentration region. For example, in Si, the fractional change in lattice parameter, $\Delta\lambda/\lambda$, with substitutional solute content n or interstitial content n' (in number of atoms per cubic centimeter), is given for C, O and P by

$$\Delta\lambda/\lambda = -6.5 \times 10^{-24}n \,, \qquad \text{for C} \qquad (7.10a)$$

$$= +4.5 \times 10^{-24}n', \qquad \text{for O} \qquad (7.10b)$$

$$= -1.2 \times 10^{-24}n \,, \qquad \text{for P} \qquad (7.10c)$$

and the number of edge dislocations in a square grid needed to satisfy the constitutional strains associated with an abrupt concentration fluctuation of magnitude $\Delta\lambda/\lambda$ is given by

$$n_D = \frac{4}{b_* d}\left(\frac{\Delta\lambda}{\lambda} - \tilde{\varepsilon}_e\right) \qquad (7.11)$$

where d is the spacing between layers across which this fluctuation repeatedly occurs and $\bar{\varepsilon}_e$ is the elastic strain.

All sources of microsegregation may give rise to such dislocations provided there are dislocation sources present in the crystal. These new dislocations would somewhat inhibit the homogenization of their generating chemical segregation patterns. Obviously rate change fluctuations like rotational striations, discussed in Chapter 4, would lead to such stresses and such dislocations. For those crystal systems that produce microscopic layer build-up at the interface (layer heights $\sim 1~\mu$m), solute partitioning at the advancing layer edges leads to an oscillating concentrating profile in the crystal with the expected type of dislocation net being that illustrated in Fig. 7.14(a) and dislocation densities $\sim 10^4$–10^5 cm^{-2} are possible. For the type of segregation produced by the cellular interface mode of crystallization, Fig. 7.14(b) illustrates the expected dislocation arrangement where dislocation densities $\sim 10^6 - 10^8$ cm^{-2} can be readily attained.

In dislocation-free Si crystals, although constitutional strains develop as a result of chemical inhomogeneities like rotational striations, there are no dislocation sources present to provide the dislocations to relieve these strains. Thus, cyclic compressive and tensile strains exist at all such chemical microheterogeneities regardless of the solute size relative to Si. Because of this, as the crystal is cooled, the excess V_{Si} congregate in the compressive regions and denude the tensile regions in an attempt to maintain a constant electrochemical potential of the vacancies. The Si_I do the reverse.

In dislocation-free Si crystals grown by the CZ technique, large quantities of C and O are picked up by the crystal. C concentrations in the range $n \sim 5 \times 10^{16}$–5×10^{17} have frequently been measured while O concentrations between 10^{17} and 3×10^{18} are common. In dislocation-free float-zone material, the O concentration is reduced to $\sim 10^{15}$ cm^{-3} and the C to $\sim 10^{16}$–10^{17} cm^{-3}. Both of these impurities have a highly favorable free energy of formation for compounding with Si. To form SiO$_2$ from Si, very large increases in free volume are needed ($\Omega_{Si} = 20$ Å^3 while $\Omega_{SiO_2} = 45.5$ Å^3) while the formation of SiC releases free volume ($\Omega_{SiC} = 10$ Å). Thus, SiC formation could act as the nucleant to allow SiO$_2$ formation synergistically which can then grow somewhat by the consumption of the supersaturated vacancies according to the reactions

$$2\,\text{Si(s)} + 2\text{C(s)} \overset{\rightarrow}{\leftarrow} 2\,\text{SiC(s)} + V_{Si} \qquad (7.12a)$$

$$\text{Si(s)} + 2\text{O}_I + V_{Si} \overset{\rightarrow}{\leftarrow} \text{SiO}_2(s) + \text{strain} \qquad (7.12b)$$

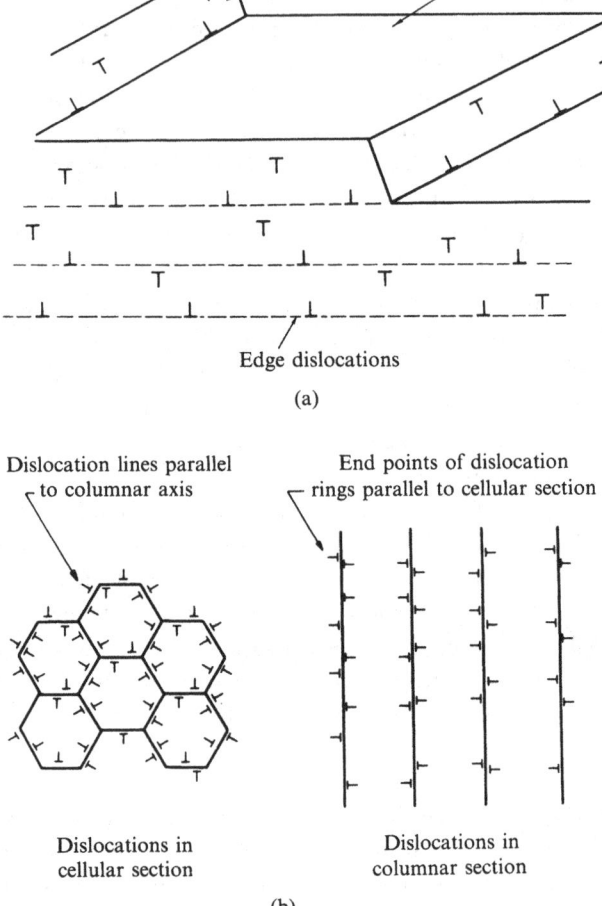

Fig. 7.14. Possible edge dislocation configuration in a crystal produced by the segregation of solute during (a) ledge movement and (b) cellular interface formation.

An additional reaction for SiO_2 particle growth after the consumption of these excess vacancies is that involving Frenkel defect formation so that Eq. (7.12) becomes

$$(1 + y)Si(s) + xV_{Si} + 2\,O_I \overset{\rightarrow}{\leftarrow} SiO_2(s) + ySi_I + strain \qquad (7.12c)$$

Thus, we can expect precipitation to occur in the cooler regions of the crystal behind the interface and microscopically in the striation pattern where the concentration fluctuation is highest. Of course, if some SiO_4 tetrahedra enter the crystal from the melt as proposed in Chapters 3 and 4, they will act as initial nuclei for the SiO_2 phase and the reaction of Eq. (7.12a) will not be needed. Since $k_0^C \approx 0.02$ and $k_0^O \sim 2.5$, the excesses for these two species will not be in exactly the same location but will be separated by ~ 1–5 μm. Further, small oxide precipitates should be present if only Eq. (7.12a) operates while large oxide precipitates with adjacent extrinsic stacking faults should be present when both Eqs. (7.12a) and (7.12c) operate.

Experimentally such microdefects have been observed to occur in a pattern that generally has the form of a spiral ramp through the crystal. Consequently, the distribution pattern assumed by such defects has gained the popular name of "swirl" or swirl defects. Swirl patterns are found to occur in both CZ and float-zone crystals irrespective of their crystallographic orientation which is consistent with the explanations of the time-dependent fluid shear flow (spiral pattern) given in Chapter 2 for the basic striation pattern. Figure 7.15 shows examples of the swirl distribution of microdefects in CZ and float-zone Si crystals.[12] Higher magnification reveals an array of shallow or empty etch pits corresponding to the microdefects in the swirl pattern. It has been noted that the presence of dislocations suppresses the formation of these microdefects in that their density is reduced while their individual size is increased in the vicinity of dislocation lines (see Fig. 7.16) which clearly points up the point defect supersaturation aspects of the microdefect formation mechanism.

Stable vacancy-oxygen complexes have been observed to form at different stages of the crystal cooling process.[13] Preferential etching and X-ray topography have revealed microdefects in two different concentration levels and two different sizes. The larger of the two types, generally referred to as A-type defects, are found to occur predominantly in regions removed from the outer surface of crystals. The smaller defects, called B-type defects, are observed over the entire crystal extending to the outer surface. The A-defects begin to form at a critical temperature T_A, whereas B-defects begin to form at $T_B > T_A$. For dislocation-free crystals grown in purified Ar (1 atm) using the pedestal technique at V between 1 and 3 mm min^{-1}, the distribution of these defects is illustrated in Fig. 7.17. The cluster concentration generally exhibits a minimum in the central area of the crystal where the magnitude of the rotational striation is least because the growth velocity fluctuation is least at that location. No A-clusters were found in a crystal surface

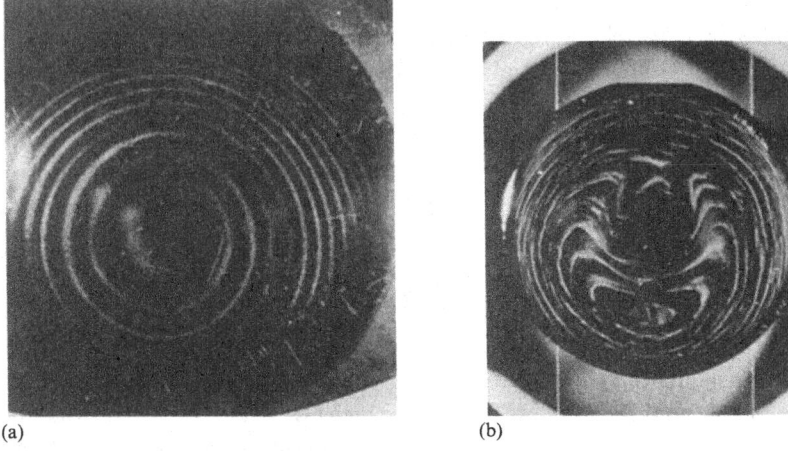

(a)   (b)

Fig. 7.15. Preferentially etched Si surfaces showing the inhomogeneous distribution of "swirl" defects, (a) {100} CZ crystal and (b) {111} float-zone crystal.[12]

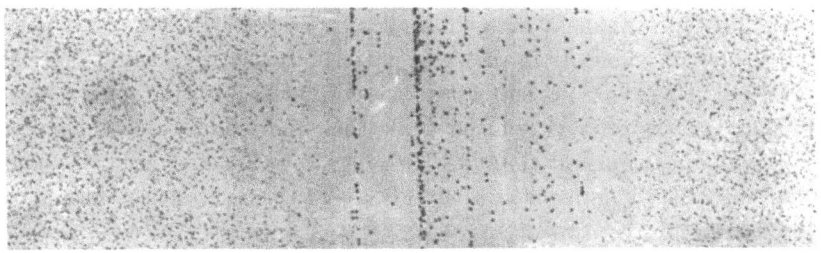

$\overline{100\ \mu\text{m}}$

Fig. 7.16. Illustrating the influence of dislocations on the distribution of microdefects.[12]

layer ~ 2 mm thick, probably because of the reduced vacancy concentration in that region. In some vacuum-grown crystals, this surface layer was completely cluster free (reduced V_{Si} and O_I species).

A-defects have been reported to be extrinsic dislocation loops roughly 1–3 μm in size and they assume a variety of complex shapes. Undecorated B-defects have not been observed by TEM although impurity decorated loops ~ 600–800 Å in size have been detected. Figure 7.18 shows relationships between the density of B-defects and the concentration of C or O in the crystal. We note that a direct relationship exists with respect to C content and a less well-defined relationship exists with

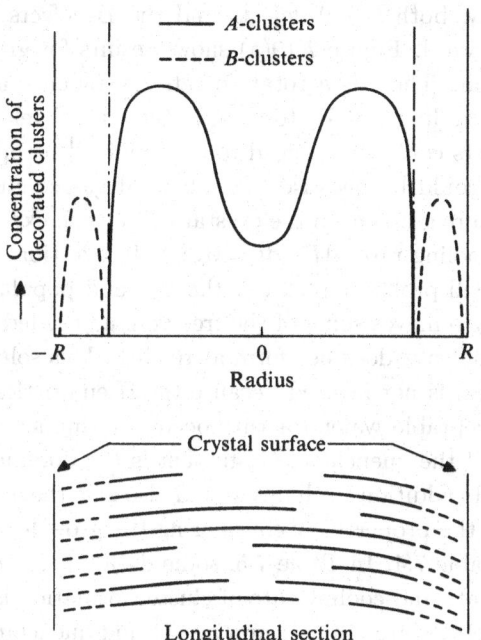

Fig. 7.17. Distribution and concentration profile of decorated micro-defects.[13]

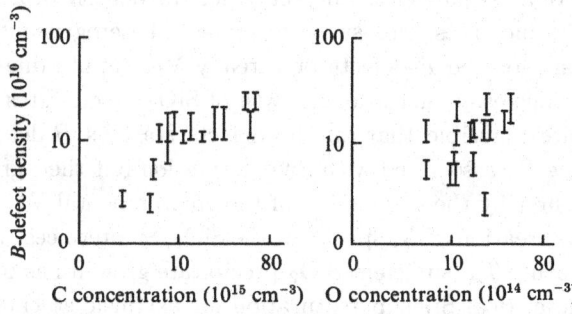

Fig. 7.18. B-defect density as a function of C and O concentrations in the crystal.[13]

respect to O content. At higher C content, B-defects are formed at higher growth rates. Thus, it seems that the B-defects at least require the operation of Eqs. (7.12) for their formation.

De Kock and others[13] have shown that, by increasing the crystal growth rate sufficiently, both the A-defects and the B-defects can be eliminated from the crystal. Figure 7.19(a) shows results for growth in pure Ar at $\omega = 30$ rpm. There is a rotation rate dependence as illustrated in Fig. 7.20. By adding 10% H_2 to the gas, the critical velocity for eliminating these defects could be reduced to 3 mm min^{-1} as indicated in Fig. 7.19(b). This could be because the enhanced gas conductivity increases the temperature gradient in the crystal which then experiences an essentially higher cooling rate ($dT_S/dt = \mathcal{G}_S V$). In addition, any interstitial H in the Si will probably seek out the V_{Si} and populate that space which essentially removes some of the free volume needed for the SiO_2 formation reaction so it does not form as readily. This solution to the swirl defect problem is not favored because the H embrittles the Si sufficiently that unacceptable wafer fracture occurs during subsequent processing. However, if the quench rate argument is the dominant factor, then the use of a He addition to the Ar would also give the enhanced $\mathcal{G}_S$. This seems to be the proper view considering the growth rate and cooling rate data of Table 7.4. In Table 7.5, some data on the relationship between growth rate and cooling rate of cluster formation is given. Since B-defects are seen for growth at 1 mm min^{-1} and quenching from $T = 1407$ °C, perhaps some grown-in supersaturation of V_{Si} species and O_I species exists in certain locations and/or perhaps some SiO_4 species enter the solid from the melt. Both possibilities would enhance the initial nucleation of SiO_2 particles. Further, since the density of B-defects increases as C_∞^C increases, this suggests either a heterogeneous nucleation mechanism for the B-defects or a ready V_{Si} supply due to SiC formation and thus easy nucleation/growth of SiO_2 precipitates.

It seems quite plausible that all the defects start as B-defects below temperature T_B associated with SiO_2 formation but their growth is greatly slowed first by the exhaustion of the environmental V_{Si} species and later by the creation of Si_I species at the SiO_2/Si interfaces. Below a second temperature T_A, sufficient SiO_2 precipitate growth has occurred to generate the critical Si_I supersaturation for extrinsic stacking fault nucleation. This fault nucleation becomes a sink for the surrounding Si_I species so their supersaturation is lowered over a large volume surrounding the extrinsic fault and the SiO_2 precipitates in that region can begin to grow more rapidly and attain a larger size. It is the stacking fault nucleation step that leads to the evolutionary transition of a small cluster of B-defects into a single A-defect with its internal sink for Si_I species. The presence of this sink in certain regions allows more rapid decrease of the O_I supersaturation so large interactive defects (A-defects) form

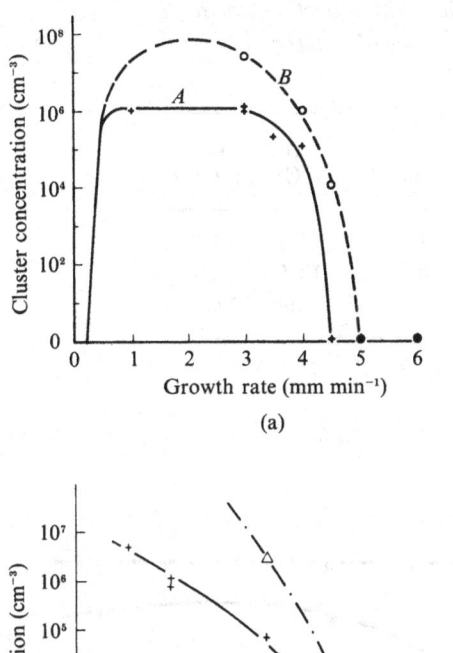

Fig. 7.19. (a) Concentration of A- and B-clusters in CZ Si as a function of V at $\omega = 30$ rpm and pure Ar gas and (b) the same but with Ar $+10\%H_2$ gas.[13]

at various locations within the rather uniform distribution of B-defects. Thus, overall, at least three growth stages can be discriminated: (1) the growth of small precipitates via the consumption of O_I and V_{Si}, (2) the continued growth of these SiO_2 precipitates but via the generation of Frenkel defects at the interface and (3) the nucleation and growth of extrinsic stacking faults as a sink for the Si_I. One of the interesting consequences of this SiO_2 formation reaction in dislocation-free Si is that it

Table 7.4. *Relationship between growth rate and temperatures of cluster formation.*[13]

Experiment	V_0 (mm min^{-1})	T_B (°C)	T_A (°C)
1	1	1407	1370
2	3	1350	1050
3	5	None	None

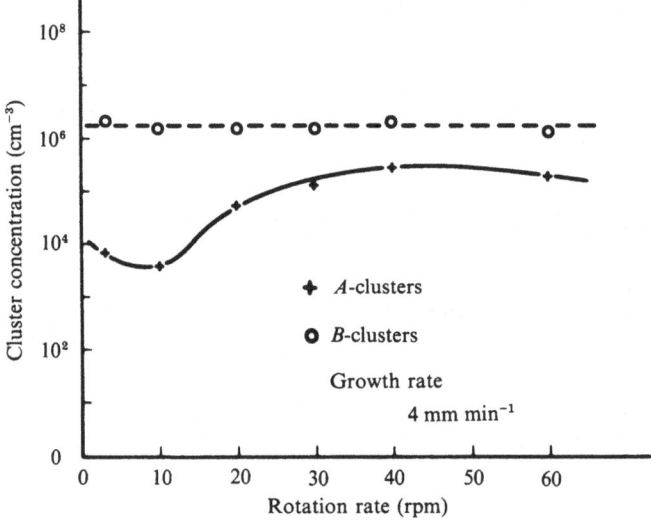

Fig. 7.20. A and B-cluster concentration in CZ Si at $V = 4$ mm min^{-1} as a function of crystal rotation rate.[13]

consumes V_{Si} species which might otherwise have supersaturated and formed dislocation loops. It also suggests that there should be specific C_∞^O and C_∞^C leading to minimum overall dislocation or stacking fault generation in the Si.

7.1.5 Inclusion-generated stress effects

As mentioned in Chapters 4 and 6, cell formation leads to sheets of liquid entrained in the crystal and these are unstable leading to an

Table 7.5. *The influence of growth rate and cooling rate on vacancy cluster formation.*[13]

Experiment	V_0 (mm/min)	Cooling from the melting point according to	Cluster formation
4	5 (no remelt)	$V_0 = 5$ mm min^{-1}	None
5	5 (no remelt)	$V_0 = 3$ mm min^{-1}	B-clusters
6	5 (no remelt)	$V_0 = 2$ mm min^{-1}	A and B-clusters
7	3	$V_0 = 3$ mm min^{-1}	A and B-clusters
8	3	$V_0 = 5$ mm min^{-1}	None

array of liquid inclusions that can migrate in the temperature gradient behind the interface. As the inclusions migrate, pressure changes of opposing kinds will occur in the droplets because of the change in local temperature along the gradient: (i) the pressure decreases along with the increasing solubility of Si in the liquid with temperature increase because the volume of a Si atom in the liquid is less than in the solid and (ii) the pressure in the droplet tends to increase with T because of the greater thermal expansion of the liquid relative to the solid. The pressure change, ΔP, caused by a temperature change, ΔT,[14] is

$$\Delta P = \left[\frac{3(\bar{\alpha}_L^* - \bar{\alpha}_S^*) + (\bar{v}_L - \bar{v}_S)(\partial C_L^{Si}/\partial T)}{B_L + 3/\mathcal{G}_S + (\bar{v}_L - \bar{v}_S)^2(C_L^{Si}/\mathcal{R}T)} \right] \Delta T \qquad (7.13a)$$

where $\bar{\alpha}^*$ refers to the linear coefficient of thermal expansion, $\bar{v}$ to the partial molar volume of Si, B_L to the liquid compressibility and C_L^{Si} to the concentration of Si in the liquid. For a typical $\mathcal{G}_S \sim 100\ °C$ cm^{-1}, at $T \sim 1200\ °C$, $\Delta P \sim -280$ MPa (tensile) for every 1 mm of migration in a Si–Aℓ alloy.[14] The increasing solubility of the Si in the liquid with increasing T overrides the counteracting thermal expansion effect.

In Si, the droplet shape is not spherical but is tetrahedral for migration in the $\langle 100 \rangle$ direction and a flat platelet when migrating in the $\langle 111 \rangle$ direction; thus, stress concentrations are expected to arise, especially for the $\langle 111 \rangle$ migration. The presence of both a stress concentration factor and a high negative pressure indicates that plastic yielding may occur around a migrating liquid drop. Any such plastic yielding should be confined to ~ 1 droplet diameter since the stress decreases as $(R/r)^3$.

Figure 7.21(a) shows a liquid inclusion of width W and height h in a

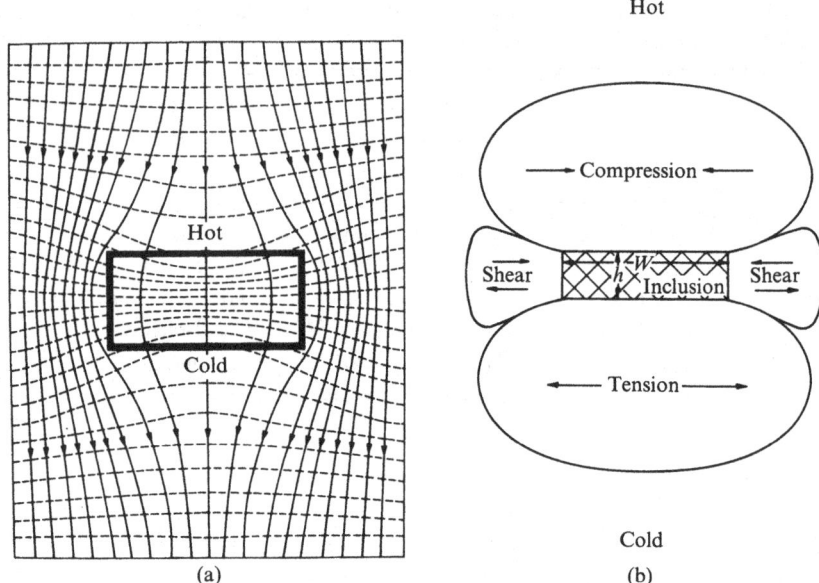

Fig. 7.21. (a) Heat flow and isotherms around a liquid inclusion in a thermal gradient $(K_L < K_S)$ and (b) stress field around a liquid inclusion in a temperature gradient for $K_L < K_S$. When $K_L > K_S$, the compression and tension fields are interchanged.[14]

temperature gradient where the thermal conductivity of the inclusion is less than that of the matrix. The region of heat flow distortion extends a distance W above and below the inclusion so that the difference in strain, $\Delta\tilde{\varepsilon}$, between a region just above or just below the droplet and a region far away from the droplet at the same average T is

$$\Delta\tilde{\varepsilon} = \bar{\alpha}_S^* W(\nabla T_L - \nabla T_S) \qquad (7.13b)$$

Because of the relative size of the regions, the surrounding crystal prevents the material near the droplet from having a strain different than the rest of the crystal, thereby generating a stress by Hooke's law of approximately

$$\sigma \approx \bar{\alpha}_S^* E_S^* W \nabla T_S[(\nabla T_L/\nabla T_S) - 1] \qquad (7.13c)$$

An examination of the isotherms of Fig. 7.21(a) shows that the region just above the droplet is hotter than areas on either side of it while the region just below the droplet is colder than areas on either side of it. The resulting stress distribution accordingly shows compression just above the droplet and tension just below it as illustrated in Fig. 7.21(b).

On either side of the droplet, the interaction of the compressive field above and tensile field below creates a shear stress of order

$$\tau = \sigma W / h \qquad (7.13d)$$

To calculate the magnitude of σ and τ, the local heat flow problem must be solved giving

$$\frac{\nabla T_L}{\nabla T_S} = \frac{K_S}{K_L + K_S(1 - F)} \qquad (7.13e)$$

where the Ks are the thermal conductivities and F is a factor that depends on the aspect ratio W/h of the inclusion.[14] The value of F increases monotonically from zero to unity as the ratio W/h increases from zero to infinity and $F = 1/3$ when $W/h = 1$. Combining these factors, we have

$$\sigma = \bar{\alpha}_S^* E_S^* W \nabla T_S \left[\frac{F(K_L - K_S)}{K_S + F(K_L - K_S)} \right] \qquad (7.13f)$$

and

$$\tau = \frac{\bar{\alpha}_S^* E_S^* W^2}{h} \nabla T_S \left[\frac{F(K_L - K_S)}{K_S + F(K_L - K_S)} \right] \qquad (7.13g)$$

For a droplet 250 μm in width migrating in the $\langle 111 \rangle$ direction in Aℓ-doped Si at $\sim$ 1200 °C, $W/h \approx 30$. For this condition, $F \approx 0.9$ and $K_L \approx 0.9 K_S$. For $\mathcal{G}_S = 100$ °C cm^{-1}, $\sigma \approx 27$ MPa and $\tau = 6.2$ MPa. Both stresses are below the yield stress of Si at 1200 °C.[14]

The foregoing results for migrating liquid inclusions also apply to solid inclusions that are not migrating. Under certain circumstances, these stresses will be sufficient to generate dislocations but in all circumstances they are sufficient to provide a binding potential for mobile solutes that influence the total character of the combined defect. Let us look more closely at such solute atmospheres at various microstructural defects.[15]

For mobile solutes which cause only dilatation of the matrix, the external work, W^j, done on the system per mole addition of a mobile component j is

$$W^j = \sigma_h \bar{V}^j \qquad (7.14a)$$

where $\sigma_h = (\sigma_{11} + \sigma_{22} + \sigma_{33})/3$ is the hydrostatic component of stress. Thus, the chemical potential of j is given by

$$\mu^j = \mu_0^j + \kappa T \ell n C^j + \sigma_h \bar{V}^j \qquad (7.14b)$$

where μ_0^j is the stress-free standard state for species j. Our interest here is in knowing the magnitude and spatial dependence of σ_h for various structural defects.

The stress fields of dislocations in various forms are well known and it is straightforward to calculate the equilibrium distribution of mobile solutes around dislocations. For the simple case of a mobile solute with only dilatational strains interacting with straight edge dislocations, we have

$$\sigma_h = -\frac{\tilde{G}b_*(1+\sigma^*)}{3\pi(1-\sigma^*)}\frac{\sin\theta}{r} \tag{7.14c}$$

where the edge dislocation lies along the z-direction in the (r, θ, z) co-ordinate representation. The concentration distribution, in the limit of infinite dilution, is

$$C^j = C_\infty^j \exp(\sigma_h \bar{v}^j / \mathcal{R}T) \tag{7.14d}$$

where C_∞^j is the concentration in the stress-free region. The average concentration $\bar{C}^j$, or the overall enhancement of solubility, is

$$\frac{\bar{C}^j}{C_\infty^j} = 1 + \frac{\phi^{j2}}{r_1^2}\left\{2\ell n\left(\frac{r_1}{r_0}\right) + \sum_{n=1}^\infty \frac{(\phi^j/r_0)^{2n}}{n[(n+1)!]^2}\right\} \tag{7.14e}$$

in a cylindrical shell around the dislocation with inner radius r_0 and outer radius r_1 ($r_1 \gg r_0$). The partial molar volume enters through the quantity ϕ; i.e.,

$$\phi^j = \frac{\tilde{G}b_*(1+\sigma^*)\bar{v}^j}{6\pi(1-\sigma^*)\mathcal{R}T} \tag{7.14f}$$

The use of the concept of partial molar strain volume, here, avoids the elastic model of an interstitial atom and the problems associated with it. These problems include the proper volume to be assigned to the atom. In the approach of Li,[15] all chemical interactions and non-linear effects of the interstitial atom are included in μ_0.

For disclinations, such as one would find in amorphous or polymeric materials, the line defect is a boundary between rotated and unrotated regions (rather than slipped and unslipped regions for the case of a dislocation). Chemical segregation atmospheres can be readily evaluated for this type of defect.[15]

Small plate-like precipitates have stress fields like dislocation loops. For example, a small patch of excess atoms behaves like a prismatic dislocation loop. Such a loop of area A lying in the xy plane and located at the origin has

$$\sigma_h = \frac{\tilde{G}b_*A(1+\sigma^*)}{6\pi(1-\sigma^*)}\frac{1}{r^3}\left(1 - \frac{3x^2}{r^2}\right) \tag{7.15a}$$

where $r^2 = x^2 + y^2 + z^2$. The atom trajectories are given by

$$r^5 = cy^2z \tag{7.15b}$$

Both are shown in Fig. 7.22(a) in the yz plane. It is seen that the flow patterns are consistent with the fact that plate-like precipitates grow in size rather than in thickness. Existing interstitial plates have a larger capacity for attracting solutes of $\bar{v} < 0$ (to the two faces) and a smaller capacity for attracting solutes of $\bar{v} > 0$ (to the edges). The result for a shear dislocation loop is given in Fig. 7.22(b). Here, we have

$$\sigma_h = \frac{\tilde{G}b_* A(1 + \sigma^*)}{6\pi(1 - \sigma^*)} \left(-\frac{3xz}{r^5} \right) \tag{7.15c}$$

with trajectories

$$r^{10} = c(z^2 - x^2)^3 \tag{7.15d}$$

It is found that shear plates do not grow in size for precipitating atoms having only dilatational strains. However, existing plates have equal capacity for attracting solutes of $\bar{v} > 0$ and $\bar{v} < 0$.

The elastic field of ellipsoidal inclusions relates to the misfit between the inclusion and the hole it occupies when both are in the stress-free state. Analysis shows that, if the misfit is purely dilatational, $\sigma_h = 0$ regardless of the inclusion shape when both the inclusion and the matrix have the same $\tilde{G}$. There is then no interaction between an inclusion of dilatational misfit and a solute of only dilatational strain.[15] However, even if an inclusion originally has no misfit with the matrix, plastic flow around the inclusion can create as large a shear misfit as the local plastic shear strain. This shear misfit introduces a hydrostatic component of stress in the matrix:

$$\sigma_h = \frac{2\tilde{G}(1 + \sigma^*)R^{3/2}e_{xy}}{3(1 - \sigma^*)} \left(\frac{xy}{r^5} \right) \tag{7.15e}$$

for a spherical inclusion of radius R at the origin with a shear misfit e_{xy}. These contours are shown in Fig. 7.23 along with their trajectories

$$r^{10} = c(y^2 - x^2)^3 \tag{7.15f}$$

Similar contours are obtained for ellipsoidal inclusions with shear misfits.

7.2 Dislocation multiplication and array formation

7.2.1 Interface penetration by slip

The influence of thermal stresses on the ultimate dislocation content of a crystal is a function of the crystal axis orientation relative to the orientations of the operating slip planes. Considering Fig. 7.24, if the slip plane makes an angle θ_a with the crystal axis, then slip in Region B causes the dislocations to intersect the outer surface rather than the interface. To shorten their line length and thus their energy, the exit

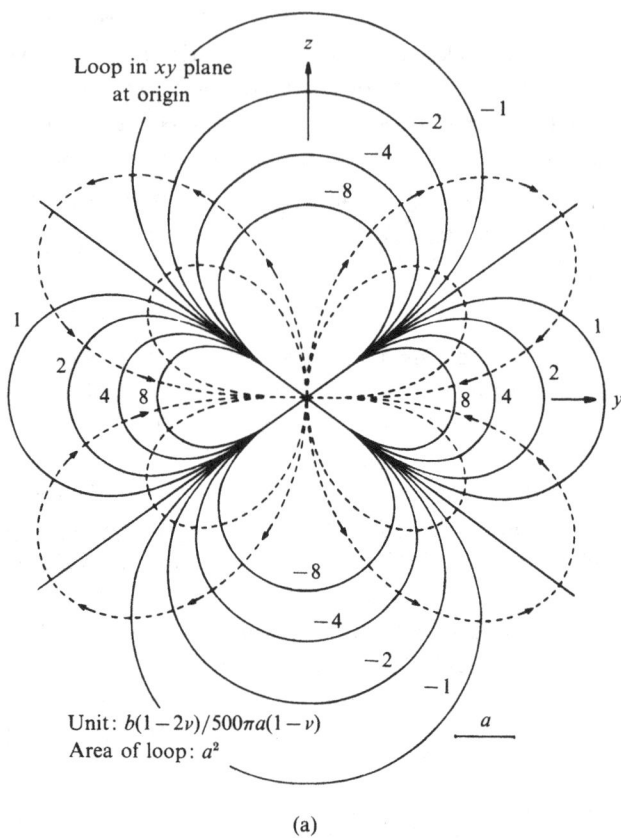

Loop in *xy* plane at origin

Unit: $b(1-2\nu)/500\pi a(1-\nu)$
Area of loop: a^2

(a)

Fig. 7.22. Dilatation field of small dislocation loops, (a) prismatic loop.[15]

end is driven up the side of the crystal and away from the interface. Slip in Region *A* causes dislocations to end in the interface and a force exists to have them move towards the interface normal direction in order to shorten their line length. They will thus propagate down the entire crystal so that, in this way, a multiplication of the dislocation density in later grown regions of the crystal occurs. For example, if Δn_D new slip dislocations penetrate the interface while the crystal grows a distance d_a (see Fig. 7.24), then at a distance $10d_a$ further along the crystal in the growth direction, the dislocation density intersecting the cross section will be $\sim 10\Delta n_D/\pi R_c^2$ where R_c is the crystal radius, provided the same temperature distribution holds, negligible annihilation occurs and the interface is flat and perpendicular to the crystal axis. If the

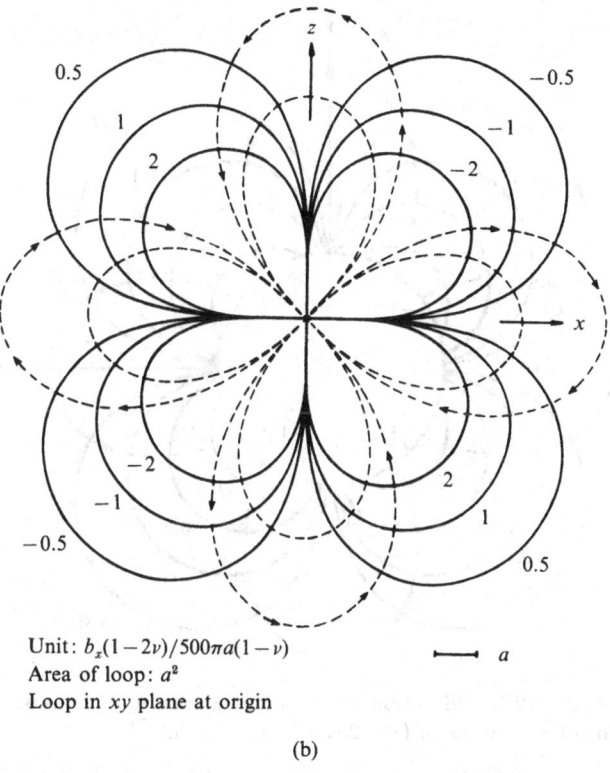

Unit: $b_x(1-2\nu)/500\pi a(1-\nu)$
Area of loop: a^2
Loop in xy plane at origin

(b)

Fig. 7.22. (b) Shear loop.[15]

interface is concave towards the liquid, the dislocation propagation is towards the central region of crystal so that the dislocation density is a function of radius (highest in center) as well as length along the crystal. If the interface is convex towards the liquid, the propagation direction of the dislocations is towards the periphery of the crystal and less overall multiplication occurs. This is a desirable condition as one would prefer to grow with this type of interface; however, for some axis orientations, like Si(111), this interface shape leads to large central facet formation that causes its own class of problems. For such orientations, one wishes to grow the crystal with a slightly concave interface and, by adjusting the thermal fields very close to the interface, make Δn_D very small.

If one is growing in a crystal orientation where the slip plane makes an angle θ_b as illustrated in Fig. 7.24, the dislocations generated in Region A' may be a large number so that Δn_D will be large and the

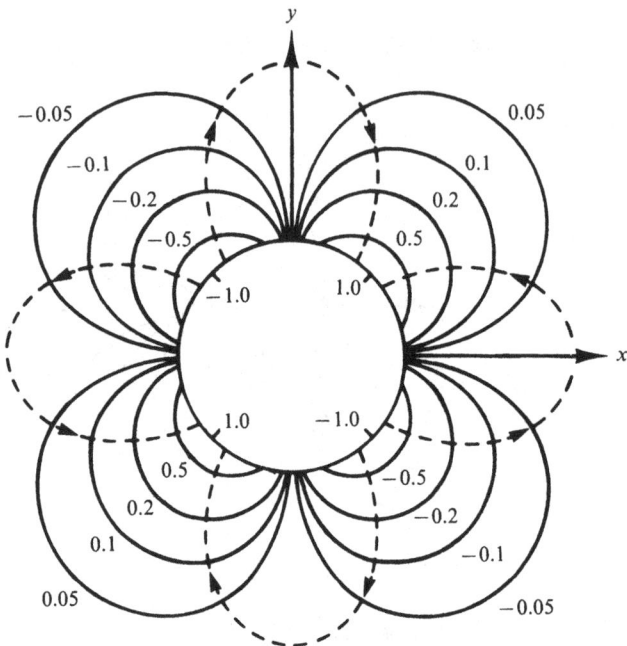

Fig. 7.23. Dilatation field around a spherical inclusion with shear misfit e_{xy} (units of $(1 - 2\sigma_*)e_{xy}/(1 - \sigma_*)$).[15]

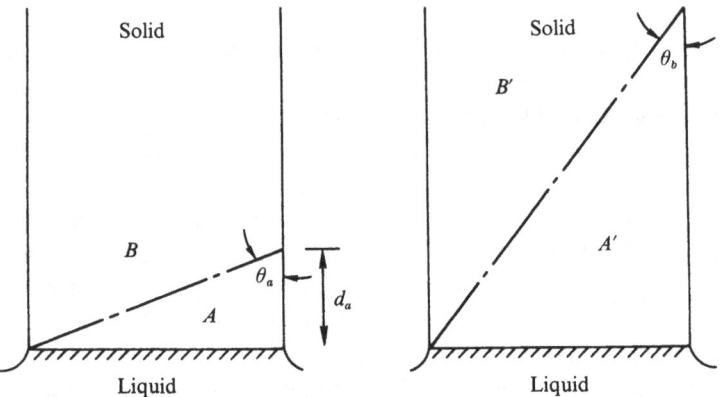

Fig. 7.24. Schematic illustration of domain A and A', for crystals of different axial orientation, wherein slip-generated dislocations may intersect the interface.

multiplication of these dislocations in subsequent regions of the crystal will be very significant. Thus, to minimize this effect, one should choose crystal axis orientations such that the slip plane makes the largest possible angle with the crystal axis. This means that, for Si, growth in the $\langle 110 \rangle$ or $\langle 211 \rangle$ is undesirable whereas growth in the $\langle 100 \rangle$ or $\langle 111 \rangle$ is preferred.

From a quantitative viewpoint, $n_D \sim 2(\bar{\alpha}^* \Delta T^* - \tilde{\varepsilon}_e)/b_*$, and as the temperature decreases σ_y increases so that $\tilde{\varepsilon}_e$ increases and eventually $(\mathrm{d}\Delta T^*/\mathrm{d}z)/\mathcal{G}_S < \mathrm{d}\sigma_y/\mathrm{d}T$. Thus, no more plastic deformation can occur so residual stresses begin to be built into the crystal below this point. In Fig. 7.24, $d_a = 2R_c \cot \theta_a$; thus, the number of dislocations penetrating the interface, Δn_D^p, from this slip source is given by

$$\Delta n_D^p \sim \frac{2}{b_*} \left[2\bar{\alpha}^* \left(\frac{\partial \Delta T^*}{\partial z} \right) R_c \cot \theta_a - \frac{\sigma_y}{E^*} \right] \tag{7.16a}$$

where

$$\frac{\partial}{\partial z} \Delta T^* \approx \frac{\partial}{\partial z} \int_0^{R_c} \frac{\partial T_S}{\partial r} \mathrm{d}r \tag{7.16b}$$

We thus see that Δn_D^p increases with crystal radius and decreases with increase of both θ_a and σ_y.

Dislocations can also be moved towards the interface by climb rather than by slip processes. The driving force for climb could come from thermal stresses, vacancy supersaturation stresses or interface image forces.

7.2.2 Interface penetration by climb

As a dislocation climbs, it will drain the surrounding region of its excess V_{Si}. We will assume that the radius, r^*, of this denuded region is just such as to allow a dislocation to climb into a fresh region in time t^*. Then, the climb velocity, V_c, is just r^*/t^*. The number of excess V_{Si} from a semicircle of radius r^* is just

$$m^* = \frac{\pi}{2} \left(\frac{r^*}{b_*} \right)^2 \left(C_{V_{Si}} - C_{V_{Si}}^* \right) = \frac{\ell}{b_*} \tag{7.17a}$$

where the dislocations climb a distance ℓ. To have a perpetuating process, we must set $\ell = r^*$ yielding

$$\frac{r^*}{b_*} = \frac{2}{\pi \left(C_{V_{Si}} - C_{V_{Si}}^* \right)} \tag{7.17b}$$

The time t^* it takes to drain the region r^* of excess V_{Si} is just given by $r^{*^2} \approx D_V t^*$ which leads to

$$V_c \approx \frac{D_V}{r^*} \approx \frac{\pi D_V \left(C_{V_{Si}} - C_{V_{Si}}^* \right)}{2b_*} \tag{7.17c}$$

If we had carried out an exact solution with Bessel functions, we would have obtained the same result but reduced by a factor of 0.2–0.4. Thus, the climb velocity, V_c, in a vacancy field is given by [11]

$$V_c \approx 0.5 \frac{D}{b_*} \left[\frac{C_{V_{Si}}(z)}{C^*_{V_{Si}}} - 1 \right] \qquad (7.17d)$$

where the self-diffusion coefficient $D = C^*_V(T)D_V$ and D_V is the vacancy diffusion coefficient. In Fig. 7.25, with $C_V(z) = C^*_V(T_M)$, a reduced velocity, $V = V_c b_*/D_0$, where $D = D_0 \exp(-U_d/\kappa T)$, is plotted versus T_M/T and we see that there is a maximum value at $\Delta T/T_M \sim 0.1$. This reduced climb velocity vanishes for small T because the vacancies become immobile and it also vanishes at the melting point because the vacancy supersaturation disappears there. One finds that the dislocations move to within ~ 0.5 cm of the interface and thereafter move at the same rate as the interface. In order for these dislocations actually to climb into the interface under this driving force, a vacancy supersaturation must exist at the interface. Thus, if $C_V(z)/C^*_V(T_M) \approx 2$ and $D(T_M) \sim 10^{-8}$–10^{-10} cm^2 s^{-1}, $V_c(T_M) \sim 10^{-1}$–10^{-3} cm s^{-1}. If the freezing velocity is less than this, these dislocations will terminate in the interface. Thus, slow freezing velocities for a fixed vacancy supersaturation favor dislocation penetration of the interface.

The image force on a dislocation parallel to and at a distance z_i from the interface is $\tilde{G}b_*^2/4(1-\nu_*)z_i \approx 10^{-5}/z_i$ dyne cm^{-1}. Thus, if $z_i \tilde{>} 1$ μm, the force attracting a dislocation to the interface will be very small. If the climb velocity V_c at $z_i \sim 1$ μm is equal to the interface velocity, image forces will take over at that point and pull the dislocations into the interface.

Dislocation annihilation does occur, but generally only over short segments of the line and even then only at high dislocation densities. For the densities of our interest, the annihilation probability is small with an upper limit of ~ 0.5.

7.2.3   Array formation

Dislocations generated in a crystal by various mechanisms will interact with each other behind the interface to form lineage boundaries and arrays in order to lower the total dislocation energy. If the arrays divide the crystal into rod-like elements such as illustrated in Fig. 7.26, they may also lower the free energy of the crystal by decreasing the excess vacancy concentration. The dislocations in the arrays will interact with

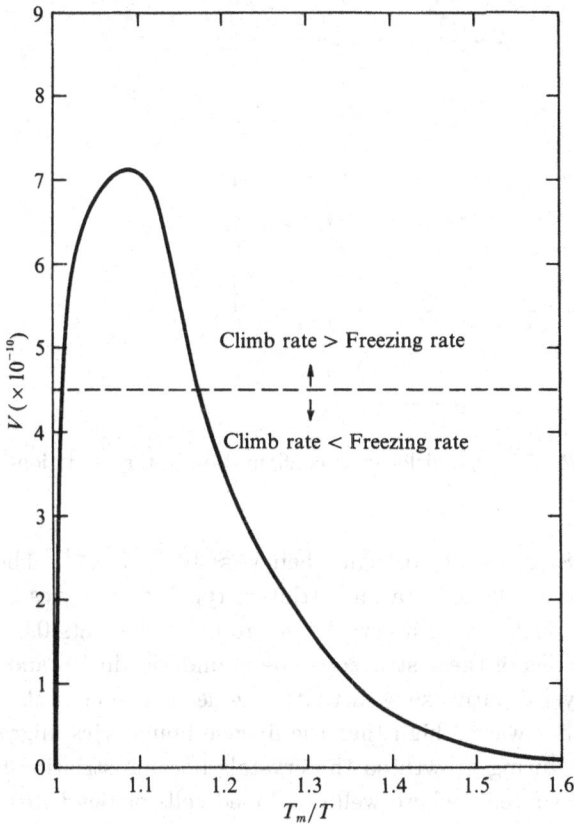

Fig. 7.25. Plot of reduced climb velocity, $V = V_c b_*/D_0$, versus the dimensionless temperature, T_M/T, for a vacancy supersaturated crystal.[11]

each other and climb, absorbing vacancies in the polygonization-type process.

An array of side λ cm (Fig. 7.26) will draw vacancies and dislocations from an area λ^2 cm^2 and, the larger is λ, the closer together are the dislocations in the array wall. Since the rate of dislocation climb is a strong function of the distance of separation, the rate of vacancy annihilation will increase as λ increases. However, when λ becomes too large, the vacancies do not have time to diffuse to a boundary from within the area λ^2. Thus, we would expect the most stable array to be the largest to which vacancies can diffuse in the time allowed which is given by

$$\lambda^2 = D_V \kappa T^2 / \Omega \mathcal{G}_S V \qquad (7.18)$$

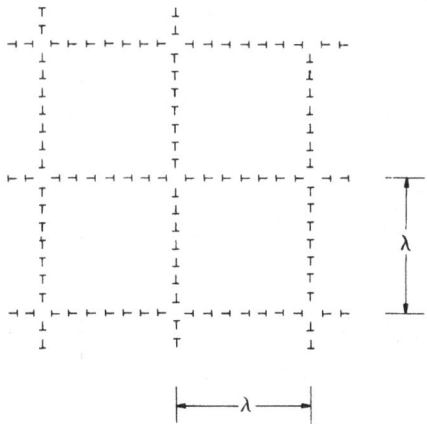

Fig. 7.26. Edge dislocation configuration in the "striation-type" array.

Equation (7.18) gives $\lambda \approx 0.1$ cm when $V \approx 10^{-3}$ cm s^{-1}. These conclusions fit the available data on "striation-type" boundaries in metals ($\Delta\theta \sim 1$–$5°$) which divide a crystal into rod-like elements 0.05–0.5 cm on a side. Studies of these striation-type boundaries in Pb and Sn as a function of crystal purity showed that, in zone-refined crystals, the dislocation mobility was so high that the lineage boundaries migrated out of the crystal during growth so the crystals became striation-free. At high impurity content, where well-developed cells or dendrites formed, no striations as such were seen probably because the dislocation mobility was too low. In this case, the lineage structure was reduced to the size scale of one or two cells, i.e., $\sim 5 \times 10^{-3}$ cm. Only in the middle range of purity is this substructure readily observed with the naked eye on etched crystals.

7.2.4 Dislocation-free crystals

To grow a dislocation-free Si seed crystal, one starts with a $\langle 111 \rangle$ or $\langle 100 \rangle$ seed containing dislocations and, after beginning growth, the seed is worked down slowly and carefully to a minimum size ($\sim$ 3 mm diameter and 30 mm long) that is safe from a strength point of view for the weight of crystal to be grown (3 mm diameter supports $\sim$ 200 kg). During this period, the interface must become strongly convex to the liquid as illustrated in Fig. 7.27 so that dislocation lines, present in the initial seed, can slide out to the sidewalls of the crystal. Subsequent growth is slowly tapered to the desired crystal diameter and

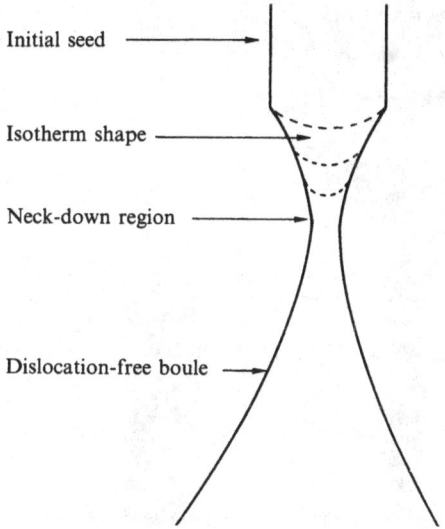

Initial seed

Isotherm shape

Neck-down region

Dislocation-free boule

Fig. 7.27. Schematic illustration of neck-down and flare-out geometries for growing dislocation-free crystals.

a dislocation-free crystal is then pulled. This is illustrated in Fig. 7.28(a) for the seeding of a float-zone crystal.[16] At this point, the crystal habit lines showing strongly on the conical surface of the crystal are used to characterize dislocation-free growth and their presence, as noted in Fig. 7.28(b) for a growing ⟨100⟩ crystal, can be taken as proof of a dislocation-free condition. As discussed in Chapter 3 of the companion book[8], the magnitude of these ridges depends on the size of the facets on the {111} layer source planes operating on the rim of the crystal in the meniscus region. The facet size will increase as the growth velocity increases and as the local temperature gradient decreases. Thus, one finds that dislocation-free Si crystal growth via the CZ method at excessive speeds causes the meniscus formed at the interface to be deformed towards a triangle from a circle for the ⟨111⟩ and towards a square from a circle for the ⟨100⟩. For a material like GaAs, which has more sluggish interface attachment kinetics than Si, the basic growth facet must be larger for the same V and G so the tendency to a triangular ⟨111⟩ or a square ⟨100⟩ crystal cross-section is even more pronounced (provided the GaAs crystal is dislocation free).

Loss of the dislocation-free condition of the crystal is thought to be indicated by the abrupt loss or change of one or more of the surface ridge-lines associated with the crystal habit. With the introduction of

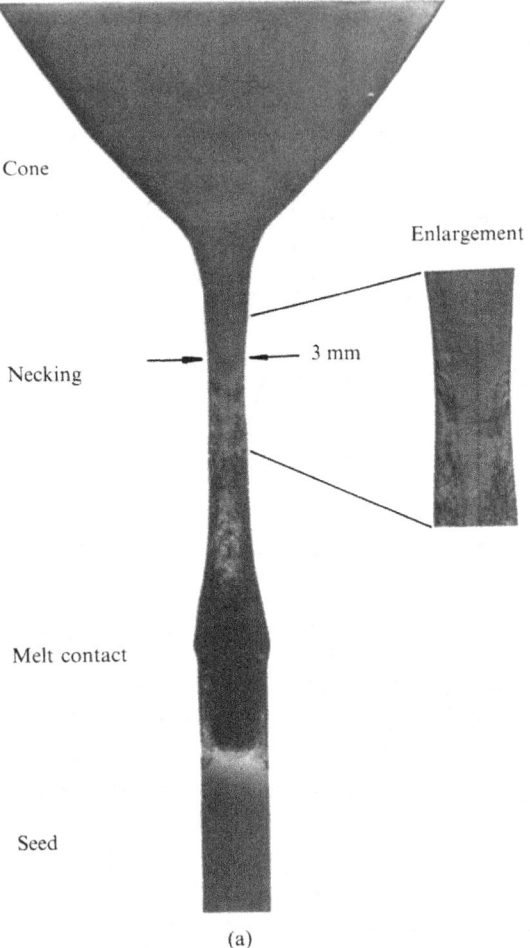

Cone

Enlargement

Necking   ⟶ 3 mm ⟵

Melt contact

Seed

(a)

Fig. 7.28. (a) X-ray topograph of seed, necking and conical part of a crystal.[16]

a suitable dislocation that can act as a layer source near one of the rim facets, the facet shrinks to a negligible size and the surface ridge generation mechanism vanishes. Thus, the abrupt loss of a surface ridge is a good indication of dislocation generation in that area. Many people feel that the generation of such dislocations is caused by SiO dust particles falling into the melt surface as a result of (a) SiO evaporation from the melt, (b) SiO deposits on the low temperature portions of the furnace, (c) fatigue cycle release of some SiO solid particles to fall onto the

Fig. 7.28. (b) Photographs of necking and conical shaping of CZ crystals for $\langle 100 \rangle$ (left) and $\langle 111 \rangle$ (right) orientations showing the brilliant crystal habit lines exhibited by dislocation-free crystals.[16]

melt surface and, (d) the natural convection flow pattern which sweeps the particles inwards towards the crystal/liquid meniscus region. The sudden introduction of dislocation layer sources will temporarily speed up the crystal growth rate which causes the crystal diameter to be reduced and, since the interface moves to a slightly higher temperature, it projects further towards the bulk melt where the convective heat transfer flux is greater so the crystal diameter should remain reduced. This explanation is consistent with the data presented in Fig. 7.29 where the abrupt increase in resistivity profile for a P-doped crystal ($k_0^P < 1$) is consistent with the smaller solute boundary layer thickness, δ_C, associated with the interface being in a region of stronger convective flow.[16] As the SiO particle induces dislocations to form, the corrugated crystal surface character changes to smooth and the growth habit fades.

7.3 Macroscopic fault formation
7.3.1 *Twins and stacking faults*
Two mechanisms are available to account for the formation of twins or twin lamellae during the growth of crystals: (i) mechanical or thermal stresses, and (ii) probability accidents that lead to en-

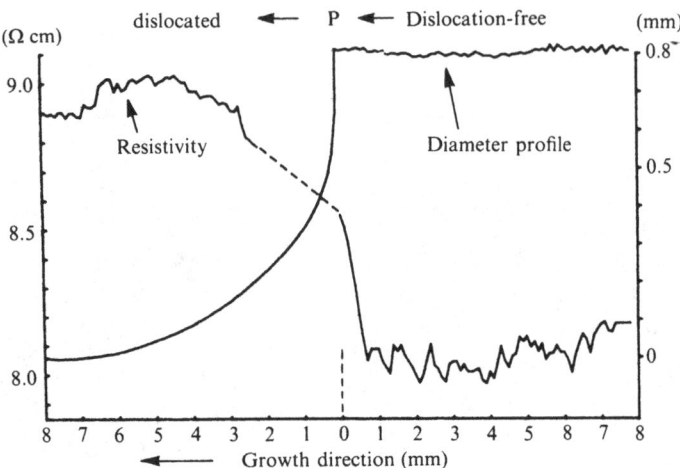

Fig. 7.29. Diameter profile change and resistivity profile change after the abrupt generation of dislocations in a CZ crystal due to an SiO particle.[16]

hanced interface attachment kinetics. The oxide crystal, $2BaO : TiO_2 : P_2O_5$(BTP), exhibits both of these mechanisms as shown in Figs. 7.30(a) and (b).[17] Here, the (100) growth twins provide reentrant corners via the interface (310)-type planes for easy and rapid growth in the [010]-direction. In the same crystal, (001) mechanical twin traces are also seen. During the horizontal growth of Ge crystals using 10° misoriented seed crystals, as illustrated in Fig. 7.31, a periodic array of twin lamellae form at the crucible walls with the density of the array increasing as V increases and $\mathcal{G}_L$ decreases.[18] The layer-flow mechanism needed for growth is illustrated in Fig. 7.32 and we see that an enhanced zone of supercooled liquid exists for the off-axis orientations both on the layer-generation side and on the layer flow side. On the layer-generation side, we require $V = V_{\langle 111 \rangle}\cos\phi$ ($\phi = 10°$ in Fig. 7.31(a)) so the increased $V_{\langle 111 \rangle}$ requires a larger facet area which means a greater undercooling. On the layer-flow side the ledge velocity must vary inversely as the ledge density so the more closely does the layer flow direction coincide with the interface isotherm tangent, the smaller is the ledge density and the greater is the supercooling needed to provide this ledge velocity.

From a theoretical viewpoint, for dislocation-free Si, twin formation occurs primarily at the facet planes on the Si in the meniscus region (see Figs. 3.5 and 3.6 of the companion volume[8]) which act as the layer

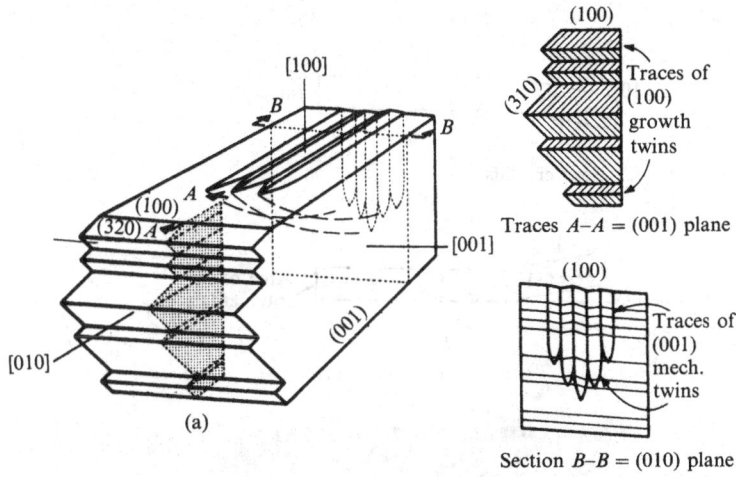

Fig. 7.30. Relative orientations of growth twins (100) and mechanical twins (001) in a $Ba_2TiP_2O_9$ crystal, (a) three-dimensional schematic drawing and (b) thin section micrograph under crossed nicols $(14\times)$.[17]

Plan view

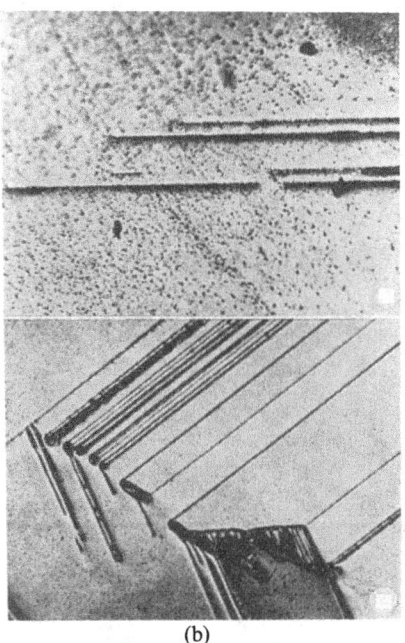

(b)

Fig. 7.31. (a) Schematic illustration of twin lamellae formation in horizontally grown Ge crystals tilted ±10° away from the ⟨111⟩ and (b) photomicrographs of internally bounded twin lamellae on decanted interfaces of Ge crystals, both in the presence of some constitutional supercooling (as evidenced by "pox" formation; (top) – small additions of Ga (300×) and (bottom) at higher Ga additions (100×).[18]

source regions for normal Si growth and which give rise to the rib-lines on the exterior of the Si crystals. Because of the supercooling needed

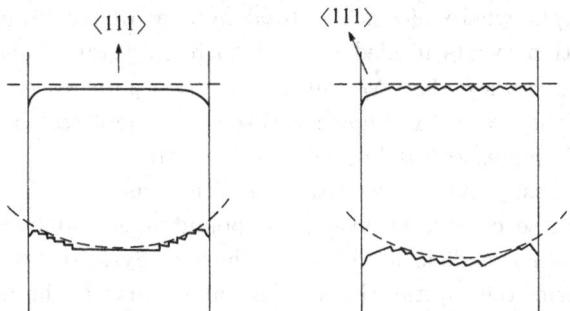

Fig. 7.32. Schematic illustration of the interface supercooling zones associated with both layer generation regions and layer-flow regions for both ⟨111⟩ and off-⟨111⟩ axis growth.

to drive these two-dimensional nucleation layer sources, especially in the presence of rotational striations and temperature fluctuations, there is always a finite probability that a two-dimensional pillbox can form in a faulted arrangement leading to the formation of a twin by lateral propagation of the faulted layer edge away from the facet region. The critical size, r^*, of this faulted pillbox is given by

$$r^* = \gamma'_\ell/(\Delta S_F \Delta T_i - \gamma_t/h) \tag{7.19a}$$

where γ'_ℓ is the ledge energy in the presence of the twin, γ_t is the twin energy and h is the height of the pillbox. The rate of formation of faulted pillboxes is given by

$$\dot{N}_f = A_F I_0 \exp[-\pi h^2 v_m \gamma'^2_\ell/\mathcal{R} T_i(h\Delta S_F \Delta T_i - \gamma_t)] \tag{7.19b}$$

where A_F is the facet area, $I_0 \sim (10^{42}/\Delta T_i^2)\exp(-\Delta G_A/\mathcal{R}T)$, ΔG_A is the activation barrier at the interface and v_m is the specific volume for Si. For Ge and Si, $\Delta S_F \approx 1.9$ J cm^{-3} K^{-1} so ΔT_i must be large or γ_t/h must be small in order that $r^* > 0$. We do not have reliable numbers for γ_t although people like to quote numbers for macroscopic twins in semiconductors in the range of 10–100 erg cm^{-2}.[19] Certainly, these numbers are an order of magnitude too large for pillbox application because h is probably not more than two atomic layers thick. Once again we see the need for atomic scale computations of the type illustrated by Fig. 8.20 of the companion book.[8]

Since useful numbers are not available, and since γ'_ℓ will be different than γ_ℓ for the unfaulted case (when $\gamma_t = 0$), there is little point in comparing $\dot{N}_f$ with $\dot{N}_{0f}$ for unfaulted pillboxes. However, we know that, to grow an untwinned Si⟨100⟩ crystal of length L, we must have

$\dot{N}_{0f}/\dot{N}_f \gg 4L/h$, where $\dot{N}_{0f}$ is the total number of two-dimensional pillbox nucleation events needed to grow such a crystal. Considering Eq. (7.19b), we can see that $\dot{N}_f$ will increase as A_F increases ($\mathcal{G}_S$ decreases), as γ'_ℓ decreases (as impurity adsorption increases), as γ_t decreases, as ΔT_i increases (as V increases, the attachment kinetics become more sluggish) and as fluctuations in V increase.

Extending these considerations to compound semiconductors, even though the crystals are dislocated and thus have ready-made layer sources in the center of the crystal, the interface must curve in the meniscus region for the reasons discussed in the companion volume.[8] However, these are strongly faceting systems and, for $\langle 100 \rangle$ crystals of GaAs or InP, four edge facets will form and expand in size as the main interface moves because the $\langle 100 \rangle$ interface layers will only grow as far as the edge of the $\{111\}$ facets because the $\{111\}$ facets have a deeper cusp in the γ-plot than do the $\{100\}$ facets. Thus, only repeated two-dimensional pillbox nucleation on these $\{111\}$-type edge facets can keep their size relatively small. This means that Eqs. (7.19) also apply to dislocated crystals that exhibit faceting; i.e., compound semiconductors. We can use the macroscopic twin energies as a guideline for γ_t so we expect that $\dot{N}_f(\text{InP}) > \dot{N}_f(\text{InAs}) > \dot{N}_f(\text{InSb or GaP}) > \dot{N}_f(\text{GaAs or GaSb}) > \dot{N}_f(\text{Si})$. We also know that, because the $(\bar{1}\bar{1}\bar{1})$ attachment kinetics are much more sluggish than the (111) attachment kinetics, $\dot{N}_f(\bar{1}\bar{1}\bar{1}) > \dot{N}_f(111)$ for any of these crystals. Likewise, for off-stoichiometry melts or impurity contaminated melts, solute build-up develops in the meniscus region leading to a lower liquidus temperature so A_F increases and γ_ℓ generally decreases. This leads to $\dot{N}_f(\text{off-stoichiometry}) > \dot{N}_f(\text{stoichiometry})$ and $\dot{N}_f(\text{impurity}) > \dot{N}_f(\text{pure})$. Finally, considering Fig. 7.32, we expect that $\dot{N}_f(\text{concave}) > \dot{N}_f(\text{plane})$ and $\dot{N}_f(\text{unsymmetrical}) > \dot{N}_f(\text{symmetrical})$. Of course, for all of these systems, as V increases and $\mathcal{G}_S$ decreases, $\dot{N}_f$ increases.

7.3.2   *Void and stray crystal formation*

As discussed in Chapter 3, potentially gaseous solutes that have small k_0 values can produce a build-up at the interface sufficient to (a) nucleate directly a bubble of gas on the interface ($N + N \rightarrow N_2$ or $O + O \rightarrow O_2$ from dissolved air in water), or (b) react in the interface region to form a new species at a sufficiently high concentration level to nucleate gas bubbles on the interface ($C + O \rightarrow CO$ in liquid Fe). The interface gas bubbles inhibit the growth of the solid beneath the cap

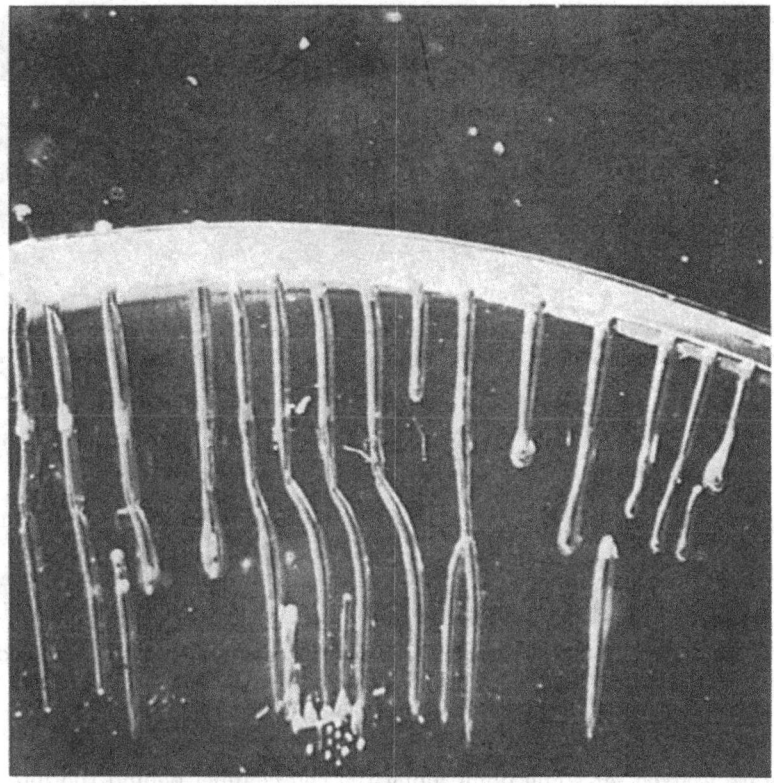

Fig. 7.33. Photomicrograph of a growing ice crystal illustrating the formation of long pores at the ice/water interface because of dissolved air build-up due to interface partitioning (15×).

of the bubble and voids are left in the solid as the interface advances. At rapid freezing rates an array of bubbles is formed whereas, at slow freezing rates, long pores are formed as the individual bubbles are fed by diffusion in the interface layer. An example of bubble formation in ice crystals is shown in Fig. 7.33.

A second prime source of void formation occurs for materials that shrink on freezing (metals, many oxides, etc.) and develop the cellular interface morphology with long narrow liquid channels penetrating into the solid. This happens because macroscopic liquid flow through these long narrow channels is required to feed the groove roots between cells. At low freezing rates this fluid flow occurs readily but at high freezing rates, the flow is inhibited by the viscosity of the liquid. Since the head of pressure is only ~ 1 atm at the macroscopic interface and the flow

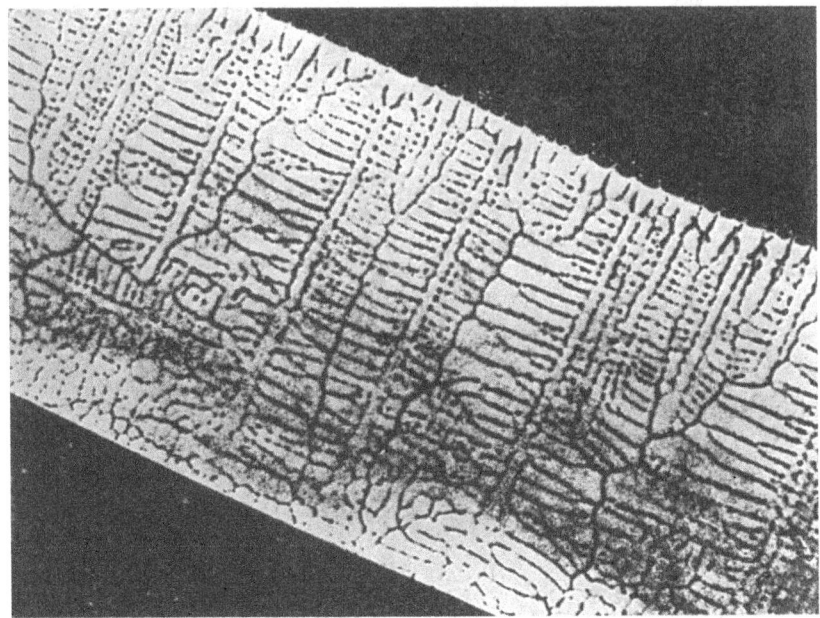

Fig. 7.34. A constrained growth Aℓ single crystal showing dendritic substructure. The dark, circular, pit-free regions surround small holes or pores that are only a few micrometers in diameter (14×).

rate is fixed, at some point along the cell channel behind the interface the local pressure will become less than zero and cavitation can occur. Cavitation will relieve the local tensile stress in the channel fluid and a void $\sim$ 1–5 μm in diameter is expected to form because this is the width of the fluid channels ($\sim$ 1–5% of the cell width). This phenomenon appears most prominently in rapidly solidified metal castings, especially Aℓ castings. Figure 7.34 shows a variety of micron size pores in an Aℓ single crystal grown dendritically.

Another void formation event comes from the area of fiber crystal growth. Although sapphire fibers can be readily grown at pull velocities $\sim$ 2 cm s^{-1}, they are unsuitable for optical applications because of the presence of microvoids. These microvoids are found to congregate within a well-defined area near the fiber center, are spherical and have a diameter $\sim$ 1 μm.[20] The microvoids were found to be arranged in a sheet-like pattern made from separate necklaces with the sheets aligned at an angle ϕ to the fiber axis for both c-axis and a-axis fibers (see Fig. 7.35). This angle increased with the pull velocity, being $\phi \sim 0°$ at $V = 1.0$ cm min^{-1}

and $\phi \sim 8\text{-}10°$ at $V = 2.3$ cm min^{-1}. At $V \approx 0.8$ cm min^{-1}, the microvoid pattern disappeared completely and the fibers were of good optical quality. These results are fully consistent with the cell formation picture. From old studies on horizontally grown Sn crystals it was found that cell boundaries are aligned parallel to the axis of heat flow for slow rates of growth and that the alignment varies from this direction as V increases to approach the dendrite direction at high V. This observation was discussed in Chapter 6 and shown to depend upon the sequential development of various layer flow systems that operate on the cell caps as the cell caps become more parabolically shaped. This cell cap shape transition occurs as the degree of constitutional supercooling to be removed is increased (see Fig. 6.11).

Two factors that strongly influence the incidence of stray crystal formation are the macroscopic interface shape and the degree of constitutional supercooling existing in the liquid ahead of the interface. As discussed earlier, in the immediate vicinity of a crucible surface for Bridgeman crystal growth, a small volume of supercooled liquid is present. Both the degree of supercooling and the volume of supercooled liquid are greater if the interface is concave rather than convex so the probability of stray crystal formation is greater for concave than for convex interfaces. In addition, for concave interfaces, stray crystals tend to grow into the existing crystal because the grain boundary tends to align itself perpendicular to the interface. When a stray crystal forms on a convex interface, it cannot grow to any appreciable size because the existing crystal encroaches on it.

If heterogeneous nuclei are present in the liquid that can operate at a supercooling of δT_c, many stray crystals will form whenever the degree of CSC exceeds δT_c. Although cell and dendrite formation reduces the interface concentration at the caps of the cells below that for the planer interface, C_i^P, and thus the degree of constitutional supercooling is reduced, sufficient supercooling to nucleate stray crystals exist when the following condition holds

$$\Delta T_K + m_L(C_i^c - C_\infty)\left[\exp\left(-\frac{V}{D}z^*\right) - 1\right] - G_L z^* \overset{\sim}{>} \delta T_c \qquad (7.20a)$$

where

$$z^* = \frac{D}{V}\ell n\left[-\frac{Vm_L}{DG_L}(C_i^c - C_\infty)\right] \qquad (7.20b)$$

is the location ahead of the interface where the maximum degree of CSC exists and where ΔT_K is the attachment kinetic undercooling needed at the interface for growth at velocity V. If this condition is satisfied, stray

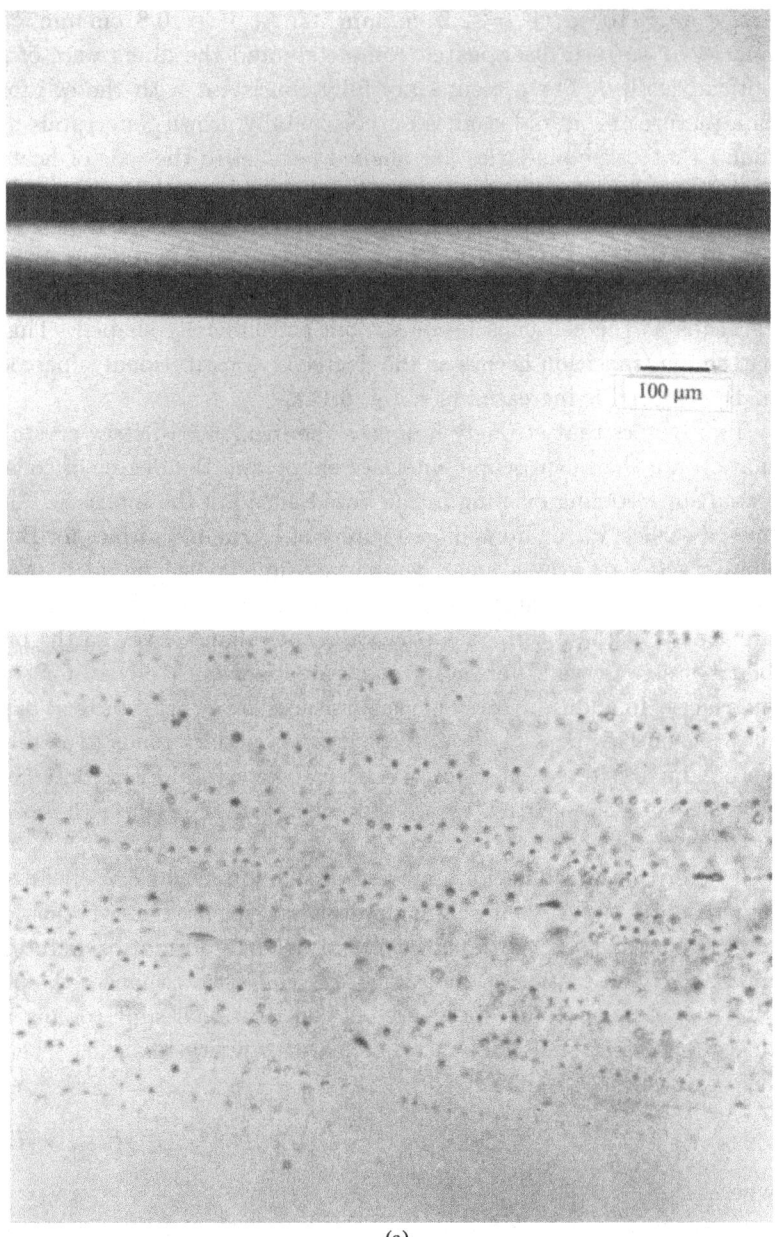

100 µm

(a)

Fig. 7.35. (a) Optical micrograph of microvoids formed in melt-pulled sapphire fibers.[20]

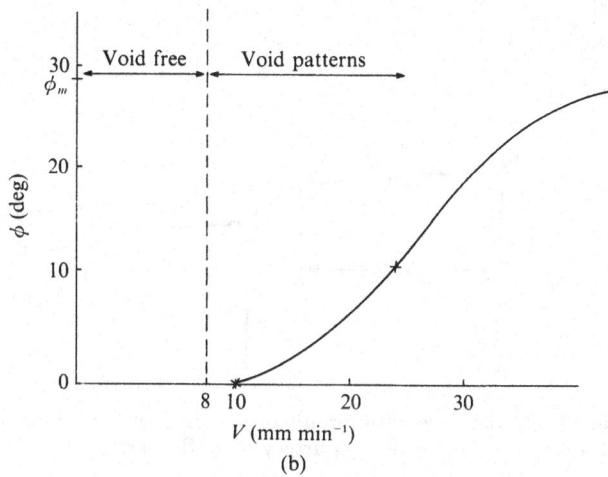

Fig. 7.35. (b) Plot of angle ϕ that the void string pattern makes with the fiber axis as a function of fiber pull velocity.

crystals will form ahead of the interface and inhibit either single crystal or columnar crystal growth. This same condition determines the onset of the equiaxed crystallization mode during ingot formation.

7.3.3 Lateral microscopic segregation trace defects

Referring to the step bunching instability of Fig. 6.16, the evaluation of this macrostep in time leads to a lateral chemical segregation trace in the crystal called a type II striation in contrast to the type I striations of Fig. 4.15 associated with fluctuations in macroscopic interface velocity. Let us try to understand the characteristics of this "trace" defect.

In the TLK picture, depending upon the interface orientation deviation, θ, from a facet plane, the step height, h, and terrace spacing λ, are randomly distributed with average values depending on the magnitude of θ. Since $V_\ell \propto h^{-1}$, generally, steps with small h catch up with steps of larger h and merge into them, leading to "step bunching". The average h as well as the average λ increase during growth due to step bunching. Continuation of this process leads to visible macrosteps on the surface. A macrostep consists of a tread and a riser with the average density of microsteps on a tread being much smaller than that on a riser (see Fig. 7.36(a)). A tracing of the riser development with time is left in the crystal because the local impurity concentration differs from that of

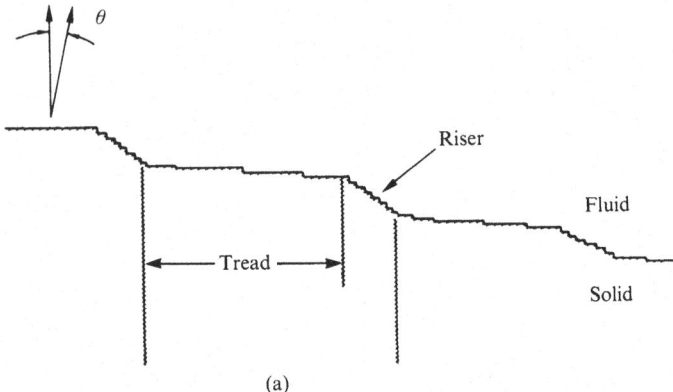

Fig. 7.36. (a) Schematic profile of a vicinal surface at orientation θ to the layer source plane. (Courtesy of E. Bauser.)

the surrounding host crystal and thus can be revealed in semiconductors by etching. The generalized cross-sectional shape of a single trace is illustrated in Fig. 7.36(b) while a three-dimensional representation of trace development is given in Fig. 7.36(c).[21] Fig. 7.36(d) illustrates the fanning out of risers R_B and R_C indicating that a decay process also exists. It is found that the angles α and ϕ steepen as the height of the riser increases.

This mechanism is not only operative during LPE crystal growth but also during CZ growth of Si and Ge. In Figs. 7.37(a) and (b), such segregation traces are illustrated for concave and convex interfaces, respectively.[22] In Fig. 7.37(c), a longitudinal section of an Sb-doped CZ Si⟨111⟩ crystal reveals these striation II traces relative to the striation I traces. In Fig. 7.38, a different microsegregation trace, called a "valley trace" due to solute trapping is illustrated.

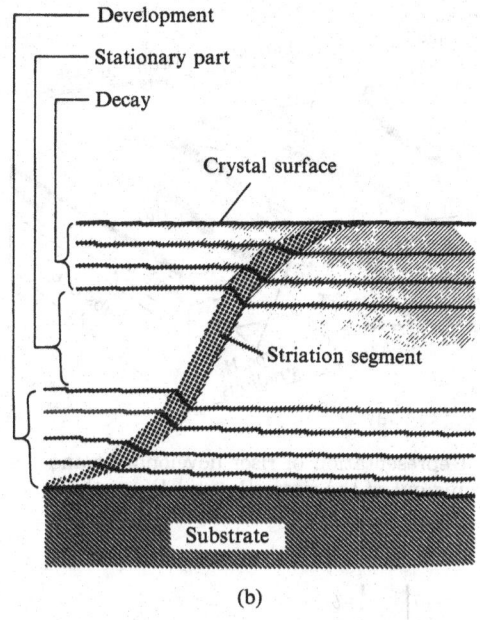

(b)

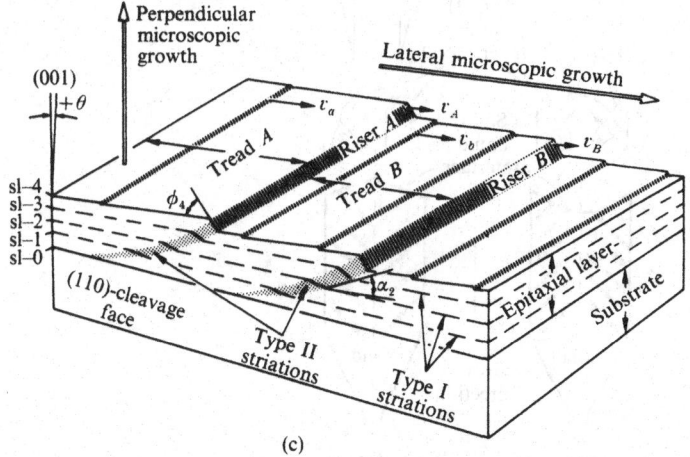

(c)

Fig. 7.36. (b) Generalized trace of the chemical segregation associated with riser height build-up and decay, (c) three-dimensional evolution of type II striation development. (Courtesy of E. Bauser.)

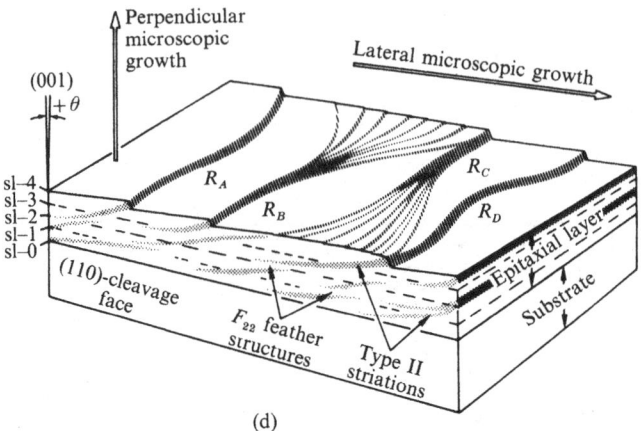

Fig. 7.36. (d) Representation of riser development and decay, indicated by the fanning out of the risers R_B and R_C at the $s\ell$-4 interface. (Courtesy of E. Bauser.)

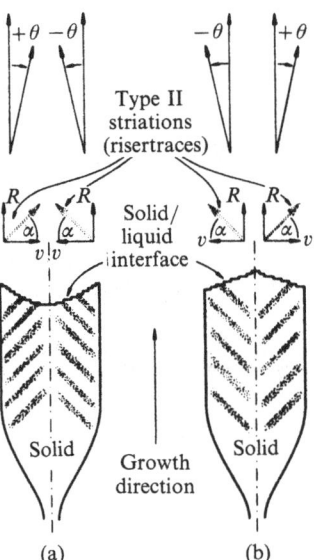

Fig. 7.37. Riser trace development during CZ growth of a Si⟨111⟩ crystal with (a) a concave solid/liquid interface and (b) a convex solid/liquid interface. (Courtesy of E. Bauser.)

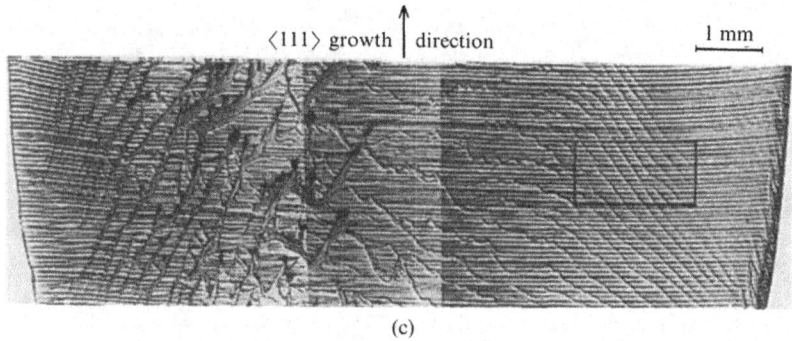

(c)

Fig. 7.37. (c) Longitudinal section of an Sb-doped CZ Si⟨111⟩ crystal showing striation II traces relative to striation I traces. (Courtesy of E. Bauser.)

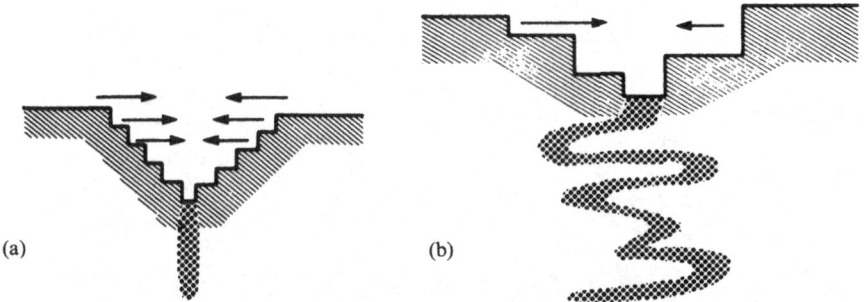

(a)   (b)

Fig. 7.38. Schematic development of "valley traces" where the direction of step motion is indicated by the arrows. At the valley bed where the steps meet and annihilate, dopants and impurities may be preferentially trapped or rejected. The trace of the valley may be straight when the step movement is symmetrical and regular, as in (a), or it may oscillate in the lateral direction as shown in (b) when the step train movement is unsymmetrical and irregular. (Courtesy of E. Bauser.)

8

Defect generation during thin film formation

8.1   Mechanics of stresses in thin films

Suppose we consider a thin film/substrate composite that is completely free of stress as in Fig. 8.1(a).[1] Because the film is under no stress, we can imagine removing it from the substrate and allowing it to be in a stress-free state. In this state, the lateral dimensions of the film will exactly match those of the substrate. However, in actuality, atomic force relaxations will produce dimensional changes in the film. If the dimensions of the film are changed in any way, then elastic strains and stresses will develop in the film when it is reattached to the substrate after some relaxation time has passed. Let us suppose that the detached film experiences a uniform volume shrinkage (dilatational transformation strain, e_T) as indicated in Fig. 8.1(c). For a pure dilatational strain, the principal strain components are $\tilde{\varepsilon}_{xx} = \tilde{\varepsilon}_{yy} = \tilde{\varepsilon}_{zz} = e_T/3$.

Let us now reattach the film to the substrate. Because the lateral dimensions of the film no longer match those of the substrate, a biaxial stress must be imposed on the film to deform it elastically, so that it again fits the dimensions of the substrate. The stress required to do this produces elastic strains $\tilde{\varepsilon}_{xx}$ and $\tilde{\varepsilon}_{yy}$ that exactly compensate these components of the transformation strain. Thus,

$$\tilde{\varepsilon} = \tilde{\varepsilon}_{xx} = \tilde{\varepsilon}_{yy} = -e_T/3 \qquad (8.1a)$$

Using Hooke's Law, this leads to a biaxial stress in the film

$$\sigma = \sigma_{xx} = \sigma_{yy} = \tilde{M}\varepsilon \qquad (8.1b)$$

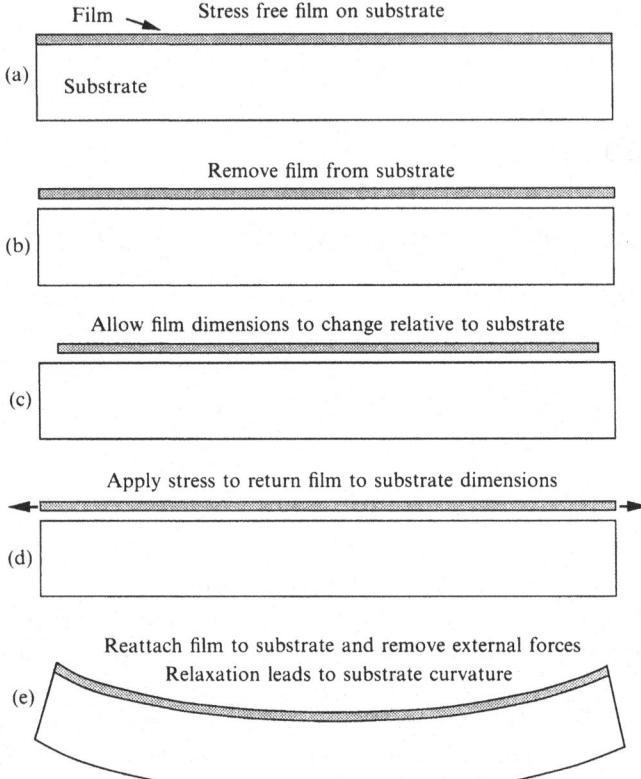

Fig. 8.1. Schematic illustration of the relationship between biaxial stress development in a thin film and the associated bending of its attached substrate.[1]

where $\tilde{M}$ is the biaxial elastic modulus of the film. For isotropic elasticity, the biaxial modulus of the film is simply

$$\tilde{M} = E^*/(1 - \sigma^*) \qquad (8.1c)$$

where E^* is Young's modulus and σ^* is Poisson's ratio.

In Fig. 8.1, the tensile forces needed to deform the film to match the dimensions of the substrate produce a biaxial stress state in the film. As long as these forces are present, the stress in the film does not change when it is reattached to the substrate. Of course, the local atomic reconstruction state in the surface of the film may not match the substrate so that thermally activated temporal atomic rearrangements will ensue. After the film is again perfectly bonded to the substrate, the edge forces are removed by superimposing forces of the opposite sign on the edges of the film. These additional forces remove the normal tractions from the edges of the film and produce shear stresses on the film/substrate interface near the edges of the film which provide the forces needed to maintain the biaxial stress in the film. These forces cause the substrate to bend elastically as indicated in Fig. 8.1.

For the case of single crystal films or for polycrystalline films with a strong crystallographic texture, the anisotropy of the elastic properties must be taken into account.[1] For a single crystal film of cubic material in which the (100)-cube plane lies parallel to the plane of the film, the biaxial elastic modulus, $\tilde{M}(001)$, is isotropic in the plane of the film and is given by

$$\tilde{M}(001) = C_{11} + C_{12} - \frac{2C_{12}^2}{C_{11}} \qquad (8.2a)$$

where C_{11} and C_{12} are the components of the stiffness matrix. Although neither E^* nor σ^* is isotropic in the (001) plane, the ratio $E^*/(1 - \sigma^*)$ is isotropic in that plane. A single crystal film with a (111) plane lying in the plane of the film is fully isotropic in that plane and the biaxial modulus, $\tilde{M}(111)$, is given by

$$\tilde{M}(111) = \frac{6\,C_{44}(C_{11} + 2C_{12})}{C_{11} + 2C_{12} + 4C_{44}} \qquad (8.2b)$$

For a single crystal film with a (011) plane parallel to the plane of the film, the elastic properties are not isotropic in the plane of the film. The elastic moduli in the mutually perpendicular [100] and [01$\bar{1}$] directions are different. If such a film is subjected to in-plane stresses such that the result is an equal biaxial strain, $\tilde{\varepsilon} = \tilde{\varepsilon}_{xx} = \tilde{\varepsilon}_{yy}$, the stresses in the two directions can be computed using the following moduli[1]

$$\tilde{M}[100] = C_{11} + C_{12} - \frac{C_{12}(C_{11} + 3C_{12} - 2C_{44})}{C_{11} + C_{12} + 2C_{44}} \qquad (8.2c)$$

and

$$\tilde{M}[01\bar{1}] = \frac{2C_{11} + 6C_{12} + 4C_{44}}{4}$$
$$- \frac{(2C_{11} + 2C_{12} - 4C_{44})(C_{11} + 3C_{12} - 2C_{44})}{4(C_{11} + C_{12} + 2C_{44})} \qquad (8.2d)$$

As stated earlier, the edge forces exerted on the substrate by the biaxial stress in the film cause the substrate to deform elastically to the shape of a spherical shell for isotropic modulus. A simple biaxial bending analysis shows the curvature, $\mathcal{K} = R^{-1}$, of the spherical shell to be given by

$$\mathcal{K} = \frac{1}{R} = \frac{6\sigma_f h_f}{\tilde{M}_s h_s^2} \qquad (8.3)$$

where R is the radius of curvature, M_s is the biaxial elastic modulus of the substrate, σ_f is the biaxial tensile stress in the film, h_f is the film thickness and h_s is the thickness of the substrate. The edge force per unit length is represented here by $\sigma_f h_f$ and we note that Eq. (8.3) can be readily inverted to yield the stress in the film. When the (011) plane of the substrate is parallel to the plane of the film (for non-cubic substrates), the substrate is not elastically isotropic in that plane and does not bend symmetrically.

It should be noted that Eq. (8.3) for the elastic bending of the substrate does not depend on the elastic properties or on any other mechanical properties of the film because the thin film approximation was used to derive Eq. (8.3). Typically, films ~ 1 μm thick are deposited onto substrates that may be 500–1000 times thicker. In such cases, the flexural modulus of the thin film/substrate composite is completely dominated by the properties of the substrate. It follows that, when multiple thin films are deposited sequentially onto a much thicker substrate, each film causes a fixed amount of bending to occur, irrespective of the order in which the films are deposited. The amount of curvature change is determined by the stress and thickness of each film. The total change of substrate curvature is simply the *algebraic* sum of the curvature changes associated with the presence of each film.[1]

As discussed above, the stresses in thin films on substrates can be viewed as arising from the misfit that must be accommodated elastically when the film is attached to the substrate. For different thermal expansion coefficients, $\bar{\alpha}^*$, thermal strains, $\tilde{\varepsilon} = -\Delta\bar{\alpha}^*\Delta T$, will develop in the film on cooling an amount ΔT. If the density of the film changes after it has been bonded to the substrate, an "intrinsic" or growth strain, $\tilde{\varepsilon} = -e_T/3$, develops in the film where e_T is the dilatational "transformation" strain associated with the change of density. A film that immediately densifies when it is attached to the substrate must be subjected to a biaxial tensile strain to match the dimensions of the substrate. Later density changes due to the transport of vacancies or interstitials into the film will change this state of strain. For epitaxial films on thick substrates, the elastic accommodation strain, $\tilde{\varepsilon} = \Delta a/a$, depends on the lattice parameters, a_f and a_s, of the film and substrate respectively ($\Delta a = a_s - a_f$). Here, all the elastic accommodation is assumed to take place in the film because the substrate, being so much thicker than the film, is essentially rigid. During the nucleation and early stages of film growth, the strain state may be quite different than during the later stages of film growth because surface and interface reconstruction effects tend to dominate these early processes.

8.2 Misfit dislocation formation in epitaxial films

There is a great deal of interest in growing semiconductor thin films on dislocation-free substrates using heteroepitaxy. Of course, any lattice mismatch between the film and the substrate must be accommodated by a uniform strain in the film which typically leads to very large biaxial stresses in the film (together with a slight bending of the substrate). One might intuitively expect these large stresses to be relaxed by plastic flow in the film, through dislocation nucleation and motion, regardless of the film thickness. However, this cannot occur in very thin films because the dislocation energy created by such relaxation processes is greater than the reduction of strain energy associated with this type of relaxation. Thus, there is a critical film thickness, h_c, below which the film is thermodynamically stable with respect to dislocation formation so that, below h_c, misfit dislocations do not form in heteroepitaxial structures. Of course, dislocations already present in the substrate can propagate into the thin film.

The equilibrium theory of misfit dislocation formation is well established.[1-3] As shown in Fig. 8.2, the excess energy per unit area of a very thin film is lowest when no dislocations are present. Such films are perfectly coherent with the substrate in their equilibrium state and have an energy per unit area given by

$$E_h = \tilde{M} h \tilde{\varepsilon}^2 \quad \text{for} \quad h < h_c \tag{8.4a}$$

where $\tilde{\varepsilon}$ is the biaxial elastic strain that must be imposed on the film to bring the lattice of the film into coincidence with that of the substrate and $\tilde{M}\tilde{\varepsilon}$ is the biaxial stress in the film. When dislocations with Burger's vector b_* and spacing λ are formed, the remaining homogeneous strain in the film is $\tilde{\varepsilon} - b_*/\lambda$ and the corresponding strain energy is reduced accordingly while E_h is reduced as misfit dislocations are introduced, the energies, E_d, of such dislocations increase the total energy of the system. The dislocation energy per unit area is given by

$$E_d = \frac{\tilde{G} b_*^2}{4\pi(1 - \sigma^*)} \frac{2}{\lambda} \ell n \left(\frac{\beta h}{b_*} \right) \tag{8.4b}$$

where $\tilde{G}$ is the shear modulus of the thin film and substrate (here assumed to be the same) and β is a numerical constant of the order of unity. The factor $2/\lambda$ represents the misfit dislocation length per unit film area. The dislocation energy depends logarithmically on the film thickness because this dimension controls the outer cut-off radius for the elastic field of the dislocations. The total energy of thin films containing

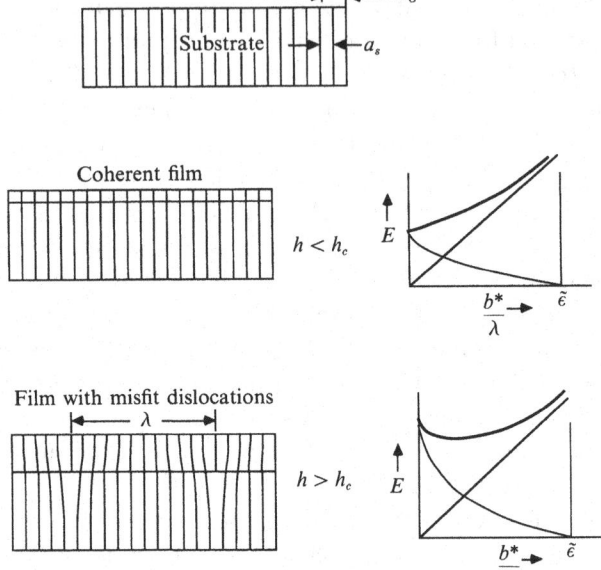

Fig. 8.2. Illustration of the equilibrium theory of misfit dislocation formation. Below a critical film thickness, h_c, the equilibrium state is dislocation free. In thicker films, dislocations form to reduce the free energy of the system.[1]

misfit dislocations is thus

$$E = \tilde{M}h\left(\tilde{\varepsilon} - \frac{b_*}{\lambda}\right)^2 + \frac{\tilde{G}b_*^2}{2\pi(1 - \sigma^*)\lambda}\ell n\left(\frac{\beta h}{b_*}\right) \qquad (8.4c)$$

We note that E_h varies linearly with h while E_d varies only logarithmically with h. As a consequence, only above a critical thickness, h_c, does the introduction of misfit dislocations lead to a decrease in the energy of the system. By minimizing E with respect to the number of misfit dislocations per unit length, λ^{-1}, h_c can be determined as the special case of $b_*/\lambda = 0$ yielding

$$\frac{h_c}{\ell n(\beta h_c/b_*)} = \frac{\tilde{G}b_*}{4\pi(1 - \sigma^*)\tilde{M}\tilde{\varepsilon}} \qquad (8.5)$$

For $h > h_c$, the equilibrium state includes misfit dislocations as indicated in Fig. 8.2. However, for $h < h_c$, the lowest energy is achieved when no misfit dislocations are present and a fully coherent epitaxial film is thermodynamically stable.

For a very thick film, the density of these misfit dislocations, n_D, per unit length of interface is given by

$$n_D = \frac{|f|}{\bar{\lambda}} \qquad (8.6a)$$

where

$$f = \frac{\lambda_f - \lambda_s}{\bar{\lambda}} = \frac{a_f - a_s}{\bar{a}} \qquad (8.6b)$$

Here, f is the misfit, λ_f and λ_s are the interplanar spacings of interest in the two crystals, a_f and a_s are the lattice parameters while $\bar{\lambda} = (\lambda_f \lambda_s)^{1/2}$ and $\bar{a} = (a_f a_s)^{1/2}$. The specific excess energy of the dislocation boundary is given by

$$E_d = \frac{1}{2} \tau_0 \bar{\lambda} |f| (A' - \ln|f|) \qquad (8.7a)$$

$$A' = \ln(\bar{\lambda}/2\pi r_D^c) \qquad (8.7b)$$

and

$$\tau_0^{-1} = \left(\frac{1 - \sigma_f^*}{\tilde{G}_f} + \frac{1 - \sigma_s^*}{\tilde{G}_s} \right) \qquad (8.7c)$$

where r_D^c is the dislocation core radius chosen in such a way as to give the correct energy of atomic misfit at the center of the dislocation. It is interesting to note that the elastic model of the two contacting crystals has them effectively in parallel so that the softer crystal predominantly determines τ_0. This dislocation array exactly cancels the far-field stress due to the disregistry and localizes the stress in the region of the interface over a film width – a few λ_D.

The picture presented here has assumed a film of homogeneous structure whereas, during film growth, we often have isolated nucleation of crystallites on the substrate followed by non-impinging crystallite growth. Such growing crystallites can develop localized stresses in the substrate and an aspect ratio such that the height exceeds h_c. If dislocations are incorporated at the interface of such crystallites then the stress fields in the surrounding substrate can increase significantly and impede the impinging crystallite stage of film growth. In such a state of non-ideal film growth, the effective value of h_c might be significantly different than that given by Eq. (8.5).

8.2.1 Some comparisons with experiment

Estimates for the stress necessary to nucleate dislocations spontaneously in an otherwise perfect crystal yield a value of approximately $\tilde{G}/30$. Once nucleated, dislocations will typically move under applied stresses of $\stackrel{\sim}{>}\tilde{G}/1000$. Numerical calculations show that the stress due to epitaxial misfit will exceed $\tilde{G}/30$ for misfits $|f|\stackrel{-}{>}0.02$ although the actual value depends on the active shear system. Since the local stress may be considerably higher than the average shear stress (precipitates, surface discontinuities, etc.) dislocations may be nucleated at these sites and subsequently glide to relieve misfit. Since climb is also an important mechanism for moving dislocations, the available vacancy concentration is another important consideration.

Si/Ge VPE:[4] The epitaxial deposition of Ge on Si substrates is one of the simplest examples of semiconductor heteroepitaxy. The mismatch here is $|f| \approx 4 \times 10^{-2}$ leading to a calculated coherent thickness of $h_c \sim 25$ Å. Thus, growth with $h \stackrel{-}{>} 100$ Å should exhibit misfit dislocation densities $\sim 7 \times 10^5$ cm^{-2} in the region of the Ge/Si interface. Vacuum evaporation was used to make thin films of Ge on Si(111) substrates at $T \sim 700$–$900\,°$C and the films were examined by transmission electron microscopy (TEM). The Ge growth on Si proceeds by the nucleation and growth of islands with a larger number of small islands at lower temperatures (typical expectation). The dislocations observed have the correct directions and Burger's vectors for misfit relieving but the density of dislocations *decreased with increasing deposition temperature*, even though thermal expansion causes the misfit to increase with increasing temperature. The density expected for relief of misfit was approached only at the lowest temperatures. At their highest temperature, the lattice parameter of the Ge islands had gone about halfway to that of Si which is consistent with the smaller dislocation density found.

Si/Si: B and Si: B/Si CVD:[4] Epitaxial composites of undoped Si on Si(111) and Si(100) substrates doped with 10^{20} cm^{-3} B, leading to an expected $|f|$ of 6×10^{-4} and $n_D \approx 2 \times 10^4$ cm^{-1}, have been investigated. Their growths were performed by decomposition of dichlorosilane (SiH$_2$Cl$_2$) and silane (SiH$_4$) at temperatures from 1050 to 1150 °C with growth rates between 1 and 2 μm min^{-1}. The substrate thickness was 200 μm and the epitaxial layer thicknesses were between 25 and 100 μm. The substrate dislocation density was $\sim 10^2$ cm^{-2}. The observed misfit dislocation density in these grown composites *never exceeded* 30 cm^{-2}, a discrepancy of almost three orders of magnitude. This result was independent of growth temperature, method of growth, growth rate, post growth annealing, thickness of the epitaxial layer or whether the back surfaces of the substrate were etched or lapped. Everything the investigators did led to the inescapable conclusion that, in some sense, 30 dislocations per square centimeter is all that is required to accommodate the net strain of the composite.

The observed radius of curvature, R, of their composites was never smaller than 5 m for epitaxial layer thicknesses of $\sim 50\mu m$ while a misfit of 6×10^{-4} should lead to $R \approx 0.2$ m. Both this large R and small n_D would be understandable if somehow $|f|$ were smaller than that expected from the impurity concentration. To test this idea, an attempt was made to etch off the substrate, to isolate the epitaxial layer for further examination. This led to a complete shattering of the epitaxial layer after most of the substrate had been removed and at least demonstrates that a considerable strain *had* been at least virtually stored in the grown layer.

Reciprocal experiments of growing 10^{20} cm^{-3} B-doped epitaxial layers from SiH$_4$ onto undoped Si substrates were also carried out. For this case, TEM showed an unresolvable density of dislocations ($n_D > 5 \times 10^3$ cm^{-1}) and a radius of curvature of ~ 0.3 m in the same sense expected for the growth of a smaller lattice parameter film on a larger lattice parameter substrate. Thus, we see that there is an obvious asymmetry between Si:B/Si and Si/Si:B composites.

Si/Aℓ$_2$O$_3$ VPE:[4] Sapphire is rhombohedral, but Si is cubic, so that for growth on any one particular sapphire orientation the misfit strain is anisotropic with mean strains of ≈ 0.1. The salient results for the early growth of Si(001) on Al$_2$O$_3$(01$\bar{1}$2) at $T \approx 1000\,°C$ are as follows:

(a) The growth proceeds via island formation, predominantly of (100) orientation, with a rotational spread of $\approx 3°$. The deposits contained a minor number of {110} islands.

(b) Individual islands appear to grow independently of one another, and the coalescence of two (100) islands is occasionally accompanied by the formation of a stacking fault at the juncture of the islands, presumably to accommodate the rotational misalignment.

(c) The minor {110} constituent is eventually covered over as the growth attains full coverage.

The end result of this complex set of events is that a large number of stacking faults and microtwins is formed and the density gradually diminishes with distance from the interface. Near the Si/Aℓ$_2$O$_3$ interface, the mean distance between faults is ≈ 100 Å. However, in deposits

with incomplete coverage, misfit dislocations have been observed to occur with the densities expected. Here, the overgrowth is free of the
complications of growth mechanisms and fault introduction.

III–V compound VPE:[5] Misfit dislocations in III–V compounds are
predominantly of the 60°-type lying along $\langle 110 \rangle$ directions with $\langle 011 \rangle$
Burger's vectors. The following relationships between strain and mis-
fit dislocations for various amounts of abrupt (non-graded) lattice mismatch in the $In_xGa_{1-x}P/GaAs$ system exist for overgrowths made using
the VPE technique:

(1) $f \overset{-}{<} 1 \times 10^{-4}$; no misfit dislocations are introduced.

(2) $f \overset{\sim}{<} 2 \times 10^{-4}$; misfit dislocations are observed along only one $\langle 011 \rangle$
    direction.

(3) $3 \times 10^{-4} < f < 1 \times 10^{-2}$; orthogonal arrays of misfit dislocations
    are observed at the interface. However, they are confined to
    within a few thousand angstroms of the interfacial region and
    do not propagate along the growth axis.

(4) $f > 1 \times 10^{-2}$; dense arrays ($> 10^{10}$ cm^{-2}) of dislocations which
    propagate away from the interface along the growth direction
    are observed.

Observation (1) is consistent with the requirement from Eq. (8.5) that
a film thickness > 2 μm is needed for dislocation introduction at this
value of $|f|$. Observation (2) is a direct consequence of the asymmetry
of the zinc blend lattice where the dislocation core energies on the (111)
versus on the ($\bar{1}\bar{1}\bar{1}$) will be different. Observation (3) shows that, within
the specified range of strain, dislocations are either introduced along
the interface plane or they get bent over into this plane whereby they
relieve most of the strain. Fig. 8.3 shows a dislocation line that extends
from the substrate, across the interface and through an early thickness
of overgrowth film. As a result of misfit between the stress-free lattice
parameters of film and substrate, there are glide forces acting on the
dislocation. These forces increase as the deposit thickness increases and
cause the dislocation to bow as shown in Fig. 8.3(b). Eventually, the
glide forces acting on the dislocation are able to overcome the forces that
oppose glide. When this happens, the length of threading dislocation
in the film and substrate moves laterally, as shown in Fig. 8.3(c) and
leaves a length of misfit-dislocation line in the interface. When the free
end of the line encounters a surface step, it can slip down the step onto
the lower terrace and in this way eventually bend to become completely

(a)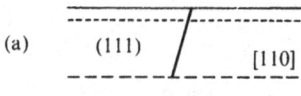

(b)

(c)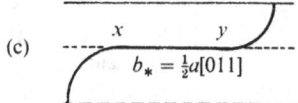

Fig. 8.3. Illustration of the generation of a length xy of complete misfit dislocation by the glide of a threading dislocation.

parallel with the growing interface. In the simple case of an elastically isotropic thick film on a thin substrate, the force, $\hat{F}_D$, tending to move the dislocation in the overgrowth to the right is

$$\hat{F}_D = 2\tilde{G}b_* h|f|\frac{(1+\sigma^*)}{(1-\sigma^*)}\cos \chi \qquad (8.7d)$$

where χ is the angle between the slip direction and that direction in the interface that is perpendicular to the line of intersection of the slip plane and the interface.

The final observation above, (4), indicates that compositional steps which contain 1% or more misfit strain should be avoided in VPE materials. The very large density of inclined dislocations probably results from misfit dislocations being introduced at the island stage of growth before a continuous film has formed. If the misfit dislocations present in each island were to remain in the interfacial plane after coalescence, the overall dislocation density would increase many fold unless an exact one-to-one match-up of dislocations occurred at each island boundary, which is highly unlikely. Since the dislocation density needed to relieve the misfit would be exceeded in this case, a more plausible happening would be for the dislocations to incline away from the interface and propagate along the growth axis.

A definite overgrowth/substrate misorientation is found in lattice-mismatched VPE layers. The misorientation, α, varies from 0.5° to

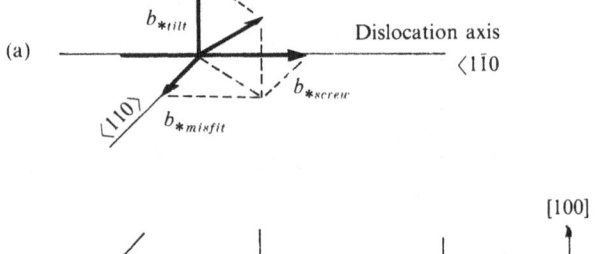

Fig. 8.4. (a) The three-dimensional Burger's vector components of a 60° dislocation. (b) A view along the dislocation axis showing the break-up of the Burger's vector into misfit and tilt accommodating components.

2.0° increasing with the lattice mismatch and the axis of rotation lies either in the substrate plane or is somewhat inclined to it ($\tilde{<}$30°).

To accommodate the different components of the interfacial stress tensor, one expects that three types of dislocations may be needed in the overgrowth layer: one set (type 1) having Burger's vectors perpendicular to the boundary plane to accommodate tilt, a second set (type 2) with Burger's vectors lying in the interface to accommodate interfacial misfit and a third set (type 3) which are screw dislocations having Burger's vectors that lie in the interface and give rise to shear stresses and/or twist boundaries. If we consider such an array of dislocations having colinear but unspecified Burger's vectors, each dislocation will, in general, have two edge components; one normal to the interface and one lying in the interface as illustrated in Fig. 8.4.[5] The type 1 dislocation will give rise to a misorientation

$$\alpha = b_{*1}\rho_1 \qquad (8.8a)$$

where b_{*1} is the Burger's vector component perpendicular to the interface while ρ_1 is the linear dislocation density in the interface. Assuming that all misfit is relieved by dislocations, the linear dislocation density may in turn be related to the misfit, $|f|$, as

$$\rho_1 = |f|/b_{*2} \qquad (8.8b)$$

where b_{*2} is the misfit-relieving Burger's vector component lying in the interface. Combining, we obtain

$$\alpha = |f| b_{*1}/b_{*2} \qquad (8.8c)$$

which shows that the misorientation is directly proportional to misfit strain. In III–V compounds, misfit relief has been shown to be accomplished largely by 60°-type dislocations that have both type 1 and type 2 Burger's vector components.

In VPE grown material, the misfit is relieved by a cross-hatched array of dislocations; however, in the same material grown by LPE at a comparable temperature, the misfit dislocations align to form a subgrain structure within the essentially single crystal epitaxial layer. The subgrains are about 10 μm in diameter and are misoriented from one another by $\tilde{<}0.2°$. However, no overall substrate/overgrowth misorientation has been observed in LPE layers. Step grading in VPE-grown materials compared to LPE-grown materials seems to be much more effective in reducing the dislocation density in the topmost regions of the layers. This is illustrated in Table 8.1 for the $In_x Ga_{1-x} As$ system. As can be clearly seen for comparable compositions and thus lattice misfit, the VPE technique produces a lower defect density. All these growths were made on GaAs substrates with dislocation densities $< 3 \times 10^3$ cm^{-2}. In this comparison, the GaAs face orientation which produced the most consistent high quality layers was 3° from the (100) toward the $\langle 110 \rangle$ direction for the VPE layers and the (111)A for the LPE layers. In LPE growth, the dislocation densities can be somewhat reduced (factor of 5–10) by simply stopping and restarting the cooling or by moving the sample from bin to bin in a multibin slider boat. These interruptions bend some dislocations over at the interface in the LPE growth with about a factor of 2 reduction at each interruption down to a minimum density of $\sim 10^5$ cm^{-2} which is still not as effective as VPE step-grading. The subgrain structure of the LPE heteroepitaxial layers appears to arise from a lattice mismatch-induced nucleation process as well as from a polygonization process acting on the high dislocation density.

Strained layer $Ge_x Si_{1-x}/(Si, Ge)$ *MBE:*[6] In Si MBE there are almost no lattice-matched candidates for heteroepitaxial growth. As such, typical heteroepitaxy experiments involve mismatches ranging from 1% (the metal silicides) to over 10% (Si on sapphire). The high interfacial energies in these mismatched systems lead to an almost universal tendency towards islanded growth. As a general rule, islanding increases with lattice mismatch and decreases with epitaxial growth temperature. Using

Table 8.1. *Comparison of LPE and VPE grown layers of*
$In_xGa_{1-x}As.$[5]

Growth method and top layer composition (% InAs)		Intermediate grading layers (% InAs)	Dislocation density in top layer (cm^{-2})	Average dislocation spacing^a (μm)	Measured diffusion length (μm)
VPE	~ 3	None	3×10^4	57	–
	6.1	None	5×10^4	44	2.6
	14.0	4.7, 6.7	5×10^6	4.5	2.3
	17.7	–	2×10^5	22.3	–
	20.5	4.7, 6.7, 9.7, 12.5, 14.7	1.5×10^7	2.5	2.7
	23	–	3×10^5	18.2	–
	28.4	2.4, 4.8, 8.5, 11.5, 14.5, 20	2×10^6	7.1	2.6
LPE	6.6	None	1.5×10^7	2.5	–
	8.1	None	5×10^6	4.5	–
	13.8	6.3	6×10^7	1.3	–
	23.8	5.7, 13.7	$> 10^8$	< 1	0.5

^{a} Taken as the square root of the reciprocal of the dislocation density.
Dashes indicate parameters that were not measured.

careful substrate preparation procedures, Bean[7] was able to produce heteroepitaxy of Si on Ge using an MBE technique with no islanding. He observed a sharp transition between island-type growth and non-island-type growth of GeSi on Si as indicated in Fig. 8.5. For temperatures below 550 °C, all GeSi compositions can be grown smoothly until ~ 400 °C is reached where the transition to polycrystalline growth occurs.

In the GeSi/Si system, defects can be avoided only if mismatch is accommodated by strain in one or both materials. Using Eq. (8.5) one would expect the introduction of misfit dislocations after only 700 Å of growth of a $Ge_{0.15}Si_{0.85}$ alloy on Si. Bean[7] attempted to avoid this equilibrium condition by finding conditions to grow layers virtually free of dislocations and with smooth near-perfect surfaces. Fig. 8.6 shows that he achieved such conditions and was able to form layer thicknesses more than an order of magnitude above the calculated equilibrium values (e.g., to 10 000 Å with 15% Ge alloys).

In a strained layer superlattice of GeSi/Si, the layers will be of com-

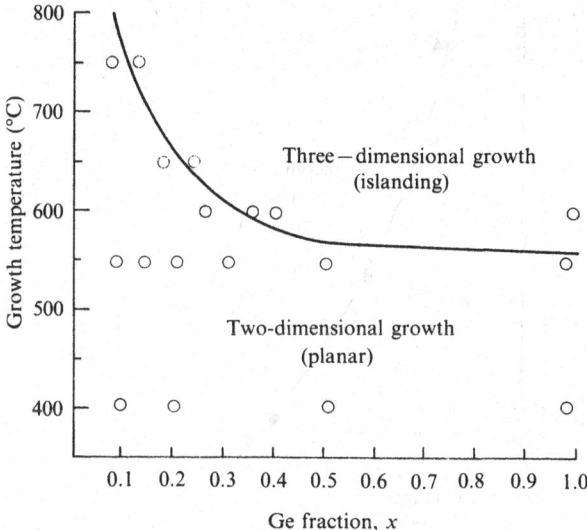

Fig. 8.5. Plot of Ge_xSi_{1-x} on Si growth morphology as a function of composition and temperature.[7]

parable thickness and the elastic strain will be distributed over both types of layer. Further, the superlattice will have an in-plane lattice constant between that of the unstrained constituents. It has been found empirically that, not only can single layer critical thicknesses exceed the calculated equilibrium limits but superlattices can be grown where the total alloy thickness is $\sim 10^2$ times the calculated single layer critical value. It is necessary only to interleave Si layers that are 3–5 times the alloy layer thickness (because Si and Ge are much stiffer than GeSi alloy layers). Even thin elemental layers retain their bulk structures effectively decoupling the strain in adjacent alloy layers. This almost completely asymmetric distribution of strain is evident not only in microscopy but also in Raman scattering measurements of such superlattices.

If superlattices are increased towards 100 periods or if elemental layers are comparable to or thinner than the alloy layers, strain will build in both layers. Bean[7] has found empirically that a superlattice will remain commensurate (strained but free of dislocations) if two requirements are met: (1) the individual alloy layers must be commensurate as defined by the critical layer thickness of Fig. 8.6 and (2) the total superlattice thickness must be less than the critical thickness for a single layer of the average superlattice concentration. For example, a

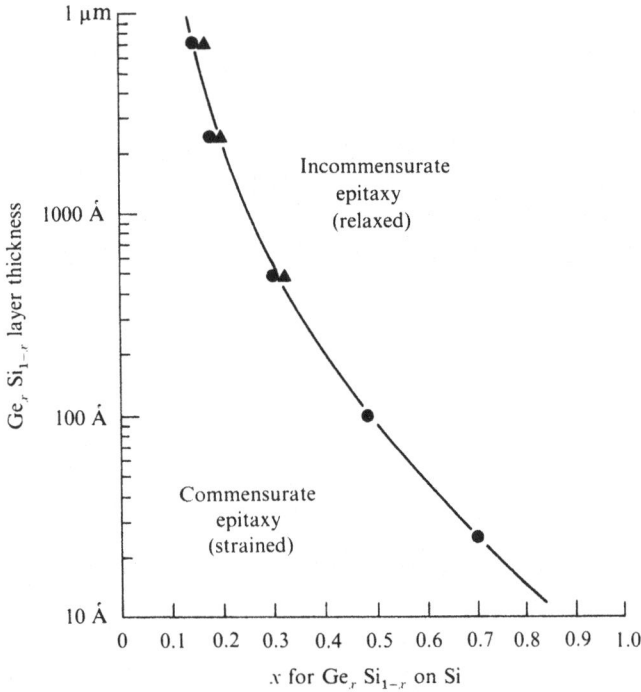

Fig. 8.6. Single layer, critical film thickness to maintain dislocation-free strained growth for Ge_xSi_{1-x} films on Ge.[7]

20 period superlattice of 250 Å Si layers and 75 Å $Ge_{0.4}Si_{0.6}$ layers is commensurate because (a) the individual $Ge_{0.4}Si_{0.6}$ layers are less than the 200 Å critical thickness for this composition and (b) the total superlattice thickness, 6500 Å is less than the critical layer thickness for a single $Ge_{0.12}Si_{0.88}$ layer (the average superlattice composition).

$Si_xGe_{1-x}/Si,Ge$ MBE: As mentioned above, heteroepitaxial films of Si–Ge on Si suggest that these films remain dislocation free at thicknesses much greater than h_c calculated from Eq. (8.5). This result is shown in Fig. 8.7(a) indicating that, for films containing 20% Ge or less, the actual film thicknesses at which misfit dislocations are observed are an order of magnitude greater than h_c calculated from Eq. (8.5).[6] Similar results are shown in Figs. 8.7(b)[1] and (c)[8] from the work of other investigators. Although there has been considerable debate about the evidence for this discrepancy, it seems that the results are genuine and that the

cause for the discrepancy is related to the kinetics of nucleation, motion and multiplication of dislocations in epitaxial semiconductor films.

A detailed study of nucleation and propagation of misfit dislocations has been made in the $Si_{1-x}Ge_x/Si$ system using a standard defect revealing etch.[9] The value of h_c for a layer of $Si_{0.875}Ge_{0.125}$ on Si is 220 Å. From the measurement of mismatch dislocation densities in layers < 5000 Å thick, the first sign of mismatch occurs at h just above h_c. Certainly, by $h = 2h_c$, a small but identifiable number of short mismatch lines were observed. For layers with a thickness in the range $h_c < h < 0.7$ μm the density and length of mismatch lines continue to increase. In the range 0.7 μm $\leq h < 1.1$ μm, the individual mismatch lines began to multiply giving rise to patches of highly relaxed material. However, the density of these patches corresponded to the density of dislocation sources intrinsic to the MBE Si ($\sim 10^3$ cm^{-2}).

In the case of a 1 μm thick $Si_{0.875}Ge_{0.125}$ layer deposited onto a Si substrate containing 10^6 dislocations per square centimeter, a relaxation of $\sim 50\%$ was observed. For comparison, an identical layer deposited on a substrate with a dislocation density of $\sim 10^3$ cm^{-2} showed no discernible relaxation by X-ray techniques. For layers thicker than 1.1 μm, the patches of relaxation begin to merge and a quantifiable measurement for the extent of relaxation can be made from X-ray rocking curves. By 2.0 μm, the patches have almost completely merged, with only a few isolated regions of unrelaxed material remaining. From these results, a dislocation dynamics model can be made for the formation of misfit dislocations in epitaxial thin films. The next section develops such a model which provides a mechanistic understanding of misfit dislocation formation in terms of the kinetics of nucleation, motion and multiplication of misfit dislocations.

8.3 Mechanisms of misfit dislocation formation

Freund[10] has shown that the most powerful way to understand the formation of misfit dislocations in epitaxial structures is to consider the energetics associated with the incremental extension of a misfit dislocation, rather than by considering the overall energy change associated with the formation of a periodic array of misfit dislocations. Following his work, we envision a dislocation that extends from the free surface of a film to the film/substrate interface and deposits a misfit dislocation at that interface as it moves. Such a situation is shown in Fig. 8.8. For simplicity, we consider the case of FCC crystals with the (001) orientation

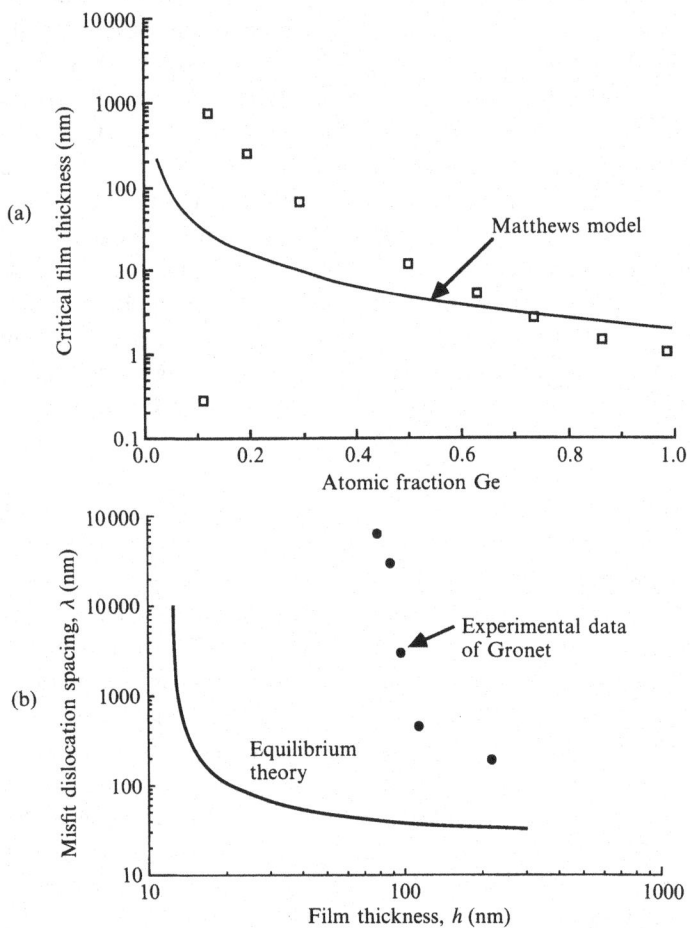

Fig. 8.7. (a) Critical film thickness as a function of misfit strain for Ge_xSi_{1-x} on Si,[1] (b) misfit dislocation spacing in $Ge_{0.2}Si_{0.8}$ films on Si ($T = 625\,^{\circ}C$, $V = 5$ nm min^{-1}) as a function of film thickness.[6]

and focus our attention on the motion of dislocations on the octahedral {111} planes. Here, the angles defining the {111} slip plane normal and the Burger's vector are $\phi = \cos^{-1}(1/\sqrt{3})$ and $\psi = \cos^{-1}(1/\sqrt{2})$, respectively.

If the dislocations shown in Fig. 8.8(a) are present in the substrate on which the film is growing, they naturally propagate through the growing film and are called "threading" dislocations.[1] Considering Fig. 8.8(b),

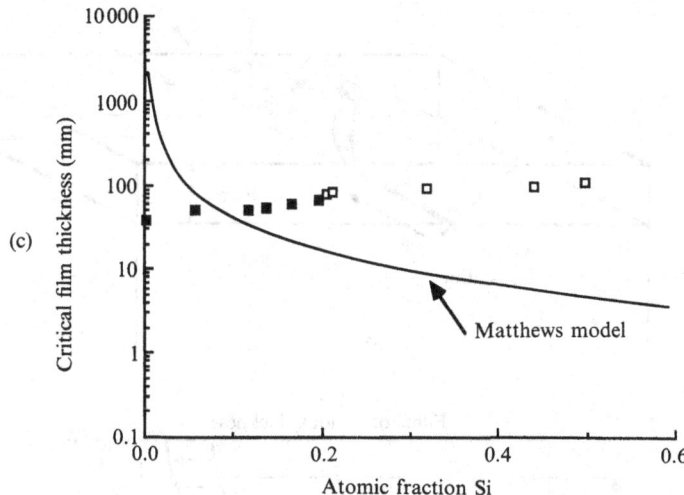

Fig. 8.7. (c) Critical film thickness as a function of misfit strain for Si_xGe_{1-x} films on Ge.[8]

the biaxial stresses in the film exert forces on the threading dislocation and cause it to move in the slip plane. The portion of the dislocation that resides in the substrate remains stationary because the forces on it are much smaller and are of opposite sign. Thus, the threading dislocation in the film bends over as it moves and eventually leaves a misfit dislocation in its wake. Continued movement of the threading dislocation extends the length of the misfit dislocation.

For dislocation-free substrates, misfit dislocations may also be formed in epitaxial films by a dislocation nucleation process at the free surface of the growing film. Fig. 8.8(c) shows how the nucleation of dislocation half-loops at the surface of the growing film eventually leads to the formation of a misfit dislocation at the film substrate interface as the two ends of the half loop move in opposite directions.

The basic idea of the theoretical model is that the work done by the stresses in the film, W_f, when the dislocation moves a unit distance must provide enough energy to deposit a unit length of misfit dislocation, E_D. For the dislocation geometry shown in Fig. 8.8(a), W_f is given by

$$W_f = \frac{\tau b_* h}{\sin \phi} = \frac{\cos \psi \cos \phi}{\sin \phi} \sigma b_* h = \frac{\sigma b_* h}{2} \qquad (8.9a)$$

For the FCC slip geometry under consideration, the Burger's vector makes an angle of 60° to the line of the misfit dislocation so the mis-

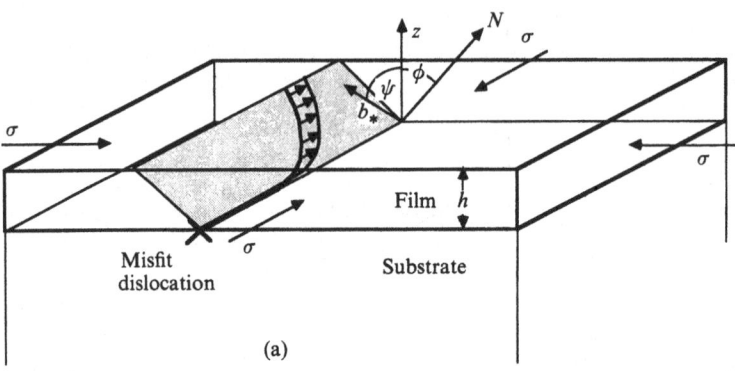

(a)

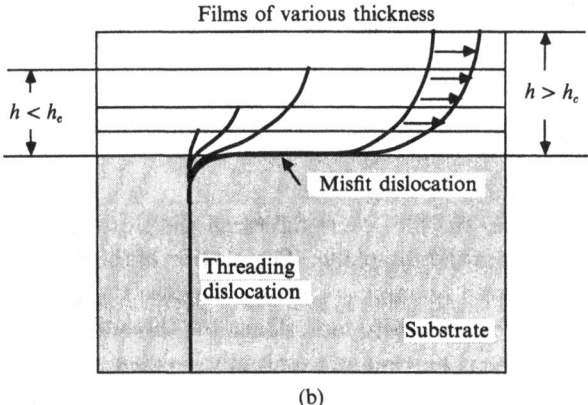

(b)

Fig. 8.8. (a) Dislocation motion in a thin film on a substrate leading to the deposition of a misfit dislocation at the interface, (b) bending of a threading dislocation in a strained film to create a misfit dislocation for films above a critical thickness. (Courtesy of W. D. Nix.)

fit dislocation deposited at the film/substrate interface is a 60° mixed dislocation and we need E_D for that case.

The energy of an edge dislocation lying at the interface between a thin film and a semiinfinite substrate is

$$E_{D\,edge} = \frac{2b_{*e}^2}{2\pi(1 - \sigma^*)} \frac{\tilde{G}_f \tilde{G}_s}{(\tilde{G}_f + \tilde{G}_s)} \ell n \left(\frac{\beta_e h}{b_{*e}} \right) \qquad (8.9b)$$

where $\beta_e = 0.701$. The result for a screw dislocation is

$$E_{D\,screw} = \frac{b_{*s}^2}{2\pi} \frac{\tilde{G}_f \tilde{G}_s}{(\tilde{G}_f + \tilde{G}_s)} \ell n \left(\frac{\beta_s h}{b_{*s}} \right) \qquad (8.9c)$$

where $\beta_s = 1.0$. We can obtain the result for a 60° mixed dislocation by

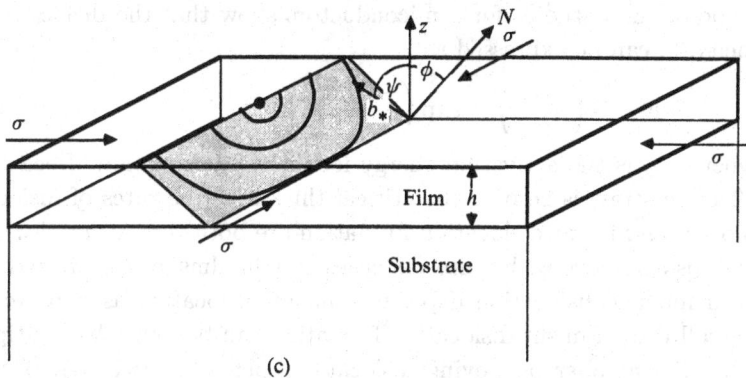

(c)

Fig. 8.8. (c) Dislocation nucleation at the surface of a strained film. The expanding half-loop eventually produces a misfit dislocation at the film/substrate interface. (Courtesy of W. D. Nix.)

noting that $b_{*e} = (\sqrt{3}/2)b_*$ and $b_{*s} = b_*/2$. By adding the energies of the edge and screw components with $\sigma^* = 0.3$, we obtain

$$E_{D\,60°} = 0.95 \frac{b_*^2}{2\pi(1-\sigma^*)} \frac{\tilde{G}_f \tilde{G}_s}{(\tilde{G}_f + \tilde{G}_s)} \ell n\left(\frac{\beta h}{b_*}\right), \qquad (8.9d)$$

where $\beta = 0.755$.

On the basis of the above, we may define an effective stress for dislocation motion in terms of the work done by the stress in the film less the work required to deposit the misfit dislocation; i.e.,

$$\tau_{eff} b_* \frac{h}{\sin \phi} = W_f - E_{D\,60°} \qquad (8.9e)$$

This net driving force is positive only at film thicknesses greater than a critical thickness and this can be found by setting the right side of Eq. (8.9e) to zero so that

$$\frac{h_c}{\ell n(\beta h_c/b_*)} = \frac{0.95 b_*}{\pi(1-\sigma^*)\sigma} \frac{\tilde{G}_f \tilde{G}_s}{(\tilde{G}_f + \tilde{G}_s)} \qquad (8.10)$$

where the biaxial stress, σ, has been used in place of $\tilde{M}\tilde{\varepsilon}$. This result is a more exact form of Eq. (8.5) for this particular orientation and it can be used to find the relaxed state of stress in a film containing the equilibrium number of misfit dislocations.[1] Thus, for $h > h_c$, we have

$$\sigma = \frac{0.95}{\pi(1-\sigma^*)h} \frac{\tilde{G}_f \tilde{G}_s}{(\tilde{G}_f + \tilde{G}_s)} \ell n\left(\frac{\beta h}{b_*}\right) \qquad (8.11)$$

The kinetics of misfit dislocation formation can be predicted by using the effective stress defined by Eq. (8.9e) and the dislocation mobility.

Experimental studies for semiconductors show that the dislocation velocity, v, can be expressed as

$$v = B \left(\frac{\tau_{eff}}{\tau_0} \right)^{1.2} \exp \left(\frac{-U_D}{\kappa T} \right) \qquad (8.12)$$

where U_D is the activation energy for dislocation motion. Because the effective stress is zero at the critical thickness, the rates of dislocation motion and misfit dislocation formation are both zero at $h = h_c$. Dislocations can move with finite velocities only in films for $h > h_c$ and, since each moving dislocation deposits a misfit dislocation as it moves, the overall rate of misfit dislocation formation can be found by multiplying v by the number of moving dislocations per unit area, n_D. Defining the misfit dislocation density, ρ_{mf}, as the line length of misfit dislocations per unit area, the rate of production of misfit dislocations can be expressed as

$$\frac{\mathrm{d}\rho_{mf}}{\mathrm{d}t} = n_D v \qquad (8.13)$$

Thus, the rate of formation of misfit dislocations depends on the density of mobile dislocations that are already there.

The simplest kinetic model is one with a fixed number of threading dislocations from the substrate and no multiplication of existing dislocations or nucleation of new ones). For this case, the rate of increase of misfit dislocation line length is governed entirely by v. Defining the spacing, λ, between the misfit dislocations as

$$\lambda = \frac{2}{\rho_{mf}} \qquad (8.14)$$

then λ will decrease during the course of film growth for $h > h_c$. Fig. 8.9 shows how λ, for an $Si_{0.8}Ge_{0.2}$ epitaxial film on Si, is expected to change during film growth. For this result, the film growth velocity was assumed to be $V = 50\,\text{Å}\,\text{min}^{-1}$ at 625 °C and v was obtained from Eq. (8.12) with $B = 7.33 \times 10^4\,\text{m s}^{-1}$, $\tau_o = 9.8\,\text{MPa}$ and $U_D = 2.2\,\text{eV}$. The curve indicating the largest misfit spacings corresponds to the lowest mobile dislocation density. For all curves, the misfit spacing tends to infinity at $h \sim 13.5$ nm, which corresponds to h_c for this system, and one notes that the curve corresponding to the largest dislocation density coincides with the equilibrium theory at large film thicknesses, as expected.

The treatment thus far is unrealistic because it doesn't allow for either dislocation multiplication or new dislocation nucleation in the course of film growth. In one substantiated multiplication mechanism,[11] two crossing misfit dislocations with the same Burger's vectors can annihilate locally and produce two new mobile dislocation segments that can

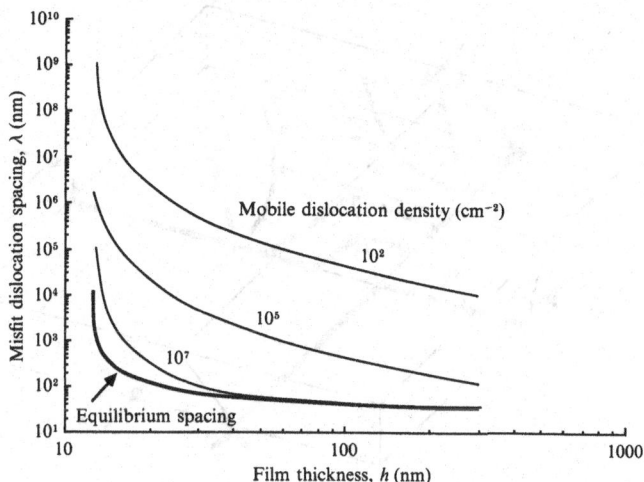

Fig. 8.9. Misfit dislocation spacing as a function of film thickness during growth of a $Si_{0.8}Ge_{0.2}$ film on Si at 625 °C with $V = 50$ Å min^{-1}. Calculations are shown for three different threading dislocation densities. (Courtesy of W. D. Nix.)

then glide and produce additional lengths of misfit dislocation. This mechanism is illustrated in Fig. 8.10. All such multiplication mechanisms can be described by using the concept of a breeding factor, δ_B, defined as the number of new dislocations produced per unit length of glide motion by each moving dislocation. Then, the multiplication of dislocations can be described by

$$\frac{dn_D}{dt} = n_D \delta_B v \qquad (8.15)$$

The effect of such dislocation multiplication on λ during film growth is shown in Fig. 8.11(a) where we have assumed a very low initial dislocation density of $n_D = 0.1$ cm^{-2} and constant breeding factors of $\delta_B = 5$, 10 and 20 dislocations per millimeter. We note that, although some dislocation motion occurs as soon as h_c is exceeded, λ remains very large until the film has reached a thickness of ~ 1000 Å.

According to the multiplication mechanism of Fig. 8.10, δ_B should be proportional to the number of such dislocation interactions produced per unit length of dislocation travel leading to

$$\delta_B = \frac{p_B}{2\lambda} \qquad (8.16)$$

where p_B is the breeding efficiency (probability that any particular dislocation intersection actually leads to a new pair of moving dislocations).

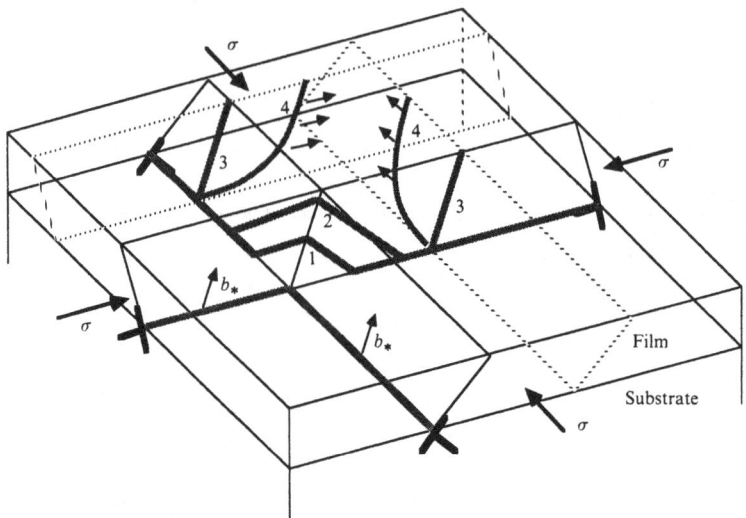

Fig. 8.10. Illustration of the Hagen–Strunk[11] mechanism of dislocation multiplication. Crossing dislocations with the same Burger's vector can react to form two new mobile dislocations as shown.[1]

The effect of this particular kind of multiplication on λ is shown in Fig. 8.11(b) for the case of a film with an initial dislocation density of $n_D = 10^3$ cm^{-2} and various breeding efficiencies. As expected, one sees that multiplication can have a profound effect on the evolution of the misfit dislocation spacing. Generally, increasing δ_B causes the misfit dislocations to form more abruptly during film growth. Indeed, the curves of Fig. 8.11 give the impression that the critical thickness for misfit dislocation formation is much greater than the true h_c of 135 Å. This helps to explain why misfit dislocations are often not observed in films that have been grown well beyond h_c.[1]

New dislocations can also be formed by half-loop nucleation, probably at the growing surface, and thus at a much higher rate for LPE than VPE at the same driving force because $\gamma_{SL} \ll \gamma_{SV}$. Such nucleation must occur in those cases where misfit dislocations are formed in epitaxial films grown on dislocation-free substrates. If $(dn_D/dt)_n$ represents the rate of nucleation of new dislocations, the total rate of formation of new dislocations can be expressed as

$$\frac{dn_D}{dt} = \left(\frac{dn_D}{dt}\right)_n + n_D \delta_B v \tag{8.17}$$

The effect of dislocation nucleation on the evolution of λ is shown in

Fig. 8.11(c) for various $(dn_D/dt)_n$. One sees that increasing the rate of nucleation causes the misfit dislocation spacing to approach the equilibrium spacing more quickly.

In all of the above calculations, stress relaxation during film growth has been taken into account. The creation of misfit dislocations is equivalent to plastic deformation in the film and relieves the elastic strain and stress in the film. The corresponding stress evolution can be described by

$$\frac{d\sigma}{dt} = -\frac{\tilde{M} n_D b_* v}{2} \tag{8.18}$$

8.4   Film formation by interface reaction

The foregoing has dealt with film formation by reaction at the top surface and has been concerned largely with misfit dislocation formation at the substrate/film interface. In many cases, the films of interest to us are those that form by reaction at the substrate/film interface; i.e., SiO_2 on Si via Si/O_2 reaction, MSi_x on Si via M/Si reaction, etc. In such cases, the dislocation structure needs to be somewhat different to ensure a perpetuating reaction because, in addition to lattice mismatch, there is a molecular volume difference to be accommodated on a continuous basis. Thus one must not only consider the diffusive transport of chemical species but also vacancies and interstitials plus the real likelihood of Frenkel defect formation at the interface.

Let us illustrate the new features by considering the formation of an SiO_2 film on Si by thermal oxidation. As discussed in Chapter 1 of the companion book,[12] O_2 diffuses interstitially through the SiO_2 to react with the Si and $\Omega_{SiO_2} \approx 45\,\text{Å}^3$ while $\Omega_{Si} = 20\,\text{Å}^3$ so that a static interface between the two phases might look like that shown in Fig. 8.12(a). Here, the extra planes of Si just look like misfit dislocations with $\lambda \sim 4.5 a_{Si}$. For oxide film thickening, Frenkel defect formation must occur at these dislocations and the Si_I species must diffuse into the two adjoining phases. These will eventually produce extrinsic stacking faults in the adjoining phases and direct experimental evidence for such structures has been found in the Si. Most of the Si_I species diffuse into the SiO_2, are internally oxidized by the incoming O_2, produce large swelling stresses and convert the crystalline SiO_2 to amorphous SiO_2 which is also observed experimentally. However, there is another option available for elimination of volume stresses and this mechanism is illustrated in Figs. 8.12(b) and (c).[13] Fig. 8.12(b) depicts a hypothetical

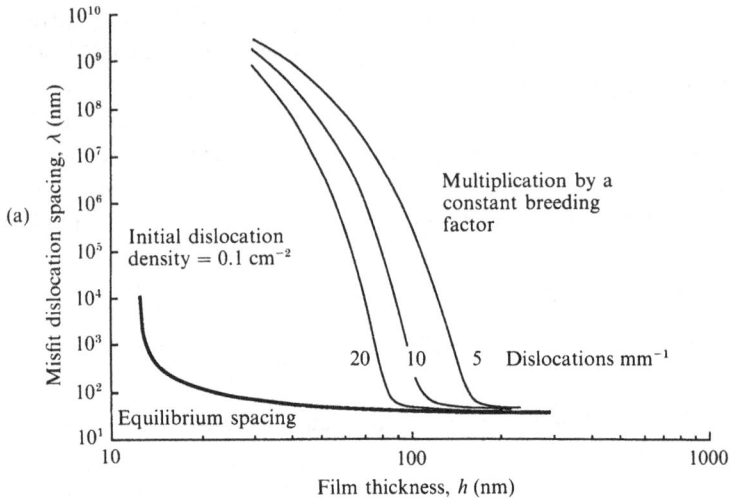

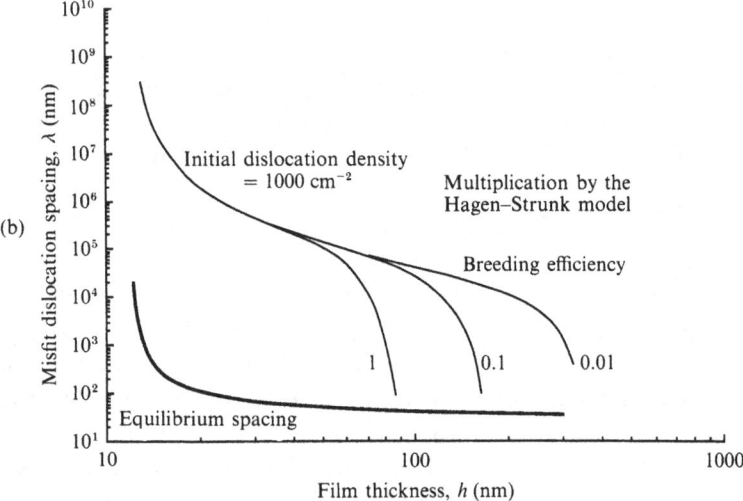

Fig. 8.11. Misfit dislocation spacing as a function of film thickness during growth of a $Si_{0.8}Ge_{0.2}$ film on Si at 625 °C with $V = 50$ Å min^{-1} for (a) three rates of dislocation multiplication, δ_B, with $n_{D_0} = 10^{-1}$ cm^{-2}, (b) Hagen–Strunk multiplication with three different breeding efficiencies, p_B, and $n_{D_0} = 10^3$ cm^{-2}. (Courtesy of W. D. Nix.)

strain-free process in which atoms evaporate from Si by a ledge mechanism and deposit, together with O, to form SiO_2. Because of the volume

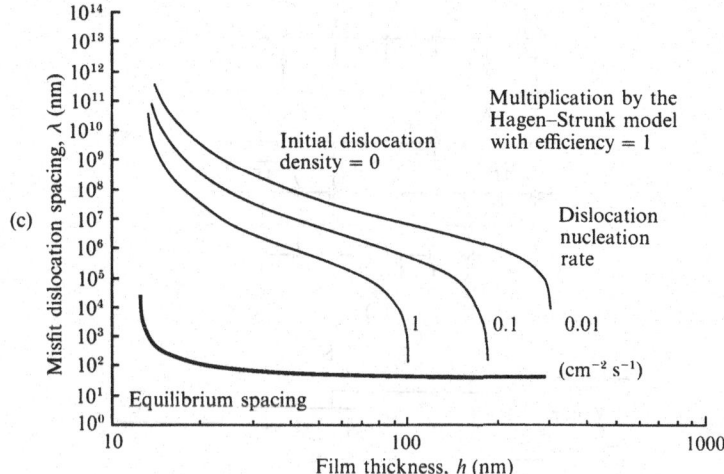

Fig. 8.11. (c) Three different nucleation rates with $n_{D_0} = 0$ and the Hagen–Strunk model with efficiency $= 1$. (Courtesy of W. D. Nix.)

differences, there must be 2.25 times as many ledges on the oxide to conserve Si (if only one set of parallel ledges is present). Now suppose that the crystals are welded together as shown in Fig. 8.12(c). The ledges at A match except for a residual step translation dislocation (fractional Burger's vector), while those at B become misorientation dislocations.

Evidently, the processes of Figs. 8.12(b) and (c) are equivalent in producing oxide growth without the development of long range elastic strains. The array of misorientation dislocations is in the form of a tilt boundary with strains limited to a distance of the order of the dislocation spacing. In Fig. 8.12(c), the growth occurs by diffusive addition of Si to the oxide at kinks in the A ledges as well as by diffusion from A ledges in the Si to misorientation dislocations B where they are diffusively added to jogs by an elementary climb step. The macroscopic displacement field in this case is purely in the direction normal to the interface. In practice, one would expect the B-dislocation array to glide a short distance away from the interface ($\sim$ 10–20 Å) for energetic reasons. To the extent that the dislocation/ledge density in the SiO_2 is inadequate, the oxide must form by some other mechanism with the attendant possibility of very large long range strain field development.

For the specific case of Si with a near surface oxide layer β-cristobalite, step translational dislocations must be present. The process corresponds to the nucleation of a circular ledge of radius r at the interface where

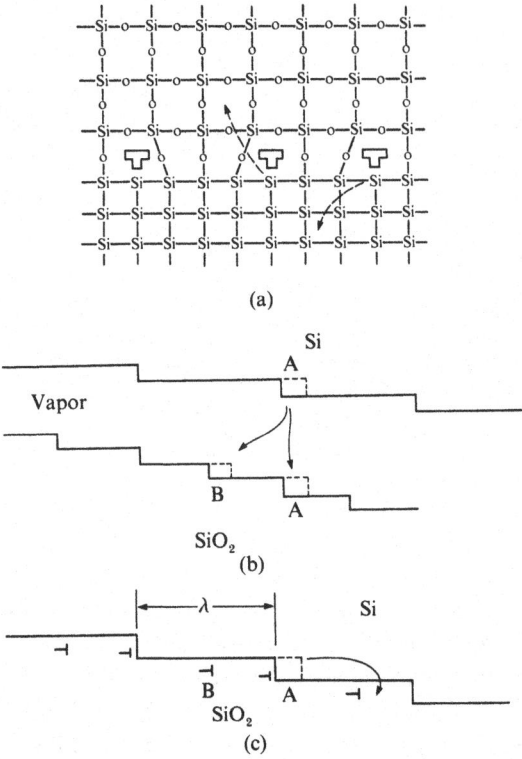

Fig 8.12. (a) Schematic illustration of lattice matching at a Si/SiO₂ interface and necessary edge dislocation representation, (b) hypothetical process of formation of oxide by detachment of Si from kink sites, vapor transport and subsequent attachment at kink sites on ledges in SiO₂ and (c) phases of (a) welded together to create ledges and tilt dislocations at the interface.

$N = \pi r^2 h / \Omega_{Si}$ atoms of Si are removed. This Si is oxidised to a slab of SiO₂ (resembling β-cristobalite) of thickness $a' = 0.356$ nm, a layer with one Si atom per unit layer thickness. A similar process could be envisaged even if the SiO₂ were amorphous since locally a discrete layer of SiO₂ would need to be added to the system. A portion of the slab of the same radius r is inserted at the location from which the Si was originally removed. The result is a circular ledge of radius r of height a' but, because $a' \neq h$, the ledge also has extrinsic dislocation character with Burger's length $b_{*\alpha} = a' - h$ and uses only N_α Si atoms. The balance of the Si atoms, $N_\beta = N - N_\alpha$, are inserted as a disc of SiO₂ of radius r' where $N_\beta = \pi r'^2 a' / \Omega_{SiO_2}$. Thus an extrinsic dislocation

Table 8.2. *Geometric parameters for ledge dislocation lines.*[a][(13)]

Plane	No. of Si layers	a (nm)	a' (nm)	h (nm)	$b_{*\alpha}$	$b_{*\beta}$	r'/r
[100]	1	0.384	0.356	0.136	0.220	−0.356	0.37
[100]	2	0.384	0.356	0.271	0.085	0.356	0.44
[100]	3	0.384	0.356	0.407	−0.052	0.356	0.57
[100]	4	0.384	0.356	0.542	−0.158	0.356	0.66
[111]	1	0.357	0.356	0.314	0.042	0.356	0.46
[111]	2	0.357	0.356	0.628	−0.271	0.356	0.67

[a] A positive value of b_* indicates an interstitial-type edge dislocation loop, a negative value indicates a vacancy type.

loop with Burger's length $b_{*\beta}$ together with its elastic strain field has been created. Table 8.2 lists the relevant step heights a and a' referred to the Si and SiO$_2$, respectively, and the Burger's length b_{*t} of the step translational dislocation.[(13)] A minimum of b_{*t} would be favored energetically because it would produce a minimum strain energy density for the dislocations. Hence, the table indicates that single layer height ledges should be observed for Si{111} surfaces while double or triple height ledges should be observed for Si{100} surfaces. Using TEM, single height ledges were observed for the {111} surfaces while {100} surfaces appear rough over a range of 2–5 atom widths. Since the images in atomic resolution TEM are integrals of ledges projected parallel to the beam, these observations are consistent with the above indications of single versus double or triple ledge heights, respectively.

In actual practice, we might expect some compromise between the different dislocation mechanisms of Fig. 8.12 so that an energetic balance is eventually made with some proportion of each type being present at the interface as indicated in Fig. 8.13(b). Of course, this minimum energy situation does not spontaneously materialize but is expected to develop via the stages of (i) a coherent highly strained SiO$_2$ layer, (ii) a partially strained SiO$_2$ layer with interface misfit dislocations generating Si$_I$ species, (iii) extrinsic fault formation in the SiO$_2$ to form the B-dislocation array and (iv) fine scale balancing of the misfit and misorientation dislocation arrays.

With some small modification, the above model can be extended

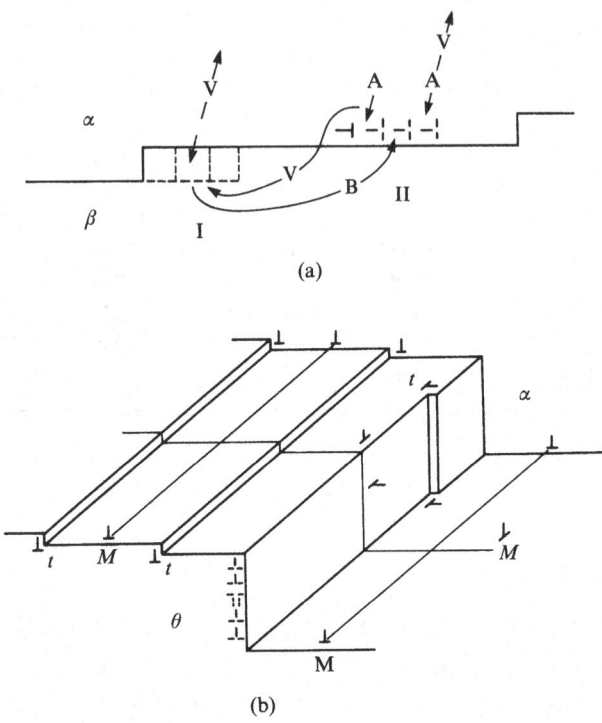

(a)

(b)

Fig. 8.13 (a) Hypothetical process for growth of α from β at steps and misorientation dislocations (tilt) and (b) a precipitate θ growing from α where the terraces are coherent while the step riser contains an intrinsic dislocation array (dashed), transformation dislocations, t, and misfit dislocations, M.[14]

to diffusive phase transformations involving substitutional solute diffusion.[14] The local process for this case is illustrated in Fig. 8.13(a) for the case of β-phase transforming to an α-phase with a change in chemical stoichiometry. For example, let us assume that the compositions near the interface are $\sim A_2B$ in α and $\sim AB_2$ in β. Let us also suppose that the diffusivity of A greatly exceeds that of B and that the partial molar volumes of A and B are the same in both α and β. Growth can then occur by the combined step and misorientation mechanism illustrated as a reversible sequence in Fig. 8.13(a) even though, in the actual case, the events represented would be average events and the atom/vacancy motions would occur by (biased) random walk. Two A atoms are added to the dislocation of II by climb, creating two vacancies. One vacancy is removed to a remote sink in α. The other moves by interface motion to effectively remove a B atom from I and also place it on the dislocation at II. This vacancy next adds an A atom at I and is then removed to a remote sink in α. The net result is the addition of an A_2B unit at II by climb, the replacement of an AB_2 unit by an A_2B unit at I by exchanging an A for a B plus the counterflow of two vacancies and two A atoms in α. For cases where misfit dislocations are present, the interstitial formation generated by interface movement becomes a natural sink for the vacancies released in the proposed process. Although the misorientation dislocation is shown symmetrically disposed to the ledges, its position could be adjusted to be closer to one of the ledges I in order to reduce the drag force associated with the relatively slow interface mobility of the B atom (postulated).

8.5   Void and columnar defects in sputtered films

Film deposition via the sputtering technique generally leads to films containing a strongly compressive stress when the atom trajectories are fairly ballistic. When higher ambient gas pressures are used, the atom mean free path is reduced so that the excess energy carried by the depositing atom is also reduced and the degree of compressive stress development in the film decreases, eventually becoming tensile in nature. It is interesting that vapor deposition of the *identical* material usually leads to tensile stress development in the film because of excess vacancy formation in the film at high deposition rates relative to the surface diffusion rate. Higher surface diffusion rates allow the depositing atoms more time to find their equilibrium positions in the film so fewer excess vacancies are formed and less tensile stress forms. Molecular dynamics calculations have also shown that the vibrational shaking of the substrate/film system due to any excess impact energy of the depositing atoms tends to order the depositing layers even when the surface diffusion coefficient is low.[15] Thus, we see three opposing forces at play for determining the state of stress in the film: (1) void formation due to both "shadowing" and high deposition rates relative to relaxation rates (leads to tensile stress), (2) atom implantation due to deposition at high impact energies (leads to compressive stress) and (3) atomic relaxation due to high surface diffusion coefficients and vibration of the surface by atomic collisions (leads to reduced stress). Let us now look a little more deeply at this "shadowing" phenomenon.

It has been known for many years that evaporated films are often composed of elongated crystals or amorphous regions oriented along a preferred direction and this column character is observed only in films with limited surface mobility. Room temperature deposition of Si, Ge and Be results in voids and low density structures, whereas evaporation of noble metals produces compact films in most cases. It has been found that a vapor beam that impinges obliquely at angle α on the film produces columns that are inclined with respect to the substrate normal by an angle β where $2\tan\beta = \tan\alpha$ provides a good fit for many systems. Konig and Helwig[16] proposed that this type of structure could be explained from the fact that any protuberances in the film surface should shadow adjacent regions from the incident beam, thereby producing voids. Henderson, Brodsky and Chaudhari[17] showed by computer simulation that the random deposition of hard spheres generally resulted in a deposit with atomic-scale surface non-uniformities and these have a sufficient slope to block neighboring regions from the beam. This simulation resulted in an amorphous columnar structure with low deposit densities.

A crude model for the growth of a grassy lawn nicely illustrates the dramatic effects of shadowing.[18] Suppose we represent a single blade of grass as a column on a lattice as depicted in Fig. 8.14. Now, further suppose that each stalk grows at a rate proportional to the amount of light it receives at the tip. On a cloudless day, this amount will be proportional to the solid angle of sky that is not blocked by neighboring stalks. This growth rule is easy to study on a computer and Fig. 8.14 shows the results for initial "seedlings" that differ in height by infinitesimally small amounts (for the case of a one-dimensional strip of grass). The shortest stalks quickly lose as they are overshadowed and choked off by their taller neighbors. The winners compete amongst themselves for the available light until a limiting morphology is achieved that exhibits elements of both regularity and randomness. In this simple model, the regularity derives from the nature of the height distribution of blades or columns. If there are N columns of height h, one finds that there are $N/2$ columns of height $2h$, $N/4$ columns of height $4h$, etc. The final morphology turns out to be *self-similar* and to exhibit "scaling" characteristics. This means that any small portion of Fig. 8.14 that is rescaled (magnified) to full size is indistinguishable from the original pattern. Remarkably, actual sputter-grown films appear to be self-similar down to some minimum size where surface diffusion becomes very important.

Some important aspects of columnar growth can be observed during the simulated deposition of hard discs (rather than hard spheres).[19] Figs. 8.15(a) and (b) show the deposits obtained with hard discs when

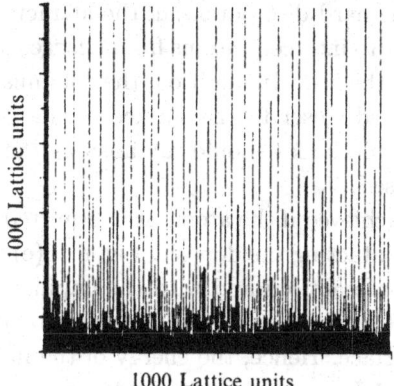

Fig. 8.14. Schematic illustration of the blade growth pattern for the grass model.[18]

the impinging disc, at an angle of incidence of 60° or 75°, was frozen at precisely the point where it first contacted a disc in the deposit. We see that the deposit consists of a series of chains that originate on the substrate and extend to the top surface of the film or are terminated when they fall into the shadow of a neighboring chain. Frequent branching events also occur. It is the shadowing by adjacent chains that produces an orientation that is more nearly perpendicular to the substrate than the incident beam direction. We can see that this is the case since impinging discs that miss the chain on the left are more likely to be high and, when they attach to the top atom of the chain, they extend it in a direction biased towards the vertical. The chain spacing is determined by a competition between chain branching and terminations, where chain branching is most probable when a column is exposed to a broad area of the incident beam (since then a large range of orientations is available for new chain segments). This occurs when the adjacent chain in the direction of the beam is either short or far away. Thus, branching provides a mechanism to increase the film density in regions where chains have terminated or are separated from one another. On the other hand, closely spaced chains have a greater probability of being terminated and thus decreasing the local density. At steady state, the average numbers of branching events and terminations must be proportional and this determines the density of the deposit. An interesting feature of the columns shown in Figs. 8.15(a) and (b) is the tendency of the chains to group themselves into columns of finite thickness. This effect is stronger at the larger angles of incidence. Further, if a Lennard–Jones (LJ) potential had been used, rather than the hard disc potential, the attractive force field of a column would cause it to intercept atoms from a larger distance and thus reduce the density of the film. In addition, the columns for the LJ case are found to be oriented closer to the surface normal (smaller β). The influence of such atomic forces should diminish as the energy of the incident beam is increased.

Dense deposits can only be achieved if the atoms are permitted to continue their motion after first contacting the deposit. Fig. 8.15(e) shows a deposit obtained[19] with LJ interactions and with the impinging atom being permitted to continue its trajectory until it comes to rest in a binding site on the deposit surface. Hence, the energy of the impinging atoms is 0.125ε, where ε is the LJ energy parameter. Although atoms of the deposit are bound close to their sites by harmonic forces, they can experience small excursions from these positions as a result of interactions with the impinging atoms. Note that inclusion of this relaxation causes a large increase in column thickness, whereas the void regions are very similar to Fig. 8.15(b), so the density of the film is correspondingly greater. The average migration distance for this value of energy was ~ 1 atomic diameter and was found to be independent of α.

The effect of a larger incident beam energy (2ε) on the structure is illustrated in Fig. 8.15(f) and such atoms have a larger migration distance (~ 2 atom diameters). Local heating of the atoms in the deposit permits the impinging atoms more time for diffusive motion. The energetic beam produces a larger column diameter, but does not significantly increase the spacing. Thus, a much larger density is produced. It is likely that further increase in the migration distance would eliminate the columnar structure, but leave a few voids. Voids of this type are frequently observed in experiments. Small changes in α should have a drastic effect on the void density.

As a result of the foregoing, a useful classification scheme for thin film microstructure, generated by the Volmer–Weber or Stranski–Krastanov mechanisms, is illustrated in Fig. 8.16. It can be argued that shadowing leads to the zone 1 morphology of low density tapered columns with domed tops. This gives way (by surface diffusion) to wider, smoother uniform columnar grains (zone 2) and finally (by bulk diffusion) to large bulk-like grains in zone 3 as the substrate temperature increases.

8.6   Stacking fault formation in epitaxial films

Structural studies of epitaxial films reveal three principal types

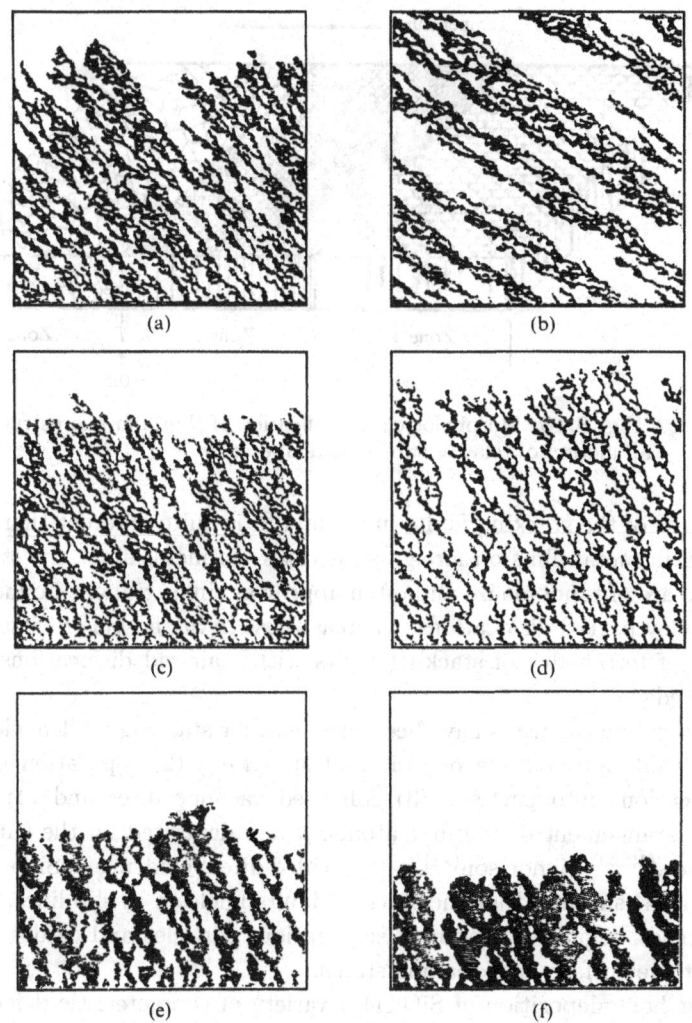

Fig. 8.15. (a) Hard disc deposits formed at beam angle $\alpha = 60°$; (b) the same for $\alpha = 75°$; (c),(d) comparison of deposits formed by hard discs at $\alpha = 45°$ (c) and LJ particles without relaxation at $\alpha = 45°$ (d), comparison of deposits formed by LJ atoms at $\alpha = 45°$ and beam energy of 0.125ε (e) and $\alpha = 45°$ and beam energy of $2.0\varepsilon(f)$. (Courtesy of G. Gilman.)

of defects: dislocations, stacking faults and "macroscopic pips". The first two are common in films of many FCC metals formed on foreign substrates and on homoepitaxial films of Ge and Si. However, the mor-

Temperature ⟶

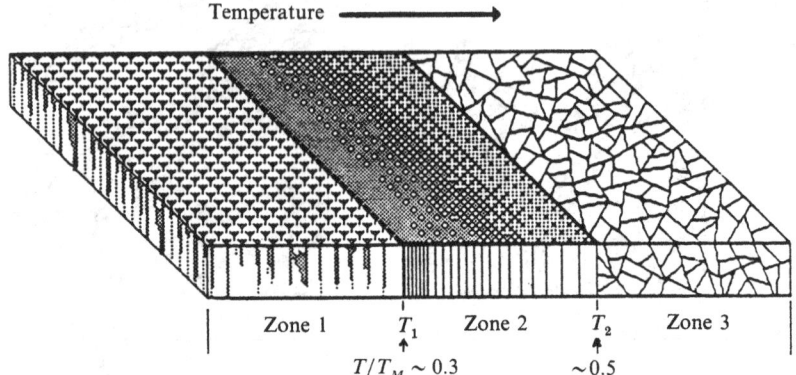

Zone 1 T_1 Zone 2 T_2 Zone 3

$T/T_M \sim 0.3$ ~ 0.5

Fig. 8.16. Morphological classification of thin film microstructure as a function of relative substrate temperature.

phology of the stacking faults in Ge and Si is quite different than those found in metal films on foreign substrates. In metal films, the stacking faults are of random size and often appear as single ribbons bounded by Shockley partial dislocations, whereas those in Si and Ge often take the form of tetrahedra of stacking faults with stair-rod dislocations along their edges.

Several mechanisms have been proposed for stacking fault nucleation: (1) misfit between the deposit and substrate, (2) separation of unit dislocations into partials, (3) collapsed vacancy discs and (4) incorrect arrangement of the first atomic layers deposited on the substrate surface.[12,20] Mendelson[21] has provided direct evidence in Si for mechanism (4) showing that the stacking faults nucleate at slip lines, microscratches, regions of impurity segregation, particles and oxides on the substrate surface and cleaning stains.

For homodeposition of Si(111) a variety of characteristic defects appear on the surface as small triangles or lines. It was concluded[20] that the defects were associated with stacking faults, that they commenced mainly at the substrate/film interface and that they propagated along one or more of the three inclined {111} planes. Some of these faults are illustrated in Fig. 8.17 along with a representation of the stacking fault sequence that gives rise to their presence. Each stacking fault is bounded by dislocation lines oriented along inclined ⟨110⟩ directions and, as the growth of the layer proceeds, the defects spread laterally and interact with one another.

To form the defect, when a new atom layer is deposited, one of the initial nucleation centers goes down in the improper sequence so that a stacking fault forms on the (111) plane parallel to the substrate. As a result, a small area is grown which is crystallographically mismatched with respect to the surrounding areas and, when these areas grow together, a mismatch boundary is formed. The addition of further atom layers causes the mismatch boundary to propagate along inclined {111} planes. This is equivalent to stacking faults on these planes as a defect forms. The shape of the mismatch boundary determines the geometrical form of the defect, while the kind of initiating stacking fault (intrinsic or extrinsic) determines the crystallography of the defect. In Fig. 8.17(a), (i) gives a triangle defect, (ii) gives a jogged triangle, (iii) gives a line, (iv) gives a prismatic form and (v) gives a V-shaped defect. In the lower left of Fig. 8.17(b), we have two similar defects which give complete cancellation of the central region (interaction of three or more such intrinsic faults leads to a central region that looks like a "pip"). In the lower right of Fig. 8.17(b), the two dissimilar defects yields incomplete cancellation. In the upper left of Fig. 8.17(b), the defect has the form of a thin triangular lamella (Fig. 8.17(a) (iii)). Finally, in the upper right of Fig. 8.17(b), the defect has the form of a regular tetrahedron.

Mendelson[21] investigated the seven different orientations illustrated in Fig. 8.18(a) and observed the different type of etch figures represented in Fig. 8.18(b). Fig. 8.19 shows a plot of the average stacking fault density versus substrate orientation. Here, ϕ is the absolute angle between the substrate orientation and a {111} when rotated about the ⟨111⟩. The plot (solid lines) shows that growth is very orientation-dependent on mechanically polished substrates but is not on chemically polished substrates. On chemically polished substrates, stacking fault nucleation is so low that orientation effects appear to be of minimal significance.

The crystallography of the fault polyhedra for the seven substrate orientations is shown in Fig. 8.20. The point 0 represents the site on the substrate at which the stacking faults nucleate. They propagate along the {111} planes and intersect the film surface in the traces designated by ℓ_1, ℓ_2 and ℓ_3. These traces join to form polyhedra which intersect the substrate surface as the polygons shown in Fig. 8.18(b).

Besides stacking faults, twinned layers could also form as the first layers are deposited. These result in stacking fault tetrahedra bounded by either extrinsic or intrinsic faults depending upon whether the number of layers in the twinned deposit is $1 + 3n$ or $2 + 3n$, respectively (n equals an integer). If the number equals $3n$, no tetrahedra form; however, the interface still has the small twinned region. It has been found experi-

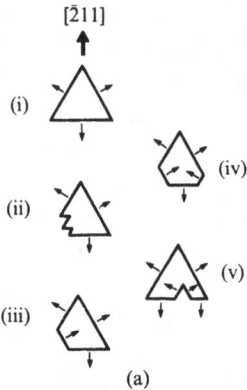

(a)

Fig. 8.17. (a) Various shapes that a fault boundary can take in the plane of the film plus the direction of boundary movement in subsequent layers.[20]

mentally that twins tend to form preferentially on Si{110} substrates. For compound semiconductor films deposited epitaxially on Ge, many twins form when the substrate orientation is {111} and {110} but not when it is {100}. It is very likely that twins and stacking faults have a common origin.

Since the stacking faults can occur when {111} layers are incorrectly deposited, all orientations other than {111} would require regions exposing {111} surfaces of sufficient dimension on which stable nuclei may form. One could thus express the stacking fault density, ρ, in terms of the area per unit thickness, A_t, of the low angle stacking fault planes or, to a first approximation, $\rho = \rho_0 A_t$, where ρ_0 is the scaling parameter to be determined experimentally. In terms of substrate orientation,

$$\rho = \frac{\rho_0 t}{\sqrt{3}(1 - \cos{^2\phi})} \qquad (8.19a)$$

where ϕ is the angle between the substrate orientation and the lowest angle {111} plane when rotated about $\langle 110 \rangle$ and t is the film thickness when the nucleus and at least one point of the substrate plane lie at different corners of a unit cell.[21] In terms of Miller indices, this is given by

$$\rho = \frac{\sqrt{3}\rho_0 P(h^2 + k^2 + \ell^2)}{3(h^2 + k^2 + \ell^2) - (Hh + Kk + L\ell)} \qquad (8.19b)$$

where $(hk\ell)$ is the substrate orientation, (HKL) is that of the particular {111} fault plane and P is an integer which relates the film thickness to the spacing between nearest $\{hk\ell\}$ planes. $P = 1, 2, 2, 6, 2, 3$ and 1 for

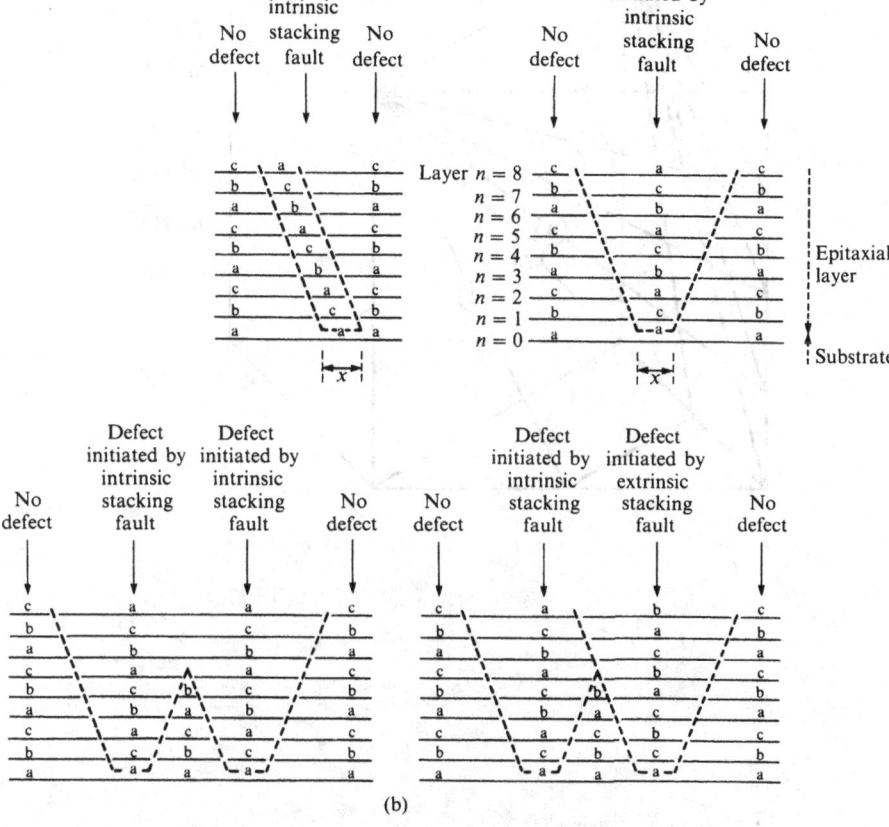

(b)

Fig. 8.17. (b) Cross-sectional views of these faults in a Si(111) film.[20]

$\{110\}, \{221\}, \{111\}, \{334\}, \{112\}, \{114\}$ and $\{100\}$, respectively. A plot of Eq. (8.19*b*), normalized to the $\{100\}$ value, is shown by the dashed line in Fig. 8.19. This agreement, from such a simple model, is encouraging.

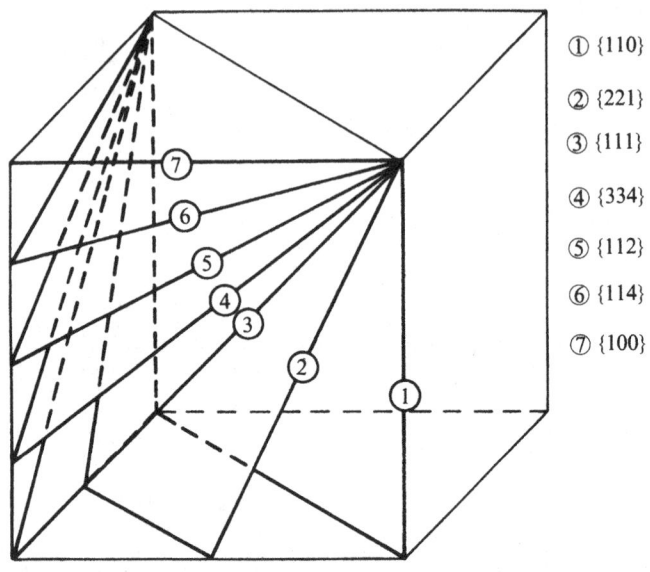

① {110}
② {221}
③ {111}
④ {334}
⑤ {112}
⑥ {114}
⑦ {100}

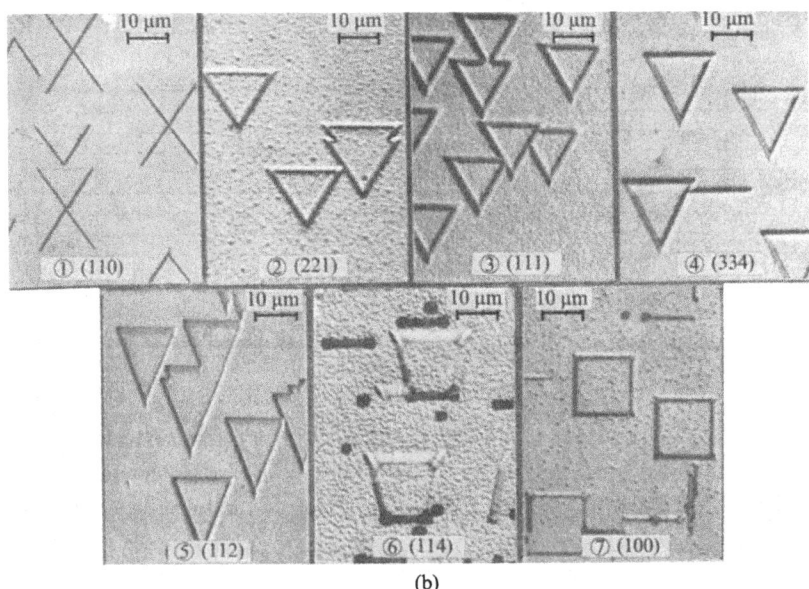

(b)

Fig. 8.18. (a) Crystallography of the seven Si substrate orientations studied in Ref. 21 and (b) etch figures of the stacking faults found in epitaxial films of these orientations. (Courtesy of S. Mendelson.)

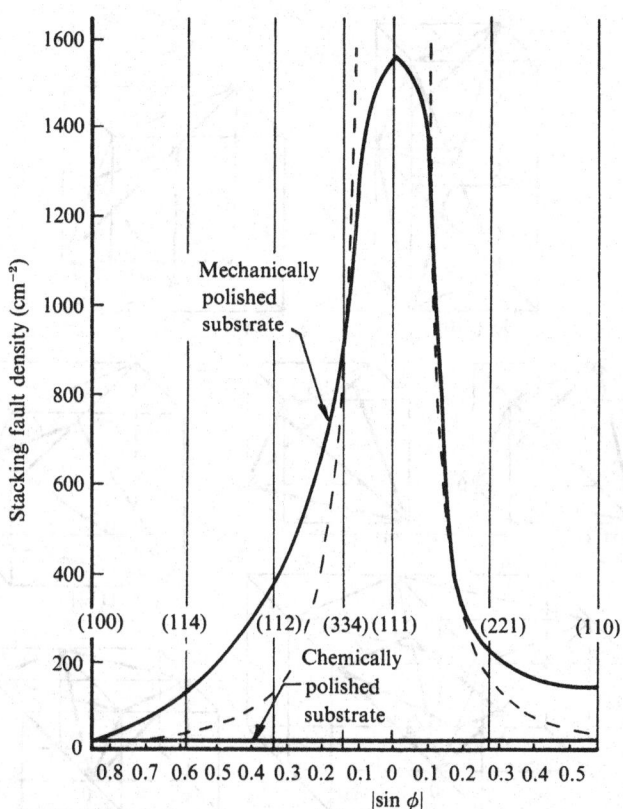

Fig. 8.19. Plot of stacking fault density versus substrate orientation. The solid curve is experimental and the dashed curve is theoretical normalized to the experimental value for {100} substrates.[21]

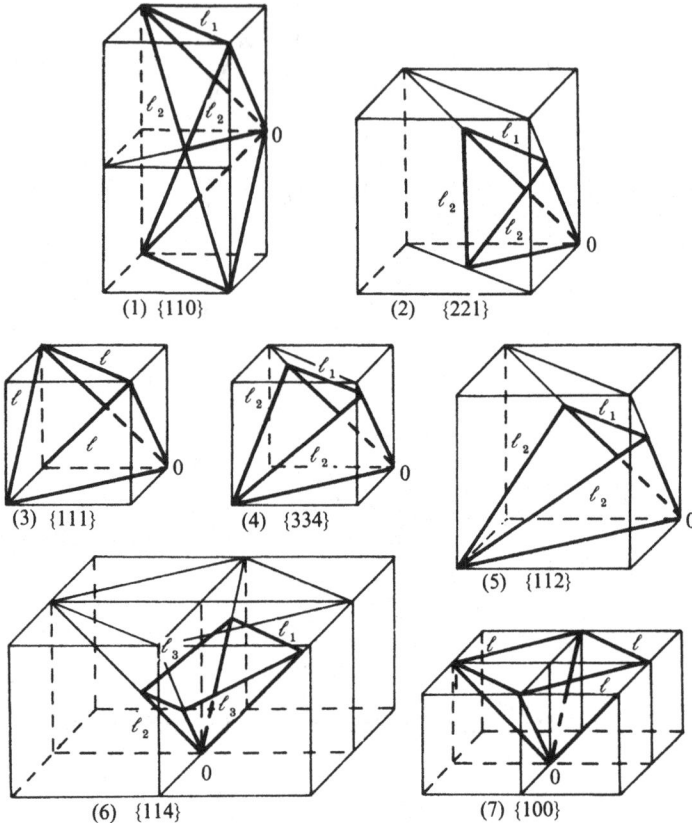

Fig. 8.20. Crystallography of stacking fault polyhedra for the seven oriented substrates.[21]

References

Chapter 1

1. W. A. Tiller, *The science of crystallization: microscopic interfacial phenomena* (Cambridge University Press, Cambridge, 1991).

Chapter 2

1. V. G. Levich, *Physicochemical hydrodynamics*, (Prentice Hall, Inc., Englewood Cliffs, NJ, 1962).
2. V. S. Arpaci and P. S. Larsen, *Convective heat transfer*, (Prentice Hall, Inc., Englewood Cliffs, NJ, 1984).
3. D. J. Tritton, *Physical fluid dynamics*, (Van Nostrand Reinhold Co, New York, 1977).
4. S. Chandrasekhar, *Hydrodynamic and hydromagnetic stability*, (Oxford University Press, London 1961).
5. G. Müller, G. Neumann and W. Weber, *J. Crystal Growth* **70** (1984) 78.
6. L. A. Dorfman, *Hydrodynamic resistance and heat loss of rotating solids*, Translator: N. Kemmer (Oliver and Boyd Ltd., London, 1963).
7. H. J. Scheel and E. O. Schulz-Dubois, *Convective transport and instability phenomena*, Eds. J. Zierep and H. Oertel, Jr. (G. Braun, Karlsruhe, 1982).
8. W. S. Liu, M. F. Wolf, D. Elwell and R. S. Feigelson, *J. Crystal Growth* **82** (1987) 589.
9. M. M. Faktor and I. Garrett, *Growth of crystals from the vapor*, (John Wiley & Sons, New York, 1975).
10. J. M. Carruthers, *J. Electrochem. Soc.* **114** (1967) 959.

11. C. H.-H. Chun and D. Schwabe "Marangoni convection in floating zones" in *Convective transport and instability phenomena*, Eds. J. Zierep and H. Oertel (G. Braun, Karlsruhe, 1982) p. 297.

12. E. Kuroda, H. Kozuka and Y. Takano, *J. Crystal Growth* **68** (1984) 613.

13. C. T. Yen "Analysis of oxygen redistribution during Czochralski crystal growth", PhD Thesis, Department of Materials Science and Engineering, Stanford University, Stanford, CA, 1991.

14. M. E. Glicksman, S. R. Coriell and G. B. McFadden, *Ann. Rev. Fluid Mech.* **18** (1986) 307.

15. N. Kobayashi, "Heat transfer in crystal growth", Ch. 3. of *Preparation and properties of solid state materials*, Vol. 6 (Marcel Dekker, Inc. New York, 1981).

16. J. C. Brice, *J. Crystal Growth* **2** (1968) 395.

17. W. R. Wilcox and R. L. Duty, *J. Heat Transfer, Trans. ASME* Ser. C**88** (1966) 47.

18. W. A. Tiller, *The science of crystallization: microscopic interfacial phenomena* (Cambridge University Press, Cambridge, 1991).

Chapter 3

1. W. A. Tiller, K. A. Jackson, J. W. Rutter and B. Chalmers, *Acta Met.* **1** (1953) 428.

2. A. Hellawell, Private communication, 1968.

3. W. A. Tiller, *The science of crystallization: microscopic interfacial phenomena*, Ch. 4 (Cambridge University Press, Cambridge, 1991).

4. W. A. Tiller and K. S. Ahn, *J. Crystal Growth*, **49** (1980) 483.

5. J. A. Burton, R. C. Prim and W. P. Slichter, *J. Chem. Phys.* **21** (1953) 1987.

6. C. T. Yen and W. A. Tiller, *J. Crystal Growth* **109** (1991) 142.

7. H. Kodera, *J. Appl. Phys. (Japan)* **2** (1963) 212.

8. C. T. Yen 'Analysis of oxygen redistribution during Czochralski crystal growth', PhD Thesis, Department of Materials Science and Engineering, Stanford University, Stanford, CA, 1991.

9. A. A. Chernov, *Soviet Phys.*, Usp (English Transl.) **4** (1964) 116.

10. J. Bloem and L. J. Giling, in *VLSI electronics: microstructure science*, Vol. 12, Ch. 3; Eds. N. G. Einspruch and H. Huff (Academic Press, New York, 1985).

11. E. Pearson, T. Halicioglu and W. A. Tiller, *J. Crystal Growth*, **83** (1987) 499.

Chapter 4

1. W. A. Tiller, *Science of crystallization: microscopic interfacial phenomena*, (Cambridge University Press, Cambridge, 1991).

2. R. W. Barton, S. A. Schwarz, W. A. Tiller and C. R. Helms, *Thin film interfaces and interactions*, Eds. J. E. E. Baglin and J. M. Poate (Electrochemical Society, Princeton, NJ, 1980) p. 104.

3. V. G. Smith, W. A. Tiller and J. W. Rutter, *Can. J. Phys.* **33** (1955) 723.
4. W. A. Tiller, K. A. Jackson, J. W. Rutter and B. Chalmer, *Acta Met.* **1** (1953) 428.
5. W. A. Tiller and R. F. Sekerka, *J. Appl. Phys.* **35** (1964) 2726.
6. R. W. Barton, "Dopant segregation at the Si/SiO$_2$ interface, studied with Auger sputter profiling," PhD Thesis, Department of Materials Science and Engineering, Stanford University, Stanford, CA, July 1981.
7. C. T. Yen, "Analysis of oxygen redistribution during Czochralski crystal growth" PhD Thesis, Department of Materials Science and Engineering, Stanford University, Stanford, CA, July 1991.
8. B. K. Jindal, V. Karelin and W. A. Tiller, *J. Electrochem. Soc.* **120** (1973) 101.
9. H. C. Gatos, *J. Electrochem. Soc.* **122** (1975) 287C.
10. A. F. Witt and H. C. Gatos, *J. Electrochem. Soc.* **115** (1968) 70.
11. K. Morizane, A. F. Witt and H. C. Gatos, *J. Electrochem. Soc.* **113** (1966) 51; **115** (1968) 747.
12. L. O. Wilson, *J. Crystal Growth* **48** (1980) 435; **48** (1980) 451.
13. C. T. Yen and W. A. Tiller, *J. Crystal Growth* 1991.
14. G. E. Lofgren, *Am. J. Sci.* **274** (1974) 243.
15. J. J. Hsieh in *Handbook on semiconductors*, Vol. 3 – *Materials properties and preparation*, Ed. S. P. Keller, (North Holland, New York, 1980).
16. F. S. Ham, *Quart. J. Appl. Math.* **17** (1959) 137.
17. G. Horvay and J. W. Cahn, *Acta Met.* **9** (1961) 694.
18. G. P. Ivanstov, *Dokl. Acad. Nauk. USSR* **58** (1947) 567.
19. J. E. McDonald, *J. Appl. Math and Phys (ZAMP)* **14** (1963) 610.
20. W. G. Pfann, *Zone melting*, (John Wiley & Sons, New York, 1957).
21. W. C. Johnston and W. A. Tiller, *Trans. AIME* **221** (1961) 331; **224** (1962) 214.
22. K. M. Kim and P. Smitama, *J. Electrochem. Soc.* **133** (1986) 1682.
23. A. Hellawell, Private communication, 1969.
24. T. R. Anthony and H. E. Cline, *J. Appl. Phys.* **47** (1976) 2316, 2325, 2332, 2550.

Chapter 5

1. J. S. Kirkaldy, *Decomposition of austenite by diffusional processes*, Eds. V. F. Zackary and H. Aaronson, (Interscience, New York, 1962) pp. 39–130.
2. W. W. Mullins and R. F. Sekerka, *J. Appl. Phys.* **34** (1963) 323.
3. C. Zener, *Trans. AIME* **167** (1946) 550.
4. W. A. Tiller, *Interfaces*, Ed. R. C. Gifkins, (Australian Inst. of Metals, Butterworths, London, 1969) pp. 257–82.
5. W. A. Tiller, *The science of crystallization: microscopic interface phenomena* (Cambridge University Press, Cambridge, 1991).

6. R. F. Xiao, J. I. D. Alexander and F. Rosenberger, *Phys. Rev.* **A38** (1988) 2447.
7. W. A. Tiller and J. W. Rutter, *Can. J. Phys.* **34** (1956) 196.
8. R. M. Sharp and A. Hellawell, *J. Crystal Growth* **6** (1970) 253, 334.
9. R. J. Schaefer and M. E. Glicksman, *Met. Trans.* **1** (1970) 1973.
10. R. F. Sekerka, *Crystal growth: an introduction*, Ed. P. Hartman, (North Holland, Amsterdam, 1973).
11. D. T. J. Hurle, *J. Crystal Growth* **5** (1969) 162.
12. K. S. Ahn, "Interface field effects on solute redistribution and interface instability during steady state planar front crystallization", PhD Thesis, Materials Science and Engineering Department, Stanford University, Stanford, CA (1981).
13. G. R. Kotler and W. A. Tiller, *Crystal growth*, Ed. H. S. Peiser (Pergammon Press, London, 1967) p. 721.
14. M. Avignon, *J. Crystal Growth* **11** (1971) 265.
15. R. J. Asaro and W. A. Tiller, *Met. Trans.* **3** (1972) 1789.

Chapter 6

1. E. Bauser and H. P. Strunk, *J. Crystal Growth* **69** (1984) 561.
2. A. Rosenberg, "The Mode of Soldification of High Purity Metals," PhD Thesis, Department of Physical Metallurgy, University of Toronto, May 1956.
3. R. Trivedi and K. Somboonsuk, *Mat. Sci. and Eng.* **65** (1984) 65.
4. W. Bardsley, J. S. Boulton and D. T. J. Hurle, *Solid-State Electron.* **5** (1962) 395.
5. L. R. Morris and W. G. Winegard, *J. Crystal Growth* **6** (1969) 61.
6. A. Rosenberg and W. A. Tiller, *Acta Met.* **5** (1957) 565.
7. W. A. Tiller, *Science of crystallization: microscopic interfacial phenomena* (Cambridge University Press, Cambridge, 1991).
8. J. D. Hunt and K. A. Jackson, *Trans. AIME* **236** (1966) 843.
9. W. A. Tiller and J. W. Rutter, *Can. J. Phys.* **34** (1956) 96.
10. M. C. Flemings, *Solidification processing* (McGraw Hill, New York, 1974).
11. G. A. Chadwick, *Progress in materials science*, Vol. 12; Ed. B. Chalmers (Pergammon Press, London, 1963).
12. A. Hellawell, Private communication, 1968.
13. K. A. Jackson and J. D. Hunt, *Trans. AIME* **236** (1966) 1129.
14. W. A. Tiller and C. T. Yen, *J. Crystal Growth* **109** (1991) 120.
15. A. Yoe, Private communication, 1968.
16. W. W. Mullins, *J. Appl. Phys.* **28** (1957) 335.
17. W. A. Tiller, *Acta Met.* **5** (1957) 56.
18. F. R. Mollard and M. C. Flemings, *Trans. AIME* **239** (1967) 1534.
19. H. E. Cline and J. D. Livingston, *Trans. AIME* **245** (1969) 1987.
20. G. P. I. Ivantsov, *Dokl. Acad. Nauk. SSSR* **58** (1947) 567.
21. D. E. Temkin, *Soviet Phys. - Dokl. Acad. Nauk. SSSR* **132** (1960) 1307.
22. G. F. Bolling and W. A. Tiller, *J. Appl. Phys.* **32** (1961) 2587.

23. M. E. Glicksman, R. J. Schaeffer and J. D. Ayers, *Metall. Trans.* A, **7** (1976) 1747; *Phil. Mag.* **32** (1975) 725.
24. W. Oldfield, *Mater. Sci. Eng.* **11** (1973) 211.
25. J. S. Langer and H. Müller-Krumbhaar, *Acta Met.* **26** (1978) 1681, 1689, 1697.
26. M. E. Glicksman, *Mater. Sci. Eng.* **65** (1984) 45.
27. G. R. Kotler and W. A. Tiller, *J. Crystal Growth* **2** (1968) 287.
28. C. S. Lindenmeyer, PhD Thesis, Division of Applied Physics, Harvard University, 1959.
29. R. Trivedi, *J. Crystal Growth* **48** (1980) 93.
30. D. R. Hamilton and R. G. Seidensticker, *J. Appl. Phys.* **34** (1963) 1450.
31. R. Wagner, Private communication, 1960.
32. H. F. John and J. W. Faust, Jr. in *Metallurgy of elemental and compound semiconductors*, Ed. R. O. Grubel (AIME/Interscience, Vol. 12, New York, 1961) p. 127.
33. B. J. Luyet, *Proceedings Royal Society*, B, **147** (1957) 434.
34. H. D. Keith and F. J. Padden, Jr., *J. Appl. Phys.* **35** (1964) 1270; **34** (1963) 2409.
35. G. T. Geering, "The role of dendrites in spherulitic crystallization", PhD Thesis, Department of Materials Science and Engineering, Stanford University, 1968.
36. W. A. Tiller and J. M. Schultz, *J. Polymer Science*, **22** (1984) 143.
37. H. Biloni, Private communication, 1988.

Chapter 7

1. P. Penning, *Philips Res. Dept.* **13** (1958) 79.
2. A. S. Jordan, A. R. Von Neida and R. Caruso, *J. Crystal Growth* **70** (1984) 555; **76** (1986) 243.
3. A. S. Jordan, R. Caruso, A. R. Von Neida and J. W. Nielson, *J. Appl. Phys.* **52** (1981) 3331.
4. J. Völkl and G. Müller, *J. Crystal Growth* **97** (1989) 136.
5. D. Maroudas and R. A. Brown, *J. Crystal Growth* 1990.
6. H. Alexander and P. Haasen, *Sol. Stat. Phys.* **22** (1968) 27.
7. B. Boley and J. Weiner, *Theory of thermal stresses* (Wiley, New York, 1960).
8. W. A. Tiller, *The science of crystallization: microscopic interfacial morphology* (Cambridge University Press, Cambridge, 1991).
9. R. J. Asaro and W. A. Tiller, *Met. Trans.* **3** (1972) 1789.
10. J. A. Van Vechten, *Handbook on semiconductors:* Vol. 3, *Materials, properties and preparation*, Ed. S. P. Keller, Series Ed. T. S. Moss (North Holland, New York, 1980) p. 82.
11. G. Schoeck and W. A. Tiller, *Phil. Mag.* **5** (1960) 43.
12. K. V. Ravi, *Imperfections and impurities in semi conductor silicon* (Wiley, New York, 1981).

13. A. J. R. de Kock, *J. Electrochem. Soc.* **118** (1971) 1851; *Acta Electronica* **16** (1973) 303; *J. Crystal Growth* **22** (1974) 311.
14. T. R. Anthony and H. E. Cline, *J. Appl. Phys.* **49** (1978) 5774.
15. J. C. M. Li in *Physical chemistry in metallurgy: proceedings of the darken conference*, Eds. R. M. Fisher, R. A. Oriani and E. T. Turkdogan (USS Research Laboratory, Monroeville, PA, 1976), p. 405.
16. T. Abe in *VLSI electronics: Microstructure science*, Vol. 12, Eds. N. G. Einspruch and H. Huff (Academic Press, New York, 1985) Ch. 1.
17. D. E. Harrison and W. A. Tiller, *J. Appl. Phys.* **33** (1962) 2451.
18. G. F. Bolling, W. A. Tiller and J. W. Rutter, *Can. J. Phys.* **34** (1956) 234.
19. H. Gottshalk, G. Patzer and H. Aleksander, *Phys. Stat. Solidi* **45** (1978) 207.
20. J. L. Nightingale, "The Growth and Optical Applications of Single Crystal Fibers", PhD Thesis, Department of Applied Physics, Stanford University, Sept. 1985.
21. E. Bauser and H. P. Strunk, *J. Crystal Growth* **69** (1984) 561.
22. E. Bauser and G. A. Rozgonyi, *Appl. Phys. Lett.* **37** (1980) 1001; *J. Electrochem. Soc.* **129** (1982) 1782.

Chapter 8

1. W. D. Nix, *Met. Trans.* A, **20A** (1989) 2217.
2. J. W. Mathews and A. E. Blakeslee, *J. Crystal Growth* **27** (1974) 118; **29** (1975) 273.
3. J. W. Mathews, *J. Vac. Sci. Technol.*, **12** (1975) 126.
4. J. Blanc, *Heteroepitaxial semiconductors for electronic devices.* Eds. G. W. Cullen and C. C. Wang (Springer Verlag, New York, 1978).
5. G. H. Olsen and M. Ettenberg, *Crystal growth: theory and technique*, Vol. 2, Ed. C. H. L. Goodman (Plenum Press, New York, 1978).
6. J. C. Bean, L. C. Feldman, A. T. Fiory, S. Nakahara and I. K. Robinson, *J. Vac. Sci. Technol.* **A2** (1984) 436.
7. J. C. Bean, *Silicon molecular beam epitaxy*, Ed. J. C. Bean (Eletrochemical Soc. INC., New Jersey, 1985) Vol. 85–7, p. 337.
8. J. Y. Tsao, B. W. Dodson, S. T. Picraux and D. M. Cornelison, *Phys. Rev. Lett.* **59** (1987) 2455.
9. C. G. Tuppen, C. J. Gibbings and M. Hockly, *J. Crystal Growth* **94** (1989) 392.
10. L. B. Freund, *J. Appl. Mech.* **54** (1987) 553.
11. W. Hagen and H. Strunk, *Appl. Phys.* **17** (1978) 85.
12. W. A. Tiller, *The science of crystallization: microscopic interfacial phenomena* (Cambridge University Press, Cambridge, 1991).
13. J. P. Hirth and W. A. Tiller, *J. Appl. Phys.* **56** (1984) 947.
14. J. P. Hirth, *Met. Trans.* 1991.
15. G. H. Gilmer, Private communication, 1989.
16. H. Konig and G. Helwig, *Optik* **6** (1950) 111.

17. D. Henderson, M. H. Brodsky and P. Chaudhari, *Appl. Phys. Lett.* **25** (1974) 641.
18. G. S. Bales, R. Bruinsina, E. A. Eklund, R. P. U. Karunasari, J. Rudnick and A. Zangwill, *Science* **249** (1990) 264.
19. G. H. Gilmer and M. H. Grabow, *SPIE* **821** (1987) 57.
20. G. Booker and R. Stickler, *J. Appl. Phys.* **33** (1962) 3281.
21. S. Mendelson, *J. Appl. Phys.* **35** (1964) 1570.

Index

Printed in the United States
By Bookmasters